清华版双语教学用书

凸优化算法

Convex Optimization Algorithms

Dimitri P. Bertsekas 著

清华大学出版社
北京

图书在版编目(CIP)数据

凸优化算法=Convex Optimization Algorithms/(美)博塞卡斯(Bertsekas,D. P.)著. —北京 :清华大学出版社,2016(2023.11 重印)
清华版双语教学用书
ISBN 978-7-302-43070-4

Ⅰ. ①凸… Ⅱ. ①博… Ⅲ. ①凸分析-最优化算法-双语教学-教材-英、汉 Ⅳ. ①O174.13 ②O242.23

中国版本图书馆 CIP 数据核字(2016)第 036014 号

责任编辑:王一玲
封面设计:常雪影
责任印制:刘海龙

出版发行:清华大学出版社
网　　址:https://www.tup.com.cn,https://www.wqxuetang.com
地　　址:北京清华大学学研大厦 A 座　　**邮　　编**:100084
社 总 机:010-83470000　　**邮　　购**:010-62786544
投稿与读者服务:010-62776969,c-service@tup.tsinghua.edu.cn
质量反馈:010-62772015,zhiliang@tup.tsinghua.edu.cn
印 装 者:涿州市般润文化传播有限公司
经　　销:全国新华书店
开　　本:155mm×235mm　　**印　　张**:36.25　　**字　　数**:623 千字
版　　次:2016 年 5 月第 1 版　　**印　　次**:2023 年 11 月第 8 次印刷
定　　价:89.00 元

产品编号:066025-01

影印版序

随着大规模资源分配、信号处理、机器学习等应用领域的快速发展，凸优化近来正引起人们日益浓厚的兴趣。本书力图给大家较为全面通俗地介绍求解大规模凸优化问题的最新算法。

本书是作者2009年出版的Convex Optimization Theory一书（原版影印本及中译本：《凸优化理论》，分别于2010年和2015年由清华大学出版社出版）的补充。不过，本书也可以单独阅读。《凸优化理论》一书侧重在凸性理论和基于对偶性的优化方面，而本书则侧重于凸优化的算法方面。本书是从《凸优化理论》原来的一章扩展而来。两本书所需要的数学基础相同，合起来内容比较完整地涵盖了有限维凸优化领域的几乎全部知识。两本书的一个共同特色是在坚持严格的数学分析基础上，十分注重对概念的直观展示。

本书几乎囊括了所有主流的凸优化算法。包括梯度法、次梯度法、多面体逼近法、邻近法和内点法等。这些方法通常依赖于代价函数和约束条件的凸性（而不一定依赖于其可微性），并与对偶性有着直接或间接的联系。作者针对具体问题的特定结构，给出了大量的例题，来充分展示算法的应用。各章的内容如下：第1章，凸优化模型概述；第2章，优化算法概述；第3章，次梯度算法；第4章，多面体逼近算法；第5章，邻近算法；第6章，其他算法问题。本书的一个特色是在强调问题之间的对偶性的同时，也十分重视建立在共轭概念上的算法之间的对偶性，这常常能为选择合适的算法实现方式提供新的灵感和计算上的便利。

本书均取材于作者过去15年在美国麻省理工学院的凸优化方面课堂教学的内容。本书和《凸优化理论》这两本书合起来可以作为一个学期的凸优化课程的教材；本书也可以用作非线性规划课程的补充材料。因为通常传统的非线性规划课程侧重于可微但非凸的内容，如Kuhn－Tucker理论、牛顿法、共轭方向法、内点法、罚函数法和增广的拉格朗日法等。

本书作者德梅萃·博塞克斯(Dimitri P. Bertsekas)教授是优化理论的国际著名学者、美国国家工程院院士，现任美国麻省理工学院电气工程与计算机科学系教授，曾在斯坦福大学工程经济系和伊利诺伊大学电气工程系任教，在优化理论、控制工程、通信工程、计算机科学等领域有丰富的科研教学经验，成果丰硕。博塞克斯教授是一位多产作者，著有 14 本专著和教科书。

赵千川 教授

2016 年 1 月于清华大学

Contents

ATHENA SCIENTIFIC

OPTIMIZATION AND COMPUTATION SERIES

1. Convex Optimization Algorithms, by Dimitri P. Bertsekas, 2015, ISBN 978-1-886529-28-1, 576 pages
2. Abstract Dynamic Programming, by Dimitri P. Bertsekas, 2013, ISBN 978-1-886529-42-7, 256 pages
3. Dynamic Programming and Optimal Control, Two-Volume Set, by Dimitri P. Bertsekas, 2012, ISBN 1-886529-08-6, 1020 pages
4. Convex Optimization Theory, by Dimitri P. Bertsekas, 2009, ISBN 978-1-886529-31-1, 256 pages
5. Introduction to Probability, 2nd Edition, by Dimitri P. Bertsekas and John N. Tsitsiklis, 2008, ISBN 978-1-886529-23-6, 544 pages
6. Convex Analysis and Optimization, by Dimitri P. Bertsekas, Angelia Nedić, and Asuman E. Ozdaglar, 2003, ISBN 1-886529-45-0, 560 pages
7. Nonlinear Programming, 2nd Edition, by Dimitri P. Bertsekas, 1999, ISBN 1-886529-00-0, 791 pages
8. Network Optimization: Continuous and Discrete Models, by Dimitri P. Bertsekas, 1998, ISBN 1-886529-02-7, 608 pages
9. Network Flows and Monotropic Optimization, by R. Tyrrell Rockafellar, 1998, ISBN 1-886529-06-X, 634 pages
10. Introduction to Linear Optimization, by Dimitris Bertsimas and John N. Tsitsiklis, 1997, ISBN 1-886529-19-1, 608 pages
11. Parallel and Distributed Computation: Numerical Methods, by Dimitri P. Bertsekas and John N. Tsitsiklis, 1997, ISBN 1-886529-01-9, 718 pages
12. Neuro-Dynamic Programming, by Dimitri P. Bertsekas and John N. Tsitsiklis, 1996, ISBN 1-886529-10-8, 512 pages
13. Constrained Optimization and Lagrange Multiplier Methods, by Dimitri P. Bertsekas, 1996, ISBN 1-886529-04-3, 410 pages
14. Stochastic Optimal Control: The Discrete-Time Case, by Dimitri P. Bertsekas and Steven E. Shreve, 1996, ISBN 1-886529-03-5, 330 pages

ABOUT THE AUTHOR

Dimitri Bertsekas studied Mechanical and Electrical Engineering at the National Technical University of Athens, Greece, and obtained his Ph.D. in system science from the Massachusetts Institute of Technology. He has held faculty positions with the Engineering-Economic Systems Department, Stanford University, and the Electrical Engineering Department of the University of Illinois, Urbana. Since 1979 he has been teaching at the Electrical Engineering and Computer Science Department of the Massachusetts Institute of Technology (M.I.T.), where he is currently the McAfee Professor of Engineering.

His teaching and research spans several fields, including deterministic optimization, dynamic programming and stochastic control, large-scale and distributed computation, and data communication networks. He has authored or coauthored numerous research papers and sixteen books, several of which are currently used as textbooks in MIT classes, including "Nonlinear Programming," "Dynamic Programming and Optimal Control," "Data Networks," "Introduction to Probability," "Convex Optimization Theory," as well as the present book. He often consults with private industry and has held editorial positions in several journals.

Professor Bertsekas was awarded the INFORMS 1997 Prize for Research Excellence in the Interface Between Operations Research and Computer Science for his book "Neuro-Dynamic Programming" (co-authored with John Tsitsiklis), the 2001 AACC John R. Ragazzini Education Award, the 2009 INFORMS Expository Writing Award, the 2014 AACC Richard Bellman Heritage Award for "contributions to the foundations of deterministic and stochastic optimization-based methods in systems and control," the 2014 Khachiyan Prize for "life-time accomplishments in optimization," and the SIAM/MOS 2015 George B. Dantzig Prize for "original research, which by its originality, breadth, and scope, is having a major impact on the field of mathematical programming." In 2001, he was elected to the United States National Academy of Engineering for "pioneering contributions to fundamental research, practice and education of optimization/control theory, and especially its application to data communication networks."

Preface

There is no royal way to geometry
(Euclid to king Ptolemy of Alexandria)

Interest in convex optimization has become intense due to widespread applications in fields such as large-scale resource allocation, signal processing, and machine learning. This book aims at an up-to-date and accessible development of algorithms for solving convex optimization problems.

The book complements the author's 2009 "Convex Optimization Theory" book, but can be read independently. The latter book focuses on convexity theory and optimization duality, while the present book focuses on algorithmic issues. The two books share mathematical prerequisites, notation, and style, and together cover the entire finite-dimensional convex optimization field. Both books rely on rigorous mathematical analysis, but also aim at an intuitive exposition that makes use of visualization where possible. This is facilitated by the extensive use of analytical and algorithmic concepts of duality, which by nature lend themselves to geometrical interpretation.

To enhance readability, the statements of definitions and results of the "theory book" are reproduced without proofs in Appendix B. Moreover, some of the theory needed for the present book, has been replicated and/or adapted to its algorithmic nature. For example the theory of subgradients for real-valued convex functions is fully developed in Chapter 3. Thus the reader who is already familiar with the analytical foundations of convex optimization need not consult the "theory book" except for the purpose of studying the proofs of some specific results.

The book covers almost all the major classes of convex optimization algorithms. Principal among these are gradient, subgradient, polyhedral approximation, proximal, and interior point methods. Most of these methods rely on convexity (but not necessarily differentiability) in the cost and constraint functions, and are often connected in various ways to duality. I have provided numerous examples describing in detail applications to specially structured problems. The reader may also find a wealth of analysis and discussion of applications in books on large-scale convex optimization, network optimization, parallel and distributed computation, signal processing, and machine learning.

The chapter-by-chapter description of the book follows:

Chapter 1: Here we provide a broad overview of some important classes of convex optimization problems, and their principal characteristics. Several problem structures are discussed, often arising from Lagrange duality theory and Fenchel duality theory, together with its special case, conic duality. Some additional structures involving a large number of additive terms in the cost, or a large number of constraints are also discussed, together with their applications in machine learning and large-scale resource allocation.

Chapter 2: Here we provide an overview of algorithmic approaches, focusing primarily on algorithms for differentiable optimization, and we discuss their differences from their nondifferentiable convex optimization counterparts. We also highlight the main ideas of the two principal algorithmic approaches of this book, iterative descent and approximation, and we illustrate their application with specific algorithms, reserving detailed analysis for subsequent chapters.

Chapter 3: Here we discuss subgradient methods for minimizing a convex cost function over a convex constraint set. The cost function may be nondifferentiable, as is often the case in the context of duality and machine learning applications. These methods are based on the idea of reduction of distance to the optimal set, and include variations aimed at algorithmic efficiency, such as ϵ-subgradient and incremental subgradient methods.

Chapter 4: Here we discuss polyhedral approximation methods for minimizing a convex function over a convex constraint set. The two main approaches here are outer linearization (also called the cutting plane approach) and inner linearization (also called the simplicial decomposition approach). We show how these two approaches are intimately connected by conjugacy and duality, and we generalize our framework for polyhedral approximation to the case where the cost function is a sum of two or more convex component functions.

Chapter 5: Here we focus on proximal algorithms for minimizing a convex function over a convex constraint set. At each iteration of the basic proximal method, we solve an approximation to the original problem. However, unlike the preceding chapter, the approximation is not polyhedral, but rather it is based on quadratic regularization, i.e., adding a quadratic term to the cost function, which is appropriately adjusted at each iteration. We discuss several variations of the basic algorithm. Some of these include combinations with the polyhedral approximation methods of the preceding chapter, yielding the class of bundle methods. Others are obtained via duality from the basic proximal algorithm, including the augmented Lagrangian method (also called method of multipliers) for constrained optimization. Finally, we discuss extensions of the proximal algorithm for finding a zero of a maximal monotone operator, and a major special case: the alternating direction method of multipliers, which is well suited for taking advantage of the structure of several types of large-scale problems.

Chapter 6: Here we discuss a variety of algorithmic topics that supplement our discussion of the descent and approximation methods of the preceding chapters. We first discuss gradient projection methods and variations with extrapolation that have good complexity properties, including Nesterov's optimal complexity algorithm. These were developed for differentiable problems, and can be extended to the nondifferentiable case by means of a smoothing scheme. Then we discuss a number of combinations of gradient, subgradient, and proximal methods that are well suited for specially structured problems. We pay special attention to incremental versions for the case where the cost function consists of the sum of a large number of component terms. We also describe additional methods, such as the classical block coordinate descent approach, the proximal algorithm with a nonquadratic regularization term, and the ϵ-descent method. We close the chapter with a discussion of interior point methods.

Our lines of analysis are largely based on differential calculus-type ideas, which are central in nonlinear programming, and on concepts of hyperplane separation, conjugacy, and duality, which are central in convex analysis. A traditional use of duality is to establish the equivalence and the connections between a pair of primal and dual problems, which may in turn enhance insight and enlarge the set of options for analysis and computation. The book makes heavy use of this type of problem duality, but also emphasizes a qualitatively different, algorithm-oriented type of duality that is largely based on conjugacy. In particular, some fundamental algorithmic operations turn out to be dual to each other, and whenever they arise in various algorithms they admit dual implementations, often with significant gains in insight and computational convenience. Some important examples are the duality between the subdifferentials of a convex function and its conjugate, the duality of a proximal operation using a convex function and an augmented Lagrangian minimization using its conjugate, and the duality between outer linearization of a convex function and inner linearization of its conjugate. Several interesting algorithms in Chapters 4-6 admit dual implementations based on these pairs of operations.

The book contains a fair number of exercises, many of them supplementing the algorithmic development and analysis. In addition a large number of theoretical exercises (with carefully written solutions) for the"theory book," together with other related material, can be obtained from the book's web page http://www.athenasc.com/convexalgorithms.html, and the author's web page http://web.mit.edu/dimitrib/www/home.html. The MIT OpenCourseWare site http://ocw.mit.edu/index.htm, also provides lecture slides and other relevant material.

The mathematical prerequisites for the book are a first course in linear algebra and a first course in real analysis. A summary of the relevant material is provided in Appendix A. Prior exposure to linear and nonlinear optimization algorithms is not assumed, although it will undoubtedly be helpful in providing context and perspective. Other than this background, the development is self-contained, with proofs provided throughout.

The present book, in conjunction with its "theory" counterpart may be used as a text for a one-semester or two-quarter convex optimization course; I have taught several variants of such a course at MIT and elsewhere over the last fifteen years. Still the book may not provide all of the convex optimization material an instructor may wish for, and it may need to be supplemented by works that aim primarily at specific types of convex optimization models, or address more comprehensively computational complexity issues. I have added representative citations for such works, which, however, are far from complete in view of the explosive growth of the literature on the subject.

The book may also be used as a supplementary source for nonlinear programming classes that are primarily focused on classical differentiable nonconvex optimization material (Kuhn-Tucker theory, Newton-like and conjugate direction methods, interior point, penalty, and augmented Lagrangian methods). For such courses, it may provide a nondifferentiable convex optimization component.

I was fortunate to have several outstanding collaborators in my research on various aspects of convex optimization: Vivek Borkar, Jon Eckstein, Eli Gafni, Xavier Luque, Angelia Nedić, Asuman Ozdaglar, John Tsitsiklis, Mengdi Wang, and Huizhen (Janey) Yu. Substantial portions of our joint research have found their way into the book. In addition, I am grateful for interactions and suggestions I received from several colleagues, including Leon Bottou, Steve Boyd, Tom Luo, Steve Wright, and particularly Mark Schmidt and Lin Xiao who read with care major portions of the book. I am also very thankful for the valuable proofreading of parts of the book by Mengdi Wang and Huizhen (Janey) Yu, and particularly by Ivan Pejcic who went through most of the book with a keen eye. I developed the book through convex optimization classes at MIT over a fifteen-year period, and I want to express appreciation for my students who provided continuing motivation and inspiration.

Finally, I would like to mention Paul Tseng, a major contributor to numerous topics in this book, who was my close friend and research collaborator on optimization algorithms for many years, and whom we unfortunately lost while he was still at his prime. I am dedicating the book to his memory.

Dimitri P. Bertsekas
dimitrib@mit.edu
January 2015

1
Convex Optimization Models: An Overview

Contents

In this chapter we provide an overview of some broad classes of convex optimization models. Our primary focus will be on large challenging problems, often connected in some way to duality. We will consider two types of duality. The first is *Lagrange duality* for constrained optimization, which is obtained by assigning dual variables to the constraints. The second is *Fenchel duality* together with its special case, conic duality, which involves a cost function that is the sum of two convex function components. Both of these duality structures arise often in applications, and in Sections 1.1 and 1.2 we provide an overview, and discuss some examples.†

In Sections 1.3 and 1.4, we discuss additional model structures involving a large number of additive terms in the cost, or a large number of constraints. These types of problems also arise often in the context of duality, as well as in other contexts such as machine learning and signal processing with large amounts of data. In Section 1.5, we discuss the exact penalty function technique, whereby we can transform a convex constrained optimization problem to an equivalent unconstrained problem.

1.1 LAGRANGE DUALITY

We start our overview of Lagrange duality with the basic case of nonlinear inequality constraints, and then consider extensions involving linear inequality and equality constraints. Consider the problem‡

$$\begin{aligned} &\text{minimize} \quad f(x) \\ &\text{subject to} \quad x \in X, \quad g(x) \leq 0, \end{aligned} \tag{1.1}$$

where X is a nonempty set,

$$g(x) = \big(g_1(x), \ldots, g_r(x)\big)',$$

and $f : X \mapsto \Re$ and $g_j : X \mapsto \Re$, $j = 1, \ldots, r$, are given functions. We refer to this as the *primal problem*, and we denote its optimal value by f^*. A vector x satisfying the constraints of the problem is referred to as *feasible*. The *dual* of problem (1.1) is given by

$$\begin{aligned} &\text{maximize} \quad q(\mu) \\ &\text{subject to} \quad \mu \in \Re^r, \end{aligned} \tag{1.2}$$

† Consistent with its overview character, this chapter contains few proofs, and refers frequently to the literature, and to Appendix B, which contains a full list of definitions and propositions (without proofs) relating to nonalgorithmic aspects of convex optimization. This list reflects and summarizes the content of the author's "Convex Optimization Theory" book [Ber09]. The proposition numbers of [Ber09] have been preserved, so all omitted proofs of propositions in Appendix B can be readily accessed from [Ber09].

‡ Appendix A contains an overview of the mathematical notation, terminology, and results from linear algebra and real analysis that we will be using.

where the dual function q is

$$q(\mu) = \begin{cases} \inf_{x \in X} L(x,\mu) & \text{if } \mu \geq 0, \\ -\infty & \text{otherwise,} \end{cases}$$

and L is the Lagrangian function defined by

$$L(x,\mu) = f(x) + \mu' g(x), \qquad x \in X, \ \mu \in \Re^r;$$

(cf. Section 5.3 of Appendix B).

Note that the dual function is extended real-valued, and that the effective constraint set of the dual problem is

$$\left\{ \mu \geq 0 \ \Big| \ \inf_{x \in X} L(x,\mu) > -\infty \right\}.$$

The optimal value of the dual problem is denoted by q^*.

The *weak duality* relation, $q^* \leq f^*$, always holds. It is easily shown by writing for all $\mu \geq 0$, and $x \in X$ with $g(x) \leq 0$,

$$q(\mu) = \inf_{z \in X} L(z,\mu) \leq L(x,\mu) = f(x) + \sum_{j=1}^{r} \mu_j g_j(x) \leq f(x),$$

so that

$$q^* = \sup_{\mu \in \Re^r} q(\mu) = \sup_{\mu \geq 0} q(\mu) \leq \inf_{x \in X,\, g(x) \leq 0} f(x) = f^*.$$

We state this formally as follows (cf. Prop. 4.1.2 in Appendix B).

Proposition 1.1.1: (Weak Duality Theorem) Consider problem (1.1). For any feasible solution x and any $\mu \in \Re^r$, we have $q(\mu) \leq f(x)$. Moreover, $q^* \leq f^*$.

When $q^* = f^*$, we say that *strong duality* holds. The following proposition gives necessary and sufficient conditions for strong duality, and primal and dual optimality (see Prop. 5.3.2 in Appendix B).

Proposition 1.1.2: (Optimality Conditions) Consider problem (1.1). There holds $q^* = f^*$, and (x^*, μ^*) are a primal and dual optimal solution pair if and only if x^* is feasible, $\mu^* \geq 0$, and

$$x^* \in \arg\min_{x \in X} L(x,\mu^*), \qquad \mu_j^* g_j(x^*) = 0, \quad j = 1,\ldots,r.$$

Both of the preceding propositions do not require any convexity assumptions on f, g, and X. However, generally the analytical and algorithmic solution process is simplified when strong duality ($q^* = f^*$) holds. This typically requires convexity assumptions, and in some cases conditions on $\mathrm{ri}(X)$, the relative interior of X, as exemplified by the following result, given in Prop. 5.3.1 in Appendix B. The result delineates the two principal cases where there is no duality gap in an inequality-constrained problem.

Proposition 1.1.3: (Strong Duality – Existence of Dual Optimal Solutions) Consider problem (1.1) under the assumption that the set X is convex, and the functions f, and $g_1, \ldots, g_r$ are convex. Assume further that f^* is finite, and that one of the following two conditions holds:

(1) There exists $\overline{x} \in X$ such that $g_j(\overline{x}) < 0$ for all $j = 1, \ldots, r$.

(2) The functions g_j, $j = 1, \ldots, r$, are affine, and there exists $\overline{x} \in \mathrm{ri}(X)$ such that $g(\overline{x}) \leq 0$.

Then $q^* = f^*$ and there exists at least one dual optimal solution. Under condition (1) the set of dual optimal solutions is also compact.

Convex Programming with Inequality and Equality Constraints

Let us consider an extension of problem (1.1), with additional linear equality constraints. It is our principal constrained optimization model under convexity assumptions, and it will be referred to as the *convex programming problem*. It is given by

$$\begin{aligned} &\text{minimize} \quad f(x) \\ &\text{subject to} \quad x \in X, \quad g(x) \leq 0, \quad Ax = b, \end{aligned} \tag{1.3}$$

where X is a convex set, $g(x) = \big(g_1(x), \ldots, g_r(x)\big)'$, $f : X \mapsto \Re$ and $g_j : X \mapsto \Re$, $j = 1, \ldots, r$, are given convex functions, A is an $m \times n$ matrix, and $b \in \Re^m$.

The preceding duality framework may be applied to this problem by converting the constraint $Ax = b$ to the equivalent set of linear inequality constraints

$$Ax \leq b, \qquad -Ax \leq -b,$$

with corresponding dual variables $\lambda^+ \geq 0$ and $\lambda^- \geq 0$. The Lagrangian function is

$$f(x) + \mu' g(x) + (\lambda^+ - \lambda^-)'(Ax - b),$$

and by introducing a dual variable

$$\lambda = \lambda^+ - \lambda^-$$

with no sign restriction, it can be written as

$$L(x, \mu, \lambda) = f(x) + \mu' g(x) + \lambda'(Ax - b).$$

The dual problem is

$$\begin{aligned} &\text{maximize} \quad \inf_{x \in X} L(x, \mu, \lambda) \\ &\text{subject to} \quad \mu \geq 0, \ \lambda \in \Re^m. \end{aligned}$$

In this manner, Prop. 1.1.3 under condition (2), together with Prop. 1.1.2, yield the following for the case where all constraint functions are linear.

Proposition 1.1.4: (Convex Programming – Linear Equality and Inequality Constraints) Consider problem (1.3).

(a) Assume that f^* is finite, that the functions g_j are affine, and that there exists $\overline{x} \in \text{ri}(X)$ such that $A\overline{x} = b$ and $g(\overline{x}) \leq 0$. Then $q^* = f^*$ and there exists at least one dual optimal solution.

(b) There holds $f^* = q^*$, and (x^*, μ^*, λ^*) are a primal and dual optimal solution pair if and only if x^* is feasible, $\mu^* \geq 0$, and

$$x^* \in \arg\min_{x \in X} L(x, \mu^*, \lambda^*), \qquad \mu_j^* g_j(x^*) = 0, \quad j = 1, \ldots, r.$$

In the special case where there are no inequality constraints:

$$\begin{aligned} &\text{minimize} \quad f(x) \\ &\text{subject to} \quad x \in X, \quad Ax = b, \end{aligned} \tag{1.4}$$

the Lagrangian function is

$$L(x, \lambda) = f(x) + \lambda'(Ax - b),$$

and the dual problem is

$$\begin{aligned} &\text{maximize} \quad \inf_{x \in X} L(x, \lambda) \\ &\text{subject to} \quad \lambda \in \Re^m. \end{aligned}$$

The corresponding result, a simpler special case of Prop. 1.1.4, is given in the following proposition.

Proposition 1.1.5: (Convex Programming – Linear Equality Constraints) Consider problem (1.4).

(a) Assume that f^* is finite and that there exists $\overline{x} \in \mathrm{ri}(X)$ such that $A\overline{x} = b$. Then $f^* = q^*$ and there exists at least one dual optimal solution.

(b) There holds $f^* = q^*$, and (x^*, λ^*) are a primal and dual optimal solution pair if and only if x^* is feasible and

$$x^* \in \arg\min_{x \in X} L(x, \lambda^*).$$

The following is an extension of Prop. 1.1.4(a) to the case where the inequality constraints may be nonlinear. It is the most general convex programming result relating to duality in this section (see Prop. 5.3.5 in Appendix B).

Proposition 1.1.6: (Convex Programming – Linear Equality and Nonlinear Inequality Constraints) Consider problem (1.3). Assume that f^* is finite, that there exists $\overline{x} \in X$ such that $A\overline{x} = b$ and $g(\overline{x}) < 0$, and that there exists $\tilde{x} \in \mathrm{ri}(X)$ such that $A\tilde{x} = b$. Then $q^* = f^*$ and there exists at least one dual optimal solution.

Aside from the preceding results, there are alternative optimality conditions for convex and nonconvex optimization problems, which are based on extended versions of the Fritz John theorem; see [BeO02] and [BOT06], and the textbooks [Ber99] and [BNO03]. These conditions are derived using a somewhat different line of analysis and supplement the ones given here, but we will not have occasion to use them in this book.

Discrete Optimization and Lower Bounds

The preceding propositions deal mostly with situations where strong duality holds ($q^* = f^*$). However, duality can be useful even when there is duality gap, as often occurs in problems that have a finite constraint set X. An example is *integer programming*, where the components of x must be integers from a bounded range (usually 0 or 1). An important special case is the linear 0-1 integer programming problem

$$\begin{aligned} &\text{minimize} \quad c'x \\ &\text{subject to} \quad Ax \le b, \quad x_i = 0 \text{ or } 1, \quad i = 1, \ldots, n, \end{aligned}$$

where $x = (x_1, \ldots, x_n)$.

A principal approach for solving discrete optimization problems with a finite constraint set is the *branch-and-bound method*, which is described in many sources; see e.g., one of the original works [LaD60], the survey [BaT85], and the book [NeW88]. The general idea of the method is that bounds on the cost function can be used to exclude from consideration portions of the feasible set. To illustrate, consider minimizing $F(x)$ over $x \in X$, and let Y_1, Y_2 be two subsets of X. Suppose that we have bounds

$$\underline{F}_1 \leq \min_{x \in Y_1} f(x), \qquad \overline{F}_2 \geq \min_{x \in Y_2} f(x).$$

Then, if $\overline{F}_2 \leq \underline{F}_1$, the solutions in Y_1 may be disregarded since their cost cannot be smaller than the cost of the best solution in Y_2. The lower bound $\underline{F}_1$ can often be conveniently obtained by minimizing f over a suitably enlarged version of Y_1, while for the upper bound $\overline{F}_2$, a value $f(x)$, where $x \in Y_2$, may be used.

Branch-and-bound is often based on weak duality (cf. Prop. 1.1.1) to obtain lower bounds to the optimal cost of restricted problems of the form

$$\begin{aligned} &\text{minimize} \quad f(x) \\ &\text{subject to} \quad x \in \tilde{X}, \qquad g(x) \leq 0, \end{aligned} \tag{1.5}$$

where $\tilde{X}$ is a subset of X; for example in the 0-1 integer case where X specifies that all x_i should be 0 or 1, $\tilde{X}$ may be the set of all 0-1 vectors x such that one or more components x_i are fixed at either 0 or 1 (i.e., are restricted to satisfy $x_i = 0$ for all $x \in \tilde{X}$ or $x_i = 1$ for all $x \in \tilde{X}$). These lower bounds can often be obtained by finding a dual-feasible (possibly dual-optimal) solution $\mu \geq 0$ of this problem and the corresponding dual value

$$q(\mu) = \inf_{x \in \tilde{X}} \big\{ f(x) + \mu' g(x) \big\}, \tag{1.6}$$

which by weak duality, is a lower bound to the optimal value of the restricted problem (1.5). In a strengthened version of this approach, the given inequality constraints $g(x) \leq 0$ may be augmented by additional inequalities that are known to be satisfied by optimal solutions of the original problem.

An important point here is that when $\tilde{X}$ is finite, the dual function q of Eq. (1.6) is concave and polyhedral. Thus solving the dual problem amounts to minimizing the polyhedral function $-q$ over the nonnegative orthant. This is a major context within which polyhedral functions arise in convex optimization.

1.1.1 Separable Problems – Decomposition

Let us now discuss an important problem structure that involves Lagrange duality and arises frequently in applications. Here x has m components,

$x = (x_1, \ldots, x_m)$, with each x_i being a vector of dimension n_i (often $n_i = 1$). The problem has the form

$$\begin{aligned} &\text{minimize} \ \sum_{i=1}^{m} f_i(x_i) \\ &\text{subject to} \ \sum_{i=1}^{m} g_{ij}(x_i) \leq 0, \quad x_i \in X_i, \ i = 1, \ldots, m, \ j = 1, \ldots, r, \end{aligned} \tag{1.7}$$

where $f_i : \Re^{n_i} \mapsto \Re$ and $g_{ij} : \Re^{n_i} \mapsto \Re^r$ are given functions, and X_i are given subsets of $\Re^{n_i}$. By assigning a dual variable μ_j to the jth constraint, we obtain the dual problem [cf. Eq. (1.2)]

$$\begin{aligned} &\text{maximize} \quad \sum_{i=1}^{m} q_i(\mu) \\ &\text{subject to} \ \ \mu \geq 0, \end{aligned} \tag{1.8}$$

where

$$q_i(\mu) = \inf_{x_i \in X_i} \left\{ f_i(x_i) + \sum_{j=1}^{r} \mu_j g_{ij}(x_i) \right\},$$

and $\mu = (\mu_1, \ldots, \mu_r)$.

Note that the minimization involved in the calculation of the dual function has been decomposed into m simpler minimizations. These minimizations are often conveniently done either analytically or computationally, in which case the dual function can be easily evaluated. This is the key advantageous structure of separable problems: it facilitates computation of dual function values (as well as subgradients as we will see in Section 3.1), and it is amenable to decomposition and distributed computation.

Let us also note that in the special case where the components x_i are one-dimensional, and the functions f_i and sets X_i are convex, there is a particularly favorable duality result for the separable problem (1.7): essentially, strong duality holds without any qualifications such as the linearity of the constraint functions, or the Slater condition of Prop. 1.1.3; see [Tse09].

Duality Gap Estimates for Nonconvex Separable Problems

The separable structure is additionally helpful when the cost and/or the constraints are not convex, and there is a duality gap. In particular, in this case *the duality gap turns out to be relatively small and can often be shown to diminish to zero relative to the optimal primal value as the number m of separable terms increases*. As a result, one can often obtain a near-optimal primal solution, starting from a dual-optimal solution, without resorting to costly branch-and-bound procedures.

The small duality gap size is a consequence of the structure of the set S of constraint-cost pairs of problem (1.7), which in the case of a separable problem, can be written as a vector sum of m sets, one for each separable term, i.e.,

$$S = S_1 + \cdots + S_m,$$

where

$$S_i = \{(g_i(x_i), f_i(x_i)) \mid x_i \in X_i\},$$

and $g_i : \Re^{n_i} \mapsto \Re^r$ is the function $g_i(x_i) = (g_{i1}(x_i), \ldots, g_{im}(x_i))$. It can be shown that the duality gap is related to how much S "differs" from its convex hull (a geometric explanation is given in [Ber99], Section 5.1.6, and [Ber09], Section 5.7). Generally, a set that is the vector sum of a large number of possibly nonconvex but roughly similar sets "tends to be convex" in the sense that any vector in its convex hull can be closely approximated by a vector in the set. As a result, the duality gap tends to be relatively small. The analytical substantiation is based on a theorem by Shapley and Folkman (see [Ber99], Section 5.1, or [Ber09], Prop. 5.7.1, for a statement and proof of this theorem). In particular, it is shown in [AuE76], and also [BeS82], [Ber82a], Section 5.6.1, under various reasonable assumptions, that the duality gap satisfies

$$f^* - q^* \leq (r+1) \max_{i=1,\ldots,m} \rho_i,$$

where for each i, ρ_i is a nonnegative scalar that depends on the structure of the functions f_i, g_{ij}, $j = 1, \ldots, r$, and the set X_i (the paper [AuE76] focuses on the case where the problem is nonconvex but continuous, while [BeS82] and [Ber82a] focus on an important class of mixed integer programming problems). This estimate suggests that as $m \to \infty$ and $|f^*| \to \infty$, the duality gap is bounded, while the "relative" duality gap $(f^* - q^*)/|f^*|$ diminishes to 0 as $m \to \infty$.

The duality gap has also been investigated in the author's book [Ber09] within the more general min common-max crossing framework (Section 4.1 of Appendix B). This framework includes as special cases minimax and zero-sum game problems. In particular, consider a function $\phi : X \times Z \mapsto \Re$ defined over nonempty subsets $X \subset \Re^n$ and $Z \subset \Re^m$. Then it can be shown that the gap between "infsup" and "supinf" of ϕ can be decomposed into the sum of two terms that can be computed separately: one term can be attributed to the lack of convexity and/or closure of ϕ with respect to x, and the other can be attributed to the lack of concavity and/or upper semicontinuity of ϕ with respect to z. We refer to [Ber09], Section 5.7.2, for the analysis.

1.1.2 Partitioning

It is important to note that there are several different ways to introduce duality in the solution of large-scale optimization problems. For example a strategy, often called *partitioning*, is to divide the variables in two subsets, and minimize first with respect to one subset while taking advantage of whatever simplification may arise by fixing the variables in the other subset.

As an example, the problem

$$\begin{aligned} &\text{minimize} \ \ F(x) + G(y) \\ &\text{subject to} \ \ Ax + By = c, \quad x \in X, \ \ y \in Y, \end{aligned}$$

can be written as

$$\begin{aligned} &\text{minimize} \ \ F(x) + \inf_{By=c-Ax,\, y\in Y} G(y) \\ &\text{subject to} \ \ x \in X, \end{aligned}$$

or

$$\begin{aligned} &\text{minimize} \ \ F(x) + p(c - Ax) \\ &\text{subject to} \ \ x \in X, \end{aligned}$$

where p is given by

$$p(u) = \inf_{By=u,\, y\in Y} G(y).$$

In favorable cases, p can be dealt with conveniently (see e.g., the book [Las70] and the paper [Geo72]).

Strategies of splitting or transforming the variables to facilitate algorithmic solution will be frequently encountered in what follows, and in a variety of contexts, including duality. The next section describes some significant contexts of this type.

1.2 FENCHEL DUALITY AND CONIC PROGRAMMING

Let us consider the Fenchel duality framework (see Section 5.3.5 of Appendix B). It involves the problem

$$\begin{aligned} &\text{minimize} \ \ f_1(x) + f_2(Ax) \\ &\text{subject to} \ \ x \in \Re^n, \end{aligned} \tag{1.9}$$

where A is an $m \times n$ matrix, $f_1 : \Re^n \mapsto (-\infty, \infty]$ and $f_2 : \Re^m \mapsto (-\infty, \infty]$ are closed proper convex functions, and we assume that there exists a feasible solution, i.e., an $x \in \Re^n$ such that $x \in \text{dom}(f_1)$ and $Ax \in \text{dom}(f_2)$.†

The problem is equivalent to the following constrained optimization problem in the variables $x_1 \in \Re^n$ and $x_2 \in \Re^m$:

$$\begin{aligned} &\text{minimize} \ \ f_1(x_1) + f_2(x_2) \\ &\text{subject to} \ \ x_1 \in \text{dom}(f_1), \ \ x_2 \in \text{dom}(f_2), \qquad x_2 = Ax_1. \end{aligned} \tag{1.10}$$

† We remind the reader that our convex analysis notation, terminology, and nonalgorithmic theory are summarized in Appendix B.

Viewing this as a convex programming problem with the linear equality constraint $x_2 = Ax_1$, we obtain the dual function as

$$\begin{aligned} q(\lambda) &= \inf_{x_1 \in \text{dom}(f_1),\ x_2 \in \text{dom}(f_2)} \big\{ f_1(x_1) + f_2(x_2) + \lambda'(x_2 - Ax_1) \big\} \\ &= \inf_{x_1 \in \Re^n} \big\{ f_1(x_1) - \lambda' A x_1 \big\} + \inf_{x_2 \in \Re^m} \big\{ f_2(x_2) + \lambda' x_2 \big\}. \end{aligned}$$

The dual problem of maximizing q over $\lambda \in \Re^m$, after a sign change to convert it to a minimization problem, takes the form

$$\begin{aligned} &\text{minimize} \quad f_1^\star(A'\lambda) + f_2^\star(-\lambda) \\ &\text{subject to} \quad \lambda \in \Re^m, \end{aligned} \tag{1.11}$$

where $f_1^\star$ and $f_2^\star$ are the conjugate functions of f_1 and f_2. We denote by f^* and q^* the corresponding optimal primal and dual values [q^* is the negative of the optimal value of problem (1.11)].

The following Fenchel duality result is given as Prop. 5.3.8 in Appendix B. Parts (a) and (b) are obtained by applying Prop. 1.1.5(a) to problem (1.10), viewed as a problem with $x_2 = Ax_1$ as the only linear equality constraint. The first equation of part (c) is a consequence of Prop. 1.1.5(b). Its equivalence with the last two equations is a consequence of the Conjugate Subgradient Theorem (Prop. 5.4.3, App. B), which states that for a closed proper convex function f, its conjugate $f^\star$, and any pair of vectors (x, y), we have

$$x \in \arg\min_{z \in \Re^n} \big\{ f(z) - z'y \big\} \quad \text{iff} \quad y \in \partial f(x) \quad \text{iff} \quad x \in \partial f^\star(y),$$

with all of these three relations being equivalent to $x'y = f(x) + f^\star(y)$. Here $\partial f(x)$ denotes the subdifferential of f at x (the set of all subgradients of f at x); see Section 5.4 of Appendix B.

Proposition 1.2.1: (Fenchel Duality) Consider problem (1.9).

(a) If f^* is finite and $\big(A \cdot \text{ri}(\text{dom}(f_1))\big) \cap \text{ri}\big(\text{dom}(f_2)\big) \neq \varnothing$, then $f^* = q^*$ and there exists at least one dual optimal solution.

(b) If q^* is finite and $\text{ri}\big(\text{dom}(f_1^\star)\big) \cap \big(A' \cdot \text{ri}(-\text{dom}(f_2^\star))\big) \neq \varnothing$, then $f^* = q^*$ and there exists at least one primal optimal solution.

(c) There holds $f^* = q^*$, and (x^*, λ^*) is a primal and dual optimal solution pair if and only if any one of the following three equivalent conditions hold:

$$x^* \in \arg\min_{x \in \Re^n} \big\{ f_1(x) - x'A'\lambda^* \big\} \quad \text{and} \quad Ax^* \in \arg\min_{z \in \Re^m} \big\{ f_2(z) + z'\lambda^* \big\}, \tag{1.12}$$

$$A'\lambda^* \in \partial f_1(x^*) \quad \text{and} \quad -\lambda^* \in \partial f_2(Ax^*), \tag{1.13}$$

$$x^* \in \partial f_1^\star(A'\lambda^*) \quad \text{and} \quad Ax^* \in \partial f_2^\star(-\lambda^*). \tag{1.14}$$

Minimax Problems

Minimax problems involve minimization over a set X of a function $\overline{F}$ of the form

$$\overline{F}(x) = \sup_{z \in Z} \phi(x, z),$$

where X and Z are subsets of $\Re^n$ and $\Re^m$, respectively, and $\phi : \Re^n \times \Re^m \mapsto \Re$ is a given function. Some (but not all) problems of this type are related to constrained optimization and Fenchel duality.

Example 1.2.1: (Connection with Constrained Optimization)

Let ϕ and Z have the form

$$\phi(x, z) = f(x) + z'g(x), \qquad Z = \{z \mid z \geq 0\},$$

where $f : \Re^n \mapsto \Re$ and $g : \Re^n \mapsto \Re^m$ are given functions. Then it is seen that

$$\overline{F}(x) = \sup_{z \in Z} \phi(x, z) = \begin{cases} f(x) & \text{if } g(x) \leq 0, \\ \infty & \text{otherwise.} \end{cases}$$

Thus minimization of $\overline{F}$ over $x \in X$ is equivalent to solving the constrained optimization problem

$$\begin{aligned} &\text{minimize} \quad f(x) \\ &\text{subject to} \quad x \in X, \quad g(x) \leq 0. \end{aligned} \tag{1.15}$$

The dual problem is to maximize over $z \geq 0$ the function

$$\underline{F}(z) = \inf_{x \in X} \{f(x) + z'g(x)\} = \inf_{x \in X} \phi(x, z),$$

and the minimax equality

$$\inf_{x \in X} \sup_{z \in Z} \phi(x, z) = \sup_{z \in Z} \inf_{x \in X} \phi(x, z) \tag{1.16}$$

is equivalent to problem (1.15) having no duality gap.

Example 1.2.2: (Connection with Fenchel Duality)

Let ϕ have the special form

$$\phi(x, z) = f(x) + z'Ax - g(z),$$

where $f : \Re^n \mapsto \Re$ and $g : \Re^m \mapsto \Re$ are given functions, and A is a given $m \times n$ matrix. Then we have

$$\overline{F}(x) = \sup_{z \in Z} \phi(x, z) = f(x) + \sup_{z \in Z} \{(Ax)'z - g(z)\} = f(x) + \hat{g}^\star(Ax),$$

where $\hat{g}^\star$ is the conjugate of the function

$$\hat{g}(z) = \begin{cases} g(z) & \text{if } z \in Z, \\ \infty & \text{otherwise.} \end{cases}$$

Thus the minimax problem of minimizing $\overline{F}$ over $x \in X$ comes under the Fenchel framework (1.9) with $f_2 = \hat{g}^\star$ and f_1 given by

$$f_1(x) = \begin{cases} f(x) & \text{if } x \in X, \\ \infty & \text{if } x \notin X. \end{cases}$$

It can also be verified that the Fenchel dual problem (1.11) is equivalent to maximizing over $z \in Z$ the function $\underline{F}(z) = \inf_{x\in X} \phi(x,z)$. Again having no duality gap is equivalent to the minimax equality (1.16) holding.

Finally note that strong duality theory is connected with minimax problems primarily when X and Z are convex sets, and ϕ is convex in x and concave in z. When Z is a finite set, there is a different connection with constrained optimization that does not involve Fenchel duality and applies without any convexity conditions. In particular, the problem

$$\begin{aligned} &\text{minimize} \quad \max\{g_1(x), \ldots, g_r(x)\} \\ &\text{subject to} \quad x \in X, \end{aligned}$$

where $g_j : \Re^n \mapsto \Re$ are any real-valued functions, is equivalent to the constrained optimization problem

$$\begin{aligned} &\text{minimize} \quad y \\ &\text{subject to} \quad x \in X, \qquad g_j(x) \le y, \quad j = 1, \ldots, r, \end{aligned}$$

where y is an additional scalar optimization variable. Minimax problems will be discussed further later, in Section 1.4, as an example of problems that may involve a large number of constraints.

Conic Programming

An important problem structure, which can be analyzed as a special case of the Fenchel duality framework is *conic programming*. This is the problem

$$\begin{aligned} &\text{minimize} \quad f(x) \\ &\text{subject to} \quad x \in C, \end{aligned} \tag{1.17}$$

where $f : \Re^n \mapsto (-\infty, \infty]$ is a closed proper convex function and C is a closed convex cone in $\Re^n$.

Indeed, let us apply Fenchel duality with A equal to the identity and the definitions

$$f_1(x) = f(x), \qquad f_2(x) = \begin{cases} 0 & \text{if } x \in C, \\ \infty & \text{if } x \notin C. \end{cases}$$

The corresponding conjugates are

$$f_1^\star(\lambda) = \sup_{x\in\Re^n}\{\lambda'x - f(x)\}, \qquad f_2^\star(\lambda) = \sup_{x\in C}\lambda'x = \begin{cases} 0 & \text{if } \lambda\in C^*, \\ \infty & \text{if } \lambda\notin C^*, \end{cases}$$

where

$$C^* = \{\lambda \mid \lambda'x \le 0,\ \forall\, x\in C\}$$

is the polar cone of C (note that $f_2^\star$ is the support function of C; cf. Section 1.6 of Appendix B). The dual problem is

$$\begin{aligned} &\text{minimize} \quad f^\star(\lambda) \\ &\text{subject to} \quad \lambda\in\hat{C}, \end{aligned} \tag{1.18}$$

where $f^\star$ is the conjugate of f and $\hat{C}$ is the negative polar cone (also called the *dual cone* of C):

$$\hat{C} = -C^* = \{\lambda \mid \lambda'x \ge 0,\ \forall\, x\in C\}.$$

Note the symmetry between primal and dual problems. The strong duality relation $f^* = q^*$ can be written as

$$\inf_{x\in C} f(x) = -\inf_{\lambda\in\hat{C}} f^\star(\lambda).$$

The following proposition translates the conditions of Prop. 1.2.1(a), which guarantees that there is no duality gap and that the dual problem has an optimal solution.

Proposition 1.2.2: (Conic Duality Theorem) Assume that the primal conic problem (1.17) has finite optimal value, and moreover $\text{ri}\big(\text{dom}(f)\big)\cap\text{ri}(C)\neq\emptyset$. Then, there is no duality gap and the dual problem (1.18) has an optimal solution.

Using the symmetry of the primal and dual problems, we also obtain that there is no duality gap and the primal problem (1.17) has an optimal solution if the optimal value of the dual conic problem (1.18) is finite and $\text{ri}\big(\text{dom}(f^\star)\big)\cap\text{ri}(\hat{C})\neq\emptyset$. It is also possible to derive primal and dual optimality conditions by translating the optimality conditions of the Fenchel duality framework [Prop. 1.2.1(c)].

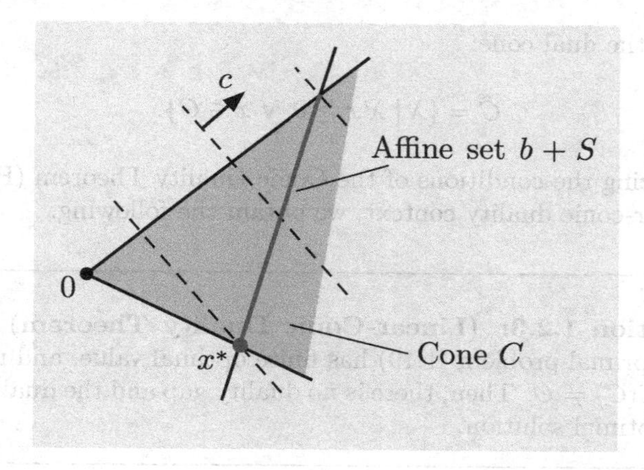

Figure 1.2.1. Illustration of a linear-conic problem: minimizing a linear function $c'x$ over the intersection of an affine set $b + S$ and a convex cone C.

1.2.1 Linear-Conic Problems

An important special case of conic programming, called *linear-conic problem*, arises when $\text{dom}(f)$ is an affine set and f is linear over $\text{dom}(f)$, i.e.,

$$f(x) = \begin{cases} c'x & \text{if } x \in b + S, \\ \infty & \text{if } x \notin b + S, \end{cases}$$

where b and c are given vectors, and S is a subspace. Then the primal problem can be written as

$$\begin{aligned} &\text{minimize} \quad c'x \\ &\text{subject to} \quad x - b \in S, \quad x \in C; \end{aligned} \tag{1.19}$$

see Fig. 1.2.1.

To derive the dual problem, we note that

$$\begin{aligned} f^\star(\lambda) &= \sup_{x-b\in S} (\lambda - c)'x \\ &= \sup_{y\in S} (\lambda - c)'(y + b) \\ &= \begin{cases} (\lambda - c)'b & \text{if } \lambda - c \in S^\perp, \\ \infty & \text{if } \lambda - c \notin S^\perp. \end{cases} \end{aligned}$$

It can be seen that the dual problem $\min_{\lambda\in\hat{C}} f^\star(\lambda)$ [cf. Eq. (1.18)], after discarding the superfluous term $c'b$ from the cost, can be written as

$$\begin{aligned} &\text{minimize} \quad b'\lambda \\ &\text{subject to} \quad \lambda - c \in S^\perp, \quad \lambda \in \hat{C}, \end{aligned} \tag{1.20}$$

where $\hat{C}$ is the dual cone:

$$\hat{C} = \{\lambda \mid \lambda' x \geq 0,\ \forall\, x \in C\}.$$

By specializing the conditions of the Conic Duality Theorem (Prop. 1.2.2) to the linear-conic duality context, we obtain the following.

Proposition 1.2.3: (Linear-Conic Duality Theorem) Assume that the primal problem (1.19) has finite optimal value, and moreover $(b+S) \cap \mathrm{ri}(C) \neq \emptyset$. Then, there is no duality gap and the dual problem has an optimal solution.

Special Forms of Linear-Conic Problems

The primal and dual linear-conic problems (1.19) and (1.20) have been placed in an elegant symmetric form. There are also other useful formats that parallel and generalize similar formats in linear programming. For example, we have the following dual problem pairs:

$$\min_{Ax=b,\ x\in C} c'x \quad \Longleftrightarrow \quad \max_{c-A'\lambda\in\hat{C}} b'\lambda, \tag{1.21}$$

$$\min_{Ax-b\in C} c'x \quad \Longleftrightarrow \quad \max_{A'\lambda=c,\ \lambda\in\hat{C}} b'\lambda, \tag{1.22}$$

where A is an $m \times n$ matrix, and $x \in \Re^n$, $\lambda \in \Re^m$, $c \in \Re^n$, $b \in \Re^m$.

To verify the duality relation (1.21), let $\overline{x}$ be any vector such that $A\overline{x} = b$, and let us write the primal problem on the left in the primal conic form (1.19) as

$$\begin{aligned} &\text{minimize} \ \ c'x \\ &\text{subject to} \ \ x - \overline{x} \in \mathrm{N}(A), \quad x \in C, \end{aligned}$$

where $\mathrm{N}(A)$ is the nullspace of A. The corresponding dual conic problem (1.20) is to solve for μ the problem

$$\begin{aligned} &\text{minimize} \ \ \overline{x}'\mu \\ &\text{subject to} \ \ \mu - c \in \mathrm{N}(A)^{\perp}, \quad \mu \in \hat{C}. \end{aligned} \tag{1.23}$$

Since $\mathrm{N}(A)^{\perp}$ is equal to $\mathrm{Ra}(A')$, the range of A', the constraints of problem (1.23) can be equivalently written as $c - \mu \in -\mathrm{Ra}(A') = \mathrm{Ra}(A')$, $\mu \in \hat{C}$, or

$$c - \mu = A'\lambda, \qquad \mu \in \hat{C},$$

for some $\lambda \in \Re^m$. Making the change of variables $\mu = c - A'\lambda$, the dual problem (1.23) can be written as

$$\begin{array}{ll} \text{minimize} & \overline{x}'(c - A'\lambda) \\ \text{subject to} & c - A'\lambda \in \hat{C}. \end{array}$$

By discarding the constant $\overline{x}'c$ from the cost function, using the fact $A\overline{x} = b$, and changing from minimization to maximization, we see that this dual problem is equivalent to the one in the right-hand side of the duality pair (1.21). The duality relation (1.22) is proved similarly.

We next discuss two important special cases of conic programming: *second order cone programming* and *semidefinite programming*. These problems involve two different special cones, and an explicit definition of the affine set constraint. They arise in a variety of applications, and their computational difficulty in practice tends to lie between that of linear and quadratic programming on one hand, and general convex programming on the other hand.

1.2.2 Second Order Cone Programming

In this section we consider the linear-conic problem (1.22), with the cone

$$C = \left\{ (x_1, \ldots, x_n) \;\middle|\; x_n \geq \sqrt{x_1^2 + \cdots + x_{n-1}^2} \right\},$$

which is known as the *second order cone* (see Fig. 1.2.2). The dual cone is

$$\hat{C} = \{ y \mid 0 \leq y'x, \ \forall\, x \in C \} = \left\{ y \;\middle|\; 0 \leq \inf_{\|(x_1,\ldots,x_{n-1})\| \leq x_n} y'x \right\},$$

and it can be shown that $\hat{C} = C$. This property is referred to as *self-duality* of the second order cone, and is fairly evident from Fig. 1.2.2. For a proof, we write

$$\begin{aligned} \inf_{\|(x_1,\ldots,x_{n-1})\| \leq x_n} y'x &= \inf_{x_n \geq 0} \left\{ y_n x_n + \inf_{\|(x_1,\ldots,x_{n-1})\| \leq x_n} \sum_{i=1}^{n-1} y_i x_i \right\} \\ &= \inf_{x_n \geq 0} \left\{ y_n x_n - \|(y_1, \ldots, y_{n-1})\|\, x_n \right\} \\ &= \begin{cases} 0 & \text{if } \|(y_1, \ldots, y_{n-1})\| \leq y_n, \\ -\infty & \text{otherwise,} \end{cases} \end{aligned}$$

where the second equality follows because the minimum of the inner product of a vector $z \in \Re^{n-1}$ with vectors in the unit ball of $\Re^{n-1}$ is $-\|z\|$. Combining the preceding two relations, we have

$$y \in \hat{C} \quad \text{if and only if} \quad 0 \leq y_n - \|(y_1, \ldots, y_{n-1})\|,$$

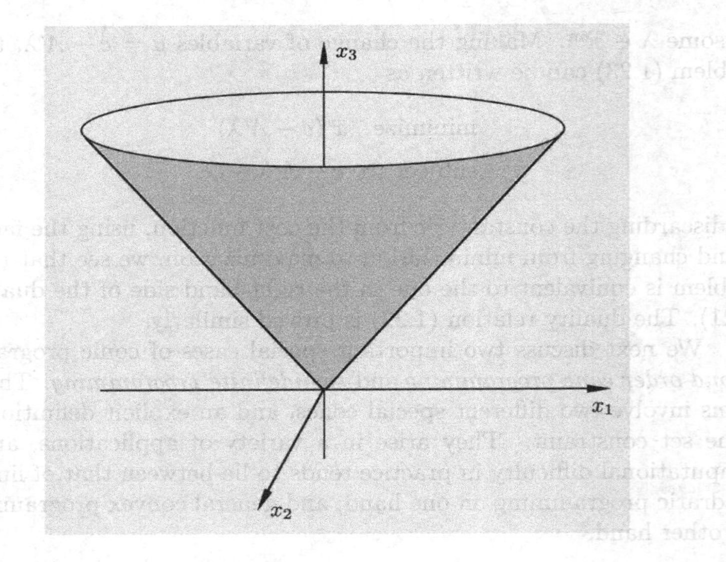

Figure 1.2.2. The second order cone

$$C = \left\{ (x_1, \ldots, x_n) \;\middle|\; x_n \geq \sqrt{x_1^2 + \cdots + x_{n-1}^2} \right\},$$

in $\Re^3$.

so $\hat{C} = C$.

The second order cone programming problem (SOCP for short) is

$$\begin{aligned} &\text{minimize} \quad c'x \\ &\text{subject to} \quad A_i x - b_i \in C_i, \ i = 1, \ldots, m, \end{aligned} \tag{1.24}$$

where $x \in \Re^n$, c is a vector in $\Re^n$, and for $i = 1, \ldots, m$, A_i is an $n_i \times n$ matrix, b_i is a vector in $\Re^{n_i}$, and C_i is the second order cone of $\Re^{n_i}$. It is seen to be a special case of the primal problem in the left-hand side of the duality relation (1.22), where

$$A = \begin{pmatrix} A_1 \\ \vdots \\ A_m \end{pmatrix}, \qquad b = \begin{pmatrix} b_1 \\ \vdots \\ b_m \end{pmatrix}, \qquad C = C_1 \times \cdots \times C_m.$$

Note that linear inequality constraints of the form $a_i'x - b_i \geq 0$ can be written as

$$\begin{pmatrix} 0 \\ a_i' \end{pmatrix} x - \begin{pmatrix} 0 \\ b_i \end{pmatrix} \in C_i,$$

where C_i is the second order cone of $\Re^2$. As a result, linear-conic problems involving second order cones contain as special cases linear programming problems.

We now observe that from the right-hand side of the duality relation (1.22), and the self-duality relation $C = \hat{C}$, the corresponding dual linear-conic problem has the form

$$\begin{aligned} &\text{maximize} \quad \sum_{i=1}^{m} b_i' \lambda_i \\ &\text{subject to} \quad \sum_{i=1}^{m} A_i' \lambda_i = c, \quad \lambda_i \in C_i, \ i = 1, \ldots, m, \end{aligned} \tag{1.25}$$

where $\lambda = (\lambda_1, \ldots, \lambda_m)$. By applying the Linear-Conic Duality Theorem (Prop. 1.2.3), we have the following.

Proposition 1.2.4: (Second Order Cone Duality Theorem) Consider the primal SOCP (1.24), and its dual problem (1.25).

(a) If the optimal value of the primal problem is finite and there exists a feasible solution $\overline{x}$ such that

$$A_i \overline{x} - b_i \in \text{int}(C_i), \qquad i = 1, \ldots, m,$$

then there is no duality gap, and the dual problem has an optimal solution.

(b) If the optimal value of the dual problem is finite and there exists a feasible solution $\overline{\lambda} = (\overline{\lambda}_1, \ldots, \overline{\lambda}_m)$ such that

$$\overline{\lambda}_i \in \text{int}(C_i), \qquad i = 1, \ldots, m,$$

then there is no duality gap, and the primal problem has an optimal solution.

Note that while the Linear-Conic Duality Theorem requires a relative interior point condition, the preceding proposition requires an interior point condition. The reason is that the second order cone has nonempty interior, so its relative interior coincides with its interior.

The SOCP arises in many application contexts, and significantly, it can be solved numerically with powerful specialized algorithms that belong to the class of interior point methods, which will be discussed in Section 6.8. We refer to the literature for a more detailed description and analysis (see e.g., the books [BeN01], [BoV04]).

Generally, SOCPs can be recognized from the presence of convex quadratic functions in the cost or the constraint functions. The following are illustrative examples. The first example relates to the field of robust optimization, which involves optimization under uncertainty described by set membership.

Example 1.2.3: (Robust Linear Programming)

Frequently, there is uncertainty about the data of an optimization problem, so one would like to have a solution that is adequate for a whole range of the uncertainty. A popular formulation of this type, is to assume that the constraints contain parameters that take values in a given set, and require that the constraints are satisfied for all values in that set. This approach is also known as a set membership description of the uncertainty and has been used in fields other than optimization, such as set membership estimation, and minimax control (see the textbook [Ber07], which also surveys earlier work).

As an example, consider the problem

$$\begin{aligned}&\text{minimize } \ c'x\\&\text{subject to } \ a_j'x \le b_j, \quad \forall\,(a_j,b_j)\in T_j, \quad j=1,\ldots,r,\end{aligned} \tag{1.26}$$

where $c \in \Re^n$ is a given vector, and T_j is a given subset of $\Re^{n+1}$ to which the constraint parameter vectors (a_j, b_j) must belong. The vector x must be chosen so that the constraint $a_j'x \le b_j$ is satisfied for all $(a_j, b_j) \in T_j$, $j = 1,\ldots,r$.

Generally, when T_j contains an infinite number of elements, this problem involves a correspondingly infinite number of constraints. To convert the problem to one involving a finite number of constraints, we note that

$$a_j'x \le b_j, \ \forall\,(a_j,b_j)\in T_j \quad \text{if and only if} \quad g_j(x)\le 0,$$

where

$$g_j(x) = \sup_{(a_j,b_j)\in T_j} \{a_j'x - b_j\}. \tag{1.27}$$

Thus, the robust linear programming problem (1.26) is equivalent to

$$\begin{aligned}&\text{minimize } \ c'x\\&\text{subject to } \ g_j(x)\le 0, \qquad j=1,\ldots,r.\end{aligned}$$

For special choices of the set T_j, the function g_j can be expressed in closed form, and in the case where T_j is an ellipsoid, it turns out that the constraint $g_j(x) \le 0$ can be expressed in terms of a second order cone. To see this, let

$$T_j = \big\{(\overline{a}_j + P_j u_j, \overline{b}_j + q_j'u_j) \mid \|u_j\| \le 1,\ u_j \in \Re^{n_j}\big\}, \tag{1.28}$$

where P_j is a given $n \times n_j$ matrix, $\overline{a}_j \in \Re^n$ and $q_j \in \Re^{n_j}$ are given vectors, and $\overline{b}_j$ is a given scalar. Then, from Eqs. (1.27) and (1.28),

$$\begin{aligned}g_j(x) &= \sup_{\|u_j\|\le 1} \big\{(\overline{a}_j + P_j u_j)'x - (\overline{b}_j + q_j'u_j)\big\}\\&= \sup_{\|u_j\|\le 1} (P_j'x - q_j)'u_j + \overline{a}_j'x - \overline{b}_j,\end{aligned}$$

and finally

$$g_j(x) = \|P_j'x - q_j\| + \overline{a}_j'x - \overline{b}_j.$$

Thus,

$$g_j(x) \le 0 \qquad \text{if and only if} \qquad (P_j'x - q_j, \overline{b}_j - \overline{a}_j'x) \in C_j,$$

where C_j is the second order cone of $\Re^{n_j+1}$; i.e., the "robust" constraint $g_j(x) \le 0$ is equivalent to a second order cone constraint. It follows that in the case of ellipsoidal uncertainty, the robust linear programming problem (1.26) is a SOCP of the form (1.24).

Example 1.2.4: (Quadratically Constrained Quadratic Problems)

Consider the quadratically constrained quadratic problem

$$\begin{aligned} &\text{minimize} && x'Q_0x + 2q_0'x + p_0 \\ &\text{subject to} && x'Q_jx + 2q_j'x + p_j \le 0, \quad j = 1, \ldots, r, \end{aligned}$$

where $Q_0, \ldots, Q_r$ are symmetric $n \times n$ positive definite matrices, $q_0, \ldots, q_r$ are vectors in $\Re^n$, and $p_0, \ldots, p_r$ are scalars. We show that the problem can be converted to the second order cone format. A similar conversion is also possible for the quadratic programming problem where Q_0 is positive definite and $Q_j = 0$, $j = 1, \ldots, r$.

Indeed, since each Q_j is symmetric and positive definite, we have

$$\begin{aligned} x'Q_jx + 2q_j'x + p_j &= \left(Q_j^{1/2}x\right)' Q_j^{1/2}x + 2\left(Q_j^{-1/2}q_j\right)' Q_j^{1/2}x + p_j \\ &= \|Q_j^{1/2}x + Q_j^{-1/2}q_j\|^2 + p_j - q_j'Q_j^{-1}q_j, \end{aligned}$$

for $j = 0, 1, \ldots, r$. Thus, the problem can be written as

$$\begin{aligned} &\text{minimize} && \|Q_0^{1/2}x + Q_0^{-1/2}q_0\|^2 + p_0 - q_0'Q_0^{-1}q_0 \\ &\text{subject to} && \|Q_j^{1/2}x + Q_j^{-1/2}q_j\|^2 + p_j - q_j'Q_j^{-1}q_j \le 0, \quad j = 1, \ldots, r, \end{aligned}$$

or, by neglecting the constant $p_0 - q_0'Q_0^{-1}q_0$,

$$\begin{aligned} &\text{minimize} && \|Q_0^{1/2}x + Q_0^{-1/2}q_0\| \\ &\text{subject to} && \|Q_j^{1/2}x + Q_j^{-1/2}q_j\| \le \left(q_j'Q_j^{-1}q_j - p_j\right)^{1/2}, \quad j = 1, \ldots, r. \end{aligned}$$

By introducing an auxiliary variable x_{n+1}, the problem can be written as

$$\begin{aligned} &\text{minimize} && x_{n+1} \\ &\text{subject to} && \|Q_0^{1/2}x + Q_0^{-1/2}q_0\| \le x_{n+1} \\ &&& \|Q_j^{1/2}x + Q_j^{-1/2}q_j\| \le \left(q_j'Q_j^{-1}q_j - p_j\right)^{1/2}, \quad j = 1, \ldots, r. \end{aligned}$$

It can be seen that this problem has the second order cone form (1.24). In particular, the first constraint is of the form $A_0x - b_0 \in C$, where C is the second order cone of $\Re^{n+1}$ and the $(n+1)$st component of $A_0x - b_0$ is x_{n+1}. The remaining r constraints are of the form $A_jx - b_j \in C$, where the $(n+1)$st component of $A_jx - b_j$ is the scalar $\big(q_j'Q_j^{-1}q_j - p_j\big)^{1/2}$.

We finally note that the problem of this example is special in that it has no duality gap, assuming its optimal value is finite, i.e., there is no need for the interior point conditions of Prop. 1.2.4. This can be traced to the fact that linear transformations preserve the closure of sets defined by quadratic constraints (see e.g., BNO03], Section 1.5.2).

1.2.3 Semidefinite Programming

In this section we consider the linear-conic problem (1.21) with C being the cone of matrices that are positive semidefinite.† This is called the *positive semidefinite cone*. To define the problem, we view the space of symmetric $n \times n$ matrices as the space $\Re^{n^2}$ with the inner product

$$< X, Y >= \text{trace}(XY) = \sum_{i=1}^{n}\sum_{j=1}^{n} x_{ij}y_{ij}.$$

The interior of C is the set of positive definite matrices.

The dual cone is

$$\hat{C} = \big\{Y \mid \text{trace}(XY) \geq 0,\ \forall\ X \in C\big\},$$

and it can be shown that $\hat{C} = C$, i.e., C is self-dual. Indeed, if $Y \notin C$, there exists a vector $v \in \Re^n$ such that

$$0 > v'Yv = \text{trace}(vv'Y).$$

Hence the positive semidefinite matrix $X = vv'$ satisfies $0 > \text{trace}(XY)$, so $Y \notin \hat{C}$ and it follows that $C \supset \hat{C}$. Conversely, let $Y \in C$, and let X be any positive semidefinite matrix. We can express X as

$$X = \sum_{i=1}^{n} \lambda_i e_i e_i',$$

where λ_i are the nonnegative eigenvalues of X, and e_i are corresponding orthonormal eigenvectors. Then,

$$\text{trace}(XY) = \text{trace}\left(Y\sum_{i=1}^{n} \lambda_i e_i e_i'\right) = \sum_{i=1}^{n} \lambda_i e_i' Y e_i \geq 0.$$

† As noted in Appendix A, throughout this book a positive semidefinite matrix is implicitly assumed to be symmetric.

It follows that $Y \in \hat{C}$ and $C \subset \hat{C}$. Thus C is self-dual, $C = \hat{C}$.

The semidefinite programming problem (SDP for short) is to minimize a linear function of a symmetric matrix over the intersection of an affine set with the positive semidefinite cone. It has the form

$$\begin{array}{ll} \text{minimize} & < D, X > \\ \text{subject to} & < A_i, X >= b_i, \quad i = 1, \ldots, m, \quad X \in C, \end{array} \tag{1.29}$$

where D, $A_1, \ldots, A_m$, are given $n \times n$ symmetric matrices, and $b_1, \ldots, b_m$, are given scalars. It is seen to be a special case of the primal problem in the left-hand side of the duality relation (1.21).

We can view the SDP as a problem with linear cost, linear constraints, and a convex set constraint. Then, similar to the case of SOCP, it can be verified that the dual problem (1.20), as given by the right-hand side of the duality relation (1.21), takes the form

$$\begin{array}{ll} \text{maximize} & b'\lambda \\ \text{subject to} & D - (\lambda_1 A_1 + \cdots + \lambda_m A_m) \in C, \end{array} \tag{1.30}$$

where $b = (b_1, \ldots, b_m)$ and the maximization is over the vector $\lambda = (\lambda_1, \ldots, \lambda_m)$. By applying the Linear-Conic Duality Theorem (Prop. 1.2.3), we have the following proposition.

Proposition 1.2.5: (Semidefinite Duality Theorem) Consider the primal SDP (1.29), and its dual problem (1.30).

(a) If the optimal value of the primal problem is finite and there exists a primal-feasible solution, which is positive definite, then there is no duality gap, and the dual problem has an optimal solution.

(b) If the optimal value of the dual problem is finite and there exist scalars $\overline{\lambda}_1, \ldots, \overline{\lambda}_m$ such that $D - (\overline{\lambda}_1 A_1 + \cdots + \overline{\lambda}_m A_m)$ is positive definite, then there is no duality gap, and the primal problem has an optimal solution.

The SDP is a fairly general problem. In particular, it can be shown that a SOCP can be cast as a SDP. Thus SDP involves a more general structure than SOCP. This is consistent with the practical observation that the latter problem is generally more amenable to computational solution. We provide some examples of problem formulation as an SDP.

Example 1.2.5: (Minimizing the Maximum Eigenvalue)

Given a symmetric $n \times n$ matrix $M(\lambda)$, which depends on a parameter vector $\lambda = (\lambda_1, \ldots, \lambda_m)$, we want to choose λ so as to minimize the maximum eigenvalue of $M(\lambda)$. We pose this problem as

$$\begin{aligned}&\text{minimize} \quad z\\&\text{subject to} \quad \text{maximum eigenvalue of } M(\lambda) \le z,\end{aligned}$$

or equivalently

$$\begin{aligned}&\text{minimize} \quad z\\&\text{subject to} \quad zI - M(\lambda) \in C,\end{aligned}$$

where I is the $n \times n$ identity matrix, and C is the semidefinite cone. If $M(\lambda)$ is an affine function of λ,

$$M(\lambda) = M_0 + \lambda_1 M_1 + \cdots + \lambda_m M_m,$$

the problem has the form of the dual problem (1.30), with the optimization variables being $(z, \lambda_1, \ldots, \lambda_m)$.

Example 1.2.6: (Semidefinite Relaxation – Lower Bounds for Discrete Optimization Problems)

Semidefinite programming provides a means for deriving lower bounds to the optimal value of several types of discrete optimization problems. As an example, consider the following quadratic problem with quadratic equality constraints

$$\begin{aligned}&\text{minimize} \quad x'Q_0x + a_0'x + b_0\\&\text{subject to} \quad x'Q_ix + a_i'x + b_i = 0, \quad i = 1, \ldots, m,\end{aligned} \tag{1.31}$$

where $Q_0, \ldots, Q_m$ are symmetric $n \times n$ matrices, $a_0, \ldots, a_m$ are vectors in $\Re^n$, and $b_0, \ldots, b_m$ are scalars.

This problem can be used to model broad classes of discrete optimization problems. To see this, consider an integer constraint that a variable x_i must be either 0 or 1. Such a constraint can be expressed by the quadratic equality $x_i^2 - x_i = 0$. Furthermore, a linear inequality constraint $a_j'x \le b_j$ can be expressed as the quadratic equality constraint $y_j^2 + a_j'x - b_j = 0$, where y_j is an additional variable.

Introducing a multiplier vector $\lambda = (\lambda_1, \ldots, \lambda_m)$, the dual function is given by

$$q(\lambda) = \inf_{x \in \Re^n} \{x'Q(\lambda)x + a(\lambda)'x + b(\lambda)\},$$

where

$$Q(\lambda) = Q_0 + \sum_{i=1}^m \lambda_i Q_i, \quad a(\lambda) = a_0 + \sum_{i=1}^m \lambda_i a_i, \quad b(\lambda) = b_0 + \sum_{i=1}^m \lambda_i b_i.$$

Let f^* and q^* be the optimal values of problem (1.31) and its dual, and note that by weak duality, we have $f^* \ge q^*$. By introducing an auxiliary scalar variable ξ, we see that the dual problem is to find a pair (ξ, λ) that solves the problem

$$\begin{aligned} &\text{maximize} \quad \xi \\ &\text{subject to} \quad q(\lambda) \geq \xi. \end{aligned}$$

The constraint $q(\lambda) \geq \xi$ of this problem can be written as

$$\inf_{x \in \Re^n} \big\{ x'Q(\lambda)x + a(\lambda)'x + b(\lambda) - \xi \big\} \geq 0,$$

or equivalently, introducing a scalar variable t and multiplying with t^2,

$$\inf_{x \in \Re^n,\, t \in \Re} \big\{ (tx)'Q(\lambda)(tx) + a(\lambda)'(tx)t + \big(b(\lambda) - \xi\big)t^2 \big\} \geq 0.$$

Writing $y = tx$, this relation takes the form of a quadratic in (y, t),

$$\inf_{y \in \Re^n,\, t \in \Re} \big\{ y'Q(\lambda)y + a(\lambda)'yt + \big(b(\lambda) - \xi\big)t^2 \big\} \geq 0,$$

or

$$\begin{pmatrix} Q(\lambda) & \frac{1}{2}a(\lambda) \\ \frac{1}{2}a(\lambda)' & b(\lambda) - \xi \end{pmatrix} \in C, \tag{1.32}$$

where C is the positive semidefinite cone. Thus the dual problem is equivalent to the SDP of maximizing ξ over all (ξ, λ) satisfying the constraint (1.32), and its optimal value q^* is a lower bound to f^*.

1.3 ADDITIVE COST PROBLEMS

In this section we focus on a structural characteristic that arises in several important contexts: a cost function f that is the sum of a large number of components $f_i : \Re^n \mapsto \Re$,

$$f(x) = \sum_{i=1}^{m} f_i(x). \tag{1.33}$$

Such cost functions can be minimized with specialized methods, called *incremental*, which exploit their additive structure, by updating x using one component function f_i at a time (see Section 2.1.5). Problems with additive cost functions can also be treated with specialized outer and inner linearization methods that approximate the component functions f_i individually (rather than approximating f); see Section 4.4.

An important special case is the cost function of the dual of a separable problem

$$\begin{aligned} &\text{maximize} \quad \sum_{i=1}^{m} q_i(\mu) \\ &\text{subject to} \quad \mu \geq 0, \end{aligned}$$

where

$$q_i(\mu) = \inf_{x_i \in X_i} \left\{ f_i(x_i) + \sum_{j=1}^{r} \mu_j g_{ij}(x_i) \right\},$$

and $\mu = (\mu_1, \ldots, \mu_r)$ [cf. Eq. (1.8)]. After a sign change to convert to minimization it takes the form (1.33) with $f_i(\mu) = -q_i(\mu)$. This is a major class of additive cost problems.

We will next describe some applications from a variety of fields. The following five examples arise in many machine learning contexts.

Example 1.3.1: (Regularized Regression)

This is a broad class of applications that relate to parameter estimation. The cost function involves a sum of terms $f_i(x)$, each corresponding to the error between some data and the output of a parametric model, with x being the vector of parameters. An example is linear least squares problems, also referred to as *linear regression* problems, where f_i has quadratic structure. Often a convex regularization function $R(x)$ is added to the least squares objective, to induce desirable properties of the solution and/or the corresponding algorithms. This gives rise to problems of the form

$$\begin{aligned} &\text{minimize} \quad R(x) + \tfrac{1}{2}\textstyle\sum_{i=1}^{m} (c_i'x - b_i)^2 \\ &\text{subject to} \quad x \in \Re^n, \end{aligned}$$

where c_i and b_i are given vectors and scalars, respectively. The regularization function R is often taken to be differentiable, and particularly quadratic. However, there are practically important examples of nondifferentiable choices (see the next example).

In statistical applications, such a problem arises when constructing a linear model for an unknown input-output relation. The model involves a vector of parameters x, to be determined, which weigh input data (the components of the vectors c_i). The inner products $c_i'x$ produced by the model are matched against the scalars b_i, which are observed output data, corresponding to inputs c_i from the true input-output relation that we try to represent. The optimal vector of parameters x^* provides the model that (in the absence of a regularization function) minimizes the sum of the squared errors $(c_i'x^* - b_i)^2$.

In a more general version of the problem, a nonlinear parametric model is constructed, giving rise to a nonlinear least squares problem of the form

$$\begin{aligned} &\text{minimize} \quad R(x) + \sum_{i=1}^{m} \big|g_i(x)\big|^2 \\ &\text{subject to} \quad x \in \Re^n, \end{aligned}$$

where $g_i : \Re^n \mapsto \Re$ are given nonlinear functions that depend on the data. This is also a common problem, referred to as *nonlinear regression*, which, however, is often nonconvex [it is convex if the functions g_i are convex *and* also nonnegative, i.e., $g_i(x) \geq 0$ for all $x \in \Re^n$].

It is also possible to use a nonquadratic function of the error between some data and the output of a linear parametric model. Thus in place of the squared error $(1/2)(c_i'x - b_i)^2$, we may use $h_i(c_i'x - b_i)$, where $h_i : \Re \mapsto \Re$ is a convex function, leading to the problem

$$\begin{aligned} &\text{minimize} \quad R(x) + \sum_{i=1}^{m} h_i(c_i'x - b_i) \\ &\text{subject to} \quad x \in \Re^n. \end{aligned}$$

Generally the choice of the function h_i is dictated by statistical modeling considerations, for which the reader may consult the relevant literature. An example is

$$h_i(c_i'x - b_i) = |c_i'x - b_i|,$$

which tends to result in a more robust estimate than least squares in the presence of large outliers in the data. This is known as the *least absolute deviations* method.

There are also constrained variants of the problems just discussed, where the parameter vector x is required to belong to some subset of $\Re^n$, such as the nonnegative orthant or a "box" formed by given upper and lower bounds on the components of x. Such constraints may be used to encode into the model some prior knowledge about the nature of the solution.

Example 1.3.2: (ℓ_1-Regularization)

A popular approach to regularized regression involves ℓ_1*-regularization*, where

$$R(x) = \gamma\|x\|_1 = \gamma \sum_{j=1}^{n} |x^j|,$$

γ is a positive scalar and x^j is the jth coordinate of x. The reason for the popularity of the ℓ_1 norm $\|x\|_1$ is that it tends to produce optimal solutions where a greater number of components x^j are zero, relative to the case of quadratic regularization (see Fig. 1.3.1). This is considered desirable in many statistical applications, where the number of parameters to include in a model may not be known a priori; see e.g., [Tib96], [DoE03], [BJM12]. The special case where a linear least squares model is used,

$$\begin{aligned} &\text{minimize} \quad \gamma\|x\|_1 + \tfrac{1}{2}\textstyle\sum_{i=1}^{m}(c_i'x - b_i)^2 \\ &\text{subject to} \quad x \in \Re^n, \end{aligned}$$

is known as the *lasso problem*.

In a generalization of the lasso problem, the ℓ_1 regularization function $\|x\|_1$ is replaced by a scaled version $\|Sx\|_1$, where S is some scaling matrix.

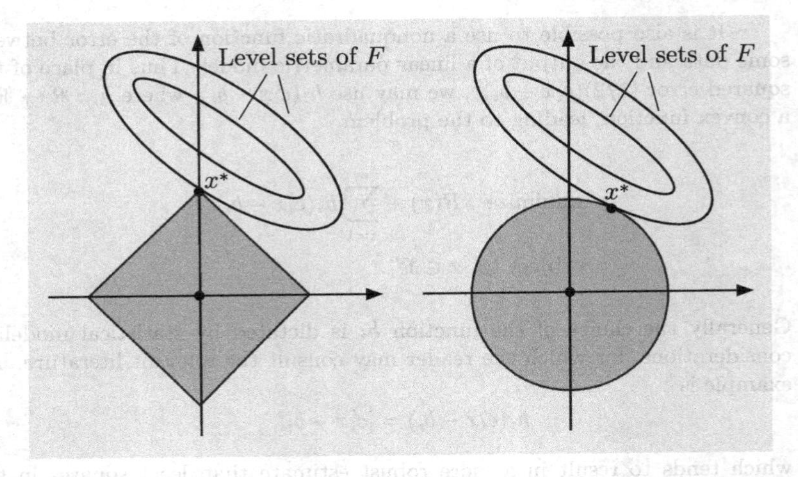

Figure 1.3.1. Illustration of the effect of ℓ_1-regularization for cost functions of the form $\gamma\|x\|_1+F(x)$, where $\gamma>0$ and $F:\Re^n\mapsto\Re$ is differentiable (figure in the left-hand side). The optimal solution x^* tends to have more zero components than in the corresponding quadratic regularization case, illustrated in the right-hand side.

The term $\|Sx\|_1$ then induces a penalty on some undesirable characteristic of the solution. For example the problem

$$\begin{aligned}&\text{minimize} \quad \gamma\sum_{i=1}^{n-1}|x_{i+1}-x_i|+\tfrac{1}{2}\textstyle\sum_{i=1}^{m}(c_i'x-b_i)^2\\&\text{subject to} \quad x\in\Re^n,\end{aligned}$$

is known as the *total variation denoising problem*; see e.g., [ROF92], [Cha04], [BeT09a]. The regularization term here encourages consecutive variables to take similar values, and tends to produce more smoothly varying solutions.

Another related example is *matrix completion with nuclear norm regularization*; see e.g., [CaR09], [CaT10], [RFP10], [Rec11], [ReR13]. Here the minimization is over all $m\times n$ matrices X, with components denoted X_{ij}. We have a set of entries M_{ij}, $(i,j)\in\Omega$, where Ω is a subset of index pairs, and we want to find X whose entries X_{ij} are close to M_{ij} for $(i,j)\in\Omega$, and has as small rank as possible, a property that is desirable on the basis of statistical considerations. The following more tractable version of the problem is solved instead:

$$\begin{aligned}&\text{minimize} \quad \gamma\|X\|_*+\tfrac{1}{2}\textstyle\sum_{(i,j)\in\Omega}(X_{ij}-M_{ij})^2\\&\text{subject to} \quad X\in\Re^{m\times n},\end{aligned}$$

where $\|X\|_*$ is the *nuclear norm* of X, defined as the sum of the singular values of X. There is substantial theory that justifies this approximation, for which we refer to the literature. It turns out that the nuclear norm is a convex function with some nice properties. In particular, its subdifferential at any X can be conveniently characterized for use in algorithms.

Let us finally note that sometimes additional regularization functions are used in conjunction with ℓ_1-type terms. An example is the sum of a quadratic and an ℓ_1-type term.

Example 1.3.3: (Classification)

In the regression problems of the preceding examples we aim to construct a parametric model that matches well an input-output relationship based on given data. Similar problems arise in a classification context, where we try to construct a parametric model for predicting whether an object with certain characteristics (also called features) belongs to a given category or not.

We assume that each object is characterized by a *feature vector* c that belongs to $\Re^n$ and a *label* b that takes the values $+1$ or -1, if the object belongs to the category or not, respectively. As illustration consider a credit card company that wishes to classify applicants as "low risk" ($+1$) or "high risk" (-1), with each customer characterized by n scalar features of financial and personal type.

We are given data, which is a set of feature-label pairs (c_i, b_i), $i = 1, \ldots, m$. Based on this data, we want to find a parameter vector $x \in \Re^n$ and a scalar $y \in \Re$ such that the sign of $c'x + y$ is a good predictor of the label of an object with feature vector c. Thus, loosely speaking, x and y should be such that for "most" of the given feature-label data (c_i, b_i) we have

$$c_i'x + y > 0, \qquad \text{if } b_i = +1,$$

$$c_i'x + y < 0, \qquad \text{if } b_i = -1.$$

In the statistical literature, $c'x + y$ is often called the *discriminant function*, and the value of

$$b_i(c_i'x + y),$$

for a given object i provides a measure of "margin" to misclassification of the object. In particular, a classification error is made for object i when $b_i(c_i'x + y) < 0$.

Thus it makes sense to formulate classification as an optimization problem where negative values of $b_i(c_i'x + y)$ are penalized. This leads to the problem

$$\begin{aligned} &\text{minimize} \quad R(x) + \sum_{i=1}^{m} h\big(b_i(c_i'x + y)\big) \\ &\text{subject to} \quad x \in \Re^n,\ y \in \Re, \end{aligned}$$

where R is a suitable regularization function, and $h : \Re \mapsto \Re$ is a convex function that penalizes negative values of its argument. It would make some sense to use a penalty of one unit for misclassification, i.e.,

$$h(z) = \begin{cases} 0 & \text{if } z \geq 0, \\ 1 & \text{if } z < 0, \end{cases}$$

but such a penalty function is discontinuous. To obtain a continuous cost function, we allow a continuous transition of h from negative to positive values, leading to a variety of nonincreasing functions h. The choice of h depends on the given application and other theoretical considerations for which we refer to the literature. Some common examples are

$$h(z) = e^{-z}, \qquad \text{(exponential loss)},$$

$$h(z) = \log\left(1 + e^{-z}\right), \qquad \text{(logistic loss)},$$

$$h(z) = \max\left\{0, 1 - z\right\}, \qquad \text{(hinge loss)}.$$

For the case of logistic loss the method comes under the methodology of *logistic regression*, and for the case of hinge loss the method comes under the methodology of *support vector machines*. As in the case of regression, the regularization function R could be quadratic, the ℓ_1 norm, or some scaled version or combination thereof. There is extensive literature on these methodologies and their applications, to which we refer for further discussion.

Example 1.3.4: (Nonnegative Matrix Factorization)

The nonnegative matrix factorization problem is to approximately factor a given nonnegative matrix B as CX, where C and X are nonnegative matrices to be determined via the optimization

$$\begin{aligned} &\text{minimize} \quad \|CX - B\|_F^2 \\ &\text{subject to} \quad C \geq 0,\ X \geq 0. \end{aligned}$$

Here $\|\cdot\|_F$ denotes the Frobenius norm of a matrix ($\|M\|_F^2$ is the sum of the squares of the scalar components of M). The matrices B, C, and X must have compatible dimensions, with the column dimension of C usually being much smaller than its row dimension, so that CX is a low-rank approximation of B. In some versions of the problem some of the nonnegativity constraints on the components of C and X may be relaxed. Moreover, regularization terms may be added to the cost function to induce sparsity or some other effect, similar to earlier examples in this section.

This problem, formulated in the 90s, [PaT94], [Paa97], [LeS99], has become a popular model for regression-type applications such as the ones of Example 1.3.1, but with the vectors c_i in the least squares objective $\sum_{i=1}^m (c_i'x - b_i)^2$ being unknown and subject to optimization. In the regression context of Example 1.3.1, we aim to (approximately) represent the data in the range space of the matrix C whose rows are the vectors c_i', and we may view C as a matrix of known basis functions. In the matrix factorization context of the present example, we aim to discover a "good" matrix C of basis functions that represents well the given data, i.e., the matrix B.

An important characteristic of the problem is that its cost function is not convex jointly in (C, X). However, it is convex in each of the matrices C and X individually, when the other matrix is held fixed. This facilitates the application of algorithms that involve alternate minimizations with respect to C and with respect to X; see Section 6.5. We refer to the literature, e.g., the papers [BBL07], [Lin07], [GoZ12], for a discussion of related algorithmic issues.

Example 1.3.5: (Maximum Likelihood Estimation)

The maximum likelihood approach is a major statistical inference methodology for parameter estimation, which is described in many sources (see e.g., the textbooks [Was04], [HTF09]). In fact in many cases, a maximum likelihood formulation is used to provide a probabilistic justification of the regression and classification models of the preceding examples.

Here we observe a sample of a random vector Z whose distribution $P_Z(\cdot;x)$ depends on an unknown parameter vector $x \in \Re^n$. For simplicity we assume that Z can take only a finite set of values, so that $P_Z(z;x)$ is the probability that Z takes the value z when the parameter vector has the value x. We estimate x based on the given sample value z, by solving the problem

$$\begin{aligned} &\text{maximize} \quad P_Z(z;x) \\ &\text{subject to } x \in \Re^n. \end{aligned} \tag{1.34}$$

The cost function $P_Z(z;\cdot)$ of this problem may either have an additive structure or may be equivalent to a problem that has an additive structure. For example the event that $Z = z$ may be the union of a large number of disjoint events, so $P_Z(z;x)$ is the sum of the probabilities of these events. For another important context, suppose that the data z consists of m independent samples $z_1,\ldots,z_m$ drawn from a distribution $P(\cdot;x)$, in which case

$$P_Z(z;x) = P(z_1;x)\cdots P(z_m;x).$$

Then the maximization (1.34) is equivalent to the additive cost minimization

$$\begin{aligned} &\text{minimize} \sum_{i=1}^{m} f_i(x) \\ &\text{subject to } x \in \Re^n, \end{aligned}$$

where

$$f_i(x) = -\log P(z_i;x).$$

In many applications the number of samples m is very large, in which case special methods that exploit the additive structure of the cost are recommended. Often a suitable regularization term is added to the cost function, similar to the preceding examples.

Example 1.3.6: (Minimization of an Expected Value - Stochastic Programming)

An important context where additive cost functions arise is the minimization of an expected value

$$\begin{aligned} &\text{minimize} \quad E\{F(x,w)\} \\ &\text{subject to } x \in X, \end{aligned}$$

where w is a random variable taking a finite but very large number of values w_i, $i = 1, \ldots, m$, with corresponding probabilities π_i. Then the cost function consists of the sum of the m functions $\pi_i F(x, w_i)$.

For example, in *stochastic programming*, a classical model of two-stage optimization under uncertainty, a vector $x \in X$ is selected, a random event occurs that has m possible outcomes $w_1, \ldots, w_m$, and another vector $y \in Y$ is selected with knowledge of the outcome that occurred (see e.g., the books [BiL97], [KaW94], [Pre95], [SDR09]). Then for optimization purposes, we need to specify a different vector $y_i \in Y$ for each outcome w_i. The problem is to minimize the expected cost

$$F(x) + \sum_{i=1}^{m} \pi_i G_i(y_i),$$

where $G_i(y_i)$ is the cost associated with the choice y_i and the occurrence of w_i, and π_i is the corresponding probability. This is a problem with an additive cost function.

Additive cost functions also arise when the expected value cost function $E\big\{F(x,w)\big\}$ is approximated by an m-sample average

$$f(x) = \frac{1}{m} \sum_{i=1}^{m} F(x, w_i),$$

where w_i are independent samples of the random variable w. The minimum of the sample average $f(x)$ is then taken as an approximation of the minimum of $E\big\{F(x,w)\big\}$.

Generally additive cost problems arise when we want to strike a balance between several types of costs by lumping them into a single cost function. The following is an example of a different character than the preceding ones.

Example 1.3.7: (Weber Problem in Location Theory)

A basic problem in location theory is to find a point x in the plane whose sum of weighted distances from a given set of points $y_1, \ldots, y_m$ is minimized. Mathematically, the problem is

$$\begin{aligned} &\text{minimize} \ \sum_{i=1}^{m} w_i \|x - y_i\| \\ &\text{subject to} \ \ x \in \Re^n, \end{aligned}$$

where $w_1, \ldots, w_m$ are given positive scalars. This problem has many variations, including constrained versions, and descends from the famous Fermat-Torricelli-Viviani problem (see [BMS99] for an account of the history of this problem). We refer to the book [DrH04] for a survey of recent research, and to the paper [BeT10] for a discussion that is relevant to our context.

The structure of the additive cost function (1.33) often facilitates the use of a distributed computing system that is well-suited for the incremental approach. The following is an illustrative example.

Example 1.3.8: (Distributed Incremental Optimization – Sensor Networks)

Consider a network of m sensors where data are collected and are used to solve some inference problem involving a parameter vector x. If $f_i(x)$ represents an error penalty for the data collected by the ith sensor, the inference problem involves an additive cost function $\sum_{i=1}^m f_i$. While it is possible to collect all the data at a fusion center where the problem will be solved in centralized manner, it may be preferable to adopt a distributed approach in order to save in data communication overhead and/or take advantage of parallelism in computation. In such an approach the current iterate x_k is passed on from one sensor to another, with each sensor i performing an incremental iteration involving just its local component f_i. The entire cost function need not be known at any one location. For further discussion we refer to representative sources such as [RaN04], [RaN05], [BHG08], [MRS10], [GSW12], and [Say14].

The approach of computing incrementally the values and subgradients of the components f_i in a distributed manner can be substantially extended to apply to general systems of asynchronous distributed computation, where the components are processed at the nodes of a computing network, and the results are suitably combined [NBB01] (see our discussion in Sections 2.1.5 and 2.1.6).

Let us finally note a constrained version of additive cost problems where the functions f_i are extended real-valued. This is essentially equivalent to constraining x to lie in the intersection of the domains

$$X_i = \mathrm{dom}(f_i),$$

resulting in a problem of the form

$$\begin{aligned} &\text{minimize} \quad \sum_{i=1}^m f_i(x) \\ &\text{subject to} \quad x \in \cap_{i=1}^m X_i, \end{aligned}$$

where each f_i is real-valued over the set X_i. Methods that are well-suited for the unconstrained version of the problem where $X_i \equiv \Re^n$ can often be modified to apply to the constrained version, as we will see in Chapter 6, where we will discuss incremental constraint projection methods. However, the case of constraint sets with many components arises independently of whether the cost function is additive or not, and has its own character, as we discuss in the next section.

1.4 LARGE NUMBER OF CONSTRAINTS

In this section we consider problems of the form

$$\begin{array}{ll} \text{minimize} & f(x) \\ \text{subject to} & x \in X, \quad g_j(x) \leq 0, \quad j = 1, \ldots, r, \end{array} \tag{1.35}$$

where the number r of constraints is very large. Problems of this type occur often in practice, either directly or via reformulation from other problems. A similar type of problem arises when the abstract constraint set X consists of the intersection of many simpler sets:

$$X = \cap_{\ell \in L} X_\ell,$$

where L is a finite or infinite index set. There may or may not be additional inequality constraints $g_j(x) \leq 0$ like the ones in problem (1.35). We provide a few examples.

Example 1.4.1: (Feasibility and Minimum Distance Problems)

A simple but important problem, which arises in many contexts and embodies important algorithmic ideas, is a classical *feasibility problem*, where the objective is to find a common point within a collection of sets X_ℓ, $\ell \in L$, where each X_ℓ is a closed convex set. In the feasibility problem the cost function is zero. A somewhat more complex problem with a similar structure arises when there is a cost function, i.e., a problem of the form

$$\begin{array}{ll} \text{minimize} & f(x) \\ \text{subject to} & x \in \cap_{\ell \in L} X_\ell, \end{array}$$

where $f : \Re^n \mapsto \Re$. An important example is the minimum distance problem, where

$$f(x) = \|x - z\|,$$

for a given vector z and some norm $\|\cdot\|$. The following example is a special case.

Example 1.4.2: (Basis Pursuit)

Consider the problem

$$\begin{array}{ll} \text{minimize} & \|x\|_1 \\ \text{subject to} & Ax = b, \end{array} \tag{1.36}$$

where $\|\cdot\|_1$ is the ℓ_1 norm in $\Re^n$, A is a given $m \times n$ matrix, and b is a vector in $\Re^m$ that consists of m given measurements. We are trying to construct a linear model of the form $Ax = b$, where x is a vector of n scalar weights for a large number n of basis functions ($m < n$). We want to satisfy exactly the measurement equations $Ax = b$, while using only a few of the basis functions in our model. Consequently, we introduce the ℓ_1 norm in the cost function of problem (1.36), aiming to delineate a small subset of basis functions, corresponding to nonzero coordinates of x at the optimal solution. This is called the *basis pursuit* problem (see, e.g., [CDS01], [VaF08]), and its underlying idea is similar to the one of ℓ_1-regularization (cf. Example 1.3.2).

It is also possible to consider a norm other than ℓ_1 in Eq. (1.36). An example is the *atomic norm* $\|\cdot\|_{\mathcal{A}}$ induced by a subset $\mathcal{A}$ that is centrally symmetric around the origin ($a \in \mathcal{A}$ if and only if $-a \in \mathcal{A}$):

$$\|x\|_{\mathcal{A}} = \inf\big\{t > 0 \mid x \in t \cdot \text{conv}(\mathcal{A})\big\}.$$

This problem, and other related problems involving atomic norms, have many applications; see for example [CRP12], [SBT12], [RSW13].

A related problem is

$$\begin{aligned} &\text{minimize} \quad \|X\|_* \\ &\text{subject to} \quad AX = B, \end{aligned}$$

where the optimization is over all $m \times n$ matrices X. The matrices A, B are given and have dimensions $\ell \times m$ and $\ell \times n$, respectively, and $\|X\|_*$ is the nuclear norm of X. This problem aims to produce a low-rank matrix X that satisfies an underdetermined set of linear equations $AX = B$ (see e.g., [CaR09], [RFP10], [RXB11]). When these equations specify that a subset of entries X_{ij}, $(i,j) \in \Omega$, are fixed at given values M_{ij},

$$X_{ij} = M_{ij}, \qquad (i,j) \in \Omega,$$

we obtain an alternative formulation of the matrix completion problem discussed in Example 1.3.2.

Example 1.4.3: (Minimax Problems)

In a minimax problem the cost function has the form

$$f(x) = \sup_{z \in Z} \phi(x,z),$$

where Z is a subset of some space and $\phi(\cdot, z)$ is a real-valued function for each $z \in Z$. We want to minimize f subject to $x \in X$, where X is a given constraint set. By introducing an artificial scalar variable y, we may transform such a problem to the general form

$$\begin{aligned} &\text{minimize} \quad y \\ &\text{subject to} \quad x \in X, \quad \phi(x,z) \le y, \quad \forall\, z \in Z, \end{aligned}$$

which involves a large number of constraints (one constraint for each z in the set Z, which could be infinite). Of course in this problem the set X may also be of the form $X = \cap_{\ell \in L} X_\ell$ as in earlier examples.

Example 1.4.4: (Basis Function Approximation for Separable Problems – Approximate Dynamic Programming)

Let us consider a large-scale separable problem of the form

$$\begin{aligned} &\text{minimize} \quad \sum_{i=1}^{m} f_i(y_i) \\ &\text{subject to} \quad \sum_{i=1}^{m} g_{ij}(y_i) \le 0, \quad \forall\, j = 1, \ldots, r, \quad y \ge 0, \end{aligned} \tag{1.37}$$

where $f_i : \Re \mapsto \Re$ are scalar functions, and the dimension m of the vector $y = (y_1, \ldots, y_m)$ is very large. One possible way to address this problem is to approximate y with a vector of the form Φx, where Φ is an $m \times n$ matrix. The columns of Φ may be relatively few, and may be viewed as basis functions for a low-dimensional approximation subspace $\{\Phi x \mid x \in \Re^n\}$. We replace problem (1.37) with the approximate version

$$\begin{aligned} &\text{minimize} \quad \sum_{i=1}^{m} f_i(\phi_i' x) \\ &\text{subject to} \quad \sum_{i=1}^{m} g_{ij}(\phi_i' x) \le 0, \quad \forall\, j = 1, \ldots, r, \\ &\qquad\qquad \phi_i' x \ge 0, \quad i = 1, \ldots, m, \end{aligned} \tag{1.38}$$

where ϕ_i' denotes the ith row of Φ, and $\phi_i' x$ is viewed as an approximation of y_i. Thus the dimension of the problem is reduced from m to n. However, the constraint set of the problem became more complicated, because the simple constraints $y_i \ge 0$ take the more complex form $\phi_i' x \ge 0$. Moreover the number m of additive components in the cost function, as well as the number of its constraints is still large. Thus the problem has the additive cost structure of the preceding section, as well as a large number of constraints.

An important application of this approach is in approximate dynamic programming (see e.g., [BeT96], [SuB98], [Pow11], [Ber12]), where the functions f_i and g_{ij} are linear. The corresponding problem (1.37) relates to the solution of the optimality condition (Bellman equation) of an infinite horizon Markovian decision problem (the constraint $y \ge 0$ may not be present in this context). Here the numbers m and r are often astronomical (in fact r can be much larger than m), in which case an exact solution cannot be obtained. For such problems, approximation based on problem (1.38) has been one of the major algorithmic approaches (see [Ber12] for a textbook presentation and references). For very large m, it may be impossible to calculate the cost function value $\sum_{i=1}^{m} f_i(\phi_i' x)$ for a given x, and one may at most be able to sample individual cost components f_i. For this reason optimization by stochastic simulation is one of the most prominent approaches in large scale dynamic programming.

Let us also mention that related approaches based on randomization and simulation have been proposed for the solution of large scale instances of classical linear algebra problems; see [BeY09], [Ber12] (Section 7.3), [DMM06], [StV09], [HMT10], [Nee10], [DMM11], [WaB13a], [WaB13b].

A large number of constraints also arises often in problems involving a graph, and may be handled with algorithms that take into account the graph structure. The following example is typical.

Example 1.4.5: (Optimal Routing in a Network – Multicommodity Flows)

Consider a directed graph that is used to transfer "commodities" from given supply points to given demand points. We are given a set W of ordered node pairs $w = (i, j)$. The nodes i and j are referred to as the *origin* and the *destination* of w, respectively, and w is referred to as an OD pair. For each w, we are given a scalar r_w referred to as the *input* of w. For example, in the context of routing of data in a communication network, r_w (measured in data units/second) is the arrival rate of traffic entering and exiting the network at the origin and the destination of w, respectively. The objective is to divide each r_w among the many paths from origin to destination in a way that the resulting total arc flow pattern minimizes a suitable cost function.

We denote:

P_w: A given set of paths that start at the origin and end at the destination of w. All arcs on each of these paths are oriented in the direction from the origin to the destination.

x_p: The portion of r_w assigned to path p, also called the *flow of path p*.

The collection of all path flows $\{x_p \mid p \in P_w,\ w \in W\}$ must satisfy the constraints

$$\sum_{p \in P_w} x_p = r_w, \qquad \forall\ w \in W, \tag{1.39}$$

$$x_p \geq 0, \qquad \forall\ p \in P_w,\ w \in W. \tag{1.40}$$

The total flow F_{ij} of arc (i, j) is the sum of all path flows traversing the arc:

$$F_{ij} = \sum_{\substack{\text{all paths } p \\ \text{containing } (i,j)}} x_p. \tag{1.41}$$

Consider a cost function of the form

$$\sum_{(i,j)} D_{ij}(F_{ij}). \tag{1.42}$$

The problem is to find a set of path flows $\{x_p\}$ that minimize this cost function subject to the constraints of Eqs. (1.39)-(1.41). It is typically assumed that D_{ij} is a convex function of F_{ij}. In data routing applications, the form of D_{ij} is often based on a queueing model of average delay, in which case D_{ij} is continuously differentiable within its domain (see e.g., [BeG92]). In a related context, arising in optical networks, the problem involves additional integer constraints on x_p, but may be addressed as a problem with continuous flow variables (see [OzB03]).

The preceding problem is known as a *multicommodity network flow problem*. The terminology reflects the fact that the arc flows consist of several different commodities; in the present example the different commodities are the data of the distinct OD pairs. This problem also arises in essentially identical form in traffic network equilibrium problems (see e.g., [FlH95], [Ber98], [Ber99], [Pat99], [Pat04]). The special case where all OD pairs have the same end node, or all OD pairs have the same start node, is known as the *single commodity network flow problem*, a much easier type of problem, for which there are efficient specialized algorithms that tend to be much faster than their multicommodity counterparts (see textbooks such as [Ber91], [Ber98]).

By expressing the total flows F_{ij} in terms of the path flows in the cost function (1.42) [using Eq. (1.41)], the problem can be formulated in terms of the path flow variables $\{x_p \mid p \in P_w,\ w \in W\}$ as

$$\begin{aligned} &\text{minimize} \quad D(x) \\ &\text{subject to} \quad \sum_{p \in P_w} x_p = r_w, \quad \forall\, w \in W, \\ &x_p \geq 0, \quad \forall\, p \in P_w,\ w \in W, \end{aligned}$$

where

$$D(x) = \sum_{(i,j)} D_{ij} \left(\sum_{\substack{\text{all paths } p \\ \text{containing } (i,j)}} x_p \right)$$

and x is the vector of path flows x_p. There is a potentially huge number of variables as well as constraints in this problem. However, by judiciously taking into account the special structure of the problem, the constraint set can be simplified and approximated by the convex hull of a small number of vectors x, and the number of variables and constraints can be reduced to a manageable size (see e.g., [BeG83], [FlH95], [OMV00], and our discussion in Section 4.2).

There are several approaches to handle a large number of constraints. One possibility, which points the way to some major classes of algorithms, is to initially discard some of the constraints, solve the corresponding less constrained problem, and later selectively reintroduce constraints that seem to be violated at the optimum. In Chapters 4-6, we will discuss methods of this type in some detail.

Another possibility is to replace constraints with penalties that assign high cost for their violation. In particular, we may replace problem (1.35) with

$$\begin{aligned} &\text{minimize} \quad f(x) + c \sum_{j=1}^{r} P\big(g_j(x)\big) \\ &\text{subject to} \quad x \in X, \end{aligned}$$

where $P(\cdot)$ is a scalar penalty function satisfying $P(u) = 0$ if $u \leq 0$, and $P(u) > 0$ if $u > 0$, and c is a positive penalty parameter. We discuss this possibility in the next section.

1.5 EXACT PENALTY FUNCTIONS

In this section we discuss a transformation that is often useful in the context of constrained optimization algorithms. We will derive a form of equivalence between a constrained convex optimization problem, and a penalized problem that is less constrained or is entirely unconstrained. The motivation is that some convex optimization algorithms do not have constrained counterparts, but can be applied to a penalized unconstrained problem. Furthermore, in some analytical contexts, it is useful to be able to work with an equivalent problem that is less constrained.

We consider the convex programming problem

$$\begin{aligned} &\text{minimize} \quad f(x) \\ &\text{subject to} \quad x \in X, \quad g_j(x) \le 0, \quad j = 1, \dots, r, \end{aligned} \tag{1.43}$$

where X is a convex subset of $\Re^n$, and $f : X \to \Re$ and $g_j : X \to \Re$ are given convex functions. We denote by f^* the primal optimal value, and by q^* the dual optimal value, i.e.,

$$q^* = \sup_{\mu \ge 0} q(\mu),$$

where

$$q(\mu) = \inf_{x \in X} \{ f(x) + \mu' g(x) \}, \qquad \forall\ \mu \ge 0,$$

with $g(x) = \big(g_1(x), \dots, g_r(x)\big)'$. We assume that $-\infty < q^* = f^* < \infty$.

We introduce a convex penalty function $P : \Re^r \mapsto \Re$, which satisfies

$$P(u) = 0, \qquad \forall\ u \le 0, \tag{1.44}$$

$$P(u) > 0, \qquad \text{if } u_j > 0 \text{ for some } j = 1, \dots, r. \tag{1.45}$$

We consider solving in place of the original problem (1.43), the "penalized" problem

$$\begin{aligned} &\text{minimize} \quad f(x) + P\big(g(x)\big) \\ &\text{subject to} \quad x \in X, \end{aligned} \tag{1.46}$$

where the inequality constraints have been replaced by the extra cost $P\big(g(x)\big)$ for their violation. Some interesting examples of penalty functions are based on the squared or the absolute value of constraint violation:

$$P(u) = \frac{c}{2} \sum_{j=1}^{r} \big(\max\{0, u_j\}\big)^2,$$

and

$$P(u) = c \sum_{j=1}^{r} \max\{0, u_j\},$$

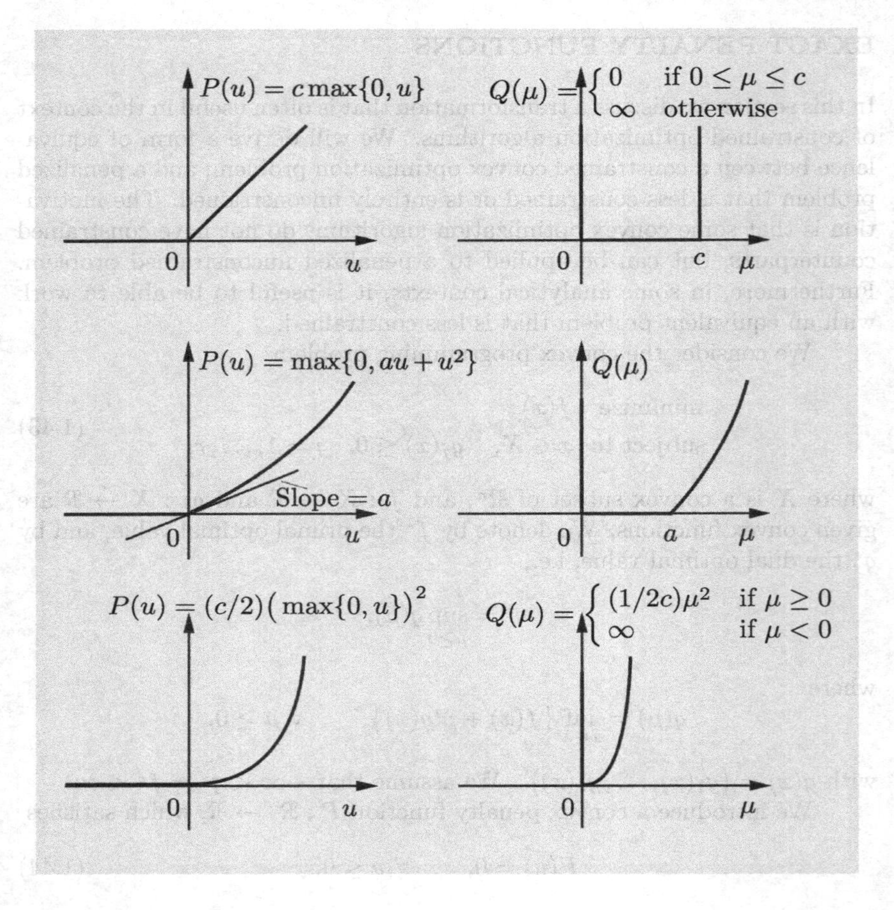

Figure 1.5.1. Illustration of various penalty functions P and their conjugate functions, denoted by Q. Because $P(u) = 0$ for $u \le 0$, we have $Q(\mu) = \infty$ for μ outside the nonnegative orthant.

where c is a positive penalty parameter. However, there are other possibilities that may be well-matched with the problem at hand.

The conjugate function of P is given by

$$Q(\mu) = \sup_{u \in \Re^r} \big\{u'\mu - P(u)\big\},$$

and it can be seen that

$$Q(\mu) \ge 0, \qquad \forall\ \mu \in \Re^r,$$

$$Q(\mu) = \infty, \qquad \text{if } \mu_j < 0 \text{ for some } j = 1, \ldots, r.$$

Figure 1.5.1 shows some examples of one-dimensional penalty functions P, together with their conjugates.

Consider the primal function of the original constrained problem,

$$p(u) = \inf_{x\in X,\, g(x)\le u} f(x), \qquad u \in \Re^r.$$

We have,

$$\begin{aligned}
\inf_{x\in X}\big\{f(x)+P\big(g(x)\big)\big\} &= \inf_{x\in X}\ \inf_{u\in\Re^r,\, g(x)\le u}\big\{f(x)+P\big(g(x)\big)\big\} \\
&= \inf_{x\in X}\ \inf_{u\in\Re^r,\, g(x)\le u}\big\{f(x)+P(u)\big\} \\
&= \inf_{x\in X,\, u\in\Re^r,\, g(x)\le u}\big\{f(x)+P(u)\big\} \\
&= \inf_{u\in\Re^r}\ \inf_{x\in X,\, g(x)\le u}\big\{f(x)+P(u)\big\} \\
&= \inf_{u\in\Re^r}\big\{p(u)+P(u)\big\},
\end{aligned}$$

where for the second equality, we use the monotonicity relation†

$$u \le v \qquad \Rightarrow \qquad P(u) \le P(v).$$

Moreover, $-\infty < q^*$ and $f^* < \infty$ by assumption, and since for any μ with $q(\mu) > -\infty$, we have

$$p(u) \ge q(\mu) - \mu' u > -\infty, \qquad \forall\, u \in \Re^r,$$

it follows that $p(0) < \infty$ and $p(u) > -\infty$ for all $u \in \Re^r$, so p is proper.

We can now apply the Fenchel Duality Theorem (Prop. 1.2.1) with the identifications $f_1 = p$, $f_2 = P$, and $A = I$. We use the conjugacy relation between the primal function p and the dual function q to write

$$\inf_{u\in\Re^r}\big\{p(u)+P(u)\big\} = \sup_{\mu\ge 0}\big\{q(\mu) - Q(\mu)\big\}, \tag{1.47}$$

so that

$$\inf_{x\in X}\big\{f(x)+P\big(g(x)\big)\big\} = \sup_{\mu\ge 0}\big\{q(\mu) - Q(\mu)\big\}; \tag{1.48}$$

see Fig. 1.5.2. Note that the conditions for application of the theorem are satisfied since the penalty function P is real-valued, so that the relative

---

† To show this relation, we argue by contradiction. If there exist u and v with $u \le v$ and $P(u) > P(v)$, then by continuity of P, there must exist $\overline{u}$ close enough to u such that $\overline{u} < v$ and $P(\overline{u}) > P(v)$. Since P is convex, it is monotonically increasing along the halfline $\big\{\overline{u} + \alpha(\overline{u} - v) \mid \alpha \ge 0\big\}$, and since $P(\overline{u}) > P(v) \ge 0$, P takes positive values along this halfline. However, since $\overline{u} < v$, this halfline eventually enters the negative orthant, where P takes the value 0 by Eq. (1.44), a contradiction.

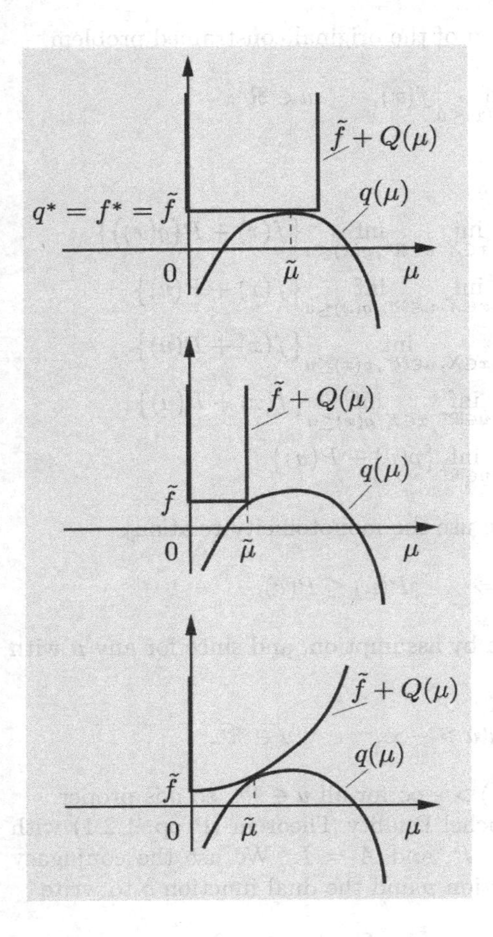

Figure 1.5.2. Illustration of the duality relation (1.48), and the optimal values of the penalized and the dual problem. Here f^* is the optimal value of the original problem, which is assumed to be equal to the optimal dual value q^*, while $\tilde{f}$ is the optimal value of the penalized problem,

$$\tilde{f} = \inf_{x \in X} \left\{ f(x) + P\big(g(x)\big) \right\}.$$

The point of contact of the graphs of the functions $\tilde{f} + Q(\mu)$ and $q(\mu)$ corresponds to the vector $\tilde{\mu}$ that attains the maximum in the relation

$$\tilde{f} = \max_{\mu \geq 0} \left\{ q(\mu) - Q(\mu) \right\}.$$

interiors of dom(p) and dom(P) have nonempty intersection. Furthermore, as part of the conclusions of part (a) of the Fenchel Duality Theorem, it follows that the supremum over $\mu \geq 0$ in Eq. (1.48) is attained.

Figure 1.5.2 suggests that in order for the penalized problem (1.46) to have the same optimal value as the original constrained problem (1.43), the conjugate Q must be "sufficiently flat" so that it is minimized by some dual optimal solution μ^*. This can be interpreted in terms of properties of subgradients, which are stated in Appendix B, Section 5.4: we must have $0 \in \partial Q(\mu^*)$ for some dual optimal solution μ^*, which by Prop. 5.4.3 in Appendix B, is equivalent to $\mu^* \in \partial P(0)$. This is part (a) of the following proposition, which was given in [Ber75a]. Parts (b) and (c) of the proposition deal with issues of equality of corresponding optimal solutions. The proposition assumes the convexity and other assumptions made in the early part in this section regarding problem (1.43) and the penalty function P.

Proposition 1.5.1: Consider problem (1.43), where we assume that $-\infty < q^* = f^* < \infty$.

(a) The penalized problem (1.46) and the original constrained problem (1.43) have equal optimal values if and only if there exists a dual optimal solution μ^* such that $\mu^* \in \partial P(0)$.

(b) In order for some optimal solution of the penalized problem (1.46) to be an optimal solution of the constrained problem (1.43), it is necessary that there exists a dual optimal solution μ^* such that

$$u'\mu^* \leq P(u), \qquad \forall\, u \in \Re^r. \tag{1.49}$$

(c) In order for the penalized problem (1.46) and the constrained problem (1.43) to have the same set of optimal solutions, it is sufficient that there exists a dual optimal solution μ^* such that

$$u'\mu^* < P(u), \qquad \forall\, u \in \Re^r \text{ with } u_j > 0 \text{ for some } j. \tag{1.50}$$

Proof: (a) We have using Eqs. (1.47) and (1.48),

$$p(0) \geq \inf_{u\in\Re^r} \big\{p(u) + P(u)\big\} = \sup_{\mu\geq 0}\big\{q(\mu) - Q(\mu)\big\} = \inf_{x\in X} \big\{f(x) + P\big(g(x)\big)\big\}. \tag{1.51}$$

Since $f^* = p(0)$, we have

$$f^* = \inf_{x\in X} \big\{f(x) + P\big(g(x)\big)\big\}$$

if and only if equality holds in Eq. (1.51). This is true if and only if

$$0 \in \arg\min_{u\in\Re^r} \big\{p(u) + P(u)\big\},$$

which by Prop. 5.4.7 in Appendix B, is true if and only if there exists some $\mu^* \in -\partial p(0)$ with $\mu^* \in \partial P(0)$ (in view of the fact that P is real-valued). Since the set of dual optimal solutions is $-\partial p(0)$ (under our assumption $-\infty < q^* = f^* < \infty$; see Example 5.4.2, [Ber09]), the result follows.

(b) If x^* is an optimal solution of both problems (1.43) and (1.46), then by feasibility of x^*, we have $P\big(g(x^*)\big) = 0$, so these two problems have equal optimal values. From part (a), there must exist a dual optimal solution $\mu^* \in \partial P(0)$, which is equivalent to Eq. (1.49), by the subgradient inequality.

(c) If x^* is an optimal solution of the constrained problem (1.43), then $P\big(g(x^*)\big) = 0$, so we have

$$f^* = f(x^*) = f(x^*) + P\big(g(x^*)\big) \geq \inf_{x\in X}\big\{f(x) + P\big(g(x)\big)\big\}.$$

The condition (1.50) implies the condition (1.49), so that by part (a), equality holds throughout in the above relation, showing that x^* is also an optimal solution of the penalized problem (1.46).

Conversely, let $x^* \in X$ be an optimal solution of the penalized problem (1.46). If x^* is feasible [i.e., satisfies in addition $g(x^*) \leq 0$], then it is an optimal solution of the constrained problem (1.43) [since $P\big(g(x)\big) = 0$ for all feasible vectors x], and we are done. Otherwise x^* is infeasible in which case $g_j(x^*) > 0$ for some j. Then, by using the given condition (1.50), it follows that there exists a dual optimal solution μ^* and an $\epsilon > 0$ such that

$$\mu^{*\prime} g(x^*) + \epsilon < P\big(g(x^*)\big).$$

Let $\tilde{x}$ be a feasible vector such that $f(\tilde{x}) \leq f^* + \epsilon$. Since $P\big(g(\tilde{x})\big) = 0$ and $f^* = \min_{x \in X}\big\{f(x) + \mu^{*\prime} g(x)\big\}$, we obtain

$$f(\tilde{x}) + P\big(g(\tilde{x})\big) = f(\tilde{x}) \leq f^* + \epsilon \leq f(x^*) + \mu^{*\prime} g(x^*) + \epsilon.$$

By combining the last two relations, it follows that

$$f(\tilde{x}) + P\big(g(\tilde{x})\big) < f(x^*) + P\big(g(x^*)\big),$$

which contradicts the hypothesis that x^* is an optimal solution of the penalized problem (1.46). This completes the proof. **Q.E.D.**

As an illustration, consider the minimization of $f(x) = -x$ over all $x \in X = \{x \mid x \geq 0\}$ with $g(x) = x \leq 0$. The dual function is

$$q(\mu) = \inf_{x \geq 0} (\mu - 1)x, \qquad \mu \geq 0,$$

so $q(\mu) = 0$ for $\mu \in [1, \infty)$ and $q(\mu) = -\infty$ otherwise. Let $P(u) = c\max\{0, u\}$, so the penalized problem is $\min_{x \geq 0}\big\{-x + c\max\{0, x\}\big\}$. Then parts (a) and (b) of the proposition apply if $c \geq 1$. However, part (c) applies only if $c > 1$. In terms of Fig. 1.5.2, the conjugate of P is $Q(\mu) = 0$ if $\mu \in [0, c]$ and $Q(\mu) = \infty$ otherwise, so when $c = 1$, Q is "flat" over an area not including an interior point of the dual optimal solution set $[1, \infty)$.

To elaborate on the idea of the preceding example, let

$$P(u) = c \sum_{j=1}^{r} \max\{0, u_j\},$$

where $c > 0$. The condition $\mu^* \in \partial P(0)$, or equivalently,

$$u'\mu^* \leq P(u), \qquad \forall\, u \in \Re^r$$

[cf. Eq. (1.49)], is equivalent to

$$\mu_j^* \leq c, \qquad \forall\, j = 1, \ldots, r.$$

Similarly, the condition $u'\mu^* < P(u)$ for all $u \in \Re^r$ with $u_j > 0$ for some j [cf. Eq. (1.50)], is equivalent to

$$\mu_j^* < c, \qquad \forall\, j = 1, \ldots, r.$$

The reader may consult the literature for other results on exact penalty functions, starting with their first proposal in the book [Zan69]. The preceding development is based on [Ber75], and focuses on convex programming problems. For additional representative references, some of which also discuss nonconvex problems, see [HaM79], [Ber82a], [Bur91], [FeM91], [BNO03], [FrT07]. In what follows we develop an exact penalty function result for the case of an abstract constraint set, which will be used in the context of incremental constraint projection algorithms in Section 6.4.4.

A Distance-Based Exact Penalty Function

Let us discuss the case of a general Lipschitz continuous (not necessarily convex) cost function and an abstract constraint set $X \subset \Re^n$. The idea is to use a penalty that is proportional to the distance from X:

$$\text{dist}(x; X) = \inf_{y \in X} \|x - y\|.$$

The next proposition from [Ber11] provides the basic result (see Fig. 1.5.3).

Proposition 1.5.2: Let $f : \Re^n \mapsto \Re$ be a function that is Lipschitz continuous with constant L over a set $Y \subset \Re^n$, i.e.,

$$\big|f(x) - f(y)\big| \leq L\|x - y\|, \qquad \forall\, x, y \in Y. \tag{1.52}$$

Let also X be a nonempty closed subset of Y, and let c be a scalar with $c > L$. Then x^* minimizes f over X if and only if x^* minimizes

$$F_c(x) = f(x) + c\,\text{dist}(x; X)$$

over Y.

Proof: For any $x \in Y$, let $\hat{x}$ denote a vector of X that is at minimum distance from x (such a vector exists by the closure of X and Weierstrass' Theorem). By using the Lipschitz assumption (1.52) and the fact $c > L$, we have

$$F_c(x) = f(x) + c\|x - \hat{x}\| = f(\hat{x}) + \big(f(x) - f(\hat{x})\big) + c\|x - \hat{x}\| \geq f(\hat{x}) = F_c(\hat{x}),$$

with strict inequality if $x \neq \hat{x}$. Thus all minima of F_c over Y must lie in X, and also minimize f over X (since $F_c = f$ on X). Conversely, all minima

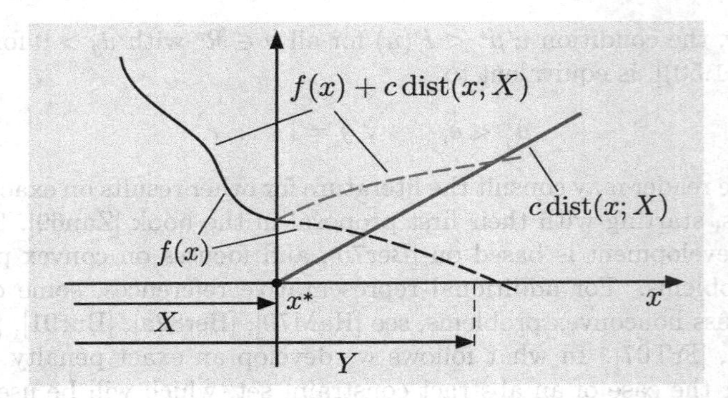

Figure 1.5.3. Illustration of Prop. 1.5.2. For c greater than the Lipschitz constant of f, the "slope" of the penalty function counteracts the "slope" of f at the optimal solution x^*.

of f over X are also minima of F_c over X (since $F_c = f$ on X), and by the preceding inequality, they are also minima of F_c over Y. **Q.E.D.**

The following proposition provides a generalization for constraints that involve the intersection of several sets.

Proposition 1.5.3: Let $f : \Re^n \mapsto \Re$ be a function, and let X_i, $i = 0, 1, \ldots, m$, be closed subsets of $\Re^n$ with nonempty intersection. Assume that f is Lipschitz continuous over X_0. Then there is a scalar $\bar{c} > 0$ such that for all $c \geq \bar{c}$, the set of minima of f over $\cap_{i=0}^m X_i$ coincides with the set of minima of

$$f(x) + c \sum_{i=1}^{m} \mathrm{dist}(x; X_i)$$

over X_0.

Proof: Let L be the Lipschitz constant for f, and let $c_1, \ldots, c_m$ be scalars satisfying

$$c_k > L + c_1 + \cdots + c_{k-1}, \qquad \forall\, k = 1, \ldots, m,$$

where $c_0 = 0$. Define

$$H_k(x) = f(x) + c_1\,\mathrm{dist}(x; X_1) + \cdots + c_k\,\mathrm{dist}(x; X_k), \qquad k = 1, \ldots, m,$$

and for $k = 0$, denote $H_0(x) = f(x)$, $c_0 = 0$. By applying Prop. 1.5.2, the set of minima of H_m over X_0 coincides with the set of minima of H_{m-1} over $X_m \cap X_0$, since c_m is greater than $L + c_1 + \cdots + c_{m-1}$, the Lipschitz constant for H_{m-1}. Similarly, for all $k = 1, \ldots, m$, the set of minima of

H_k over $\left(\cap_{i=k+1}^m X_i\right) \cap X_0$ coincides with the set of minima of H_{k-1} over $\left(\cap_{i=k}^m X_i\right) \cap X_0$. Thus, for $k = 1$, we obtain that the set of minima of H_m over X_0 coincides with the set of minima of H_0, which is f, over $\cap_{i=0}^m X_i$. Let

$$X^* \subset \cap_{i=0}^m X_i$$

be this set of minima. For $c \geq c_m$, we have $F_c \geq H_m$, while F_c coincides with H_m on X^*. Hence X^* is the set of minima of F_c over X_0. **Q.E.D.**

We finally note that exact penalty functions, and particularly the distance function $\text{dist}(x; X_i)$, are often relatively convenient in various contexts where difficult constraints complicate the algorithmic solution. As an example, see Section 6.4.4, where incremental proximal methods for highly constrained problems are discussed.

1.6 NOTES, SOURCES, AND EXERCISES

There is a very extensive literature on convex optimization, and in this section we will restrict ourselves to noting some books, research monographs, and surveys. In subsequent chapters, we will discuss in greater detail the literature that relates to the specialized content of these chapters.

Books relating primarily to duality theory are Rockafellar [Roc70], Stoer and Witzgall [StW70], Ekeland and Temam [EkT76], Bonnans and Shapiro [BoS00], Zalinescu [Zal02], Auslender and Teboulle [AuT03], and Bertsekas [Ber09].

The books by Rockafellar and Wets [RoW98], Borwein and Lewis [BoL00], and Bertsekas, Nedić, and Ozdaglar [BNO03] straddle the boundary between convex and variational analysis, a broad spectrum of topics that integrate classical analysis, convexity, and optimization of both convex and nonconvex (possibly nonsmooth) functions.

The book by Hiriart-Urruty and Lemarechal [HiL93] focuses on convex optimization algorithms. The books by Rockafellar [Roc84] and Bertsekas [Ber98] have a more specialized focus on network optimization algorithms and monotropic programming problems, which will be discussed in Chapters 4 and 6. The book by Ben-Tal and Nemirovski [BeN01] focuses on conic and semidefinite programming [see also the 2005 class notes by Nemirovski (on line), and the representative survey papers by Alizadeh and Goldfarb [AlG03], and Todd [Tod01]]. The book by Wolkowicz, Saigal, and Vanderberghe [WSV00] contains a collection of survey articles on semidefinite programming. The book by Boyd and Vanderberghe [BoV04] describes many applications, and contains a lot of related material and references. The book by Ben-Tal, El Ghaoui, and Nemirovski [BGN09] focuses on robust optimization; see also the survey by Bertsimas, Brown, and Caramanis [BBC11]. The book by Bauschke and Combettes [BaC11] develops the connection of convex analysis and monotone operator theory in infinite dimensional spaces. The book by Rockafellar and Wets [RoW98] also has a substantial finite-dimensional treatment of this subject. The books by Cottle, Pang, and Stone [CPS92], and Facchinei and Pang [FaP03] focus on complementarity and variational inequality problems. The books by Palomar and Eldar [PaE10], and Vetterli, Kovacevic, and Goyal [VKG14], and the surveys in the May 2010 issue of the IEEE Signal Processing Magazine describe applications of convex optimization in communications and signal processing. The books by Hastie, Tibshirani, and Friedman [HTF09], and Sra, Nowozin, and Wright [SNW12] describe applications of convex optimization in machine learning.

EXERCISES

1.1 (Support Vector Machines and Duality)

Consider the classification problem associated with a support vector machine,

$$\begin{aligned}&\text{minimize} \quad \tfrac{1}{2}\|x\|^2 + \beta \textstyle\sum_{i=1}^m \max\big\{0, 1 - b_i(c_i'x + y)\big\}\\ &\text{subject to} \quad x \in \Re^n,\ y \in \Re,\end{aligned}$$

with quadratic regularization, where β is a positive regularization parameter (cf. Example 1.3.3).

(a) Write the problem in the equivalent form

$$\begin{aligned}&\text{minimize} \quad \tfrac{1}{2}\|x\|^2 + \beta \textstyle\sum_{i=1}^m \xi_i\\ &\text{subject to} \quad x \in \Re^n,\ y \in \Re,\\ &\qquad\qquad 0 \le \xi_i, \quad 1 - b_i(c_i'x + y) \le \xi_i,\ i = 1, \ldots, m.\end{aligned}$$

Associate dual variables $\mu_i \ge 0$ with the constraints $1 - b_i(c_i'x + y) \le \xi_i$, and show that the dual function is given by

$$q(\mu) = \begin{cases} \hat{q}(\mu) & \text{if } \sum_{j=1}^m \mu_j b_j = 0,\ 0 \le \mu_i \le \beta,\ i = 1, \ldots, m,\\ -\infty & \text{otherwise,}\end{cases}$$

where

$$\hat{q}(\mu) = \sum_{i=1}^m \mu_i - \tfrac{1}{2} \textstyle\sum_{i=1}^m \sum_{j=1}^m b_i b_j c_i' c_j \mu_i \mu_j.$$

Does the dual problem, viewed as the equivalent quadratic program

$$\begin{aligned}&\text{minimize} \quad \tfrac{1}{2} \textstyle\sum_{i=1}^m \sum_{j=1}^m b_i b_j c_i' c_j \mu_i \mu_j - \sum_{i=1}^m \mu_i\\ &\text{subject to} \quad \sum_{j=1}^m \mu_j b_j = 0, \quad 0 \le \mu_i \le \beta,\ i = 1, \ldots, m,\end{aligned}$$

always have a solution? Is the solution unique? *Note*: The dual problem may have high dimension, but it has a generally more favorable structure than the primal. The reason is the simplicity of its constraint set, which makes it suitable for special types of quadratic programming methods, and the two-metric projection and coordinate descent methods of Section 2.1.2.

(b) Consider an alternative formulation where the variable y is set to 0, leading to the problem

$$\begin{aligned} &\text{minimize} \quad \tfrac{1}{2}\|x\|^2 + \beta \textstyle\sum_{i=1}^m \max\{0, 1 - b_i c_i' x\} \\ &\text{subject to} \quad x \in \Re^n. \end{aligned}$$

Show that the dual problem should be modified so that the constraint $\sum_{j=1}^m \mu_j b_j = 0$ is not present, thus leading to a bound-constrained quadratic dual problem.

Note: The literature of the support vector machine field is extensive. Many of the nondifferentiable optimization methods to be discussed in subsequent chapters have been applied in connection to this field; see e.g., [MaM01], [FeM02], [SmS04], [Bot05], [Joa06], [JFY09], [JoY09], [SSS07], [LeW11].

1.2 (Minimizing the Sum or the Maximum of Norms [LVB98])

Consider the problems

$$\begin{aligned} &\text{minimize} \quad \sum_{i=1}^p \|F_i x + g_i\| \\ &\text{subject to} \quad x \in \Re^n, \end{aligned} \tag{1.53}$$

and

$$\begin{aligned} &\text{minimize} \quad \max_{i=1,\ldots,p} \|F_i x + g_i\| \\ &\text{subject to} \quad x \in \Re^n, \end{aligned}$$

where F_i and g_i are given matrices and vectors, respectively. Convert these problems to second order cone form and derive the corresponding dual problems.

1.3 (Complex l_1 and l_∞ Approximation [LVB98])

Consider the complex l_1 approximation problem

$$\begin{aligned} &\text{minimize} \quad \|Ax - b\|_1 \\ &\text{subject to} \quad x \in C^n, \end{aligned}$$

where C^n is the set of n-dimensional vectors whose components are complex numbers, and A and b are given matrix and vector with complex components. Show that it is a special case of problem (1.53) and derive the corresponding dual problem. Repeat for the complex l_∞ approximation problem

$$\begin{aligned} &\text{minimize} \quad \|Ax - b\|_\infty \\ &\text{subject to} \quad x \in C^n. \end{aligned}$$

1.4

The purpose of this exercise is to show that the SOCP can be viewed as a special case of SDP.

(a) Show that a vector $x \in \Re^n$ belongs to the second order cone if and only if the matrix

$$x_n I + \begin{pmatrix} 0 & 0 & \cdots & 0 & x_1 \\ 0 & 0 & \cdots & 0 & x_2 \\ \vdots & \vdots & \vdots & \vdots & \vdots \\ 0 & 0 & \cdots & 0 & x_{n-1} \\ x_1 & x_2 & \cdots & x_{n-1} & 0 \end{pmatrix}$$

is positive semidefinite. *Hint*: We have that for any positive definite symmetric $n \times n$ matrix A, vector $b \in \Re^n$, and scalar d, the matrix

$$\begin{pmatrix} A & b \\ b' & c \end{pmatrix}$$

is positive definite if and only if

$$c - b'A^{-1}b > 0.$$

(b) Use part (a) to show that the primal SOCP can be written in the form of the dual SDP.

1.5 (Explicit Form of a Second Order Cone Problem)

Consider the SOCP (1.24).

(a) Partition the $n_i \times (n+1)$ matrices $(A_i \;\; b_i)$ as

$$(A_i \;\; b_i) = \begin{pmatrix} D_i & d_i \\ p_i' & q_i \end{pmatrix}, \qquad i = 1, \ldots, m,$$

where D_i is an $(n_i - 1) \times n$ matrix, $d_i \in \Re^{n_i - 1}$, $p_i \in \Re^n$, and $q_i \in \Re$. Show that

$$A_i x - b_i \in C_i \qquad \text{if and only if} \qquad \|D_i x - d_i\| \le p_i' x - q_i,$$

so we can write the SOCP (1.24) as

$$\begin{aligned} &\text{minimize} \quad c'x \\ &\text{subject to} \quad \|D_i x - d_i\| \le p_i' x - q_i, \; i = 1, \ldots, m. \end{aligned}$$

(b) Similarly partition λ_i as

$$\lambda_i = \begin{pmatrix} \mu_i \\ \nu_i \end{pmatrix}, \qquad i = 1, \ldots, m,$$

where $\mu_i \in \Re^{n_i-1}$ and $\nu_i \in \Re$. Show that the dual problem (1.25) can be written in the form

$$\begin{aligned} &\text{maximize} \quad \sum_{i=1}^{m} (d_i'\mu_i + q_i\nu_i) \\ &\text{subject to} \quad \sum_{i=1}^{m} (D_i'\mu_i + \nu_i p_i) = c, \quad \|\mu_i\| \le \nu_i, \ i = 1, \ldots, m. \end{aligned}$$

(c) Show that the primal and dual interior point conditions for strong duality (Prop. 1.2.4) hold if there exist primal and dual feasible solutions $\overline{x}$ and $(\overline{\mu}_i, \overline{\nu}_i)$ such that

$$\|D_i\overline{x} - d_i\| < p_i'\overline{x} - q_i, \qquad i = 1, \ldots, m,$$

and

$$\|\overline{\mu}_i\| < \overline{\nu}_i, \qquad i = 1, \ldots, m,$$

respectively.

1.6 (Separable Conic Problems)

Consider the problem

$$\begin{aligned} &\text{minimize} \quad \sum_{i=1}^{m} f_i(x_i) \\ &\text{subject to} \quad x \in S \cap C, \end{aligned}$$

where $x = (x_1, \ldots, x_m)$ with $x_i \in \Re^{n_i}$, $i = 1, \ldots, m$, and $f_i : \Re^{n_i} \mapsto (-\infty, \infty]$ is a proper convex function for each i, and S and C are a subspace and a cone of $\Re^{n_1+\cdots+n_m}$, respectively. Show that a dual problem is

$$\begin{aligned} &\text{maximize} \quad \sum_{i=1}^{m} q_i(\lambda_i) \\ &\text{subject to} \quad \lambda \in \hat{C} + S^{\perp}, \end{aligned}$$

where $\lambda = (\lambda_1, \ldots, \lambda_m)$, $\hat{C}$ is the dual cone of C, and

$$q_i(\lambda_i) = \inf_{z_i \in \Re} \big\{ f_i(z_i) - \lambda_i' z_i \big\}, \qquad i = 1, \ldots, m.$$

1.7 (Weber Points)

Consider the problem of finding a circle of minimum radius that contains r points $y_1, \ldots, y_r$ in the plane, i.e., find x and z that minimize z subject to $\|x - y_j\| \le z$ for all $j = 1, \ldots, r$, where x is the center of the circle under optimization.

(a) Introduce multipliers μ_j, $j = 1, \ldots, r$, for the constraints, and show that the dual problem has an optimal solution and there is no duality gap.

(b) Show that calculating the dual function at some $\mu \ge 0$ involves the computation of a Weber point of $y_1, \ldots, y_r$ with weights $\mu_1, \ldots, \mu_r$, i.e., the solution of the problem

$$\min_{x \in \Re^2} \sum_{j=1}^{r} \mu_j \|x - y_j\|$$

(see Example 1.3.7).

1.8 (Inconsistent Convex Systems of Inequalities)

Let $g_j : \Re^n \mapsto \Re$, $j = 1, \ldots, r$, be convex functions over the nonempty convex set $X \subset \Re^n$. Show that the system

$$g_j(x) < 0, \qquad j = 1, \ldots, r,$$

has no solution within X if and only if there exists a vector $\mu \in \Re^r$ such that

$$\sum_{j=1}^{r} \mu_j = 1, \qquad \mu \ge 0,$$

$$\mu' g(x) \ge 0, \qquad \forall\, x \in X.$$

Note: This is an example of what is known as a *theorem of the alternative*. There are many results of this type, with a long history, such as the Farkas Lemma, and the theorems of Gordan, Motzkin, and Stiemke, which address the feasibility (possibly strict) of linear inequalities. They can be found in many sources, including Section 5.6 of [Ber09]. *Hint*: Consider the convex program

$$\begin{aligned} &\text{minimize} \ \ y \\ &\text{subject to} \ \ x \in X, \quad y \in \Re, \qquad g_j(x) \le y, \quad j = 1, \ldots, r. \end{aligned}$$

2

Optimization Algorithms: An Overview

Contents

In this book we are primarily interested in optimization algorithms, as opposed to "modeling," i.e., the formulation of real-world problems as mathematical optimization problems, or "theory," i.e., conditions for strong duality, optimality conditions, etc. In our treatment, we will mostly focus on guaranteeing convergence of algorithms to desired solutions, and the associated rate of convergence and complexity analysis. We will also discuss special characteristics of algorithms that make them suitable for particular types of large scale problem structures, and distributed (possibly asynchronous) computation. In this chapter we provide an overview of some broad classes of optimization algorithms, their underlying ideas, and their performance characteristics.

Iterative algorithms for minimizing a function $f : \Re^n \mapsto \Re$ over a set X generate a sequence $\{x_k\}$, which will hopefully converge to an optimal solution. In this book we focus on iterative algorithms for the case where X is convex, and f is either convex or is nonconvex but differentiable. Most of these algorithms involve one or both of the following two ideas, which will be discussed in Sections 2.1 and 2.2, respectively:

(a) *Iterative descent*, whereby the generated sequence $\{x_k\}$ is feasible, i.e., $\{x_k\} \subset X$, and satisfies

$$\phi(x_{k+1}) < \phi(x_k) \qquad \text{if and only if } x_k \text{ is not optimal,}$$

where ϕ is a *merit function*, that measures the progress of the algorithm towards optimality, and is minimized only at optimal points, i.e.,

$$\arg\min_{x\in X} \phi(x) = \arg\min_{x\in X} f(x).$$

Examples are $\phi(x) = f(x)$ and $\phi(x) = \inf_{x^*\in X^*} \|x - x^*\|$, where X^* is the set of optimal points, assumed nonempty. In some cases, iterative descent may be the primary idea, but modifications or approximations are introduced for a variety of reasons. For example one may modify an iterative descent method to make it suitable for distributed asynchronous computation, or to deal with random or nonrandom errors, but in the process lose the iterative descent property. In this case, the analysis is appropriately modified, but often maintains important aspects of its original descent-based character.

(b) *Approximation*, whereby the generated sequence $\{x_k\}$ need not be feasible, and is obtained by solving at each k an approximation to the original optimization problem, i.e.,

$$x_{k+1} \in \arg\min_{x\in X_k} F_k(x),$$

where F_k is a function that approximates f and X_k is a set that approximates X. These may depend on the prior iterates $x_0, \ldots, x_k$, as well as other parameters. Key ideas here are that minimization of F_k over X_k should be easier than minimization of f over X, and that x_k should be a good starting point for obtaining x_{k+1} via some (possibly special purpose) method. Of course, the approximation of f by F_k and/or X by X_k should improve as k increases, and there should be some convergence guarantees as $k \to \infty$. We will summarize the main approximation ideas of this book in Section 2.2.

A major class of problems that we aim to solve is dual problems, which by their nature involve nondifferentiable optimization. The fundamental reason is that the negative of a dual function is typically a conjugate function, which is closed and convex, but need not be differentiable. Moreover nondifferentiable cost functions naturally arise in other contexts, such as exact penalty functions, and machine learning with ℓ_1 regularization. Accordingly many of the algorithms that we discuss in this book do not require cost function differentiability for their application.

Still, however, differentiability plays a major role in problem formulations and algorithms, so it is important to maintain a close connection between differentiable and nondifferentiable optimization approaches. Moreover, nondifferentiable problems can often be converted to differentiable ones by using a smoothing scheme (see Section 2.2.5). We consequently summarize in Section 2.1 some of the main ideas of iterative algorithms that rely on differentiability, such as gradient and Newton methods, and their incremental variants. We return to some of these ideas in Sections 6.1-6.3, but for most of the remainder of the book we focus primarily on convex possibly nondifferentiable cost functions.

Since the present chapter has an overview character, our discussion will not be supplemented by complete proofs; in many cases we will provide just intuitive explanations and refer to the literature for a more detailed analysis. In subsequent chapters we will treat various types of algorithms in greater detail. In particular, in Chapter 3, we discuss descent-type iterative methods that use subgradients. In Chapters 4 and 5, we discuss primarily the approximation approach, focusing on two types of algorithms and their combinations: polyhedral approximation and proximal, respectively. In Chapter 6, we discuss a number of additional methods, which extend and combine the ideas of the preceding chapters.

2.1 ITERATIVE DESCENT ALGORITHMS

Iterative algorithms generate sequences $\{x_k\}$ according to

$$x_{k+1} = G_k(x_k),$$

where $G_k : \Re^n \mapsto \Re^n$ is some function that may depend on k, and x_0 is some starting point. In a more general context, G_k may depend on some preceding iterates $x_{k-1}, x_{k-2}, \ldots$. We are typically interested in the convergence of the generated sequence $\{x_k\}$ to some desirable point. We are also interested in questions of rate of convergence, such as for example the number of iterations needed to bring a measure of error to within a given tolerance, or asymptotic bounds on some measure of error as the number of iterations increases.

A *stationary* iterative algorithm is obtained when G_k does not depend on k, i.e.,

$$x_{k+1} = G(x_k).$$

This algorithm aims to solve a fixed point problem: finding a solution of the equation $x = G(x)$. A classical optimization example is the gradient iteration

$$x_{k+1} = x_k - \alpha \nabla f(x_k), \tag{2.1}$$

which aims at satisfying the optimality condition $\nabla f(x) = 0$ for an unconstrained minimum of a differentiable function $f : \Re^n \mapsto \Re$. Here α is a positive stepsize parameter that is used to ensure that the iteration makes progress towards the solution set of the corresponding problem. Another example is the iteration

$$x_{k+1} = x_k - \alpha(Qx_k - b) = (I - \alpha Q)x_k + \alpha b, \tag{2.2}$$

which aims at solution of the linear system $Qx = b$, where Q is a matrix that has eigenvalues with positive real parts (so that the matrix $I - \alpha Q$ has eigenvalues within the unit circle for sufficiently small $\alpha > 0$, and the iteration is convergent to the unique solution). If f is the quadratic function $f(x) = \frac{1}{2}x'Qx - b'x$, where Q is positive definite symmetric, then we have $\nabla f(x_k) = Qx_k - b$ and the gradient iteration (2.1) can be written in the form (2.2).

Convergence of the stationary iteration $x_{k+1} = G(x_k)$ can be ascertained in a number of ways. The most common is to verify that G is a *contraction mapping* with respect to some norm, i.e., for some $\rho < 1$, and some norm $\|\cdot\|$ (not necessarily the Euclidean norm), we have

$$\big\|G(x) - G(y)\big\| \leq \rho\|x - y\|, \qquad \forall\, x, y \in \Re^n.$$

Then it can be shown that G has a unique fixed point x^*, and $x_k \to x^*$, starting from any $x_0 \in \Re^n$; this is the well-known Banach Fixed Point Theorem (see Prop. A.4.1 in Section A.4 of Appendix A, where the contraction and other approaches for convergence analysis are discussed). An example is the mapping

$$G(x) = (I - \alpha Q)x + \alpha b$$

of the linear iteration (2.2), when the eigenvalues of $I - \alpha Q$ lie strictly within the unit circle.

The case where G is a contraction mapping provides an example of convergence analysis based on a descent approach: at each iteration we have

$$\|x_{k+1} - x^*\| \leq \rho\|x_k - x^*\|, \tag{2.3}$$

so the distance $\|x - x^*\|$ is decreased with each iteration at a nonsolution point x. Moreover, in this case we obtain an estimate of the convergence rate: $\|x_k - x^*\|$ is decreased at least as fast as the geometric progression $\{\rho^k\|x_0 - x^*\|\}$; this is called *linear* or *geometric* convergence.†

Many optimization algorithms involve a contraction mapping as described above. There are also other types of convergent fixed point iterations, which do not require that G is a contraction mapping. In particular, there are cases where G is a *nonexpansive mapping* [$\rho = 1$ in Eq. (2.3)], and there is sufficient structure in G to ensure a form of improvement of an appropriate figure of merit at each iteration; the proximal algorithm, introduced in Section 2.2.3 and discussed in detail in Chapter 5, is an important example of this type.

There are also many cases of *nonstationary* iterations of the form

$$x_{k+1} = G_k(x_k),$$

whose convergence analysis is difficult or impossible with a contraction or nonexpansive mapping approach. An example is unconstrained minimization of a differentiable function f with a gradient method of the form

$$x_{k+1} = x_k - \alpha_k \nabla f(x_k), \tag{2.4}$$

where the stepsize α_k is not constant. Still many of these algorithms admit a convergence analysis based on a descent approach, whereby we introduce a function ϕ that measures the progress of the algorithm towards optimality, and show that

$$\phi(x_{k+1}) < \phi(x_k) \qquad \text{if and only if } x_k \text{ is not optimal.}$$

Two common cases are when $\phi(x) = f(x)$ or $\phi(x) = \text{dist}(x, X^*)$, the Euclidean minimum distance of x from the set X^* of minima of f. For example convergence of the gradient algorithm (2.4) is often analyzed by showing that for all k,

$$f(x_{k+1}) \leq f(x_k) - \gamma_k\big\|\nabla f(x_k)\big\|^2,$$

where γ_k is a positive scalar that depends on α_k and some characteristics of f, and is such that $\sum_{k=0}^{\infty} \gamma_k = \infty$; this brings to bear the convergence

† Generally, we say that a nonnegative scalar sequence $\{\beta_k\}$ *converges (at least) linearly or geometrically* if there exist scalars $\gamma > 0$ and $\rho \in (0, 1)$ such that $\beta_k \leq \gamma\rho^k$ for all k. For a discussion of different definitions of linear and other types of convergence rate, see [OrR70], [Ber82a], and [Ber99].

methodology of Section A.4 in Appendix A and guarantees that either $\nabla f(x_k) \to 0$ or $f(x_k) \to -\infty$.

In what follows in this section we will provide an overview of iterative optimization algorithms that rely on some form of descent for their validity, we discuss some of their underlying motivation, and we raise various issues that will be discussed later. We will also provide in the exercises a sampling of some related convergence analysis, while deferring to subsequent chapters a more detailed theoretical development. Moreover, in the present section we focus in greater detail on the differentiable cost function case and the potential benefits of differentiability. Our focus in subsequent chapters will be primarily on nondifferentiable problems.

2.1.1 Differentiable Cost Function Descent – Unconstrained Problems

A natural iterative descent approach to minimizing a real-valued function $f : \Re^n \mapsto \Re$ over a set X is based on cost improvement: starting with a point $x_0 \in X$, construct a sequence $\{x_k\} \subset X$ such that

$$f(x_{k+1}) < f(x_k), \qquad k = 0, 1, \ldots,$$

unless x_k is optimal for some k, at which time the method stops.

In this context it is useful to consider the directional derivative of f at a point x in a direction d. For a differentiable f, it is given by

$$f'(x; d) = \lim_{\alpha \downarrow 0} \frac{f(x + \alpha d) - f(x)}{\alpha} = \nabla f(x)' d, \tag{2.5}$$

(cf. Section A.3 of Appendix A). From this formula it follows that if d_k is a *descent direction* at x_k, in the sense that

$$f'(x_k; d_k) < 0,$$

we may reduce the cost by moving from x_k along d_k with a small enough positive stepsize α. In the unconstrained case where $X = \Re^n$, this leads to an algorithm of the form

$$x_{k+1} = x_k + \alpha_k d_k, \tag{2.6}$$

where d_k is a descent direction at x_k and α_k is a positive scalar stepsize. If no descent direction can be found at x_k, i.e., $f'(x_k; d) \geq 0$, for all $d \in \Re^n$, from Eq. (2.5) it follows that x_k must satisfy the necessary condition for optimality

$$\nabla f(x_k) = 0.$$

Gradient Methods for Differentiable Unconstrained Minimization

For the case where f is differentiable and $X = \Re^n$, there are many popular descent algorithms of the form (2.6). An important example is the classical gradient method, where we use $d_k = -\nabla f(x_k)$ in Eq. (2.6):

$$x_{k+1} = x_k - \alpha_k \nabla f(x_k).$$

Since for differentiable f we have

$$f'(x_k; d) = \nabla f(x_k)'d,$$

it follows that

$$-\frac{\nabla f(x_k)}{\|\nabla f(x_k)\|} = \arg \min_{\|d\|\leq 1} f'(x_k; d)$$

[assuming $\nabla f(x_k) \neq 0$]. Thus the gradient method is the descent algorithm of the form (2.6) that uses the direction that yields the greatest rate of cost improvement. For this reason it is also called the method of *steepest descent.*

Let us now discuss the convergence rate of the steepest descent method, assuming that f is twice continuously differentiable. With proper stepsize choice, it can be shown that the method has a linear rate, assuming that it generates a sequence $\{x_k\}$ that converges to a vector x^* such that $\nabla f(x^*) = 0$ and $\nabla^2 f(x^*)$ is positive definite. For example, if α_k is a sufficiently small constant $\alpha > 0$, the corresponding iteration

$$x_{k+1} = x_k - \alpha \nabla f(x_k), \tag{2.7}$$

can be shown to be contractive within a sphere centered at x^*, so it converges linearly.

To get a sense of this, assume for convenience that f is quadratic,† so by adding a suitable constant to f, we have

$$f(x) = \tfrac{1}{2}(x - x^*)'Q(x - x^*), \qquad \nabla f(x) = Q(x - x^*),$$

---

† Convergence analysis using a quadratic model is commonly used in nonlinear programming. The rationale is that behavior of an algorithm for a positive definite quadratic cost function is typically a correct predictor of its behavior for a twice differentiable cost function in the neighborhood of a minimum where the Hessian matrix is positive definite. Since the gradient is zero at that minimum, the positive definite quadratic term dominates the other terms in the Taylor series expansion, and the asymptotic behavior of the method does not depend on terms of order higher than two.

This time-honored line of analysis underlies some of the most widely used unconstrained optimization methods, such as Newton, quasi-Newton, and conjugate direction methods, which will be briefly discussed later. However, the rationale for these methods is weakened when the Hessian is singular at the minimum, since in this case third order terms may become significant. For this reason, when considering algorithmic options for a given differentiable optimization problem, it is important to consider (in addition to its cost function structure) whether the problem is "singular or "nonsingular."

where Q is the positive definite symmetric Hessian of f. Then for a constant stepsize α, the steepest descent iteration (2.7) can be written as

$$x_{k+1} - x^* = (I - \alpha Q)(x_k - x^*).$$

For $\alpha < 2/\lambda_{max}$, where λ_{max} is the largest eigenvalue of Q, the matrix $I - \alpha Q$ has eigenvalues strictly within the unit circle, and is a contraction with respect to the Euclidean norm. It can be shown (cf. Exercise 2.1) that the optimal modulus of contraction can be achieved with the stepsize choice

$$\alpha^* = \frac{2}{M + m},$$

where M and m are the minimum and maximum eigenvalues of Q. With this stepsize, we obtain the linear convergence rate estimate

$$\|x_{k+1} - x^*\| \leq \left(\frac{\frac{M}{m} - 1}{\frac{M}{m} + 1} \right) \|x_k - x^*\|. \tag{2.8}$$

Thus the convergence rate of steepest descent may be estimated in terms of the *condition number of* Q, the ratio M/m of largest to smallest eigenvalue. As the condition number increases to ∞ (i.e., the problem is increasingly "ill-conditioned") the modulus of contraction approaches 1, and the convergence can be very slow. This is the dominant characteristic of the behavior of gradient methods for the class of twice differentiable problems with positive definite Hessian. This class of problems is very broad, so condition number issues often become the principal consideration when implementing gradient methods in practice.

Choosing an appropriate constant stepsize may require some preliminary experimentation. Another possibility is the *line minimization rule*, which uses some specialized line search algorithm to determine

$$\alpha_k \in \arg\min_{\alpha \geq 0} f\big(x_k - \alpha \nabla f(x_k)\big).$$

With this rule, when the steepest descent method converges to a vector x^* such that $\nabla f(x^*) = 0$ and $\nabla^2 f(x^*)$ is positive definite, its convergence rate is also linear, but not faster than the one of Eq. (2.8), which is associated with an optimally chosen constant stepsize (see [Ber99], Section 1.3).

If the method converges to an optimal point x^* where the Hessian matrix $\nabla^2 f(x^*)$ is singular or does not exist, the convergence rate that we can guarantee is typically slower than linear. For example, with a properly chosen constant stepsize, and under some reasonable conditions (Lipschitz continuity of ∇f), we can show that

$$f(x_k) - f^* \leq \frac{c(x_0)}{k}, \qquad k = 1, 2, \ldots, \tag{2.9}$$

where f^* is the optimal value of f and $c(x_0)$ is a constant that depends on the initial point x_0 (see Section 6.1).

For problems where ∇f is continuous but cannot be assumed Lipschitz continuous at or near the minimum, it is necessary to use a stepsize rule that can produce time-varying stepsizes. For example in the scalar case where $f(x) = |x|^{3/2}$, the steepest descent method with any constant stepsize oscillates around the minimum $x^* = 0$, because the gradient grows too fast around x^*. However, the line minimization rule as well as other rules, such as the Armijo rule to be discussed shortly, guarantee a satisfactory form of convergence (see the end-of-chapter exercises and the discussion of Section 6.1).

On the other hand, with additional assumptions on the structure of f, we can obtain a faster convergence than the $O(1/k)$ estimate on the cost function error of Eq. (2.9). In particular, the rate of convergence to a singular minimum depends on the order of growth of the cost function near that minimum; see [Dun81], which shows that if f is convex, has a unique minimum x^*, and satisfies the growth condition

$$\beta\|x - x^*\|^\gamma \leq f(x) - f(x^*), \qquad \forall\ x \text{ such that } f(x) \leq f(x_0),$$

for some scalars $\beta > 0$ and $\gamma > 2$, then for the method of steepest descent with the Armijo rule and other related rules we have

$$f(x_k) - f(x^*) = O\left(\frac{1}{k^{\frac{\gamma}{\gamma-2}}}\right). \tag{2.10}$$

Thus for example, with a quartic order of growth of f ($\gamma = 4$), an $O(1/k^2)$ estimate is obtained for the cost function error after k iterations. The paper [Dun81] provides a more comprehensive analysis of the convergence rate of gradient-type methods based on order of growth conditions, including cases where the convergence rate is linear and faster than linear.

Scaling

To improve the convergence rate of the steepest descent method one may "scale" the gradient $\nabla f(x_k)$ by multiplication with a positive definite symmetric matrix D_k, i.e., use a direction $d_k = -D_k\nabla f(x_k)$, leading to the algorithm

$$x_{k+1} = x_k - \alpha_k D_k \nabla f(x_k); \tag{2.11}$$

cf. Fig. 2.1.1. Since for $\nabla f(x_k) \neq 0$ we have

$$f'(x_k; d_k) = -\nabla f(x_k)' D_k \nabla f(x_k) < 0,$$

it follows that we still have a cost descent method, as long as the positive stepsize α_k is sufficiently small so that $f(x_{k+1}) < f(x_k)$.

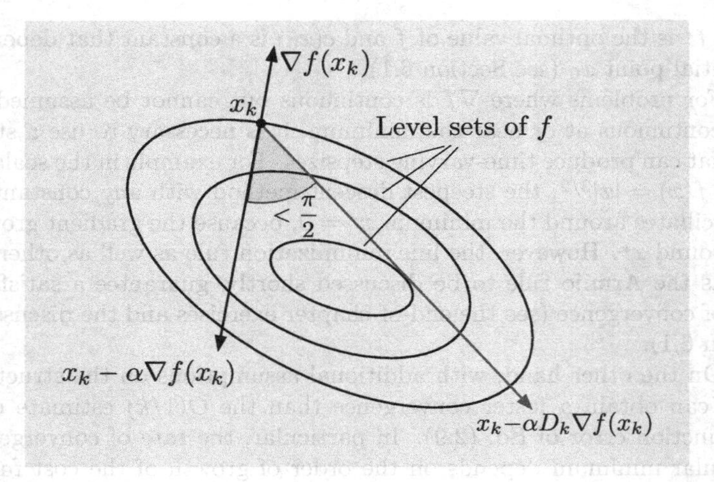

Figure 2.1.1. Illustration of descent directions. Any direction of the form

$$d_k = -D_k \nabla f(x_k),$$

where D_k is a positive definite matrix, is a descent direction because $d_k' \nabla f(x_k) = -d_k' D_k d_k < 0$. In this case d_k makes an angle less than $\pi/2$ with $-\nabla f(x_k)$.

Scaling is a major concept in the algorithmic theory of nonlinear programming. It is motivated by the idea of modifying the "effective condition number" of the problem through a linear change of variables of the form $x = D_k^{1/2} y$. In particular, the iteration (2.11) may be viewed as a steepest descent iteration

$$y_{k+1} = y_k - \alpha \nabla h_k(y_k)$$

for the equivalent problem of minimizing the function $h_k(y) = f(D_k^{1/2} y)$. For a quadratic problem, where $f(x) = \frac{1}{2}x'Qx - b'x$, the condition number of h_k is the ratio of largest to smallest eigenvalue of the matrix $D_k^{1/2} Q D_k^{1/2}$ (rather than Q).

Much of unconstrained nonlinear programming methodology deals with ways to compute "good" scaling matrices D_k, i.e., matrices that result in fast convergence rate. The "best" scaling in this sense is attained with

$$D_k = \left(\nabla^2 f(x_k)\right)^{-1},$$

assuming that the inverse above exists and is positive definite, which asymptotically leads to an "effective condition number" of 1. This is Newton's method, which will be discussed shortly. A simpler alternative is to use a diagonal approximation to the Hessian matrix $\nabla^2 f(x_k)$, i.e., the diagonal matrix D_k that has the inverse second partial derivatives

$$\left(\frac{\partial^2 f(x_k)}{(\partial x^i)^2}\right)^{-1}, \qquad i = 1, \ldots, n,$$

along the diagonal. This often improves the performance of the classical gradient method dramatically, by providing automatic scaling of the units in which the components x^i of x are measured, and also facilitates the choice of stepsize – good values of α_k are often chose to 1 (see the subsequent discussion of Newton's method and sources such as [Ber99], Section 1.3).

The nonlinear programming methodology also prominently includes quasi-Newton methods, which construct scaling matrices iteratively, using gradient information collected during the algorithmic process (see nonlinear programming textbooks such as [Pol71], [GMW81], [Lue84], [DeS96], [Ber99], [Fle00], [NoW06], [LuY08]). Some of these methods approximate the full inverse Hessian of f, and eventually attain the fast convergence rate of Newton's method. Other methods use a limited number of gradient vectors from previous iterations (have "limited memory") to construct a relatively crude but still effective approximation to the Hessian of f, and attain a convergence rate that is considerably faster than the one of the unscaled gradient method; see [Noc80], [NoW06].

Gradient Methods with Extrapolation

A variant of the gradient method, known as *gradient method with momentum*, involves extrapolation along the direction of the difference of the preceding two iterates:

$$x_{k+1} = x_k - \alpha_k \nabla f(x_k) + \beta_k (x_k - x_{k-1}), \tag{2.12}$$

where β_k is a scalar in $[0, 1)$, and we define $x_{-1} = x_0$. When α_k and β_k are chosen to be constant scalars α and β, respectively, the method is known as the *heavy ball method* [Pol64]; see Fig. 2.1.2. This is a sound method with guaranteed convergence under a Lipschitz continuity assumption on ∇f. It can be shown to have faster convergence rate than the corresponding gradient method where α_k is constant and $\beta_k \equiv 0$ (see [Pol87], Section 3.2.1, or [Ber99], Section 1.3). In particular, for a positive definite quadratic problem, and with optimal choices of the constants α and β, the convergence rate of the heavy ball method is linear, and is governed by the formula (2.8) but with $\sqrt{M/m}$ in place of M/m. This is a substantial improvement over the steepest descent method, although the method can still be very slow. Simple examples also suggest that with a momentum term, the steepest descent method is less prone to getting trapped at "shallow" local minima, and deals better with cost functions that are alternately very flat and very steep along the path of the algorithm.

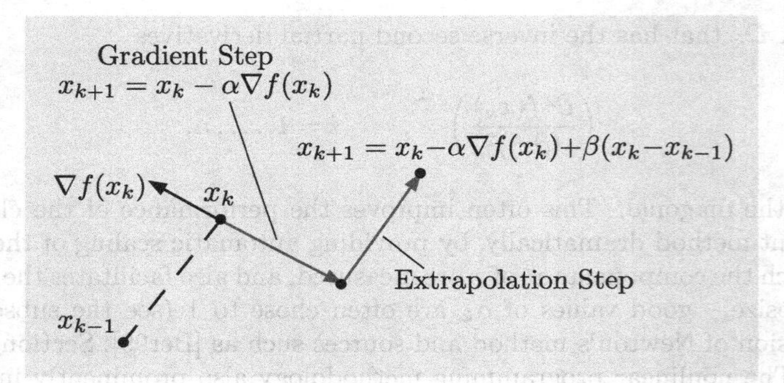

Figure 2.1.2. Illustration of the heavy ball method (2.12), where $\alpha_k \equiv \alpha$ and $\beta_k \equiv \beta$.

A method with similar structure as (2.12), proposed in [Nes83], has received a lot of attention because it has optimal iteration complexity properties under certain conditions, including Lipschitz continuity of ∇f. As we will see in Section 6.2, it improves on the $O(1/k)$ error estimate (2.9) of the gradient method by a factor of $1/k$. The iteration of this method, when applied to unconstrained minimization of a differentiable function f is commonly described in two steps: first an extrapolation step, to compute

$$y_k = x_k + \beta_k(x_k - x_{k-1})$$

with β_k chosen in a special way so that $\beta_k \to 1$, and then a gradient step with constant stepsize α, and gradient calculated at y_k,

$$x_{k+1} = y_k - \alpha\nabla f(y_k).$$

Compared to the method (2.12), it reverses the order of gradient calculation and extrapolation, and uses $\nabla f(y_k)$ in place of $\nabla f(x_k)$.

Conjugate Gradient Methods

There is an interesting connection between the extrapolation method (2.12) and the *conjugate gradient method* for unconstrained differentiable optimization. This is a classical method, with an extensive theory, and the distinctive property that it minimizes an n-dimensional convex quadratic cost function in at most n iterations, each involving a single line minimization. Fast progress is often obtained in much less than n iterations, depending on the eigenvalue structure of the quadratic cost [see e.g., [Ber82a] (Section 1.3.4), or [Lue84] (Chapter 8)]. The method can be implemented in several different ways, for which we refer to textbooks such as [Lue84], [Ber99]. It is a member of the more general class of *conjugate direction methods*, which involve a sequence of exact line searches along directions that are orthogonal with respect to some generalized inner product.

It turns out that if the parameters α_k and β_k in iteration (2.12) are chosen optimally *for each* k so that

$$(\alpha_k, \beta_k) \in \arg \min_{\alpha \in \Re,\, \beta \in \Re} f\big(x_k - \alpha \nabla f(x_k) + \beta(x_k - x_{k-1})\big), \qquad k = 0, 1, \ldots, \tag{2.13}$$

with $x_{-1} = x_0$, the resulting method is an implementation of the conjugate gradient method (see e.g., [Ber99], Section 1.6). By this we mean that if f is a convex quadratic function, *the method (2.12) with the stepsize choice (2.13) generates exactly the same iterates as the conjugate gradient method*, and hence minimizes f in at most n iterations. Finding the optimal parameters according to Eq. (2.13) requires solution of a two-dimensional optimization problem in α and β, which may be impractical in the absence of special structure. However, this optimization is facilitated in some important special cases, which also favor the use of other types of conjugate direction methods.†

There are several other ways to implement the conjugate gradient method, all of which generate identical iterates for quadratic cost functions, but may differ substantially in their behavior for nonquadratic ones. One of them, which resembles the preceding extrapolation methods, is the *method of parallel tangents* or PARTAN, first proposed in the paper [SBK64]. In particular, each iteration of PARTAN involves extrapolation and *two one-dimensional line minimizations*. At the typical iteration, given x_k, we obtain x_{k+1} as follows:

(1) We find a vector y_k that minimizes f over the line

$$\big\{y = x_k - \gamma \nabla f(x_k) \mid \gamma \geq 0\big\}.$$

(2) We generate x_{k+1} by minimizing f over the line that passes through x_{k-1} and y_k.

† Examples of favorably structured problems for conjugate direction methods include cost functions of the form $f(x) = h(Ax)$, where A is a matrix such that the calculation of the vector $y = Ax$ for a given x is far more expensive than the calculation of $h(y)$ and its gradient and Hessian (assuming it exists). Several of the applications described in Sections 1.3 and 1.4 are of this type; see also the papers [NaZ05] and [GoS10], where the application of the subspace minimization method (2.13) and PARTAN are discussed. For such problems, calculation of a stepsize by line minimization along a direction d, as in various types of conjugate direction methods, is relatively inexpensive. In particular, calculation of values, first, and second derivatives of the function $g(\alpha) \equiv f(x + \alpha d) = h(Ax + \alpha Ad)$ requires just two expensive operations: the one-time calculation of the matrix-vector products Ax and Ad. Similarly, minimization over a subspace that passes through x and is spanned by m directions $d_1, \ldots, d_m$, requires the one-time calculation of the matrix-vector products Ax and $Ad_1, \ldots, Ad_m$.

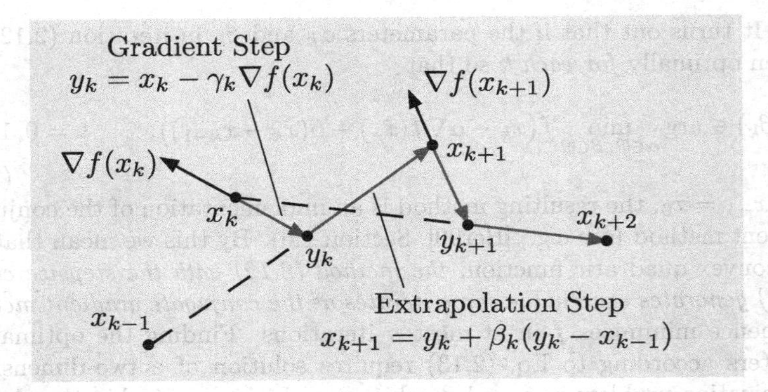

Figure 2.1.3. Illustration of the two-step method

$$y_k = x_k - \gamma_k \nabla f(x_k), \qquad x_{k+1} = y_k + \beta_k(y_k - x_{k-1}).$$

By writing the method equivalently as

$$x_{k+1} = x_k - \gamma_k(1 + \beta_k)\nabla f(x_k) + \beta_k(x_k - x_{k-1}),$$

we see that the heavy ball method (2.12) with constant parameters α and β is obtained when $\gamma_k \equiv \alpha/(1 + \beta)$ and $\beta_k \equiv \beta$. The PARTAN method is obtained when γ_k and β_k are chosen by line minimization, in which case the corresponding parameter α_k of iteration (2.12) is $\alpha_k = \gamma_k(1 + \beta_k)$.

This iteration is a special case of the gradient method with momentum (2.12), corresponding to special choices of α_k and β_k. To see this, observe that we can write iteration (2.12) as a two-step method:

$$y_k = x_k - \gamma_k \nabla f(x_k), \qquad x_{k+1} = y_k + \beta_k(y_k - x_{k-1}),$$

where

$$\gamma_k = \frac{\alpha_k}{1 + \beta_k}.$$

Thus starting from x_k, the parameter β_k is determined by the second line search of PARTAN as the optimal stepsize along the line that passes through x_{k-1} and y_k, and then α_k is determined as $\gamma_k(1 + \beta_k)$, where γ_k is the optimal stepsize along the line

$$\{x_k - \gamma \nabla f(x_k) \mid \gamma \geq 0\}$$

(cf. Fig. 2.1.3).

The salient property of PARTAN is that when f is convex quadratic it is mathematically equivalent to the conjugate gradient method (it generates exactly the same iterates and terminates in at most n iterations). For this it is essential that the line minimizations are exact, which may be difficult to guarantee in practice. However, PARTAN seems to be quite resilient to line minimization errors relative to other conjugate gradient implementations. Note that PARTAN ensures at least as large cost reduction at each iteration, as the steepest descent method, since the latter method omits the second line minimization. Thus even for nonquadratic cost functions it tends to perform faster than steepest descent, and often considerably so. We refer to [Lue84], [Pol87], and [Ber99], Section 1.6, for further discussion. These books also address additional issues for the conjugate gradient and other conjugate direction methods, such as alternative implementations, scaling (also called preconditioning), one-dimensional line search algorithms, and rate of convergence.

Newton's Method

In Newton's method the descent direction is

$$d_k = -\big(\nabla^2 f(x_k)\big)^{-1}\nabla f(x_k),$$

provided $\nabla^2 f(x_k)$ exists and is positive definite, so the iteration takes the form

$$x_{k+1} = x_k - \alpha_k\big(\nabla^2 f(x_k)\big)^{-1}\nabla f(x_k).$$

If $\nabla^2 f(x_k)$ is not positive definite, some modification is necessary. There are several possible modifications of this type, for which the reader may consult nonlinear programming textbooks. The simplest one is to add to $\nabla^2 f(x_k)$ a small positive multiple of the identity. Generally, when f is convex, $\nabla^2 f(x_k)$ is positive semidefinite (Prop. 1.1.10 in Appendix B), and this facilitates the implementation of reliable Newton-type algorithms.

The idea in Newton's method is to minimize at each iteration the quadratic approximation of f around the current point x_k given by

$$\tilde{f}_k(x) = f(x_k) + \nabla f(x_k)'(x - x_k) + \tfrac{1}{2}(x - x_k)'\nabla^2 f(x_k)(x - x_k).$$

By setting the gradient of $\tilde{f}_k(x)$ to zero,

$$\nabla f(x_k) + \nabla^2 f(x_k)(x - x_k) = 0,$$

and solving for x, we obtain as next iterate the minimizing point

$$x_{k+1} = x_k - \big(\nabla^2 f(x_k)\big)^{-1}\nabla f(x_k). \tag{2.14}$$

This is the Newton iteration corresponding to a stepsize $\alpha_k = 1$. It follows that, assuming $\alpha_k = 1$, *Newton's method finds the global minimum of a positive definite quadratic function in a single iteration.*

Newton's method typically converges very fast asymptotically, assuming that it converges to a vector x^* such that $\nabla f(x^*) = 0$ and $\nabla^2 f(x^*)$

is positive definite, and that a stepsize $\alpha_k = 1$ is used, at least after some iteration. For a simple argument, we may use Taylor's theorem to write

$$0 = \nabla f(x^*) = \nabla f(x_k) + \nabla^2 f(x_k)'(x^* - x_k) + o(\|x_k - x^*\|).$$

By multiplying this relation with $(\nabla^2 f(x_k))^{-1}$ we have

$$x_k - x^* - (\nabla^2 f(x_k))^{-1} \nabla f(x_k) = o(\|x_k - x^*\|),$$

so for the Newton iteration with stepsize $\alpha_k = 1$ we obtain

$$x_{k+1} - x^* = o(\|x_k - x^*\|),$$

or, for $x_k \neq x^*$,

$$\lim_{k \to \infty} \frac{\|x_{k+1} - x^*\|}{\|x_k - x^*\|} = \lim_{k \to \infty} \frac{o(\|x_k - x^*\|)}{\|x_k - x^*\|} = 0,$$

implying convergence that is faster than linear (also called *superlinear*). This argument can also be used to show *local convergence* to x^* with $\alpha_k \equiv 1$, that is, convergence assuming that x_0 is sufficiently close to x^*.

In implementations of Newton's method, some stepsize rule is often used to ensure cost reduction, but the rule is typically designed so that near convergence we have $\alpha_k = 1$, to ensure that a superlinear convergence rate is attained [assuming $\nabla^2 f(x^*)$ is positive definite at the limit x^*]. Methods that approximate Newton's method also use a stepsize close to 1, and modify the stepsize based on the results of the computation (see sources on nonlinear programming, such as [Ber99], Section 1.4).

The price for the fast convergence of Newton's method is the overhead required to calculate the Hessian matrix, and to solve the linear system of equations

$$\nabla^2 f(x_k) d_k = -\nabla f(x_k)$$

in order to find the Newton direction. There are many iterative algorithms that are patterned after Newton's method, and aim to strike a balance between fast convergence and high overhead (e.g., quasi-Newton, conjugate direction, and others, extensive discussions of which may be found in nonlinear programming textbooks such as [GMW81], [DeS96], [Ber99], [Fle00], [BSS06], [NoW06], [LuY08]).

We finally note that for some problems the special structure of the Hessian matrix can be exploited to facilitate the implementation of Newton's method. For example the Hessian matrix of the dual function of the separable convex programming problem of Section 1.1, when it exists, has particularly favorable structure; see [Ber99], Section 6.1. The same is true for optimal control problems that involve a discrete-time dynamic system and a cost function that is additive over time; see [Ber99], Section 1.9.

Stepsize Rules

There are several methods to choose the stepsize α_k in the scaled gradient iteration (2.11). For example, α_k may be chosen by *line minimization*:

$$\alpha_k \in \arg\min_{\alpha \geq 0} f\big(x_k - \alpha D_k \nabla f(x_k)\big).$$

This can typically be implemented only approximately, with some iterative one-dimensional optimization algorithm; there are several such algorithms (see nonlinear programming textbooks such as [GMW81], [Ber99], [BSS06], [NoW06], [LuY08]).

Our analysis in subsequent chapters of this book will mostly focus on two cases: when α_k is chosen to be *constant*,

$$\alpha_k = \alpha, \qquad k = 0, 1, \ldots,$$

and when α_k is chosen to be *diminishing* to 0, while satisfying the conditions†

$$\sum_{k=0}^{\infty} \alpha_k = \infty, \qquad \sum_{k=0}^{\infty} \alpha_k^2 < \infty. \tag{2.15}$$

A convergence analysis for these two stepsize rules is given in the end-of-chapter exercises, and also in Chapter 3, in the context of subgradient methods, as well as in Section 6.1.

We emphasize the constant and diminishing stepsize rules because they are the ones that most readily generalize to nondifferentiable cost functions. However, other stepsize rules, briefly discussed in this chapter, are also important, particularly for differentiable problems, and are used widely. One possibility is the line minimization rule discussed earlier. There are also other rules, which are simple and are based on successive reduction of α_k, until a form of descent is achieved that guarantees convergence. One of these, the *Armijo rule* (first proposed in [Arm66], and sometimes called *backtracking rule*), is popular in unconstrained minimization algorithm implementations. It is given by

$$\alpha_k = \beta^{m_k} s_k,$$

where m_k is the first nonnegative integer m for which

$$f(x_k) - f\big(x_k - \beta^m s_k D_k \nabla f(x_k)\big) \geq \sigma \beta^m s_k \nabla f(x_k)' D_k \nabla f(x_k),$$

† The condition $\sum_{k=0}^{\infty} \alpha_k = \infty$ is needed so that the method can approach the minimum from arbitrarily far, and the condition $\sum_{k=0}^{\infty} \alpha_k^2 < \infty$ is needed so that $\alpha_k \to 0$ and also for technical reasons relating to the convergence analysis (see Section 3.2). If f is a positive definite quadratic, the steepest descent method with a diminishing stepsize α_k satisfying $\sum_{k=0}^{\infty} \alpha_k = \infty$ can be shown to converge to the optimal solution, but at a rate that is slower than linear.

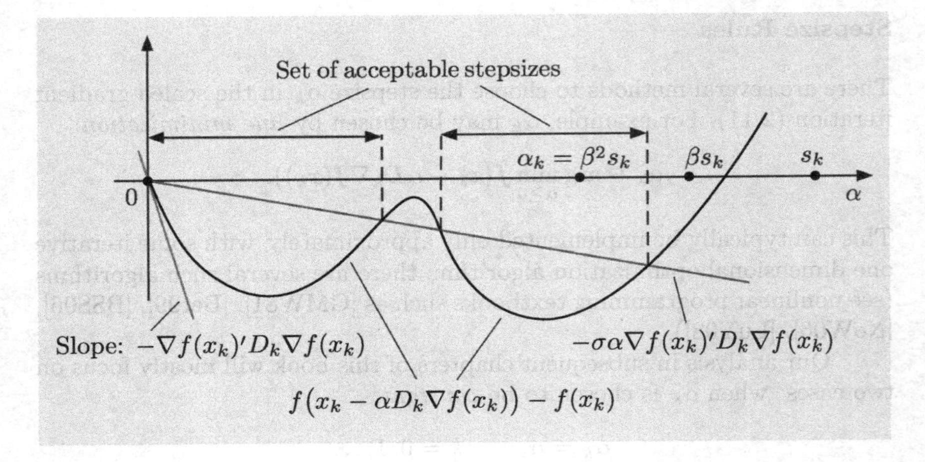

Figure 2.1.4. Illustration of the successive points tested by the Armijo rule along the descent direction $d_k = -D_k\nabla f(x_k)$. In this figure, α_k is obtained as $\beta^2 s_k$ after two unsuccessful trials. Because $\sigma \in (0,1)$, the set of acceptable stepsizes begins with a nontrivial interval interval when $d_k \neq 0$. This implies that if $d_k = 0$, the Armijo rule will find an acceptable stepsize with a finite number of stepsize reductions.

where $\beta \in (0,1)$ and $\sigma \in (0,1)$ are some constants, and $s_k > 0$ is positive initial stepsize, chosen to be either constant or through some simplified search or polynomial interpolation. In other words, starting with an initial trial s_k, the stepsizes $\beta^m s_k$, $m = 0, 1, \ldots$, are tried successively until the above inequality is satisfied for $m = m_k$; see Fig. 2.1.4. We will explore the convergence properties of this rule in the exercises.

Aside from guaranteeing cost function descent, successive reduction rules have the additional benefit of adapting the size of the stepsize α_k to the search direction $-D_k\nabla f(x_k)$, particularly when the initial stepsize s_k is chosen by some simplified search process. We refer to nonlinear programming sources for detailed discussions.

Note that the diminishing stepsize rule does not guarantee cost function descent at each iteration, although it reduces the cost function value once the stepsize becomes sufficiently small. There are also some other rules, often called *nonmonotonic*, which do not explicitly try to enforce cost function descent and have achieved some success, but are based on ideas that we will not discuss in this book; see [GLL86], [BaB88], [Ray93], [Ray97], [BMR00], [DHS06]. An alternative approach to enforce descent without explicitly using stepsizes is based on the trust region methodology for which we refer to book sources such as [Ber99], [CGT00], [Fle00], [NoW06].

2.1.2 Constrained Problems – Feasible Direction Methods

Let us now consider minimizing a differentiable cost function f over a closed convex subset X of $\Re^n$. In a natural form of the cost function descent approach, we may consider generating a feasible sequence $\{x_k\} \subset X$ with an iteration of the form

$$x_{k+1} = x_k + \alpha_k d_k, \tag{2.16}$$

while enforcing cost improvement. However, this is now more complicated because it is not enough for d_k to be a descent direction at x_k. It must also be a *feasible direction* in the sense that $x_k + \alpha d_k$ must belong to X for small enough $\alpha > 0$, in order for the new iterate x_{k+1} to belong to X with suitably small choice of α_k. By multiplying d_k with a positive constant if necessary, this essentially restricts d_k to be of the form $\bar{x}_k - x_k$ for some $\bar{x}_k \in X$ with $\bar{x}_k \neq x_k$. Thus, if f is differentiable, for a feasible descent direction, it is sufficient that

$$d_k = \bar{x}_k - x_k, \qquad \text{for some } \bar{x}_k \in X \text{ with } \nabla f(x_k)'(\bar{x}_k - x_k) < 0.$$

Methods of the form (2.16), where d_k is a feasible descent direction were introduced in the 60s (see e.g., the books [Zou60], [Zan69], [Pol71], [Zou76]), and have been used extensively in applications. We refer to them as *feasible direction methods*, and we give examples of some of the most popular ones.

Conditional Gradient Method

The simplest feasible direction method is to find at iteration k,

$$\bar{x}_k \in \arg\min_{x \in X} \nabla f(x_k)'(x - x_k), \tag{2.17}$$

and set

$$d_k = \bar{x}_k - x_k$$

in Eq. (2.16); see Fig. 2.1.5. Clearly $\nabla f(x_k)'(\bar{x}_k - x_k) \leq 0$, with equality holding only if $\nabla f(x_k)'(x - x_k) \geq 0$ for all $x \in X$, which is a necessary condition for optimality of x_k.

This is the *conditional gradient method* (also known as the *Frank-Wolfe algorithm*) proposed in [FrW56] for convex programming problems with linear constraints, and for more general problems in [LeP65]. The method has been used widely in many contexts, as it is theoretically sound, quite simple, and often convenient. In particular, when X is a polyhedral set, computation of $\bar{x}_k$ requires the solution of a linear program. In some important cases, this linear program has special structure, which results in great simplifications, e.g., in the multicommodity flow problem of Example

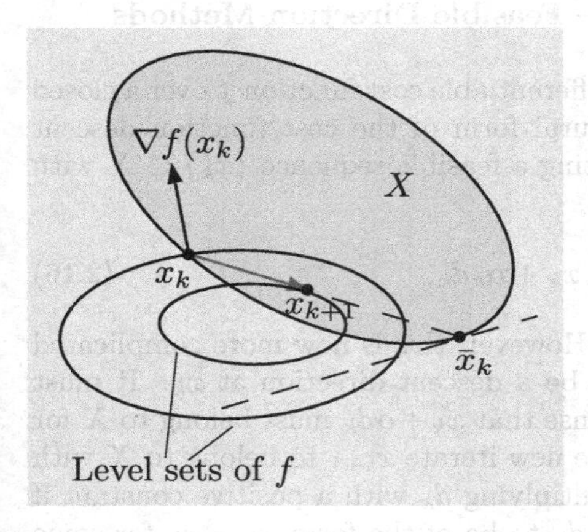

Figure 2.1.5. Illustration of the conditional gradient iteration at x_k. We find $\bar{x}_k$, a point of X that lies farthest along the negative gradient direction $-\nabla f(x_k)$. We then set

$$x_{k+1} = x_k + \alpha_k(\bar{x}_k - x_k),$$

where α_k is a stepsize from $(0, 1]$ (the figure illustrates the case where α_k is chosen by line minimization).

1.4.5 (see the book [BeG92], or the surveys [FlH95], [Pat01]). There has been intensified interest in the conditional gradient method, thanks to applications in machine learning; see e.g., [Cla10], [Jag13], [LuT13], [RSW13], [FrG14], [HJN14], and the references quoted there.

However, the conditional gradient method often tends to converge very slowly relative to its competitors (its asymptotic convergence rate can be slower than linear even for positive definite quadratic programming problems); see [CaC68], [Dun79], [Dun80]. For this reason, other methods with better practical convergence rate properties are often preferred.

One of these methods, is the *simplicial decomposition algorithm* (first proposed independently in [CaG74] and [Hol74]), which will be discussed in detail in Chapter 4. This method is not a feasible direction method of the form (2.16), but instead it is based on multidimensional optimizations over approximations of the constraint set by convex hulls of finite numbers of points. When X is a polyhedral set, it converges in a finite number of iterations, and while this number can potentially be very large, the method often attains practical convergence in very few iterations. Generally, simplicial decomposition can provide an attractive alternative to the conditional gradient method because it tends to be well-suited for the same type of problems [it also requires solution of linear cost subproblems of the form (2.17); see the discussion of Section 4.2].

Somewhat peculiarly, the practical performance of the conditional gradient method tends to improve in highly constrained problems. An explanation for this is given in the papers [Dun79], [DuS83], where it is shown among others that the convergence rate of the method is linear when the cost function is positive definite quadratic, and the constraint set is not polyhedral but rather has a "positive curvature" property (for example it is a sphere). When there are many linear constraints, the constraint set tends to have very many closely spaced extreme points, and has this "positive curvature" property in an approximate sense.

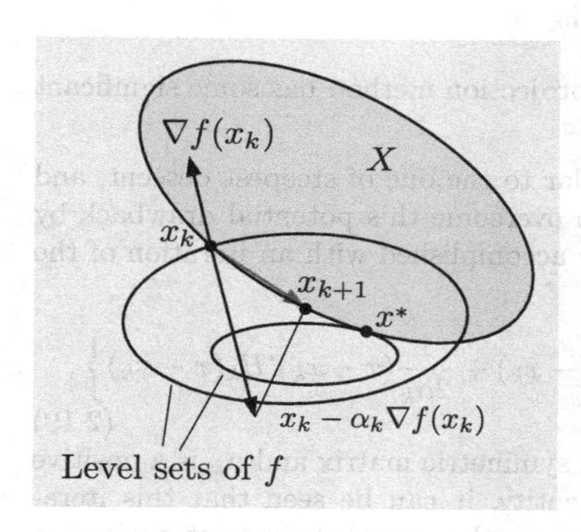

Figure 2.1.6. Illustration of the gradient projection iteration at x_k. We move from x_k along the direction $-\nabla f(x_k)$ and project $x_k - \alpha_k \nabla f(x_k)$ onto X to obtain x_{k+1}. We have

$$\nabla f(x_k)'(x_{k+1} - x_k) \leq 0,$$

and unless $x_{k+1} = x_k$, in which case x_k minimizes f over X, the angle between $\nabla f(x_k)$ and $(x_{k+1} - x_k)$ is strictly greater than 90 degrees, and we have

$$\nabla f(x_k)'(x_{k+1} - x_k) < 0.$$

Gradient Projection Method

Another major feasible direction method, which generally achieves a faster convergence rate than the conditional gradient method, is the *gradient projection method* (originally proposed in [Gol64], [LeP65]), which has the form

$$x_{k+1} = P_X\big(x_k - \alpha_k \nabla f(x_k)\big), \tag{2.18}$$

where $\alpha_k > 0$ is a stepsize and $P_X(\cdot)$ denotes projection on X (the projection is well defined since X is closed and convex; see Fig. 2.1.6).

To get a sense of the validity of the method, note that from the Projection Theorem (Prop. 1.1.9 in Appendix B), we have

$$\nabla f(x_k)'(x_{k+1} - x_k) \leq 0,$$

and by the optimality condition for convex functions (cf. Prop. 1.1.8 in Appendix B), the inequality is strict unless x_k is optimal. Thus $x_{k+1} - x_k$ defines a feasible descent direction at x_k, and based on this fact, we can show the descent property $f(x_{k+1}) < f(x_k)$ when α_k is sufficiently small.

The stepsize α_k is chosen similar to the unconstrained gradient method, i.e., constant, diminishing, or through some kind of reduction rule to ensure cost function descent and guarantee convergence to the optimum; see the convergence analysis of Section 6.1, and [Ber99], Section 2.3, for a detailed discussion and references. Moreover the convergence rate estimates given earlier for unconstrained steepest descent in the positive definite quadratic cost case [cf. Eq. (2.8)] and in the singular case [cf. Eqs. (2.9) and (2.10)] generalize to the gradient projection method under various stepsize rules (see Exercise 2.1 for the former case and [Dun81] for the latter case).

Two-Metric Projection Methods

Despite its simplicity, the gradient projection method has some significant drawbacks:

(a) Its rate of convergence is similar to the one of steepest descent, and is often slow. It is possible to overcome this potential drawback by a form of scaling. This can be accomplished with an iteration of the form

$$x_{k+1} \in \arg\min_{x \in X} \left\{ \nabla f(x_k)'(x - x_k) + \frac{1}{2\alpha_k}(x - x_k)' H_k (x - x_k) \right\}, \tag{2.19}$$

where H_k is a positive definite symmetric matrix and α_k is a positive stepsize. When H_k is the identity, it can be seen that this iteration gives the same iterate x_{k+1} as the unscaled gradient projection iteration (2.18). When $H_k = \nabla^2 f(x_k)$ and $\alpha_k = 1$, we obtain a constrained form of Newton's method (see nonlinear programming sources for analysis; e.g., [Ber99]).

(b) Depending on the nature of X, the projection operation may involve substantial overhead. The projection is simple when H_k is the identity (or more generally, is diagonal), and X consists of simple lower and/or upper bounds on the components of x:

$$X = \{(x^1, \ldots, x^n) \mid \underline{b}^i \leq x^i \leq \bar{b}^i,\ i = 1, \ldots, n\}. \tag{2.20}$$

This is an important special case where the use of gradient projection is convenient. Then the projection decomposes to n scalar projections, one for each $i = 1, \ldots, n$: the ith component of x_{k+1} is obtained by projection of the ith component of $x_k - \alpha_k \nabla f(x_k)$,

$$\big(x_k - \alpha_k \nabla f(x_k)\big)^i,$$

onto the interval of corresponding bounds $[\underline{b}^i, \bar{b}^i]$, and is very simple. However, for general nondiagonal scaling the overhead for solving the quadratic programming problem (2.19) is substantial even if X has a simple bound structure of Eq. (2.20).

To overcome the difficulty with the projection overhead, a scaled projection method known as *two-metric projection method* has been proposed for the case of the bound constraints (2.20) in [Ber82a], [Ber82b]. It has a similar form to the scaled gradient method (2.11), and it is given by

$$x_{k+1} = P_X\big(x_k - \alpha_k D_k \nabla f(x_k)\big). \tag{2.21}$$

It is thus a natural and simple adaptation of unconstrained Newton-like methods to bound-constrained optimization, including quasi-Newton methods. The main difficulty here is that an arbitrary positive definite matrix

D_k will not necessarily yield a descent direction. However, it turns out that if some of the off-diagonal terms of D_k that correspond to components of x_k that are at their boundary are set to zero, one can obtain descent (see Exercise 2.8). Furthermore, one can select D_k as the inverse of a partially diagonalized version of the Hessian matrix $\nabla^2 f(x_k)$ and attain the fast convergence rate of Newton's method (see [Ber82a], [Ber82b], [GaB84]).

The idea of simple two-metric projection with partial diagonalization may be generalized to more complex constraint sets, and it has been adapted in [Ber82b], and subsequent papers such as [GaB84], [Dun91], [LuT93b], to problems of the form

$$\begin{aligned} &\text{minimize} \quad f(x) \\ &\text{subject to} \quad \underline{b} \leq x \leq \bar{b}, \quad Ax = c, \end{aligned}$$

where A is an $m \times n$ matrix, and $\underline{b}, \bar{b} \in \Re^n$ and $c \in \Re^m$ are given vectors. For example the algorithm (2.21) can be easily modified when the constraint set involves bounds on the components of x together with a few linear constraints, e.g., problems involving a simplex constraint such as

$$\begin{aligned} &\text{minimize} \quad f(x) \\ &\text{subject to} \quad 0 \leq x, \quad a'x = c, \end{aligned}$$

where $a \in \Re^n$ and $c \in \Re$, or a Cartesian product of simplexes. For an example of a Newton algorithm of this type, applied to the multicommodity flow problem of Example 1.4.5, see [BeG83]. For representative applications in related large-scale contexts we refer to the papers [Dun91], [LuT93b], [FJS98], [Pyt98], [GeM05], [OJW05], [TaP13], [WSK14].

The advantage that the two-metric projection approach can offer is to identify quickly the constraints that are active at an optimal solution. After this happens, the method reduces essentially to an unconstrained scaled gradient method (possibly Newton method, if D_k is a partially diagonalized Hessian matrix), and attains a fast convergence rate. This property has also motivated variants of the two-metric projection method for problems involving ℓ_1-regularization, such as the ones of Example 1.3.2; see [SFR09], [Sch10], [GKX10], [SKS12], [Lan14].

Block Coordinate Descent

The preceding methods require the computation of the gradient and possibly the Hessian of the cost function at each iterate. An alternative descent approach that does not require derivatives or other direction calculations is the classical *block coordinate descent* method, which we will briefly describe here and consider further in Section 6.5. The method applies to the problem

$$\begin{aligned} &\text{minimize} \quad f(x) \\ &\text{subject to} \quad x \in X, \end{aligned}$$

where $f : \Re^n \mapsto \Re$ is a differentiable function, and X is a Cartesian product of closed convex sets $X_1, \ldots, X_m$:

$$X = X_1 \times X_2 \times \cdots \times X_m.$$

The vector x is partitioned as

$$x = (x^1, x^2, \ldots, x^m),$$

where each x^i belongs to $\Re^{n_i}$, so the constraint $x \in X$ is equivalent to

$$x^i \in X_i, \qquad i = 1, \ldots, m.$$

The most common case is when $n_i = 1$ for all i, so the components x^i are scalars. The method involves minimization with respect to a single component x^i at each iteration, with all other components kept fixed.

In an example of such a method, given the current iterate $x_k = (x_k^1, \ldots, x_k^m)$, we generate the next iterate $x_{k+1} = (x_{k+1}^1, \ldots, x_{k+1}^m)$, according to the "cyclic" iteration

$$x_{k+1}^i \in \arg \min_{\xi \in X_i} f(x_{k+1}^1, \ldots, x_{k+1}^{i-1}, \xi, x_k^{i+1}, \ldots, x_k^m), \quad i = 1, \ldots, m. \quad (2.22)$$

Thus, at each iteration, the cost is minimized with respect to each of the "block coordinate" vectors x^i, taken one-at-a-time in cyclic order.

Naturally, the method makes practical sense only if it is possible to perform this minimization fairly easily. This is frequently so when each x^i is a scalar, but there are also other cases of interest, where x^i is a multi-dimensional vector. Moreover, the method can take advantage of special structure of f; an example of such structure is a form of "sparsity," where f is the sum of component functions, and for each i, only a relatively small number of the component functions depend on x^i, thereby simplifying the minimization (2.22). The following is an example of a classical algorithm that can be viewed as a special case of block coordinate descent.

Example 2.1.1 (Parallel Projections Algorithm)

We are given m closed convex sets $X_1, \ldots, X_m$ in $\Re^n$, and we want to find a point in their intersection. This problem can equivalently be written as

$$\begin{aligned} &\text{minimize} \quad \sum_{i=1}^m \|y^i - x\|^2 \\ &\text{subject to} \quad x \in \Re^n, \ \ y^i \in X_i, \ i = 1, \ldots, m, \end{aligned}$$

where the variables of the optimization are $x, y^1, \ldots, y^m$ (the optimal solutions of this problem are the points in the intersection $\cap_{i=1}^m X_i$, if this intersection is nonempty). A block coordinate descent algorithm iterates on each of the vectors $y^1, \ldots, y^m$ in parallel according to

$$y_{k+1}^i = P_{X_i}(x_k), \qquad i = 1, \ldots, m,$$

and then iterates with respect to x according to

$$x_{k+1} = \frac{y_{k+1}^1 + \cdots + y_{k+1}^m}{m},$$

which minimizes the cost function with respect to x when each y^i is fixed at y_{k+1}^i.

Here is another example where the coordinate descent method takes advantage of decomposable structure.

Example 2.1.2 (Hierarchical Decomposition)

Consider an optimization problem of the form

$$\begin{aligned} &\text{minimize} \quad \sum_{i=1}^m \big(h_i(y^i) + f_i(x, y^i)\big) \\ &\text{subject to} \quad x \in X, \qquad y^i \in Y_i, \quad i = 1, \ldots, m, \end{aligned}$$

where X and Y_i, $i = 1, \ldots, m$, are closed, convex subsets of corresponding Euclidean spaces, and h_i, f_i are given functions, assumed differentiable. This problem is associated with a paradigm of optimization of a system consisting of m subsystems, with a cost function $h_i + f_i$ associated with the operations of the ith subsystem. Here y^i is viewed as a vector of local decision variables that influences the cost of the ith subsystem only, and x is viewed as a vector of global or coordinating decision variables that affects the operation of all the subsystems.

The block coordinate descent method has the form

$$y_{k+1}^i \in \arg\min_{y^i \in Y_i} \big\{h_i(y^i) + f_i(x_k, y^i)\big\}, \qquad i = 1, \ldots, m,$$

$$x_{k+1} \in \arg\min_{x \in X} \sum_{i=1}^m f_i(x, y_{k+1}^i).$$

The method has a natural real-life interpretation: at each iteration, each subsystem optimizes its own cost, viewing the global variables as fixed at their current values, and then the coordinator optimizes the overall cost for the current values of the local variables (without having to know the "local" cost functions h_i of the subsystems).

In the absence of special structure of f, differentiability is essential for the validity of the coordinate descent method; this can be verified with simple examples. In our convergence analysis of Chapter 6, we will also require a form of strict convexity of f along each block component, as first suggested in the book [Zan69] (subtle examples of nonconvergence have been constructed in the absence of a property of this kind [Pow73]). We note, however, there are some interesting cases where nondifferentiabilities with special structure can be dealt with; see Section 6.5.

There are several variants of the method, which incorporate various descent algorithms in the solution of the block minimizations (2.22). Another type of variant is one where the block components are iterated in an irregular order instead of a fixed cyclic order. In fact there is a substantial theory of asynchronous distributed versions of coordinate descent, for which we refer to the parallel and distributed algorithms book [BeT89a], and the sources quoted there; see also the discussion in Sections 2.1.6 and 6.5.2.

2.1.3 Nondifferentiable Problems – Subgradient Methods

We will now briefly consider the minimization of a convex nondifferentiable cost function $f : \Re^n \mapsto \Re$ (optimization of a nonconvex and nondifferentiable function is a far more complicated subject, which we will not address in this book). It is possible to generalize the steepest descent approach so that when f is nondifferentiable at x_k, we use a direction d_k that minimizes the directional derivative $f'(x_k; d)$ subject to $\|d\| \leq 1$,

$$d_k \in \arg\min_{\|d\| \leq 1} f'(x_k; d).$$

Unfortunately, this minimization (or more generally finding a descent direction) may involve a nontrivial computation. Moreover, there is a worrisome theoretical difficulty: the method may get stuck far from the optimum, depending on the stepsize rule. An example is given in Fig. 2.1.7, where the stepsize is chosen using the minimization rule

$$\alpha_k \in \arg\min_{\alpha \geq 0} f(x_k + \alpha d_k).$$

In this example, *the algorithm fails even though it never encounters a point where f is nondifferentiable,* which suggests that convergence questions in convex optimization are delicate and should not be treated lightly. The problem here is a lack of continuity: the steepest descent direction may undergo a large/discontinuous change close to the convergence limit. By contrast, this would not happen if f were continuously differentiable at the limit, and in fact the steepest descent method with the minimization stepsize rule has sound convergence properties when used for differentiable functions.

Because the implementation of cost function descent has the limitations outlined above, a different kind of descent approach, based on the notion of subgradient, is often used when f is nondifferentiable. The theory of subgradients of extended real-valued functions is outlined in Section 5.4 of Appendix B, as developed in the textbook [Ber09]. The properties of

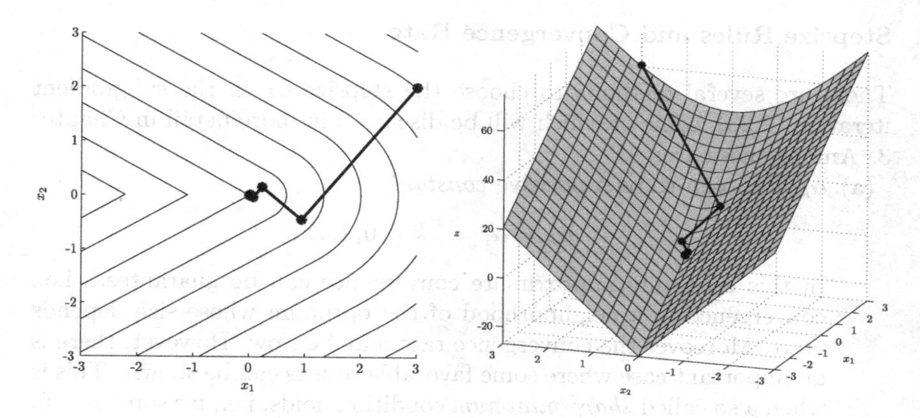

Figure 2.1.7. An example of failure of the steepest descent method with the line minimization stepsize rule for a convex nondifferentiable cost function [Wol75]. Here we have the two-dimensional cost function

$$f(x_1,x_2)=\begin{cases}5(9x_1^2+16x_2^2)^{1/2} & \text{if } x_1>|x_2|,\\ 9x_1+16|x_2| & \text{if } x_1\le|x_2|,\end{cases}$$

shown in the figure. Consider the method that moves in the direction of steepest descent from the current point, with the stepsize determined by cost minimization along that direction (this can be done analytically). Suppose that the algorithm starts anywhere within the set

$$\{(x_1,x_2)\mid x_1>|x_2|>(9/16)^2|x_1|\}.$$

The generated iterates are shown in the figure, and it can be verified that they converge to the nonoptimal point $(0,0)$.

subgradients of real-valued convex functions will also be discussed in detail in Section 3.1.

In the most common subgradient method (first proposed and analyzed in the mid 60s by Shor in a series of papers, and later in the books [Sho85], [Sho98]), an arbitrary subgradient g_k of f at x_k is used in an iteration of the form

$$x_{k+1}=x_k-\alpha_k g_k, \tag{2.23}$$

where α_k is a positive stepsize. The method, together with its many variations, will be discussed extensively in this book, starting with Chapter 3. We will see that while it may not yield a cost reduction for any value of α_k it has another descent property, which enhances the convergence process: at any nonoptimal point x_k, it satisfies

$$\operatorname{dist}(x_{k+1},X^*)<\operatorname{dist}(x_k,X^*)$$

for a sufficiently small stepsize α_k, where $\operatorname{dist}(x,X^*)$ denotes the Euclidean minimum distance of x from the optimal solution set X^*.

Stepsize Rules and Convergence Rate

There are several methods to choose the stepsize α_k in the subgradient iteration (2.23), some of which will be discussed in more detail in Chapter 3. Among these are:

(a) α_k is chosen to be a positive *constant*,

$$\alpha_k = \alpha, \qquad k = 0, 1, \ldots.$$

In this case only approximate convergence can be guaranteed, i.e., convergence to a neighborhood of the optimum whose size depends on α. Moreover the convergence rate may be slow. However, there is an important case where some favorable results can be shown. This is when a so called *sharp minimum* condition holds, i.e., for some $\beta > 0$,

$$f^* + \beta \min_{x^* \in X^*} \|x - x^*\| \leq f(x), \qquad \forall\, x \in X, \tag{2.24}$$

where f^* is the optimal value (see Exercise 3.10). We will prove in Prop. 5.1.6 that this condition holds when f and X are polyhedral, as for example in dual problems arising in integer programming.

(b) α_k is chosen to be *diminishing* to 0, while satisfying the conditions

$$\sum_{k=0}^{\infty} \alpha_k = \infty, \qquad \sum_{k=0}^{\infty} \alpha_k^2 < \infty.$$

Then exact convergence can be guaranteed, but the convergence rate is sublinear, even for polyhedral problems, and typically very slow.

There are also more sophisticated stepsize rules, which are based on estimation of f^* (see Section 3.2, and [BNO03] for a detailed account). Still, unless the condition (2.24) holds, the convergence rate can be very slow relative to other methods. On the other hand in the presence of special structure, such as in additive cost problems, incremental versions of subgradient methods (see Section 2.1.5) may perform satisfactorily.

2.1.4 Alternative Descent Methods

Aside from methods that are based on gradients or subgradients, like the ones of the preceding sections, there are some other approaches to effect cost function descent. A major approach, which applies to any convex cost function is the *proximal algorithm*, to be discussed in detail in Chapter 5. This algorithm embodies both the cost improvement and the approximation ideas. In its basic form, it approximates the minimization of a closed proper convex function $f : \Re^n \mapsto (-\infty, \infty]$ with another minimization that involves a quadratic term. It is given by

$$x_{k+1} \in \arg \min_{x \in \Re^n} \left\{ f(x) + \frac{1}{2c_k} \|x - x_k\|^2 \right\}, \tag{2.25}$$

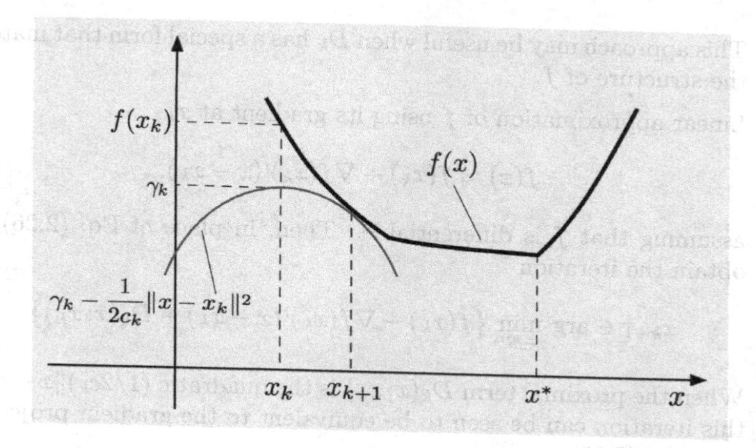

Figure 2.1.8. Illustration of the proximal algorithm (2.25) and its descent property. The minimum of $f(x)+\frac{1}{2c_k}\|x-x_k\|^2$ is attained at the unique point x_{k+1} at which the graph of the quadratic function $-\frac{1}{2c_k}\|x-x_k\|^2$, raised by the amount

$$\gamma_k = f(x_{k+1}) + \frac{1}{2c_k}\|x_{k+1} - x_k\|^2,$$

just touches the graph of f. Since $\gamma_k < f(x_k)$, it follows that $f(x_{k+1}) < f(x_k)$, unless x_k minimizes f, which happens if and only if $x_{k+1} = x_k$.

where x_0 is an arbitrary starting point and c_k is a positive scalar parameter (see Fig. 2.1.8). One of the motivations for the algorithm is that it "regularizes" the minimization of f: the quadratic term in Eq. (2.25) when added to f makes it strictly convex with compact level sets, so it has a unique minimum (cf. Prop. 3.1.1 and Prop. 3.2.1 in Appendix B).

The algorithm has an inherent descent character, which facilitates its combination with other algorithmic schemes. To see this note that since $x = x_{k+1}$ gives a lower value of $f(x) + \frac{1}{2c_k}\|x - x_k\|^2$ than $x = x_k$, we have

$$f(x_{k+1}) + \frac{1}{2c_k}\|x_{k+1} - x_k\|^2 \leq f(x_k).$$

It follows that $\{f(x_k)\}$ is monotonically nonincreasing; see also Fig. 2.1.8.

There are several variations of the proximal algorithm, which will be discussed in Chapters 5 and 6. Some of these variations involve modification of the proximal minimization problem of Eq. (2.25), motivated by the need for a convenient solution of this problem. Here are some examples:

(a) The use of a nonquadratic proximal term $D_k(x; x_k)$ in Eq. (2.25), in place of $(1/2c_k)\|x - x_k\|^2$, i.e., the iteration

$$x_{k+1} \in \arg\min_{x\in\Re^n} \{f(x) + D_k(x; x_k)\}. \tag{2.26}$$

This approach may be useful when D_k has a special form that matches the structure of f.

(b) Linear approximation of f using its gradient at x_k

$$f(x) \approx f(x_k) + \nabla f(x_k)'(x - x_k),$$

assuming that f is differentiable. Then, in place of Eq. (2.26), we obtain the iteration

$$x_{k+1} \in \arg \min_{x \in \Re^n} \{f(x_k) + \nabla f(x_k)'(x - x_k) + D_k(x; x_k)\}.$$

When the proximal term $D_k(x; x_k)$ is the quadratic $(1/2c_k)\|x - x_k\|^2$, this iteration can be seen to be equivalent to the gradient projection iteration (2.18):

$$x_{k+1} = P_X(x_k - c_k \nabla f(x_k)),$$

but there are other choices of D_k that lead to interesting methods, known as *mirror descent algorithms*.

(c) The *proximal gradient algorithm*, which applies to the problem

$$\begin{aligned} &\text{minimize} \quad f(x) + h(x) \\ &\text{subject to } x \in \Re^n, \end{aligned}$$

where $f : \Re^n \mapsto \Re$ is a differentiable convex function, and $h : \Re^n \mapsto (-\infty, \infty]$ is a closed proper convex function. This algorithm combines ideas from the gradient projection method and the proximal method. It replaces f with a linear approximation in the proximal minimization, i.e.,

$$x_{k+1} \in \arg \min_{x \in \Re^n} \left\{ \nabla f(x_k)'(x - x_k) + h(x) + \frac{1}{2\alpha_k}\|x - x_k\|^2 \right\}, \tag{2.27}$$

where $\alpha_k > 0$ is a parameter. Thus when f is a linear function, we obtain the proximal algorithm for minimizing $f + h$. When h is the indicator function of a closed convex set, we obtain the gradient projection method. Note that there is an alternative/equivalent way to write the algorithm (2.27):

$$z_k = x_k - \alpha_k \nabla f(x_k), \quad x_{k+1} \in \arg \min_{x \in \Re^n} \left\{ h(x) + \frac{1}{2\alpha_k}\|x - z_k\|^2 \right\}, \tag{2.28}$$

as can be verified by expanding the quadratic

$$\|x - z_k\|^2 = \left\|x - x_k + \alpha_k \nabla f(x_k)\right\|^2.$$

Thus the method alternates gradient steps on f with proximal steps on h. The advantage that this method may have over the proximal algorithm is that the proximal step in Eq. (2.28) is executed with h rather than with $f+h$, and this may be significant if h has simple/favorable structure (e.g., h is the ℓ_1 norm or a distance function to a simple constraint set), while f has unfavorable structure. Under relatively mild assumptions, it can be shown that the method has a cost function descent property, provided the stepsize α is sufficiently small (see Section 6.3).

In Section 6.7, we will also discuss another descent approach, called *ϵ-descent*, which aims to avoid the difficulties due to the discontinuity of the steepest descent direction (cf. Fig. 2.1.7). This is done by obtaining a descent direction via projection of the origin on an ϵ-subdifferential, an enlarged version of the subdifferential. The method is theoretically interesting and will be used to establish conditions for strong duality in *extended monotropic programming*, an important class of problems with partially separable structure, to be discussed in Section 4.4.

Finally, we note that there are a few types of descent methods that we will not discuss at all, either because they are based on ideas that do not connect well with convexity, or because they are not well suited for the type of large-scale problems that we emphasize in this book. Included are direct search methods that do not use derivatives, such as the Nelder-Mead simplex algorithm [DeT91], [Tse95], [LRW98], [NaT02], feasible direction methods such as reduced gradient and gradient projection methods based on manifold suboptimization [GiM74], [GMW81], [MoT89], and sequential quadratic programming methods [Ber82a], [Ber99], [NoW06]. Some of these methods have extensive literature and applications, but are beyond our scope.

2.1.5 Incremental Algorithms

An interesting form of approximate gradient, or more generally subgradient method, is an *incremental* variant, which applies to minimization over a closed convex set X of an additive cost function of the form

$$f(x) = \sum_{i=1}^{m} f_i(x),$$

where the functions $f_i : \Re^n \mapsto \Re$ are either differentiable or convex and nondifferentiable. We mentioned several contexts where cost functions of this type arise in Section 1.3. The idea of the incremental approach is to sequentially take steps along the subgradients of the component functions f_i, with intermediate adjustment of x after processing each f_i.

Incremental methods are interesting when m is very large, so a full subgradient step is very costly. For such problems one hopes to make progress with approximate but much cheaper incremental steps. Incremental methods are also well-suited for problems where m is large and the component functions f_i become known sequentially, over time. Then one may be able to operate on each component as it reveals itself, without waiting for the other components to become known, i.e., in *an on-line fashion*.

In a common type of incremental subgradient method, *an iteration is viewed as a cycle of m subiterations*. If x_k is the vector obtained after k cycles, the vector x_{k+1} obtained after one more cycle is

$$x_{k+1} = \psi_{m,k}, \tag{2.29}$$

where starting with

$$\psi_{0,k} = x_k,$$

we obtain $\psi_{m,k}$ after the m steps

$$\psi_{i,k} = P_X(\psi_{i-1,k} - \alpha_k g_{i,k}), \qquad i = 1, \ldots, m, \tag{2.30}$$

with $g_{i,k}$ being a subgradient of f_i at $\psi_{i-1,k}$ [or the gradient $\nabla f_i(\psi_{i-1,k})$ in the differentiable case].

In a *randomized* version of the method, given x_k at iteration k, an index i_k is chosen from the set $\{1, \ldots, m\}$ randomly, and the next iterate x_{k+1} is generated by

$$x_{k+1} = P_X(x_k - \alpha_k g_{i_k}), \qquad i = 1, \ldots, m, \tag{2.31}$$

where g_{i_k} is a subgradient of f_{i_k} at x_k. Here it is important that all indexes are chosen with equal probability. It turns out that there is a rate of convergence advantage for this and other types of randomization, as we will discuss in Section 6.4.2. We will ignore for the moment the possibility of randomizing the component selection, and assume cyclic selection as in Eqs. (2.29)-(2.30).

In the present section we will explain the ideas underlying incremental methods by focusing primarily on the case where the component functions f_i are differentiable. We will thus consider methods that compute at each step a component gradient ∇f_i and possibly Hessian matrix $\nabla^2 f_i$. We will discuss the case where f_i may be nondifferentiable in Section 6.4, after the analysis of nonincremental subgradient methods to be given in Section 3.2.

Incremental Gradient Method

Assume that the component functions f_i are differentiable. We refer to the method

$$x_{k+1} = \psi_{m,k}, \tag{2.32}$$

where starting with $\psi_{0,k} = x_k$, we generate $\psi_{m,k}$ after the m steps

$$\psi_{i,k} = P_X\big(\psi_{i-1,k} - \alpha_k \nabla f_i(\psi_{i-1,k})\big), \qquad i = 1, \ldots, m, \tag{2.33}$$

[cf. (2.29)-(2.30)], as the *incremental gradient method*. A well known and important example of such a method is the following. Together with its many variations, it is widely used in computerized imaging; see e.g., the book [Her09].

Example 2.1.3: (Kaczmarz Method)

Let

$$f_i(x) = \frac{1}{2\|c_i\|^2}(c_i'x - b_i)^2, \qquad i = 1, \ldots, m,$$

where c_i are given nonzero vectors in $\Re^n$ and b_i are given scalars, so we have a linear least squares problem. The constant term $1/(2\|c_i\|^2)$ multiplying each of the squared functions $(c_i'x - b_i)^2$ serves a scaling purpose: with its inclusion, all the components f_i have a Hessian matrix

$$\nabla^2 f_i(x) = \frac{1}{\|c_i\|^2} c_i c_i'$$

with trace equal to 1. This type of scaling is often used in least squares problems (see [Ber99] for explanations). The incremental gradient method (2.32)-(2.33) takes the form $x_{k+1} = \psi_{m,k}$, where $\psi_{m,k}$ is obtained after the m steps

$$\psi_{i,k} = \psi_{i-1,k} - \frac{\alpha_k}{\|c_i\|^2}(c_i'\psi_{i-1,k} - b_i)c_i, \qquad i = 1, \ldots, m, \tag{2.34}$$

starting with $\psi_{0,k} = x_k$ (see Fig. 2.1.9).

The stepsize α_k may be chosen in a number of different ways, but if α_k is chosen identically equal to 1, $\alpha_k \equiv 1$, we obtain the Kaczmarz method, which dates to 1937 [Kac37]; see Fig. 2.1.9(a). The interpretation of the iteration (2.34) in this case is very simple: *$\psi_{i,k}$ is obtained by projecting $\psi_{i,k-1}$ onto the hyperplane defined by the single equation $c_i'x = b_i$*. Indeed from Eq. (2.34) with $\alpha_k = 1$, it is easily verified that $c_i'\psi_{i,k} = b_i$ and that $\psi_{i,k} - \psi_{i,k-1}$ is orthogonal to the hyperplane, since it is proportional to its normal c_i. (There are also other related methods involving alternating projections on subspaces or other convex sets, one of them attributed to von Neumann from 1933; see Section 6.4.4.)

If the system of equations $c_i'x = b_i$, $i = 1, \ldots, m$, is consistent, i.e., has a unique solution x^*, then the unique minimum of $\sum_{i=1}^m f_i(x)$ is x^*. In this case it turns out that for a constant stepsize $\alpha_k \equiv \alpha$, with $0 < \alpha < 2$, the method converges to x^*. The convergence process is illustrated in Fig. 2.1.9(b) for the case $\alpha_k \equiv 1$: the distance $\|\psi_{i,k} - x^*\|$ is guaranteed not to increase for any i within cycle k, and to strictly decrease for at least one i, so x_{k+1} will be closer to x^* than x_k (assuming $x_k \neq x^*$). Generally, the order in which the equations are taken up for iteration can affect significantly the

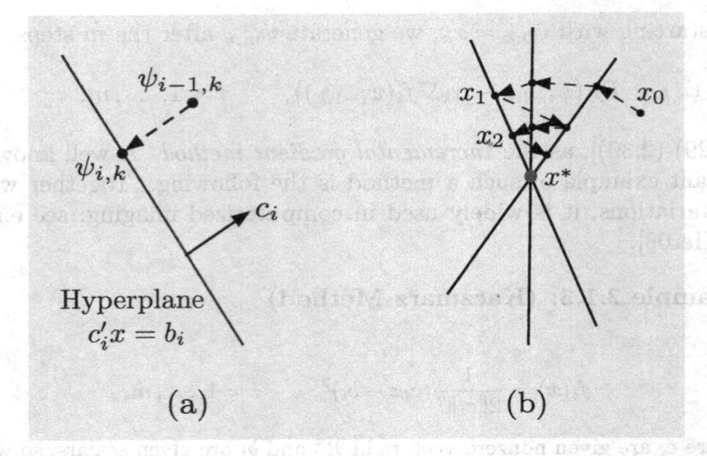

Figure 2.1.9. Illustration of the Kaczmarz method (2.34) with unit stepsize $\alpha_k \equiv 1$: (a) $\psi_{i,k}$ is obtained by projecting $\psi_{i-1,k}$ onto the hyperplane defined by the single equation $c_i'x = b_i$. (b) The convergence process for the case where the system of equations $c_i'x = b_i$, $i = 1, \ldots, m$, is consistent and has a unique solution x^*. Here $m = 3$, and x_k is the vector obtained after k cycles through the equations. Each incremental iteration decreases the distance to x^*, unless the current iterate lies on the hyperplane defined by the corresponding equation.

performance. In particular, faster convergence can be shown if the order is randomized in a special way; see [StV09].

If the system of equations

$$c_i'x = b_i, \qquad i = 1, \ldots, m,$$

is inconsistent, the method does not converge with a constant stepsize; see Fig. 2.1.10. In this case a diminishing stepsize α_k is necessary for convergence to an optimal solution. These convergence properties will be discussed further later in this section, and in Chapters 3 and 6.

Convergence Properties of Incremental Methods

The motivation for the incremental approach is faster convergence. In particular, we hope that far from the solution, a single cycle of the incremental gradient method will be as effective as several (as many as m) iterations of the ordinary gradient method (think of the case where the components f_i are similar in structure). Near a solution, however, the incremental method may not be as effective. Still, the frequent superiority of the incremental method when far from convergence can be a decisive advantage for problems where solution accuracy is not of paramount importance.

To be more specific, we note that there are two complementary performance issues to consider in comparing incremental and nonincremental methods:

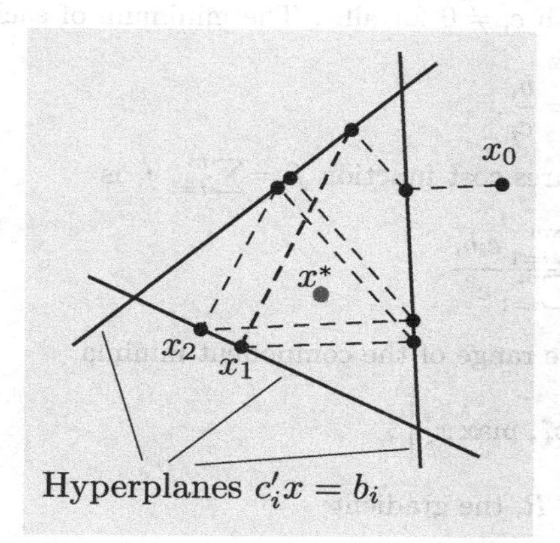

Figure 2.1.10. Illustration of the Kaczmarz method (2.34) with $\alpha_k \equiv 1$ for the case where the system of equations $c_i'x = b_i$, $i = 1, \ldots, m$, is inconsistent. In this figure there are three equations with corresponding hyperplanes as shown. The method approaches a neighborhood of the optimal solution, and then oscillates. A similar behavior would occur if the stepsize α_k were a constant $\alpha \in (0, 1)$, except that the size of the oscillation would diminish with α.

(a) *Progress when far from convergence.* Here the incremental method can be much faster. For an extreme case let $X = \Re^n$ (no constraints), and take m large and all components f_i identical to each other. Then an incremental iteration requires m times less computation than a classical gradient iteration, but gives exactly the same result, when the stepsize is appropriately scaled to be m times larger. While this is an extreme example, it reflects the essential mechanism by which incremental methods can be much superior: far from the minimum a single component gradient will point to "more or less" the right direction, at least most of the time.

(b) *Progress when close to convergence.* Here the incremental method can be inferior. As a case in point, assume that all components f_i are differentiable functions. Then the nonincremental gradient projection method can be shown to converge with a constant stepsize under reasonable assumptions, as we will see in Section 6.1. However, the incremental method requires a diminishing stepsize, and its ultimate rate of convergence can be much slower. When the component functions f_i are nondifferentiable, both the nonincremental and the incremental subgradient methods require a diminishing stepsize. The nonincremental method tends to require a smaller number of iterations, but each of the iterations involves all the components f_i and thus larger computation overhead, so that on balance, in terms of computation time, the incremental method tends to perform better.

As an illustration consider the following example.

Example 2.1.4:

Consider a scalar linear least squares problem where the components f_i have the form

$$f_i(x) = \tfrac{1}{2}(c_i x - b_i)^2, \qquad x \in \Re,$$

where c_i and b_i are given scalars with $c_i \neq 0$ for all i. The minimum of each of the components f_i is

$$x_i^* = \frac{b_i}{c_i},$$

while the minimum of the least squares cost function $f = \sum_{i=1}^m f_i$ is

$$x^* = \frac{\sum_{i=1}^m c_i b_i}{\sum_{i=1}^m c_i^2}.$$

It can be seen that x^* lies within the range of the component minima

$$R = \left[\min_i x_i^*, \, \max_i x_i^*\right],$$

and that for all x *outside* the region R, the gradient

$$\nabla f_i(x) = (c_i x - b_i)c_i$$

has the same sign as $\nabla f(x)$ (see Fig. 2.1.11). As a result, when outside the region R, the incremental gradient method

$$\psi_i = \psi_{i-1} - \alpha_k(c_i\psi_{i-1} - b_i)c_i$$

approaches x^* at each step, provided the stepsize α_k is small enough. In fact it is sufficient that

$$\alpha_k \leq \min_i \frac{1}{c_i^2}.$$

However, *for x inside the region R, the ith step of a cycle of the incremental gradient method need not make progress.* It will approach x^* (for small enough stepsize α_k) only if the current point ψ_{i-1} does not lie in the interval connecting x_i^* and x^*. This induces an oscillatory behavior within R, and as a result, the incremental gradient method will typically not converge to x^* unless $\alpha_k \to 0$.

Let us now compare the incremental gradient method with the nonincremental version, which takes the form

$$x_{k+1} = x_k - \alpha_k \sum_{i=1}^m (c_i x_k - b_i)c_i.$$

It can be shown that this method converges to x^* for any constant stepsize $\alpha_k \equiv \alpha$ satisfying

$$0 < \alpha \leq \frac{1}{\sum_{i=1}^m c_i^2}.$$

On the other hand, for x outside the region R, an iteration of the nonincremental method need not make more progress towards the solution than a single step of the incremental method. In other words, with comparably intelligent stepsize choices, *far from the solution (outside R), a single cycle through the entire set of component functions by the incremental method is roughly as effective as m iterations by the nonincremental method, which require m times as many component gradient calculations.*

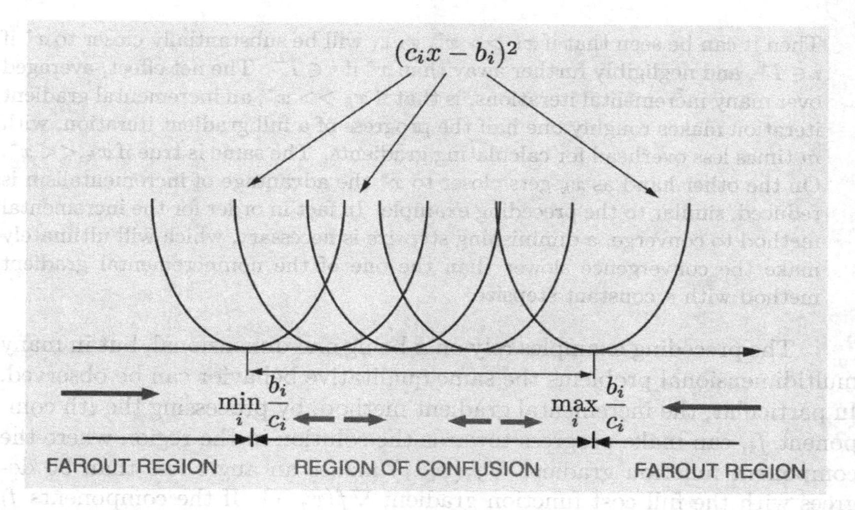

Figure 2.1.11. Illustrating the advantage of incrementalism when far from the optimal solution. The region of component minima

$$R = \left[\min_i x_i^*, \max_i x_i^*\right],$$

is labeled as the "region of confusion." It is the region where the method does not have a clear direction towards the optimum. The ith step in an incremental gradient cycle is a gradient step for minimizing $(c_ix - b_i)^2$, so if x lies outside the region of component minima $R = \left[\min_i x_i^*, \max_i x_i^*\right]$, (labeled as the "farout region") and the stepsize is small enough, progress towards the solution x^* is made.

Example 2.1.5:

The preceding example assumes that each component function f_i has a minimum, so that the range of component minima is defined. In cases where the components f_i have no minima, a similar phenomenon may occur. As an example consider the case where f is the sum of increasing and decreasing convex exponentials, i.e.,

$$f_i(x) = a_i e^{b_i x}, \qquad x \in \Re,$$

where a_i and b_i are scalars with $a_i > 0$ and $b_i \neq 0$. Let

$$I^+ = \{i \mid b_i > 0\}, \qquad I^- = \{i \mid b_i < 0\},$$

and assume that I^+ and I^- have roughly equal numbers of components. Let also x^* be the minimum of $\sum_{i=1}^m f_i$.

Consider the incremental gradient method that given the current point, call it x_k, chooses some component f_{i_k} and iterates according to the incremental iteration

$$x_{k+1} = x_k - \alpha_k \nabla f_{i_k}(x_k).$$

Then it can be seen that if $x_k >> x^*$, x_{k+1} will be substantially closer to x^* if $i \in I^+$, and negligibly further away than x^* if $i \in I^-$. The net effect, averaged over many incremental iterations, is that if $x_k >> x^*$, an incremental gradient iteration makes roughly one half the progress of a full gradient iteration, with m times less overhead for calculating gradients. The same is true if $x_k << x^*$. On the other hand as x_k gets closer to x^* the advantage of incrementalism is reduced, similar to the preceding example. In fact in order for the incremental method to converge, a diminishing stepsize is necessary, which will ultimately make the convergence slower than the one of the nonincremental gradient method with a constant stepsize.

The preceding examples rely on x being one-dimensional, but in many multidimensional problems the same qualitative behavior can be observed. In particular, the incremental gradient method, by processing the ith component f_i, can make progress towards the solution in the region where the component function gradient $\nabla f_i(\psi_{i-1})$ makes an angle less than 90 degrees with the full cost function gradient $\nabla f(\psi_{i-1})$. If the components f_i are not "too dissimilar," this is likely to happen in a region of points that are not too close to the optimal solution set.

Stepsize Selection

The choice of the stepsize α_k plays an important role in the performance of incremental gradient methods. On close examination, it turns out that the iterate differential $x_k - x_{k+1}$ corresponding to a full cycle of the incremental gradient method, and the corresponding vector $\alpha_k \nabla f(x_k)$ of its nonincremental counterpart differ by an error that is proportional to the stepsize (see the discussion in Exercises 2.6 and 2.10). For this reason a diminishing stepsize is essential for convergence to a minimizing point of f. However, it turns out that a peculiar form of convergence also typically occurs for the incremental gradient method if the stepsize α_k is a constant but sufficiently small α. In this case, the iterates converge to a "limit cycle," whereby the ith iterates ψ_i within the cycles converge to a different limit than the jth iterates ψ_j for $i \neq j$. The sequence $\{x_k\}$ of the iterates obtained at the end of cycles converges, except that the limit obtained *need not be optimal* even if f is convex. The limit tends to be close to an optimal point when the constant stepsize is small [for analysis of the case where the components f_i are quadratic, see Exercise 2.13(a), [BeT96] (Section 3.2), and [Ber99] (Section 1.5), where a linear convergence rate is also shown].

In practice, it is common to use a constant stepsize for a (possibly prespecified) number of iterations, then decrease the stepsize by a certain factor, and repeat, up to the point where the stepsize reaches a prespecified minimum. An alternative possibility is to use a stepsize α_k that diminishes to 0 at an appropriate rate [cf. Eq. (2.15)]. In this case convergence can be shown under reasonable conditions; see Exercise 2.10.

Still another possibility is to use an adaptive stepsize rule, whereby the stepsize is reduced (or increased) when the progress of the method indicates that the algorithm is oscillating because it operates within (or outside, respectively) the region of confusion. There are formal ways to implement such stepsize rules with sound convergence properties (see [Gri94], [Tse98], [MYF03]). One of the ideas is to look at a batch of incremental updates $\psi_i, \ldots, \psi_{i+M}$, for some relatively large $M \leq m$, and compare $\|\psi_i - \psi_{i+M}\|$ with $\sum_{\ell=1}^{M} \|\psi_{1+\ell-1} - \psi_{i+\ell}\|$. If the ratio of these two numbers is "small" this suggests that the method is oscillating.

Incremental gradient and subgradient methods have a rich theory, which includes convergence and rate of convergence analysis, optimization and randomization issues of the component order selection, and distributed computation aspects. Moreover they admit interesting combinations with other methods, such as the proximal algorithm. We will more fully discuss their properties and extensions in Chapter 6, Section 6.4.

Aggregated Gradient Methods

Another variant of incremental gradient is the *incremental aggregated gradient method*, which has the form

$$x_{k+1} = P_X \left(x_k - \alpha_k \sum_{\ell=0}^{m-1} \nabla f_{i_{k-\ell}}(x_{k-\ell}) \right), \tag{2.35}$$

where f_{i_k} is the new component function selected for iteration k. Here, the component indexes i_k may either be selected in a cyclic order [$i_k = (k \text{ modulo } m) + 1$], or according to some randomization scheme, consistently with Eq. (2.31). Also for $k < m$, the summation should go up to $\ell = k$, and α should be replaced by a corresponding larger value, such as $\alpha_k = m\alpha/(k+1)$. This method, first proposed in [BHG08], computes the gradient incrementally, one component per iteration, but in place of the single component gradient, it uses an approximation to the total cost gradient $\nabla f(x_k)$, which is the sum of the component gradients computed in the past m iterations.

There is analytical and experimental evidence that by aggregating the component gradients one may be able to attain a faster asymptotic convergence rate, by ameliorating the effect of approximating the full gradient with component gradients; see the original paper [BHG08], which provides an analysis for quadratic problems, the paper [SLB13], which provides a more general convergence and convergence rate analysis, and extensive computational results, and the papers [Mai13], [Mai14], [DCD14], which describe related methods. The expectation of faster convergence should be tempered, however, because in order for the effect of aggregating the component gradients to fully manifest itself, at least one pass (and possibly quite a few more) through the components must be made, which may be too long if m is very large.

A drawback of this aggregated gradient method is that it requires that the most recent component gradients be kept in memory, so that when a component gradient is reevaluated at a new point, the preceding gradient of the same component is discarded from the sum of gradients of Eq. (2.35). There have been alternative implementations of the incremental aggregated gradient method idea that ameliorate this memory issue, by recalculating the full gradient periodically and replacing an old component gradient by a new one, once it becomes available; see [JoZ13], [ZMJ13], [XiZ14]. For an example, instead of the gradient sum

$$s_k = \sum_{\ell=0}^{m-1} \nabla f_{i_{k-\ell}}(x_{k-\ell}),$$

in Eq. (2.35), such a method may use $\tilde{s}_k$, updated according to

$$\tilde{s}_k = \nabla f_{i_k}(x_k) - \nabla f_{i_k}(\tilde{x}_k) + \tilde{s}_{k-1},$$

where $\tilde{s}_0$ is the full gradient computed at the start of the current cycle, and $\tilde{x}_k$ is the point at which this full gradient has been calculated. Thus to obtain $\tilde{s}_k$ one only needs to compute the difference of the two gradients

$$\nabla f_{i_k}(x_k) - \nabla f_{i_k}(\tilde{x}_k)$$

and add it to the current approximation of the full gradient $\tilde{s}_{k-1}$. This bypasses the need for extensive memory storage, and with proper implementation, typically leads to small degradation in performance. In particular, convergence with a sufficiently small constant stepsize, with an attendant superior convergence rate over the incremental gradient method, has been shown.

Incremental Gradient Method with Momentum

There is an incremental version of the gradient method with momentum or heavy ball method, discussed in Section 2.1.1 [cf. Eq. (2.12)]. It is given by

$$x_{k+1} = x_k - \alpha_k \nabla f_{i_k}(x_k) + \beta_k(x_k - x_{k-1}), \tag{2.36}$$

where f_{i_k} is the component function selected for iteration k, β_k is a scalar in $[0, 1)$, and we define $x_{-1} = x_0$; see e.g., [MaS94], [Tse98]. As noted earlier, special nonincremental methods with similarities to the one above have optimal iteration complexity properties under certain conditions; cf. Section 6.2. However, there have been no proposals of incremental versions of these optimal complexity methods.

The heavy ball method (2.36) is related with the aggregated gradient method (2.35) when $\beta_k \approx 1$. In particular, when $\alpha_k \equiv \alpha$ and $\beta_k \equiv \beta$, the sequence generated by Eq. (2.36) satisfies

$$x_{k+1} = x_k - \alpha \sum_{\ell=0}^{k} \beta^\ell \nabla f_{i_{k-\ell}}(x_{k-\ell}) \tag{2.37}$$

[both iterations (2.35) and (2.37) involve different types of diminishing dependence on past gradient components]. Thus, the heavy ball iteration (2.36) provides an approximate implementation of the incremental aggregated gradient method (2.35), while it does not have the memory storage issue of the latter.

A further way to intertwine the ideas of the aggregated gradient method (2.35) and the heavy ball method (2.36) for the unconstrained case ($X = \Re^n$) is to form an infinite sequence of components

$$f_1, f_2, \ldots, f_m, f_1, f_2, \ldots, f_m, f_1, f_2, \ldots, \tag{2.38}$$

and group together blocks of successive components into batches. One way to implement this idea is to add p preceding gradients (with $1 < p < m$) to the current component gradient in iteration (2.36), thus iterating according to

$$x_{k+1} = x_k - \alpha_k \sum_{\ell=0}^{p} \nabla f_{i_{k-\ell}}(x_{k-\ell}) + \beta_k(x_k - x_{k-1}). \tag{2.39}$$

Here f_{i_k} is the component function selected for iteration k using the order of the sequence (2.38). This essentially amounts to reformulating the problem by redefining the components as sums of $p+1$ successive components and applying an approximation of the incremental heavy ball method (2.36). The advantage of the method (2.39) over the aggregated gradient method is that it requires keeping in memory only p previous component gradients, and p can be chosen according to the memory limitations of the given computational environment. Generally in incremental methods, grouping together several components f_i, a process sometimes called *batching*, tends to reduce the size of the region of confusion (cf. Fig. 2.1.11), and with a small region of confusion, the incentive for aggregating the component gradients diminishes (see [Ber97] and [FrS12], for different implementations and analysis of this idea). The process of batching can also be implemented adaptively, based on some form of heuristic detection that the method has entered the region of confusion.

Stochastic Subgradient Methods

Incremental subgradient methods are related to methods that aim to minimize an expected value

$$f(x) = E\{F(x, w)\},$$

where w is a random variable, and $F(\cdot, w) : \Re^n \mapsto \Re$ is a convex function for each possible value of w. The stochastic subgradient method for minimizing f over a closed convex set X is given by

$$x_{k+1} = P_X\big(x_k - \alpha_k g(x_k, w_k)\big), \tag{2.40}$$

where w_k is a sample of w and $g(x_k, w_k)$ is a subgradient of $F(\cdot, w_k)$ at x_k. This method has a rich theory and a long history, particularly for the case where $F(\cdot, w)$ is differentiable for each value of w (for representative references, see [PoT73], [Lju77], [KuC78], [TBA86], [Pol87], [BeT89a], [BeT96], [Pfl96], [LBB98], [BeT00], [KuY03], [Bot05], [BeL07], [Mey07], [Bor08], [BBG09], [Ben09], [NJL09], [Bot10], [BaM11], [DHS11], [ShZ12], [FrG13], [NSW14]). It is strongly related to the classical algorithmic field of *stochastic approximation*; see the books [KuC78], [BeT96], [KuY03], [Spa03], [Mey07], [Bor08], [BPP13].

If we view the expected value cost $E\{F(x, w)\}$ as a weighted sum of cost function components, we see that the stochastic subgradient method (2.40) is related to the incremental subgradient method

$$x_{k+1} = P_X(x_k - \alpha_k g_{i,k}) \tag{2.41}$$

for minimizing a finite sum $\sum_{i=1}^{m} f_i$, when randomization is used for component selection [cf. Eq. (2.31)]. An important difference is that the former method involves sequential sampling of cost components $F(x, w)$ from an infinite population under some statistical assumptions, while in the latter the set of cost components f_i is predetermined and finite. However, it is possible to view the incremental subgradient method (2.41), with uniform randomized selection of the component function f_i (i.e., with i_k chosen to be any one of the indexes $1, \ldots, m$, with equal probability $1/m$, and independently of preceding choices), as a stochastic subgradient method.

Despite the apparent similarity of the incremental and the stochastic subgradient methods, the view that the problem

$$\begin{aligned} &\text{minimize} \quad f(x) = \sum_{i=1}^{m} f_i(x) \\ &\text{subject to} \quad x \in X, \end{aligned} \tag{2.42}$$

can simply be treated as a special case of the problem

$$\begin{aligned} &\text{minimize} \quad f(x) = E\{F(x, w)\} \\ &\text{subject to} \quad x \in X, \end{aligned}$$

is questionable.

One reason is that once we convert the finite sum problem to a stochastic problem, we preclude the use of methods that exploit the finite sum structure, such as the aggregated gradient methods we discussed earlier. Under certain conditions, these methods offer more attractive convergence rate guarantees than incremental and stochastic gradient methods, and can be very effective for many problems, as we have noted.

Another reason is that the finite-component problem (2.42) is often genuinely deterministic, and to view it as a stochastic problem at the outset may mask some of its important characteristics, such as the number m of cost components, or the sequence in which the components are ordered and processed. These characteristics may potentially be algorithmically exploited. For example, with insight into the problem's structure, one may be able to discover a special deterministic or partially randomized order of processing the component functions that is superior to a uniform randomized order.

Example 2.1.6:

Consider the one-dimensional problem

$$\text{minimize} \quad f(x) = \frac{1}{2}\sum_{i=1}^{m}(x - w_i)^2$$
$$\text{subject to} \quad x \in \Re,$$

where the scalars w_i are given by

$$w_i = \begin{cases} 1 & \text{if } i\text{: odd,} \\ -1 & \text{if } i\text{: even.} \end{cases}$$

Assuming that m is an even number, the optimal solution is $x^* = 0$.

An incremental gradient method with the commonly used diminishing stepsize $\alpha_k = 1/(k+1)$ chooses a component index i_k at iteration k, and updates x_k according to

$$x_{k+1} = x_k - \frac{1}{k+1}(x_k - w_{i_k}),$$

starting with some initial iterate x_0. It is then easily verified by induction that

$$x_k = \frac{x_0}{k} + \frac{w_{i_0} + \cdots + w_{i_{k-1}}}{k}, \qquad k = 1, 2, \ldots.$$

Thus the iteration error, which is x_k (since $x^* = 0$), consists of two terms. The first is the error term x_0/k, which is independent of the method of selecting i_k, and the second is the error term

$$e_k = \frac{w_{i_0} + \cdots + w_{i_{k-1}}}{k},$$

which depends on the selection method for i_k.

If i_k is chosen by independently randomizing with equal probability $1/2$ over the odd and even cost components, then e_k will be a random variable whose variance can be calculated to be $1/2k$. Thus the standard deviation of the error x_k will be of order $O(1/\sqrt{k})$. If on the other hand i_k is chosen by the deterministic order, which alternates between the odd and even components, we will have $e_k = 1/k$ for the odd iterations and $e_k = 0$ for the even iterations, so the error x_k will be of order $O(1/k)$, much smaller than the one for the randomized order. Of course, this is a favorable deterministic order, and we may obtain much worse results with an unfavorable deterministic order (such as selecting first all the odd components and then all the even components). However, the point here is that if we take the view that we are minimizing an expected value, we are disregarding at the outset information about the problem's structure that could be algorithmically useful.

A related experimental observation is that by suitably mixing the deterministic and the stochastic order selection methods we may produce better practical results. As an example, a popular technique for incremental methods, called *random reshuffling*, is to process the component functions f_i in cycles, with each component selected once in each cycle, and to re-order randomly the components after each cycle. This alternative order selection scheme has the nice property of allocating exactly one computation slot to each component in an m-slot cycle (m incremental iterations). By comparison, choosing components by uniform sampling allocates one computation slot to each component *on the average*, but some components may not get a slot while others may get more than one. A nonzero variance in the number of slots that any fixed component gets within a cycle, may be detrimental to performance, and suggests that reshuffling randomly the order of the component functions after each cycle works better. While it seems difficult to establish this fact analytically, a justification is suggested by the view of the incremental gradient method as a gradient method with error in the computation of the gradient (see Exercise 2.10). The error has apparently greater variance in the uniform sampling method than in the random reshuffling method. Heuristically, if the variance of the error is larger, the direction of descent deteriorates, suggesting slower convergence. For some experimental evidence, see [Bot09], [ReR13].

Let us also note that in Section 6.4 we will compare more formally various component selection orders in incremental methods. Our analysis will indicate that in the absence of problem-specific knowledge that can be exploited to select a favorable deterministic order, a uniform randomized order (each component f_i chosen with equal probability $1/m$ at each iteration, independently of preceding choices) has superior worst-case complexity.

Our conclusion is that in incremental methods, it may be beneficial to search for a favorable order for processing the component functions f_i, exploiting whatever problem-specific information may be available, rather than ignore all prior information and apply a uniform randomized order of the type commonly used in stochastic gradient methods. However, if a favorable order cannot be found, a randomized order is usually better than a fixed deterministic order, although there is no guarantee that this will be so for a given practical problem; for example a fixed deterministic order has been reported to be considerably faster on some benchmark test problems without any attempt to order the components favorably [Bot09].

Incremental Newton Methods

We will now consider an incremental version of Newton's method for unconstrained minimization of an additive cost function of the form

$$f(x) = \sum_{i=1}^{m} f_i(x),$$

where the functions $f_i : \Re^n \mapsto \Re$ are convex and twice continuously differentiable. Consider the quadratic approximation $\tilde{f}_i$ of a function f_i at a vector $\psi \in \Re^n$, i.e., the second order Taylor expansion of f_i at ψ:

$$\tilde{f}_i(x;\psi) = \nabla f_i(\psi)'(x-\psi) + \tfrac{1}{2}(x-\psi)'\nabla^2 f_i(\psi)(x-\psi), \quad \forall\, x,\psi \in \Re^n.$$

Similar to Newton's method, which minimizes a quadratic approximation at the current point of the cost function [cf. Eq. (2.14)], the incremental form of Newton's method minimizes a sum of quadratic approximations of components. Similar to the incremental gradient method, we view an iteration as a cycle of m subiterations, each involving a single additional component f_i, and its gradient and Hessian at the current point within the cycle. In particular, if x_k is the vector obtained after k cycles, the vector x_{k+1} obtained after one more cycle is

$$x_{k+1} = \psi_{m,k},$$

where starting with $\psi_{0,k} = x_k$, we obtain $\psi_{m,k}$ after the m steps

$$\psi_{i,k} \in \arg\min_{x\in\Re^n} \sum_{\ell=1}^{i} \tilde{f}_\ell(x;\psi_{\ell-1,k}), \qquad i=1,\ldots,m. \tag{2.43}$$

If all the functions f_i are quadratic, it can be seen that the method finds the solution in a single cycle.† The reason is that when f_i is quadratic, each $f_i(x)$ differs from $\tilde{f}_i(x;\psi)$ by a constant, which does not depend on x. Thus the difference

$$\sum_{i=1}^{m} f_i(x) - \sum_{i=1}^{m} \tilde{f}_i(x;\psi_{i-1,k})$$

† Here we assume that the m quadratic minimizations (2.43) to generate $\psi_{m,k}$ have a solution. For this it is sufficient that the first Hessian matrix $\nabla^2 f_1(x_0)$ be positive definite, in which case there is a unique solution at every iteration. A simple possibility to deal with this requirement is to add to f_1 a small positive definite quadratic term, such as $\frac{\epsilon}{2}\|x-x_0\|^2$. Another possibility is to lump together several of the component functions (enough to ensure that the sum of their quadratic approximations at x_0 is positive definite), and to use them in place of f_1. This is generally a good idea and leads to smoother initialization, as it ensures a relatively stable behavior of the algorithm for the initial iterations.

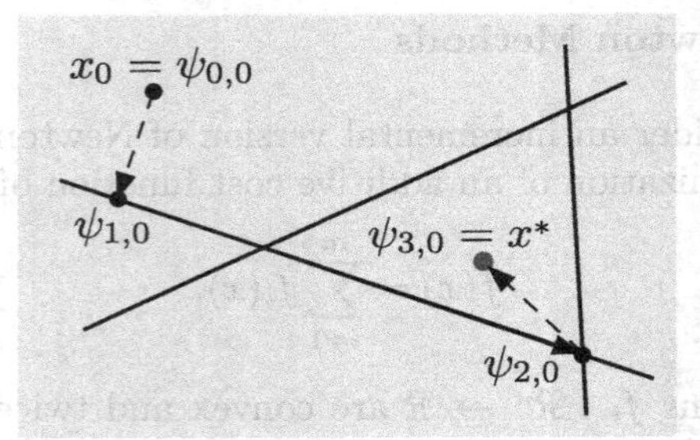

Figure 2.1.12. Illustration of the incremental Newton method for the case of a two-dimensional linear least squares problem with $m = 3$ cost function components (compare with the Kaczmarz method, cf. Fig. 2.1.10).

is a constant that is independent of x, and minimization of either sum in the above expression gives the same result.

As an example, consider a linear least squares problem, where

$$f_i(x) = \tfrac{1}{2}(c_i'x - b_i)^2, \qquad i = 1, \ldots, m.$$

Then the ith subiteration within a cycle minimizes

$$\sum_{\ell=1}^{i} f_\ell(x),$$

and when $i = m$, the solution of the problem is obtained (see Fig. 2.1.12). This convergence behavior should be compared with the one for the Kaczmarz method (cf. Fig. 2.1.10).

It is important to note that the quadratic minimizations of Eq. (2.43) can be carried out efficiently. For simplicity, let as assume that $\tilde{f}_1(x;\psi)$ is a positive definite quadratic, so that for all i, $\psi_{i,k}$ is well defined as the unique solution of the minimization problem in Eq. (2.43). We will show that the incremental Newton method (2.43) can be implemented in terms of the incremental update formula

$$\psi_{i,k} = \psi_{i-1,k} - D_{i,k}\nabla f_i(\psi_{i-1,k}), \tag{2.44}$$

where $D_{i,k}$ is given by

$$D_{i,k} = \left(\sum_{\ell=1}^{i} \nabla^2 f_\ell(\psi_{\ell-1,k})\right)^{-1}, \tag{2.45}$$

and is generated iteratively as

$$D_{i,k} = \left(D_{i-1,k}^{-1} + \nabla^2 f_i(\psi_{i,k})\right)^{-1}. \tag{2.46}$$

Indeed, from the definition of the method (2.43), the quadratic function $\sum_{\ell=1}^{i-1} \tilde{f}_\ell(x; \psi_{\ell-1,k})$ is minimized by $\psi_{i-1,k}$ and its Hessian matrix is $D_{i-1,k}^{-1}$, so we have

$$\sum_{\ell=1}^{i-1} \tilde{f}_\ell(x; \psi_{\ell-1,k}) = \tfrac{1}{2}(x - \psi_{\ell-1,k})' D_{i-1,k}^{-1}(x - \psi_{\ell-1,k}) + \text{constant}.$$

Thus, by adding $\tilde{f}_i(x; \psi_{i-1,k})$ to both sides of this expression, we obtain

$$\begin{aligned}\sum_{\ell=1}^{i} \tilde{f}_\ell(x; \psi_{\ell-1,k}) &= \tfrac{1}{2}(x - \psi_{\ell-1,k})' D_{i-1,k}^{-1}(x - \psi_{\ell-1,k}) + \text{constant} \\ &\quad + \tfrac{1}{2}(x - \psi_{i-1,k})' \nabla^2 f_i(\psi_{i-1,k})(x - \psi_{i-1,k}) + \nabla f_i(\psi_{i-1,k})'(x - \psi_{i-1,k}).\end{aligned}$$

Since by definition $\psi_{i,k}$ minimizes this function, we obtain Eqs. (2.44)-(2.46).

The update formula (2.46) for the matrix $D_{i,k}$ can often be efficiently implemented by using convenient formulas for the inverse of the sum of two matrices. In particular, if f_i is given by

$$f_i(x) = h_i(a_i'x - b_i),$$

for some twice differentiable convex function $h_i : \Re \mapsto \Re$, vector a_i, and scalar b_i, we have

$$\nabla^2 f_i(\psi_{i-1,k}) = \nabla^2 h_i(\psi_{i-1,k})\, a_i a_i',$$

and the update formula (2.46) can be written as

$$D_{i,k} = D_{i-1,k} - \frac{D_{i-1,k} a_i a_i' D_{i-1,k}}{\nabla^2 h_i(\psi_{i-1,k})^{-1} + a_i' D_{i-1,k} a_i};$$

this is the well-known Sherman-Morrison formula for the inverse of the sum of an invertible matrix and a rank-one matrix (see the matrix inversion formula in Section A.1 of Appendix A).

We have considered so far a single cycle of the incremental Newton method. One algorithmic possibility for cycling through the component functions multiple times, is to simply create a larger set of components by concatenating multiple copies of the original set, that is, by forming what we refer to as *the extended set of components*

$$f_1, f_2, \ldots, f_m,\ f_1, f_2, \ldots, f_m,\ f_1, f_2, \ldots.$$

The incremental Newton method, when applied to the extended set, asymptotically resembles a scaled incremental gradient method with diminishing stepsize of the type described earlier. Indeed, from Eq. (2.45)], the matrix

$D_{i,k}$ diminishes roughly in proportion to $1/k$. From this it follows that the asymptotic convergence properties of the incremental Newton method are similar to those of an incremental gradient method with diminishing stepsize of order $O(1/k)$. Thus its convergence rate is slower than linear.

To accelerate the convergence of the method one may employ a form of restart, so that $D_{i,k}$ does not converge to 0. For example $D_{i,k}$ may be reinitialized and increased in size at the beginning of each cycle. For problems where f has a unique nonsingular minimum x^* [one for which $\nabla^2 f(x^*)$ is nonsingular], one may design incremental Newton schemes with restart that converge linearly to within a neighborhood of x^* (and even superlinearly if x^* is also a minimum of all the functions f_i, so there is no region of confusion). Alternatively, the update formula (2.46) may be modified to

$$D_{i,k} = \left(\lambda_k D_{i-1,k}^{-1} + \nabla^2 f_\ell(\psi_{i,k})\right)^{-1}, \tag{2.47}$$

by introducing a fading factor $\lambda_k \in (0,1)$, which can be used to accelerate the practical convergence rate of the method (see [Ber96] for an analysis of schemes where $\lambda_k \to 1$; in cases where λ_k is some constant $\lambda < 1$, linear convergence to within a neighborhood of the optimum may be shown).

The following example provides some insight regarding the behavior of the method when the cost function f has a very large number of cost components, as is the case when f is defined as the average of a very large number of random samples.

Example 2.1.7: (Infinite Number of Cost Components)

Consider the problem

$$\begin{aligned} &\text{minimize} \quad f(x) \overset{\text{def}}{=} \lim_{m\to\infty} \frac{1}{m} \sum_{i=1}^{m} F(x, w_i) \\ &\text{subject to} \quad x \in \Re^n, \end{aligned}$$

where $\{w_k\}$ is a given sequence from some set, and each function $F(\cdot, w_i) : \Re^n \mapsto \Re$ is positive semidefinite quadratic. We assume that f is well-defined (i.e., the limit above exists for each $x \in \Re^n$), and is a positive definite quadratic. This type of problem arises in linear regression models (cf. Example 1.3.1) involving an infinite amount of data that is obtained through random sampling.

The natural extension of the incremental Newton's method, applied to the infinite set of components $F(\cdot, w_1), F(\cdot, w_2), \ldots$, generates the sequence $\{x_k^*\}$ where

$$x_k^* \in \arg\min_{x\in\Re^n} f_k(x) \overset{\text{def}}{=} \frac{1}{k} \sum_{i=1}^{k} F(x, w_i).$$

Since f is positive definite and the same is true for f_k, when k is large enough, we have $x_k^* \to x^*$, where x^* is the minimum of f. The rate of convergence is determined strictly by the rate at which the vectors x_k^* approach x^*, or equivalently by the rate at which f_k approaches f. It is impossible to achieve a faster rate of convergence with an algorithm that is nonanticipative in the sense that it uses just the first k cost components in the first k iterations.

By contrast, if we were to apply the natural extension of the incremental gradient method to this problem, the convergence rate could be much worse. There would be an error due to the difference $(x_k^* - x^*)$, but also an additional error due to the difference $(x_k^* - x_k)$ between x_k^* and the kth iterate x_k of the incremental gradient method, which is generally diminishing quite slowly, possibly more slowly than $(x_k^* - x^*)$. The same is true for other gradient-type methods based on incremental computations, including the aggregated gradient methods discussed earlier.

Incremental Newton Method with Diagonal Approximation

Generally, with proper implementation, the incremental Newton method is often substantially faster than the incremental gradient method, in terms of numbers of iterations (there are theoretical results suggesting this property for stochastic versions of the two methods; see the end-of-chapter references). However, in addition to computation of second derivatives, the incremental Newton method involves greater overhead per iteration due to matrix-vector calculations in Eqs. (2.44), (2.46), and (2.47), so it is suitable only for problems where n, the dimension of x, is relatively small.

One way to remedy in part this difficulty is to approximate $\nabla^2 f_i(\psi_{i,k})$ by a diagonal matrix, and recursively update a diagonal approximation of $D_{i,k}$ using Eqs. (2.46) or (2.47). One possibility, inspired by similar diagonal scaling schemes for nonincremental gradient methods, is to set to 0 the off-diagonal components of $\nabla^2 f_i(\psi_{i,k})$. In this case, the iteration (2.44) becomes a diagonally scaled version of the incremental gradient method, and involves comparable overhead per iteration (assuming the required diagonal second derivatives are easily computed). As an additional option, one may multiply the diagonal components with a stepsize parameter that is close to 1 and add a small positive constant (to bound them away from 0). Ordinarily, for the convex problems considered here, this method should require little experimentation with stepsize selection.

Incremental Newton Methods with Constraints

The incremental Newton method can also be adapted to constrained problems of the form

$$\begin{aligned} &\text{minimize} \quad \sum_{i=1}^{m} f_i(x) \\ &\text{subject to} \quad x \in X, \end{aligned}$$

where $f_i : \Re^n \mapsto \Re$ are convex, twice continuously differentiable convex functions. If X has a relatively simple form, such as upper and lower bounds on the variables, one may use a two-metric implementation, such as the ones discussed earlier, whereby the matrix $D_{i,k}$ is partially diagonalized before it is applied to the iteration

$$\psi_{i,k} = P_X\big(\psi_{i-1,k} - D_{i,k}\nabla f_i(\psi_{i-1,k})\big),$$

[cf. Eqs. (2.21) and (2.44)].

For more complicated constraint sets of the form

$$X = \cap_{i=1}^m X_i,$$

where each X_i is a relatively simple component constraint set (such as a halfspace), there is another possibility. This is to apply an incremental projected Newton iteration, with projection on a single individual component X_i, i.e., an iteration of the form

$$\psi_{i,k} \in \arg\min_{\psi\in X_i} \big\{\nabla f_i(\psi_{i-1,k})'(\psi-\psi_{i-1,k}) + \tfrac{1}{2}(\psi-\psi_{i-1,k})'H_{i,k}(\psi-\psi_{i-1,k})\big\},$$

where

$$H_{i,k} = \sum_{\ell=1}^{i} \nabla^2 f_\ell(\psi_{\ell-1,k}).$$

Note that each component X_i can be relatively simple, in which case the quadratic optimization problem above may be simple despite the fact that $H_{i,k}$ is nondiagonal. Depending on the problem's special structure, one may also use efficient methods that pass information from the solution of one quadratic subproblem to the next.

A similar method may also be used for problems of the form

$$\begin{aligned} &\text{minimize} \quad R(x) + \sum_{i=1}^m f_i(x) \\ &\text{subject to} \quad x \in X = \cap_{i=1}^m X_i, \end{aligned}$$

where $R(x)$ is a regularization function that is a multiple of either the ℓ_1 or the ℓ_2 norm. Then the incremental projected Newton iteration takes the form

$$\begin{aligned} \psi_{i,k} \in \arg\min_{\psi\in X_i} \big\{&R(\psi) + \nabla f_i(\psi_{i-1,k})'(\psi - \psi_{i-1,k}) \\ &+ \tfrac{1}{2}(\psi - \psi_{i-1,k})'H_{i,k}(\psi - \psi_{i-1,k})\big\}. \end{aligned}$$

When X_i is a polyhedral set, this problem is a quadratic program.

The idea of incremental projection on constraint set components is complementary to the idea of using gradient and possibly Hessian information from single cost function components at each iteration, and will be discussed in more detail in Section 6.4.4, in the context of incremental subgradient and incremental proximal methods. Several variations are possible, whereby the cost function component and constraint set component selected at each iteration may be chosen according to special deterministic or randomized rules. We refer to the papers [Ned11], [Ber11], [WaB13a] for a discussion of these incremental methods, their variations, and their convergence analysis.

Incremental Gauss-Newton Method – The Extended Kalman Filter

We will next consider an algorithm that operates similar to the incremental Newton method, but is specialized for the nonlinear least squares problem

$$\begin{aligned} &\text{minimize} \quad \sum_{i=1}^{m} \big\|g_i(x)\big\|^2 \\ &\text{subject to} \quad x \in \Re^n, \end{aligned}$$

where $g_i : \Re^n \mapsto \Re^{n_i}$ are some possibly nonlinear functions (cf. Example 1.3.1). As noted in Section 1.3, this is a common problem in practice.

We introduce a function $\tilde{g}_i$ that represents a linear approximation of g_i at a vector $\psi \in \Re^n$:

$$\tilde{g}_i(x;\psi) = \nabla g_i(\psi)'(x-\psi) + g_i(\psi), \qquad \forall\ x,\psi \in \Re^n,$$

where $\nabla g_i(\psi)$ is the $n \times n_i$ gradient matrix of g_i evaluated at ψ. Similar to the incremental gradient and Newton methods, we view an iteration as a cycle of m subiterations, each requiring linearization of a single additional component at the current point within the cycle. In particular, if x_k is the vector obtained after k cycles, the vector x_{k+1} obtained after one more cycle is

$$x_{k+1} = \psi_{m,k}, \tag{2.48}$$

where starting with $\psi_{0,k} = x_k$, we obtain $\psi_{m,k}$ after the m steps

$$\psi_{i,k} \in \arg\min_{x \in \Re^n} \sum_{\ell=1}^{i} \big\|\tilde{g}_\ell(x;\psi_{\ell-1,k})\big\|^2, \qquad i = 1,\ldots,m. \tag{2.49}$$

If all the functions g_i are linear, we have $\tilde{g}_i(x;\psi) = g_i(x)$, and the method solves the problem exactly in a single cycle. It then becomes identical to the incremental Newton method.

When the functions g_i are nonlinear the algorithm differs from the incremental Newton method because it does not involve second derivatives of g_i. It may be viewed instead as an incremental version of the Gauss-Newton method, a classical nonincremental scaled gradient method for solving nonlinear least squares problems (see e.g., [Ber99], Section 1.5). It is also known as the *extended Kalman filter*, and has found extensive application in state estimation and control of dynamic systems, where it was introduced in the mid-60s (it was also independently proposed in [Dav76]).

The implementation issues of the extended Kalman filter are similar to the ones of the incremental Newton method. This is because both methods solve similar linear least squares problems at each iteration [cf. Eqs. (2.43) and (2.49)]. The convergence behaviors of the two methods are also similar: they asymptotically operate as scaled forms of incremental gradient methods with diminishing stepsize. Both methods are primarily well-suited for problems where the dimension of x is much smaller than the number of components in the additive cost function, so that the associated matrix-vector operations are not overly costly. Moreover their practical convergence rate can be accelerated by introducing a fading factor [cf. Eq. (2.47)]. We refer to [Ber96], [MYF03] for convergence analysis, variations, and computational experimentation.

2.1.6 Distributed Asynchronous Iterative Algorithms

We will now consider briefly distributed asynchronous counterparts of some of the algorithms discussed earlier in this section. We have in mind a situation where an iterative algorithm, such as a gradient method or a coordinate descent method, is parallelized by separating it into several local algorithms operating concurrently at different processors. The main characteristic of an asynchronous algorithm is that the local algorithms do not have to wait at predetermined points for predetermined information to become available. We thus allow some processors to execute more iterations than others, we allow some processors to communicate more frequently than others, and we allow the communication delays to be substantial and unpredictable.

Let us consider for simplicity the problem of unconstrained minimization of a differentiable function $f : \Re^n \mapsto \Re$. Out of the iterative algorithms of Sections 2.1.1-2.1.3, there are three types that are suitable for asynchronous distributed computation. Their asynchronous versions are as follows:

(a) *Gradient methods*, where we assume that the ith coordinate x^i is updated at a subset of times $\mathcal{R}_i \subset \{0, 1, \ldots\}$, according to

$$x_{k+1}^i = \begin{cases} x_k^i & \text{if } k \notin \mathcal{R}_i, \\ x_k^i - \alpha_k \dfrac{\partial f\left(x_{\tau_{i1}(k)}^1, \ldots, x_{\tau_{in}(k)}^n\right)}{\partial x^i} & \text{if } k \in \mathcal{R}_i, \end{cases} \qquad i = 1, \ldots, n,$$

where α_k is a positive stepsize. Here $\tau_{ij}(k)$ is the time at which the jth coordinate used in this update was computed, and the difference $k - \tau_{ij}(k)$ is commonly called the *communication delay* from j to i at time k. In a distributed setting, each coordinate x^i (or block of coordinates) may be updated by a separate processor, on the basis of values of coordinates made available by other processors, with some delay.

(b) *Coordinate descent methods*, where for simplicity we consider a block size of one; cf. Eq. (2.22). We assume that the ith scalar coordinate is updated at a subset of times $\mathcal{R}_i \subset \{0, 1, \ldots\}$, according to

$$x_{k+1}^i \in \arg\min_{\xi \in \Re} f\left(x_{\tau_{i1}(k)}^1, \ldots, x_{\tau_{i,i-1}(k)}^{i-1}, \xi, x_{\tau_{i,i+1}(k)}^{i+1}, \ldots, x_{\tau_{in}(k)}^n\right),$$

and is left unchanged ($x_{k+1}^i = x_k^i$) if $k \notin \mathcal{R}_i$. The meanings of the subsets of updating times $\mathcal{R}_i$ and indexes $\tau_{ij}(k)$ are the same as in the case of gradient methods. Also the distributed environment where the method can be applied is similar to the case of the gradient method. Another practical setting that may be modeled well by this iteration is when all computation takes place at a single computer, but any number of coordinates may be simultaneously updated at a time, with the order of coordinate selection possibly being random.

(c) *Incremental gradient methods* for the case where

$$f(x) = \sum_{i=1}^m f_i(x).$$

Here the ith component is used to update x at a subset of times $\mathcal{R}_i$:

$$x_{k+1} = x_k - \alpha_k \nabla f_i\left(x_{\tau_{i1}(k)}^1, \ldots, x_{\tau_{in}(k)}^n\right), \qquad k \in \mathcal{R}_i,$$

where we assume that a single component gradient ∇f_i is used at each time (i.e., $\mathcal{R}_i \cap \mathcal{R}_j = \emptyset$ for $i \neq j$). The meaning of $\tau_{ij}(k)$ is the same as in the preceding cases, and the gradient ∇f_i can be replaced by a subgradient in the case of nondifferentiable f_i. Here the entire vector x is updated at a central computer, based on component gradients ∇f_i that are computed at other computers and are communicated with some delay to the central computer. For validity of these methods, it is essential that all the components f_i are used in the iteration with the same asymptotic frequency, $1/m$ (see [NBB01]). For this type of asynchronous implementation to make sense, the computation of ∇f_i must be substantially more time-consuming than the update of x_k using the preceding incremental iteration.

An interesting fact is that some asynchronous algorithms, called *totally asynchronous*, can tolerate arbitrarily large delays $k - \tau_{ij}(k)$, while other algorithms, called *partially asynchronous*, are not guaranteed to work unless there is an upper bound on these delays. The convergence mechanisms at work in each of these two cases are genuinely different and so are their analyses (see [BeT89a], where totally and partially asynchronous algorithms, and various special cases including gradient and coordinate descent methods, are discussed in Chapters 6 and 7, respectively).

The totally asynchronous algorithms are valid only under special conditions, which guarantee that any progress in the computation of the individual processors, is consistent with progress in the collective computation. For example to show convergence of the (synchronous) stationary iteration

$$x_{k+1} = G(x_k)$$

it is sufficient to show that G is a contraction mapping with respect to some norm (see Section A.4 of Appendix A), but for asynchronous convergence it turns out that one needs the contraction to be with respect to the sup-norm $\|x\|_\infty = \max_{i=1,\dots,n} |x^i|$ or a weighted sup-norm (see Section 6.5.2). To guarantee totally asynchronous convergence of a gradient method with a constant and sufficiently small stepsize $\alpha_k \equiv \alpha$, a diagonal dominance condition is required; see the paper [Ber83]. In the special case of a quadratic cost function

$$f(x) = \tfrac{1}{2}x'Qx + b'x$$

this condition is that the Hessian matrix Q is diagonally dominant, i.e., has components q_{ij} such that

$$q_{ii} > \sum_{j \neq i}^{n} |q_{ij}|, \qquad i = 1, \dots, n.$$

Without this diagonal dominance condition, totally asynchronous convergence is unlikely to be guaranteed (for examples see [BeT89a], Section 6.3.2).

The partially asynchronous algorithms do not need a weighted sup-norm contraction structure, but typically require either a diminishing or a stepsize that is small and inversely proportional to the size of the delays. The idea is that when the delays are bounded and the stepsize is small enough, the asynchronous algorithm resembles its synchronous counterpart sufficiently closely, so that the convergence properties of the latter are maintained (see [BeT89a], particularly Sections 7.1 and 7.5; also the convergence analysis of gradient methods with errors in the exercises). This mechanism for convergence is similar to the one for incremental methods. For this reason, incremental gradient, gradient projection, and coordinate descent methods are natural candidates for partially asynchronous implementation; see [BeT89a], Chapter 7, and [NBB01].

For further discussion of the implementation and convergence analysis of partially asynchronous algorithms, we refer to the paper [TBA86], to the books [BeT89a] (Chapters 6 and 7) and [Bor08] for deterministic and stochastic gradient, and coordinate descent methods, and to the paper [NBB01] for incremental gradient and subgradient methods. For recent related work on distributed partially asynchronous algorithms of the gradient and coordinate descent type, see [NeO09a], [RRW11], and for partially asynchronous implementations of the Kaczmarz method, see [LWS14].

2.2 APPROXIMATION METHODS

Approximation methods for minimizing a convex function $f : \Re^n \mapsto \Re$ over a convex set X, are based on replacing f and X with approximations F_k

and X_k, respectively, at each iteration k, and finding

$$x_{k+1} \in \arg\min_{x \in X_k} F_k(x).$$

At the next iteration, F_{k+1} and X_{k+1} are generated by refining the approximation, based on the new point x_{k+1}, and possibly on the earlier points $x_k, \ldots, x_0$. Of course such a method makes sense only if the approximating problems are simpler than the original. There is a great variety of approximation methods, with different aims, and suitable for different circumstances. The present section provides a brief overview and orientation, while Chapters 4-6 provide a detailed analysis.

2.2.1 Polyhedral Approximation

In polyhedral approximation methods, F_k is a polyhedral function that approximates f and X_k is a polyhedral set that approximates X. The idea is that the approximate problem is polyhedral, so it may be easier to solve than the original problem. The methods include mechanisms for progressively refining the approximation, thereby obtaining a solution of the original problem in the limit. In some cases, only one of f and X is polyhedrally approximated.

In Chapter 4, we will discuss the two main approaches for polyhedral approximation: *outer linearization* (also called the *cutting plane* approach) and *inner linearization* (also called the *simplicial decomposition* approach). As the name suggests, outer linearization approximates epi(f) and X from without, $F_k(x) \leq f(x)$ for all x, and $X_k \supset X$, using intersections of finite numbers of halfspaces. By contrast, inner linearization approximates epi(f) and X from within, $F_k(x) \geq f(x)$ for all x, and $X_k \subset X$, using convex hulls of finite numbers of halflines or points. Figure 2.2.1 illustrates outer and inner linearization of convex sets and functions.

We will show in Sections 4.3 and 4.4 that these two approaches are intimately connected by conjugacy and duality: *the dual of an outer approximating problem is an inner approximating problem involving the conjugates of F_k and the indicator function of X_k, and reversely.* In fact, using this duality, outer and inner approximations may be combined in the same algorithm.

One of the major applications of the cutting plane approach is in *Dantzig-Wolfe decomposition*, an important method for solving large scale problems with special structure, including the separable problems of Section 1.1.1 (see e.g., [BeT97], [Ber99]). Simplicial decomposition also finds many important applications in problems with special structure; e.g., in high-dimensional problems with a constraint set X such that minimization of a linear function over X is relatively simple. This is exactly the same structure that favors the use of the conditional gradient method discussed in Section 2.1.2 (see Chapter 4). A prominent example is the multicommodity flow problem of Example 1.4.5.

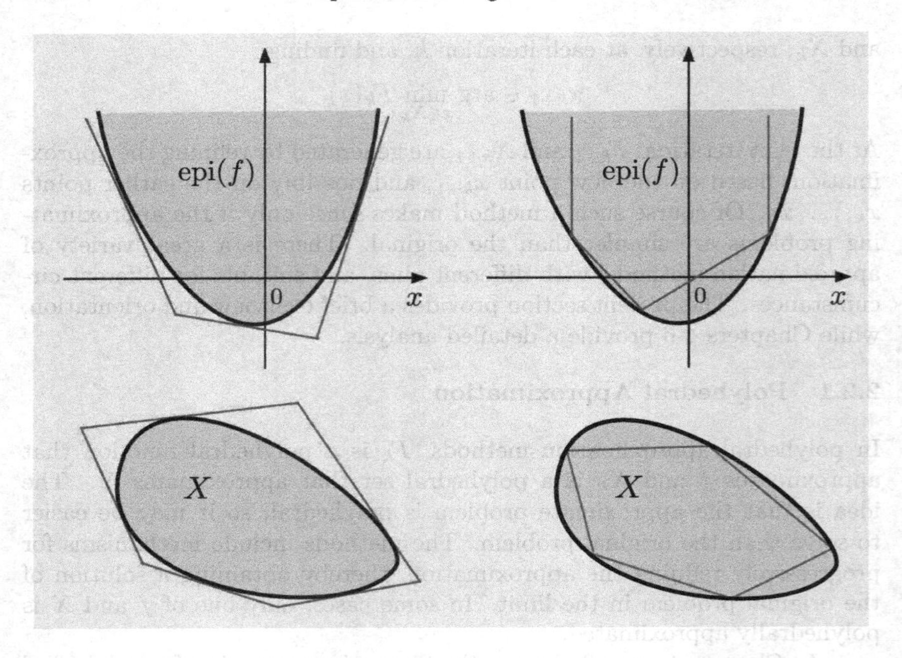

Figure 2.2.1. Illustration of outer and inner linearization of a convex function f and a convex set X using hyperplanes and convex hulls.

2.2.2 Penalty, Augmented Lagrangian, and Interior Point Methods

Generally in optimization problems, the presence of constraints complicates the algorithmic solution, and limits the range of available algorithms. For this reason it is natural to try to eliminate constraints by using approximation of the corresponding indicator functions. In particular, we may replace constraints by penalty functions that prescribe a high cost for their violation. We discussed in Section 1.5 such an approximation scheme, which uses exact nondifferentiable penalty functions. In this section we focus on differentiable penalty functions that are not necessarily exact.

To illustrate this approach, let us consider the equality constrained problem

$$\begin{aligned} &\text{minimize} \quad f(x) \\ &\text{subject to} \quad x \in X, \quad a_i'x = b_i, \quad i = 1, \ldots, m. \end{aligned} \tag{2.50}$$

We replace this problem with a penalized version

$$\begin{aligned} &\text{minimize} \quad f(x) + c_k \sum_{i=1}^{m} P(a_i'x - b_i) \\ &\text{subject to} \quad x \in X, \end{aligned} \tag{2.51}$$

where $P(\cdot)$ is a scalar penalty function satisfying

$$P(u) = 0 \quad \text{if} \quad u = 0,$$

and

$$P(u) > 0 \quad \text{if} \quad u \neq 0.$$

The scalar c_k is a positive penalty parameter, so by increasing c_k to ∞, the solution x_k of the penalized problem tends to decrease the constraint violation, thereby providing an increasingly accurate approximation to the original problem. An important practical point here is that c_k should be increased gradually, using the optimal solution of each approximating problem to start the algorithm that solves the next approximating problem. Otherwise serious numerical problems occur due to "ill-conditioning."

A common choice for P is the quadratic penalty function

$$P(u) = \tfrac{1}{2}u^2,$$

in which case the penalized problem (2.51) takes the form

$$\begin{aligned} &\text{minimize} \quad f(x) + \frac{c_k}{2}\|Ax - b\|^2 \\ &\text{subject to} \quad x \in X, \end{aligned} \tag{2.52}$$

where $Ax = b$ is a vector representation of the system of equations $a_i'x = b_i$, $i = 1, \ldots, m$.

An important enhancement of the penalty function approach is the augmented Lagrangian methodology, where we add a linear term to $P(u)$, involving a multiplier vector $\lambda_k \in \Re^m$. Then in place of problem (2.52), we solve the problem

$$\begin{aligned} &\text{minimize} \quad f(x) + \lambda_k'(Ax - b) + \frac{c_k}{2}\|Ax - b\|^2 \\ &\text{subject to} \quad x \in X. \end{aligned} \tag{2.53}$$

After a minimizing vector x_k is obtained, the multiplier vector λ_k is updated by some formula that aims to approximate an optimal dual solution. A common choice that we will discuss in Chapter 5 is

$$\lambda_{k+1} = \lambda_k + c_k(Ax_k - b). \tag{2.54}$$

This is also known as the *first order augmented Lagrangian method* (also called *first order method of multipliers*). It is a major general purpose, highly reliable, constrained optimization method, which applies to nonconvex problems as well. It has a rich theory, with a strong connection to duality, and many variations that are aimed at increased efficiency, involving for example second order multiplier updates and inexact minimization of the augmented Lagrangian. In the convex programming setting of this book, augmented Lagrangian methods embody additional favorable structure. Among others, convergence is guaranteed for *any* nondecreasing sequence $\{c_k\}$ (for nonconvex problems, c_k must exceed a certain positive threshold). Moreover there is no requirement that $c_k \to \infty$, which is needed for penalty methods that do not involve multiplier updates, and is often the cause of numerical problems.

Generally, penalty and augmented Lagrangian methods can be used for inequality as well as equality constraints. The penalty function is modified to reflect penalization for violation of inequalities. For example the inequality constraint analog of the quadratic penalty $P(u) = \frac{1}{2}u^2$ is

$$P(u) = \tfrac{1}{2}\big(\max\{0, u\}\big)^2.$$

We will consider these possibilities in greater detail in Section 5.2.1.

The penalty methods just discussed are known as *exterior penalty methods*: they approximate the indicator function of the constraint set from without. Another type of algorithm involves approximation from within, which leads to the so called *interior point methods*. These are important methods that find application in a broad variety of problems, including linear programming. They will be discussed in Section 6.8.

2.2.3 Proximal Algorithm, Bundle Methods, and Tikhonov Regularization

The proximal algorithm, briefly discussed in Section 2.1.4, aims to minimize a closed proper convex function $f : \Re^n \mapsto (-\infty, \infty]$, and is given by

$$x_{k+1} \in \arg\min_{x \in \Re^n} \left\{ f(x) + \frac{1}{2c_k}\|x - x_k\|^2 \right\}, \tag{2.55}$$

[cf. Eq. (2.25)], where x_0 is an arbitrary starting point and c_k is a positive scalar parameter. As the parameter c_k tends to ∞, the quadratic regularization term becomes insignificant and the proximal minimization (2.55) approximates more closely the minimization of f, hence the connection of the proximal algorithm with the approximation approach.

We will discuss the proximal algorithm in much more detail in Chapter 5, including dual and polyhedral approximation versions. Among others, we will show that when f is the dual function of the constrained optimization problem (2.50), the proximal algorithm, via Fenchel duality, becomes equivalent to the multiplier iteration of the augmented Lagrangian method [cf. Eq. (2.54)]. Since any closed proper convex function can be viewed as the dual function of an appropriate convex constrained optimization problem, it follows that *the proximal algorithm (2.55) is essentially equivalent to the augmented Lagrangian method*: the two algorithms are dual sides of the same coin.

There are also variants of the proximal algorithm where f in Eq. (2.55) is approximated by a polyhedral or other function. One possibility is *bundle methods*, which involve a combination of the proximal and polyhedral approximation ideas. The motivation here is to simplify the proximal minimization subproblem (2.25), replacing it for example with a quadratic programming problem. Some of these methods may be viewed as regularized versions of Dantzig-Wolfe decomposition (see Section 4.3).

Another approximation approach that bears similarity to the proximal algorithm is *Tikhonov regularization*, which approximates the minimization of f with the minimization

$$x_{k+1} \in \arg \min_{x \in \Re^n} \left\{ f(x) + \frac{1}{2c_k} \|x\|^2 \right\}. \tag{2.56}$$

The quadratic regularization term makes the cost function of the preceding problem strictly convex, and guarantees that it has a unique minimum. Sometimes the quadratic term in Eq. (2.56) is scaled and a term $\|Sx\|^2$ is used instead, where S is a suitable scaling matrix. The difference with the proximal algorithm (2.55) is that x_k does not enter directly the minimization to determine x_{k+1}, so the method relies for its convergence on increasing c_k to ∞. By contrast this is not necessary for the proximal algorithm, which is generally convergent even when c_k is left constant (as we will see in Section 5.1), and is typically much faster. Similar to the proximal algorithm, there is a dual and essentially equivalent algorithm to Tikhonov regularization. This is the penalty method that consists of sequential minimization of the quadratically penalized cost function (2.52) for a sequence $\{c_k\}$ with $c_k \to \infty$.

2.2.4 Alternating Direction Method of Multipliers

The proximal algorithm embodies fundamental ideas that lead to a variety of other interesting methods. In particular, when properly generalized (see Section 5.1.4), it contains as a special case the *alternating direction method of multipliers* (ADMM for short), a method that resembles the augmented Lagrangian method, but is well-suited for some important classes of problems with special structure.

The starting point for the ADMM is the minimization problem of the Fenchel duality context:

$$\begin{aligned} &\text{minimize} \quad f_1(x) + f_2(Ax) \\ &\text{subject to} \quad x \in \Re^n, \end{aligned} \tag{2.57}$$

where A is an $m \times n$ matrix, $f_1 : \Re^n \mapsto (-\infty, \infty]$ and $f_2 : \Re^m \mapsto (-\infty, \infty]$ are closed proper convex functions. We convert this problem to the equivalent constrained minimization problem

$$\begin{aligned} &\text{minimize} \quad f_1(x) + f_2(z) \\ &\text{subject to} \quad x \in \Re^n,\ z \in \Re^m,\ \ Ax = z, \end{aligned} \tag{2.58}$$

and we introduce its augmented Lagrangian function

$$L_c(x, z, \lambda) = f_1(x) + f_2(z) + \lambda'(Ax - z) + \frac{c}{2}\|Ax - z\|^2,$$

where c is a positive parameter.

The ADMM, given the current iterates $(x_k, z_k, \lambda_k) \in \Re^n \times \Re^m \times \Re^m$, generates a new iterate $(x_{k+1}, z_{k+1}, \lambda_{k+1})$ by first minimizing the augmented Lagrangian with respect to x, then with respect to z, and finally performing a multiplier update:

$$x_{k+1} \in \arg \min_{x \in \Re^n} L_c(x, z_k, \lambda_k), \tag{2.59}$$

$$z_{k+1} \in \arg \min_{z \in \Re^m} L_c(x_{k+1}, z, \lambda_k), \tag{2.60}$$

$$\lambda_{k+1} = \lambda_k + c(Ax_{k+1} - z_{k+1}). \tag{2.61}$$

The important advantage that the ADMM may offer over the augmented Lagrangian method, is that it does not involve a joint minimization with respect to x and z. Thus the complications resulting from the coupling of x and z in the penalty term $\|Ax - z\|^2$ of the augmented Lagrangian are eliminated. This property can be exploited in special applications, for which the ADMM is structurally well suited, as we will discuss in Section 5.4. On the other hand the ADMM may converge more slowly than the augmented Lagrangian method, so the flexibility it provides must be weighted against this potential drawback.

In Chapter 5, we will see that the proximal algorithm for minimization can be viewed as a special case of a generalized proximal algorithm for finding a solution of an equation involving a multivalued monotone operator. While we will not fully develop the range of algorithms that are based on this generalization, we will show that both the augmented Lagrangian method and the ADMM are special cases of the generalized proximal algorithm, corresponding to two different multivalued monotone operators. Because of the differences of these two operators, some of the properties of the two methods are quite different. For example, contrary to the case of the augmented Lagrangian method (where c_k is often taken to be increasing with k in order to accelerate convergence), there seems to be no generally good way to adjust c in ADMM from one iteration to the next. Moreover, even when both methods have a linear convergence rate, the performance of the two methods may differ markedly in practice. Still there is more than a superficial connection between the two methods, which can be understood within the context of their common proximal algorithm ancestry.

In Section 6.3, we will also see another connection of ADMM with proximal-related methods, and particularly the proximal gradient method, which we briefly discussed in Section 2.1.4 [cf. Eq. (2.27)]. It turns out that both the ADMM and the proximal gradient method can be viewed as instances of splitting algorithms for finding a zero of the sum of two monotone operators. The idea is to decouple the two operators within an iteration: one operator is treated by a proximal-type algorithm, while the other is treated by a proximal-type or a gradient algorithm. In so doing, the complications that arise from coupling of the two operators are mitigated.

2.2.5 Smoothing of Nondifferentiable Problems

Generally speaking, differentiable cost functions are preferable to nondifferentiable ones, because algorithms for the former are better developed and are more effective than algorithms for the latter. Thus there is an incentive to eliminate nondifferentiabilities by "smoothing" their corners. It turns out that penalty functions and smoothing are closely related, reflecting the fact that constraints and nondifferentiabilities are also closely related. As an example of this connection, the unconstrained minimax problem

$$\begin{aligned} &\text{minimize} \quad \max\{f_1(x),\dots,f_m(x)\} \\ &\text{subject to} \quad x \in \Re^n, \end{aligned} \tag{2.62}$$

where $f_1,\dots,f_m$ are differentiable functions can be converted to the differentiable constrained problem

$$\begin{aligned} &\text{minimize} \quad z \\ &\text{subject to} \quad f_j(x) \le z, \quad j = 1,\dots,m, \end{aligned} \tag{2.63}$$

where z is an artificial scalar variable. When a penalty or augmented Lagrangian method is applied to the constrained problem (2.63), we will show that a smoothing method is obtained for the minimax problem (2.62).

We will now describe a technique (first given in [Ber75b], and generalized in [Ber77]) to obtain smoothing approximations. Let $f : \Re^n \mapsto (-\infty,\infty]$ be a closed proper convex function with conjugate denoted by $f^\star$. For fixed $c > 0$ and $\lambda \in \Re^n$, define

$$f_{c,\lambda}(x) = \inf_{u\in\Re^n} \left\{ f(x-u) + \lambda' u + \frac{c}{2}\|u\|^2 \right\}, \qquad x \in \Re^n. \tag{2.64}$$

The conjugates of $\phi_1(u) = f(x-u)$ and $\phi_2(u) = \lambda' u + \frac{c}{2}\|u\|^2$ are $\phi_1^\star(y) = f^\star(-y) + y'x$ and $\phi_2^\star(y) = \frac{1}{2c}\|y-\lambda\|^2$, so by using the Fenchel duality formula $\inf_{u\in\Re^n}\big\{\phi_1(u)+\phi_2(u)\big\} = \sup_{y\in\Re^n}\big\{-\phi_1^\star(-y) - \phi_2^\star(y)\big\}$, we have

$$f_{c,\lambda}(x) = \sup_{y\in\Re^n} \left\{ y'x - f^\star(y) - \frac{1}{2c}\|y-\lambda\|^2 \right\}, \qquad x \in \Re^n. \tag{2.65}$$

It can be seen that $f_{c,\lambda}$ approximates f in the sense that

$$\lim_{c\to\infty} f_{c,\lambda}(x) = f^{\star\star}(x) = f(x), \qquad \forall\, x, \lambda \in \Re^n;$$

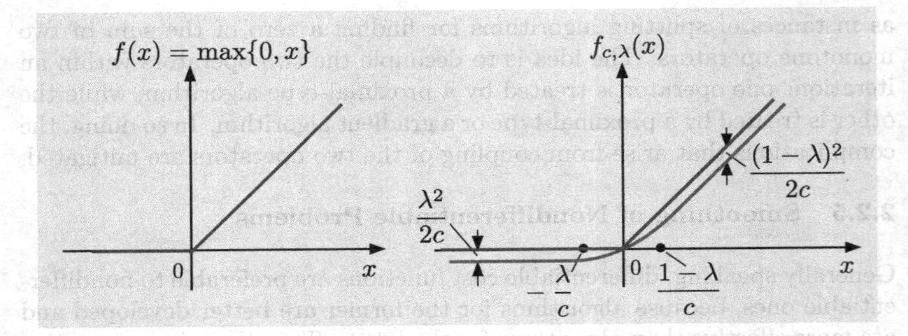

Figure 2.2.2. Illustration of smoothing of the function $f(x) = \max\{0, x\}$. Note that as $c \to \infty$, we have $f_{c,\lambda}(x) \to f(x)$ for all $x \in \Re$, regardless of the value of λ.

the double conjugate f^{**} is equal to f by the Conjugacy Theorem (Prop. 1.6.1 in Appendix B). Furthermore, it can be shown using the optimality conditions of the Fenchel Duality Theorem Prop. 1.2.1(c) (see also [Ber77]) that $f_{c,\lambda}$ is convex and differentiable as a function of x for fixed c and λ, and that the gradient $\nabla f_{c,\lambda}(x)$ at any $x \in \Re^n$ can be obtained in two ways:

(i) As the vector $\lambda + cu$, where u is the unique vector attaining the infimum in Eq. (2.64).

(ii) As the unique vector y that attains the supremum in Eq. (2.65).

The smoothing approach consists of replacing unsmoothed functions f with their smooth approximations $f_{c,\lambda}$, wherever they occur within the cost and constraint functions of a given problem. Note that there may be several functions f that are being smoothed simultaneously, and each occurrence of f may use different λ and c. In this way we obtain a differentiable problem that approximates the original.

An an example consider a common source of nondifferentiability:

$$f(x) = \max\{0, x\}, \qquad x \in \Re.$$

It can be verified using Eqs. (2.64) and (2.65) that

$$f_{c,\lambda}(x) = \begin{cases} x - \dfrac{(1-\lambda)^2}{2c} & \text{if } \frac{1-\lambda}{c} \le x, \\ \lambda x + \dfrac{c}{2}x^2 & \text{if } -\frac{\lambda}{c} \le x \le \frac{1-\lambda}{c}, \\ -\dfrac{\lambda^2}{2c} & \text{if } x \le -\frac{\lambda}{c}; \end{cases}$$

see Fig. 2.2.2. The function $f(x) = \max\{0, x\}$ may also be used as a building block to construct more complicated nondifferentiable functions, such as for example

$$\max\{x_1, x_2\} = x_1 + \max\{0, x_1 - x_2\};$$

see [Ber82a], Ch. 3.

Smoothing and Augmented Lagrangians

The smoothing technique just described can also be combined with the augmented Lagrangian method. As an example, let $f : \Re^n \mapsto (-\infty, \infty]$ be a closed proper convex function with conjugate denoted by $f^\star$. Let $F : \Re^n \mapsto \Re$ be another convex function, and let X be a closed convex set. Consider the problem

$$\begin{aligned} &\text{minimize} \quad F(x) + f(x) \\ &\text{subject to } x \in X, \end{aligned}$$

and the equivalent problem

$$\begin{aligned} &\text{minimize} \quad F(x) + f(x-u) \\ &\text{subject to } x \in X,\ u = 0. \end{aligned}$$

Applying the augmented Lagrangian method (2.53)-(2.54) to the latter problem leads to minimizations of the form

$$(x_{k+1}, u_{k+1}) \in \arg \min_{x\in X,\, u\in\Re^n} \left\{ F(x) + f(x-u) + \lambda_k' u + \frac{c_k}{2}\|u\|^2 \right\}.$$

By first minimizing over $u \in \Re^n$, these minimizations yield

$$x_{k+1} \in \arg \min_{x\in X} \big\{ F(x) + f_{c_k,\lambda_k}(x) \big\},$$

where f_{c_k,λ_k} is the smoothed function

$$f_{c_k,\lambda_k}(x) = \inf_{u\in\Re^n} \left\{ f(x-u) + \lambda_k' u + \frac{c_k}{2}\|u\|^2 \right\},$$

[cf. Eq. (2.64)]. The corresponding multiplier update (2.54) is

$$\lambda_{k+1} = \lambda_k + c_k u_{k+1},$$

where

$$u_{k+1} \in \arg \min_{u\in\Re^n} \left\{ f(x_{k+1} - u) + {\lambda_k}' u + \frac{c_k}{2}\|u\|^2 \right\}.$$

The preceding technique can be extended so that it applies to general convex/concave minimax problems. Let Z be a nonempty convex subset of $\Re^m$, respectively, and $\phi : \Re^n \times Z \mapsto \Re$ is a function such that $\phi(\cdot, z) : \Re^n \mapsto \Re$ is convex for each $z \in Z$, and $-\phi(x, \cdot) : Z \mapsto \Re$ is convex and closed for each $x \in \Re^n$. Consider the problem

$$\begin{aligned} &\text{minimize} \quad \sup_{z\in Z} \phi(x, z) \\ &\text{subject to } x \in X, \end{aligned}$$

where X is a nonempty closed convex subset of $\Re^n$. Consider also the equivalent problem

$$\begin{aligned} &\text{minimize} \quad f(x,y) \\ &\text{subject to} \quad x \in X, \ y = 0, \end{aligned}$$

where f is the function

$$f(x,y) = \sup_{z\in Z}\{\phi(x,z) - y'z\}, \qquad x \in \Re^n, \ y \in \Re^m,$$

which is closed and convex, being the supremum of closed and convex functions. The augmented Lagrangian minimization (2.53) for this problem takes the form

$$x_{k+1} \in \arg\min_{x\in X} f_{c_k,\lambda_k}(x),$$

where $f_{c,\lambda} : \Re^n \mapsto \Re$ is the differentiable function given by

$$f_{c,\lambda}(x) = \min_{y\in\Re^m}\left\{f(x,y) + \lambda'y + \frac{c}{2}\|y\|^2\right\}, \qquad x \in \Re^n.$$

The corresponding multiplier update (2.54) is

$$\lambda_{k+1} = \lambda_k + c_k y_{k+1},$$

where

$$y_{k+1} \in \arg\min_{y\in\Re^m}\left\{f(x_{k+1},y) + \lambda_k'y + \frac{c_k}{2}\|y\|^2\right\}.$$

This method of course makes sense only in the case where the function f has a convenient form that facilitates the preceding minimization.

For further discussion of the relations and combination of smoothing with the augmented Lagrangian method, see [Ber75b], [Ber77], [Pap81], and for a detailed textbook analysis, [Ber82a], Ch. 3. There have also been many variations of smoothing ideas and applications in different contexts; see [Ber73], [Geo77], [Pol79], [Pol88], [BeT89b], [PiZ94], [Nes05], [Che07], [OvG14]. In Section 6.2, we will also see an application of smoothing as an analytical device, in the context of complexity analysis.

Exponential Smoothing

We have used so far a quadratic penalty function as the basis for smoothing. It is also possible to use other types of penalty functions. A simple penalty function, which often leads to convenient formulas is the exponential, which will also be discussed further in Section 6.6. The advantage of the exponential function over the quadratic is that it produces twice differentiable approximating functions. This may be significant when Newton's method is used to solve the smoothed problem.

As an example, a smooth approximation of the function

$$f(x) = \max\{f_1(x), \ldots, f_m(x)\}$$

is given by

$$f_{c,\lambda}(x) = \frac{1}{c} \ln \left\{ \sum_{i=1}^{m} \lambda^i e^{c f_i(x)} \right\}, \tag{2.66}$$

where $c > 0$, and $\lambda = (\lambda^1, \ldots, \lambda^m)$ is a vector with

$$\sum_{i=1}^{m} \lambda^i = 1, \qquad \lambda^i > 0, \quad \forall\, i = 1, \ldots, m.$$

There is an augmented Lagrangian method associated with the approximation (2.66). It involves minimizing over $x \in \Re^n$ the function $f_{c_k,\lambda_k}(x)$ for a given c_k and λ_k to obtain an approximation x_k to the minimum of f. This approximation is refined by setting $c_{k+1} \geq c_k$ and

$$\lambda^i_{k+1} = \frac{\lambda^i_k e^{c_k f_i(x_k)}}{\sum_{j=1}^{m} \lambda^j_k e^{c_k f_j(x_k)}}, \qquad i = 1, \ldots, m, \tag{2.67}$$

and by repeating the process.† The generated sequence $\{x_k\}$ can be shown to converge to the minimum of f under mild assumptions, based on general convergence properties of augmented Lagrangian methods that use nonquadratic penalty functions; see [Ber82a], Ch. 5, for a detailed development.

Example 2.2.1: (Smoothed ℓ_1 Regularization)

Consider the ℓ_1-regularized least squares problem

$$\begin{aligned} &\text{minimize} \quad \gamma \sum_{j=1}^{n} |x^j| + \frac{1}{2} \sum_{i=1}^{m} (a_i' x - b_i)^2 \\ &\text{subject to} \quad x \in \Re^n, \end{aligned}$$

† Sometimes the use of the exponential in Eq. (2.67) and other related formulas, such as (2.66), may lead to very large numbers and computer overflow. In this case one may use a translation device to avoid such numbers, e.g., multiplying numerator and denominator in Eq. (2.67) by $e^{-\beta_k}$ where

$$\beta_k = \max_{i=1,\ldots,m} \{c_k f_i(x_k)\}.$$

A similar idea works for Eq. (2.66).

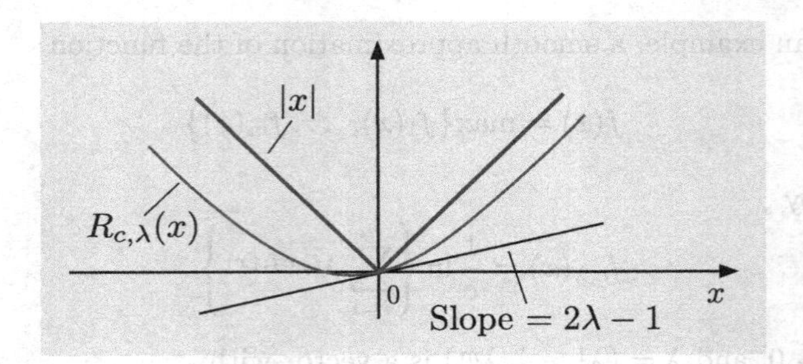

Figure 2.2.3. Illustration of the exponentially smoothed version

$$R_{c,\lambda}(x) = \frac{1}{c} \ln \left\{ \lambda e^{cx} + (1 - \lambda) e^{-cx} \right\}$$

of the absolute value function $|x|$. The approximation becomes asymptotically exact as $c \to \infty$ for any fixed value of the multiplier $\lambda \in (0, 1)$. Also by adjusting the multiplier λ within the range $(0, 1)$, we can attain better approximation for x positive or negative. As $\lambda \to 1$ (or $\lambda \to 0$) the approximation becomes asymptotically exact for $x \geq 0$ (or $x \leq 0$, respectively).

where a_i and b_i are given vectors and scalars, respectively (cf. Example 1.3.1). The nondifferentiable ℓ_1 penalty may be smoothed by writing each term $|x^j|$ as $\max\{x^j, -x^j\}$ and by smoothing it using Eq. (2.66), i.e., replacing it by

$$R_{c,\lambda^j}(x^j) = \frac{1}{c} \ln \left\{ \lambda^j e^{cx^j} + (1 - \lambda^j) e^{-cx^j} \right\},$$

where c and λ^j are scalars satisfying $c > 0$ and $\lambda^j \in (0, 1)$ (see Fig. 2.2.3). We may then consider an exponential type of augmented Lagrangian method, whereby we minimize over $\Re^n$ the twice differentiable function

$$\gamma \sum_{j=1}^{n} R_{c_k, \lambda_k^j}(x^j) + \frac{1}{2} \sum_{i=1}^{m} (a_i' x - b_i)^2, \tag{2.68}$$

to obtain an approximation x_k to the optimal solution. This approximation is refined by setting $c_{k+1} \geq c_k$ and

$$\lambda_{k+1}^j = \frac{\lambda_k^j e^{c_k x_k^j}}{\lambda_k^j e^{c_k x_k^j} + (1 - \lambda_k^j) e^{-c_k x_k^j}}, \qquad j = 1, \ldots, n, \tag{2.69}$$

[cf. Eq. (2.67)], and by repeating the process. Note that the minimization of the exponentially smoothed cost function (2.68) can be carried out efficiently by incremental methods, such as the incremental gradient and Newton methods of Section 2.1.5.

As Fig. 2.2.3 suggests, the adjustment of the multiplier λ^j can selectively reduce the error

$$|x^j| - R_{c,\lambda^j}(x^j)$$

depending on whether good approximation for positive or negative x^j is desired. For this reason *it is not necessary to increase c_k to infinity*; the multiplier iteration (2.69) is sufficient for convergence even with c_k kept constant at some positive value (see [Ber82a], Ch. 5).

2.3 NOTES, SOURCES, AND EXERCISES

Section 2.1: Textbooks on nonlinear programming with a substantial algorithmic content are good sources for the overview material on unconstrained and constrained differentiable optimization in this chapter, e.g., Zangwill [Zan69], Polak [Pol71], Hestenes [Hes75], Zoutendijk [Zou76], Shapiro [Sha79], Gill, Murray, and Wright [GMW81], Luenberger [Lue84], Poljak [Pol87], Dennis and Schnabel [DeS96], Bertsekas [Ber99], Kelley [Kel99], Fletcher [Fle00], Nesterov [Nes04], Bazaraa, Shetty, and Sherali [BSS06], Nocedal and Wright [NoW06], Ruszczynski [Rus06], Griva, Nash, and Sofer [GNS08], Luenberger and Ye [LuY08]. References for specific algorithmic nondifferentiable optimization topics will be given in subsequent chapters.

Incremental gradient methods have a long history, particularly for the unconstrained case ($X = \Re^n$), starting with the Widrow-Hoff least mean squares (LMS) method [WiH60], which stimulated much subsequent research. They have also been used widely, under the generic name "backpropagation methods," for training of neural networks, which involves nonquadratic/nonconvex differentiable cost components. There is an extensive literature on this subject, and for some representative works, we refer to the papers by Rumelhart, Hinton, and Williams [RHW86], [RHW88], Becker and LeCun [BeL88], Vogl et al. [VMR88], and Saarinen, Bramley, and Cybenko [SBC91], and the books by Bishop [Bis95], Bertsekas and Tsitsiklis [BeT96], and Haykin [Hay08]. Some of this literature overlaps with the literature on stochastic gradient methods, which we noted in Section 2.1.5.

Deterministic convergence analyses of several variants of incremental gradient methods were given in the 90s under various assumptions and for a variety of stepsize rules; see Luo [Luo91], Grippo [Gri94], [Gri00], Luo and Tseng [LuT94a], Mangasarian and Solodov [MaS94], Bertsekas [Ber97], Solodov [Sol98], Tseng [Tse98], and Bertsekas and Tsitsiklis [BeT00]. Recent theoretical work on incremental gradient methods has focused, among others, on the aggregated gradient methods discussed in Section 2.1.5, on extensions to nondifferentiable and constrained problems, on combinations with the proximal algorithm, and on constrained versions where the constraints are treated by incremental projection (see Section 6.4 and the references cited there).

The incremental Newton and related methods have been considered by several authors in a stochastic approximation framework; e.g., by Sakrison [Sak66], Venter [Ven67], Fabian [Fab73], Poljak and Tsypkin [PoT80], [PoT81], and more recently by Bottou and LeCun [BoL05], and Bhatnagar, Prasad, and Prashanth [BPP13]. Among others, these references quantify the convergence rate advantage that stochastic Newton methods have over stochastic gradient methods. Deterministic incremental Newton methods have received little attention (for a recent work see Gurbuzbalaban, Ozdaglar, and Parrilo [GOP14]). However, they admit an analysis that is similar to a deterministic analysis of the extended Kalman filter, the incremental version of the Gauss-Newton method (see Bertsekas [Ber96], and Moriyama, Yamashita, and Fukushima [MYF03]). There are also many stochastic analyses of the extended Kalman filter in the literature of estimation and control of dynamic systems.

Let us also note another approach to accelerate the theoretical convergence rate of incremental gradient methods, which involves using a larger than $O(1/k)$ stepsize and averaging the iterates (for analysis of the corresponding stochastic gradient methods, see Ruppert [Rup85], and Poljak and Juditsky [PoJ92], and for a textbook account, Kushner and Yin [KuY03]).

Section 2.2: The nonlinear programming textbooks cited earlier contain a lot of material on approximation methods. In particular, the literature on polyhedral approximation is extensive. It dates to the early days of nonlinear and convex programming, and it involves applications in data communication and transportation networks, and large-scale resource allocation. This literature will be reviewed in Chapter 4.

The research monographs by Fiacco and MacCormick [FiM68], and Bertsekas [Ber82a] focus on penalty and augmented Lagrangian methods, respectively. The latter book also contains a lot of material on smoothing methods and the proximal algorithm, including cases where nonquadratic regularization is involved, leading in turn to nonquadratic penalty terms in the augmented Lagrangian (e.g., logarithmic regularization and exponential penalty).

The proximal algorithm was proposed in the early 70s by Martinet [Mar70], [Mar72]. The literature on the algorithm and its extensions, spurred by the influential paper by Rockafellar [Roc76a], is voluminous, and reflects the central importance of proximal ideas in convex optimization and other problems. The ADMM, an important special case of the proximal context, was proposed by Glowinskii and Morocco [GIM75], and Gabay and Mercier [GaM76], and was further developed by Gabay [Gab79], [Gab83]. We refer to Section 5.4 for a detailed discussion of this algorithm, its applications, and its connections to more general operator splitting methods. Recent work involving proximal ideas has focused on combinations with other algorithms, such as gradient, subgradient, and coordinate descent methods. Some of these combined methods will be discussed in detail in Chapters 5 and 6.

EXERCISES

2.1 (Convergence Rate of Steepest Descent and Gradient Projection for a Quadratic Cost Function)

Let f be the quadratic cost function,

$$f(x) = \tfrac{1}{2}x'Qx - b'x,$$

where Q is a symmetric positive definite matrix, and let m and M be the minimum and maximum eigenvalues of Q, respectively. Consider the minimization of f over a closed convex set X and the gradient projection mapping

$$G(x) = P_X\big(x - \alpha\nabla f(x)\big)$$

with constant stepsize $\alpha < 2/M$.

(a) Show that G is a contraction mapping and we have

$$\big\|G(x) - G(y)\big\| \leq \max\big\{|1-\alpha m|, |1-\alpha M|\big\}\|x-y\|, \qquad \forall\, x, y \in \Re^n,$$

and its unique fixed point is the unique minimum x^* of f over X. *Solution*: First note the nonexpansive property of the projection

$$\big\|P_X(x) - P_X(y)\big\| \leq \|x-y\|, \qquad \forall\, x, y \in \Re^n;$$

(use a Euclidean geometric argument, or see Section 3.2 for a proof). Use this property and the gradient formula $\nabla f(x) = Qx - b$ to write

$$\begin{aligned}\big\|G(x) - G(y)\big\| &= \big\|P_X\big(x - \alpha\nabla f(x)\big) - P_X\big(y - \alpha\nabla f(y)\big)\big\| \\ &\leq \big\|\big(x - \alpha\nabla f(x)\big) - \big(y - \alpha\nabla f(y)\big)\big\| \\ &= \big\|(I - \alpha Q)(x-y)\big\| \\ &\leq \max\big\{|1-\alpha m|, |1-\alpha M|\big\}\|x-y\|,\end{aligned}$$

where m and M are the minimum and maximum eigenvalues of Q. Clearly x^* is a fixed point of G if and only if $x^* = P_X\big(x^* - \alpha\nabla f(x^*)\big)$, which by the projection theorem, is true if and only if the necessary and sufficient condition for optimality $\nabla f(x^*)'(x - x^*) \geq 0$ for all $x \in X$ is satisfied. *Note*: In a generalization of this convergence rate estimate to the case of a nonquadratic strongly convex differentiable function f, the maximum eigenvalue M is replaced by the Lipschitz constant of ∇f and the minimum eigenvalue m is replaced by the modulus of strong convexity of f; see Section 6.1.

(b) Show that the value of α that minimizes the bound of part (a) is

$$\alpha^* = \frac{2}{M+m},$$

in which case

$$\big\|G(x) - G(y)\big\| \le \left(\frac{M/m - 1}{M/m + 1}\right) \|x - y\|.$$

Note: The linear convergence rate estimate,

$$\|x_{k+1} - x^*\| \le \left(\frac{M/m - 1}{M/m + 1}\right) \|x_k - x^*\|,$$

that this contraction property implies for steepest descent with constant stepsize is sharp, in the sense that there exist starting points x_0 for which the preceding inequality holds as an equation for all k (see [Ber99], Section 2.3).

2.2 (Descent Inequality)

This exercise deals with an inequality that is fundamental for the convergence analysis of gradient methods. Let X be a convex set, and let $f : \Re^n \mapsto \Re$ be a differentiable function such that for some constant $L > 0$, we have

$$\big\|\nabla f(x) - \nabla f(y)\big\| \le L\|x - y\|, \qquad \forall\, x, y \in X.$$

Show that

$$f(y) \le f(x) + \nabla f(x)'(y - x) + \frac{L}{2}\|y - x\|^2, \qquad \forall\, x, y \in X. \tag{2.70}$$

Proof: Let t be a scalar parameter and let $g(t) = f\big(x + t(y - x)\big)$. The chain rule yields $(dg/dt)(t) = \nabla f\big(x + t(y - x)\big)'(y - x)$. Thus, we have

$$\begin{aligned}
f(y) - f(x) &= g(1) - g(0) \\
&= \int_0^1 \frac{dg}{dt}(t)\, dt \\
&= \int_0^1 (y - x)'\nabla f\big(x + t(y - x)\big)\, dt \\
&\le \int_0^1 (y - x)'\nabla f(x)\, dt + \left|\int_0^1 (y - x)'\big(\nabla f\big(x + t(y - x)\big) - \nabla f(x)\big)\, dt\right| \\
&\le \int_0^1 (y - x)'\nabla f(x)\, dt + \int_0^1 \|y - x\| \cdot \|\nabla f\big(x + t(y - x)\big) - \nabla f(x)\| dt \\
&\le (y - x)'\nabla f(x) + \|y - x\| \int_0^1 Lt\|y - x\|\, dt \\
&= (y - x)'\nabla f(x) + \frac{L}{2}\|y - x\|^2.
\end{aligned}$$

2.3 (Convergence of Steepest Descent with Constant Stepsize)

Let $f : \Re^n \mapsto \Re$ be a differentiable function such that for some constant $L > 0$, we have

$$\big\|\nabla f(x) - \nabla f(y)\big\| \le L\|x - y\|, \qquad \forall\, x, y \in \Re^n. \tag{2.71}$$

Consider the sequence $\{x_k\}$ generated by the steepest descent iteration

$$x_{k+1} = x_k - \alpha \nabla f(x_k),$$

where $0 < \alpha < \frac{2}{L}$. Show that if $\{x_k\}$ has a limit point, then $\nabla f(x_k) \to 0$, and every limit point $\overline{x}$ of $\{x_k\}$ satisfies $\nabla f(\overline{x}) = 0$. *Proof*: We use the descent inequality (2.70) to show that the cost function is reduced at each iteration according to

$$\begin{aligned} f(x_{k+1}) &= f\big(x_k - \alpha \nabla f(x_k)\big) \\ &\le f(x_k) + \nabla f(x_k)'\big(-\alpha \nabla f(x_k)\big) + \frac{\alpha^2 L}{2}\|\nabla f(x_k)\|^2 \\ &= f(x_k) - \alpha\left(1 - \frac{\alpha L}{2}\right)\|\nabla f(x_k)\|^2. \end{aligned}$$

Thus if there exists a limit point $\overline{x}$ of $\{x_k\}$, we have $f(x_k) \to f(\overline{x})$ and $\nabla f(x_k) \to 0$. This implies that $\nabla f(\overline{x}) = 0$, since $\nabla f(\cdot)$ is continuous by Eq. (2.71).

2.4 (Armijo/Backtracking Stepsize Rule)

Consider minimization of a continuously differentiable function $f : \Re^n \mapsto \Re$, using the iteration

$$x_{k+1} = x_k + \alpha_k d_k,$$

where d_k is a descent direction. Given fixed scalars β, and σ, with $0 < \beta < 1$, $0 < \sigma < 1$, and s_k with $\inf_{k \ge 0} s_k > 0$, the stepsize α_k is determined as follows: we set $\alpha_k = \beta^{m_k} s_k$, where m_k is the first nonnegative integer m for which

$$f(x_k) - f(x_k + \beta^m s_k d_k) \ge -\sigma \beta^m s_k \nabla f(x_k)' d_k.$$

Assume that there exist positive scalars c_1, c_2 such that for all k we have

$$c_1\big\|\nabla f(x_k)\big\|^2 \le -\nabla f(x_k)' d_k, \qquad \|d_k\|^2 \le c_2 \big\|\nabla f(x_k)\big\|^2. \tag{2.72}$$

(a) Show that the stepsize α_k is well-defined, i.e., that it will be determined after a finite number of reductions if $\nabla f(x_k) \ne 0$. *Proof*: We have for all $s > 0$

$$f(x_k + s d_k) - f(x_k) = s \nabla f(x_k)' d_k + o(s).$$

Thus the test for acceptance of a stepsize $s > 0$ is written as

$$s\nabla f(x_k)'d_k + o(s) \le \sigma s \nabla f(x_k)'d_k,$$

or using Eq. (2.72),

$$\frac{o(s)}{s} \le (1-\sigma)c_1 \big\|\nabla f(x_k)\big\|^2,$$

which is satisfied for s in some interval $(0, \bar{s}_k]$. Thus the test will be passed for all m for which $\beta^m s_k \le \bar{s}_k$.

(b) Show that every limit point $\overline{x}$ of the generated sequence $\{x_k\}$ satisfies $\nabla f(\overline{x}) = 0$. *Proof*: Assume, to arrive at a contradiction, that there is a subsequence $\{x_k\}_\mathcal{K}$ that converges to some $\overline{x}$ with $\nabla f(\overline{x}) \neq 0$. Since $\{f(x_k)\}$ is monotonically nonincreasing, $\{f(x_k)\}$ either converges to a finite value or diverges to $-\infty$. Since f is continuous, $f(\overline{x})$ is a limit point of $\{f(x_k)\}$, so it follows that the entire sequence $\{f(x_k)\}$ converges to $f(\overline{x})$. Hence,

$$f(x_k) - f(x_{k+1}) \to 0.$$

By the definition of the Armijo rule and the descent property $\nabla f(x_k)'d_k \le 0$ of the direction d_k, we have

$$f(x_k) - f(x_{k+1}) \ge -\sigma\alpha_k \nabla f(x_k)'d_k \ge 0,$$

so by combining the preceding two relations,

$$\alpha_k \nabla f(x_k)'d_k \to 0. \tag{2.73}$$

From the left side of Eq. (2.72) and the hypothesis $\nabla f(\overline{x}) \neq 0$, it follows that

$$\limsup_{\substack{k\to\infty\\ k\in\mathcal{K}}} \nabla f(x_k)'d_k < 0, \tag{2.74}$$

which together with Eq. (2.73) implies that

$$\{\alpha_k\}_\mathcal{K} \to 0.$$

Since s_k, the initial trial value for α_k, is bounded away from 0, s_k will be reduced at least once for all $k \in \mathcal{K}$ that are greater than some iteration index $\overline{k}$. Thus we must have for all $k \in \mathcal{K}$ with $k > \overline{k}$,

$$f(x_k) - f\big(x_k + (\alpha_k/\beta)d_k\big) < -\sigma(\alpha_k/\beta)\nabla f(x_k)'d_k. \tag{2.75}$$

From the right side of Eq. (2.72), $\{d_k\}_\mathcal{K}$ is bounded, and it follows that there exists a subsequence $\{d_k\}_{\overline{\mathcal{K}}}$ of $\{d_k\}_\mathcal{K}$ such that

$$\{d_k\}_{\overline{\mathcal{K}}} \to \overline{d},$$

where $\overline{d}$ is some vector. From Eq. (2.75), we have

$$\frac{f(x_k) - f(x_k + \overline{\alpha}_k d_k)}{\overline{\alpha}_k} < -\sigma \nabla f(x_k)' d_k, \qquad \forall\, k \in \overline{\mathcal{K}},\ k \geq \overline{k},$$

where $\overline{\alpha}_k = \alpha_k/\beta$. By using the mean value theorem, this relation is written as

$$-\nabla f(x_k + \tilde{\alpha}_k d_k)' d_k < -\sigma \nabla f(x_k)' d_k, \qquad \forall\, k \in \overline{\mathcal{K}},\ k \geq \overline{k},$$

where $\tilde{\alpha}_k$ is a scalar in the interval $[0, \overline{\alpha}_k]$. Taking limits in the preceding relation we obtain

$$-\nabla f(\overline{x})' \overline{d} \leq -\sigma \nabla f(\overline{x})' \overline{d},$$

or

$$0 \leq (1 - \sigma) \nabla f(\overline{x})' \overline{d}.$$

Since $\sigma < 1$, it follows that

$$0 \leq \nabla f(\overline{x})' \overline{d},$$

a contradiction of Eq. (2.74).

2.5 (Convergence of Steepest Descent to a Single Limit [BGI95])

Let $f : \Re^n \mapsto \Re$ be a differentiable convex function, and assume that for some $L > 0$, we have

$$\big\|\nabla f(x) - \nabla f(y)\big\| \leq L\,\|x - y\|, \qquad \forall\, x, y \in \Re^n.$$

Let X^* be the set of minima of f, and assume that X^* is nonempty. Consider the steepest descent method

$$x_{k+1} = x_k - \alpha_k \nabla f(x_k).$$

Show that $\{x_k\}$ converges to a minimizing point of f under each of the following two stepsize rule conditions:

(i) For some $\epsilon > 0$, we have

$$\epsilon \leq \alpha_k \leq \frac{2(1-\epsilon)}{L}, \qquad \forall\, k.$$

(ii) $\alpha_k \to 0$ and $\sum_{k=0}^{\infty} \alpha_k = \infty$.

Notes: The original source [BGI95] also shows convergence to a single limit for a variant of the Armijo rule. This should be contrasted with a result of [Gon00], which shows that the steepest descent method with the exact line minimization rule may produce a sequence with multiple limit points (all of which are of course optimal), even for a convex cost function. There is also a "local capture" theorem that applies to gradient methods for *nonconvex* continuously differentiable cost functions f and an isolated local minimum of f (a local minimum x^* that is unique within a neighborhood of x^*). Under mild conditions it asserts that there is an open sphere S_{x^*} centered at x^* such that once the generated sequence $\{x_k\}$ enters S_{x^*}, it converges to x^* (see [Ber82a], Prop. 1.12, or [Ber99], Prop. 1.2.5 and the references given there). *Abbreviated Proof*: Consider the stepsize rule (i). From the descent inequality (Exercise 2.2), we have for all k

$$f(x_{k+1}) \leq f(x_k) - \alpha_k \left(1 - \frac{\alpha_k L}{2}\right) \left\|\nabla f(x_k)\right\|^2 \leq f(x_k) - \epsilon^2 \left\|\nabla f(x_k)\right\|^2,$$

so $\{f(x_k)\}$ is monotonically nonincreasing and converges. Adding the preceding relation for all values of k and taking the limit as $k \to \infty$, we obtain for all $x^* \in X^*$,

$$f(x^*) \leq f(x_0) - \epsilon^2 \sum_{k=0}^{\infty} \left\|\nabla f(x_k)\right\|^2.$$

It follows that $\sum_{k=0}^{\infty} \left\|\nabla f(x_k)\right\|^2 < \infty$ and $\nabla f(x_k) \to 0$, and also

$$\sum_{k=0}^{\infty} \|x_{k+1} - x_k\|^2 < \infty, \tag{2.76}$$

since $\nabla f(x_k) = (x_k - x_{k+1})/\alpha_k$. Moreover any limit point of $\{x_k\}$ belongs to X^*, since $\nabla f(x_k) \to 0$ and f is convex.

Using the convexity of f, we have for all $x^* \in X^*$,

$$\begin{aligned}\|x_{k+1} - x^*\|^2 - \|x_k - x^*\|^2 - \|x_{k+1} - x_k\|^2 &= -2(x^* - x_k)'(x_{k+1} - x_k) \\ &= 2\alpha_k (x^* - x_k)'\nabla f(x_k) \\ &\leq 2\alpha_k \big(f(x^*) - f(x_k)\big) \\ &\leq 0,\end{aligned}$$

so that

$$\|x_{k+1} - x^*\|^2 \leq \|x_k - x^*\|^2 + \|x_{k+1} - x_k\|^2, \qquad \forall\, x^* \in X^*. \tag{2.77}$$

We now use Eqs. (2.76) and (2.77), and the Fejér Convergence Theorem (Prop. A.4.6 in Appendix A). From part (a) of that theorem it follows that $\{x_k\}$ is bounded, and hence it has a limit point $\overline{x}$, which must belong to X^* as shown earlier. Using this fact and part (b) of the theorem, it follows that $\{x_k\}$ converges to $\overline{x}$.

The proof for the case of the stepsize rule (ii) is similar. Using the assumptions $\alpha_k \to 0$ and $\sum_{k=0}^{\infty} \alpha_k = \infty$, and the descent inequality, we show that $\nabla f(x_k) \to 0$, that $\{f(x_k)\}$ converges, and that Eq. (2.76) holds. From this point, the preceding proof applies.

2.6 (Convergence of Gradient Method with Errors [BeT00])

Consider the problem of unconstrained minimization of a differentiable function $f : \Re^n \mapsto \Re$. Let $\{x_k\}$ be a sequence generated by the method

$$x_{k+1} = x_k - \alpha_k\big(\nabla f(x_k) + w_k\big),$$

where α_k is a positive stepsize, and w_k is an error vector satisfying for some positive scalars p and q,

$$\|w_k\| \le \alpha_k\big(q + p\|\nabla f(x_k)\|\big), \qquad k = 0, 1, \ldots. \tag{2.78}$$

Assume that for some constant $L > 0$, we have

$$\big\|\nabla f(x) - \nabla f(y)\big\| \le L\|x - y\|, \qquad \forall\, x, y \in \Re^n,$$

and that

$$\sum_{k=0}^{\infty} \alpha_k = \infty, \qquad \sum_{k=0}^{\infty} \alpha_k^2 < \infty. \tag{2.79}$$

Show that either $f(x_k) \to -\infty$ or else $f(x_k)$ converges to a finite value and $\lim_{k\to\infty} \nabla f(x_k) = 0$. Furthermore, every limit point $\overline{x}$ of $\{x_k\}$ satisfies $\nabla f(\overline{x}) = 0$. *Abbreviated Proof*: The descent inequality (2.70) yields

$$f(x_{k+1}) \le f(x_k) - \alpha_k \nabla f(x_k)'\big(\nabla f(x_k) + w_k\big) + \frac{\alpha_k^2 L}{2}\big\|\nabla f(x_k) + w_k\big\|^2.$$

Using Eq. (2.78), we have

$$\begin{aligned} -\nabla f(x_k)'\big(\nabla f(x_k) + w_k\big) &\le -\big\|\nabla f(x_k)\big\|^2 + \big\|\nabla f(x_k)\big\|\,\|w_k\| \\ &\le -\big\|\nabla f(x_k)\big\|^2 + \alpha_k q\big\|\nabla f(x_k)\big\| + \alpha_k p\big\|\nabla f(x_k)\big\|^2, \end{aligned}$$

and

$$\begin{aligned} \frac{1}{2}\big\|\nabla f(x_k) + w_k\big\|^2 &\le \big\|\nabla f(x_k)\big\|^2 + \|w_k\|^2 \\ &\le \big\|\nabla f(x_k)\big\|^2 + \alpha_k^2\big(q^2 + 2pq\big\|\nabla f(x_k)\big\| + p^2\big\|\nabla f(x_k)\big\|^2\big). \end{aligned}$$

Combining the preceding three relations and collecting terms, it follows that

$$\begin{aligned} f(x_{k+1}) \le f(x_k) &- \alpha_k(1 - \alpha_k L - \alpha_k p - \alpha_k^3 p^2 L)\big\|\nabla f(x_k)\big\|^2 \\ &+ \alpha_k^2(q + 2\alpha_k^2 pqL)\big\|\nabla f(x_k)\big\| + \alpha_k^4 q^2 L. \end{aligned}$$

Since $\alpha_k \to 0$, we have for some positive constants c and d, and all k sufficiently large

$$f(x_{k+1}) \le f(x_k) - \alpha_k c\big\|\nabla f(x_k)\big\|^2 + \alpha_k^2 d\big\|\nabla f(x_k)\big\| + \alpha_k^4 q^2 L.$$

Using the inequality $\|\nabla f(x_k)\| \le 1 + \|\nabla f(x_k)\|^2$, the above relation yields for all k sufficiently large

$$f(x_{k+1}) \le f(x_k) - \alpha_k(c - \alpha_k d)\|\nabla f(x_k)\|^2 + \alpha_k^2 d + \alpha_k^4 q^2 L.$$

By applying the Supermartingale Convergence Theorem (Prop. A.4.4 in Appendix A), using also the assumption (2.79), it follows that either $f(x_k) \to -\infty$ or else $f(x_k)$ converges to a finite value and $\sum_{k=0}^{\infty} \alpha_k \|\nabla f(x_k)\|^2 < \infty$. In the latter case, in view of the assumption $\sum_{k=0}^{\infty} \alpha_k = \infty$, we must have $\liminf_{k\to\infty} \|\nabla f(x_k)\| = 0$. This implies that $\nabla f(x_k) \to 0$; for a detailed proof of this last step see [BeT00]. This reference also provides a stochastic version of the result of this exercise. This result, however, requires a different line proof, which does not rely on supermartingale convergence arguments.

2.7 (Steepest Descent Direction for Nondifferentiable Cost Functions [BeM71])

Let $f : \Re^n \mapsto \Re$ be a convex function, and let us view the steepest descent direction at x as the solution of the problem

$$\begin{aligned} &\text{minimize} \quad f'(x; d) \\ &\text{subject to} \quad \|d\| \le 1. \end{aligned} \tag{2.80}$$

Show that this direction is $-g^*$, where g^* is the vector of minimum norm in $\partial f(x)$. *Abbreviated Solution*: From Prop. 5.4.8 in Appendix B, $f'(x; \cdot)$ is the support function of the nonempty and compact subdifferential $\partial f(x)$, i.e.,

$$f'(x; d) = \max_{g \in \partial f(x)} d'g, \qquad \forall\, x, d \in \Re^n.$$

Since the sets $\{d \mid \|d\| \le 1\}$ and $\partial f(x)$ are convex and compact, and the function $d'g$ is linear in each variable when the other variable is fixed, by the Saddle Point Theorem of Prop. 5.5.3 in Appendix B, it follows that

$$\min_{\|d\| \le 1} \max_{g \in \partial f(x)} d'g = \max_{g \in \partial f(x)} \min_{\|d\| \le 1} d'g,$$

and that a saddle point exists. For any saddle point (d^*, g^*), g^* maximizes the function $\min_{\|d\| \le 1} d'g = -\|g\|$ over $\partial f(x)$, so g^* is the unique vector of minimum norm in $\partial f(x)$. Moreover, d^* minimizes $\max_{g \in \partial f(x)} d'g$ or equivalently $f'(x; d)$ [by Eq. (2.80)] subject to $\|d\| \le 1$ (so it is a direction of steepest descent), and minimizes $d'g^*$ subject to $\|d\| \le 1$, so it has the form

$$d^* = -\frac{g^*}{\|g^*\|}$$

[except if $0 \in \partial f(x)$, in which case $d^* = 0$].

2.8 (Two-Metric Projection Methods for Bound Constraints [Ber82a], [Ber82b])

Consider the minimization of a continuously differentiable function $f : \Re^n \mapsto \Re$ over the set

$$X = \{(x^1, \dots, x^n) \mid \underline{b}^i \le x^i \le \bar{b}^i,\ i = 1, \dots, n\},$$

where $\underline{b}^i$ and $\bar{b}^i$, $i = 1, \dots, n$, are given scalars. The two-metric projection method for this problem has the form

$$x_{k+1} = P_X\big(x_k - \alpha_k D_k \nabla f(x_k)\big),$$

where D_k is a positive definite symmetric matrix.

(a) Construct an example of f and D_k, where x_k does not minimize f over X and $f(x_{k+1}) > f(x_k)$ for all $\alpha_k > 0$.

(b) For given $x_k \in X$, let $I_k = \{i \mid x_k^i = \underline{b}^i \text{ with } \partial f(x_k)/\partial x^i > 0 \text{ or } x_k^i = \bar{b}^i \text{ with } \partial f(x_k)/\partial x^i < 0\}$. Assume that D_k is diagonal with respect to I_k in the sense that $(D_k)_{ij} = (D_k)_{ji} = 0$ for all $i \in I_k$ and $j = 1, \dots, n$, with $j \neq i$. Show that if x_k is not optimal, there exists $\bar{\alpha}_k > 0$ such that $f(x_{k+1}) < f(x_k)$ for all $\alpha_k \in (0, \bar{\alpha}_k]$.

(c) Assume that the nondiagonal portion of D_k is the inverse of the corresponding portion of $\nabla^2 f(x_k)$. Argue informally that the method can be reasonably expected to have superlinear convergence rate.

2.9 (Incremental Methods – Computational Exercise)

This exercise deals with the (perhaps approximate) solution of a system of linear inequalities $c_i'x \le b_i$, $i = 1, \dots, m$, where $c_i \in \Re^n$ and $b_i \in \Re$ are given.

(a) Consider a variation of the Kaczmarz algorithm that operates in cycles as follows. At the end of cycle k, we set $x_{k+1} = \psi_{m,k}$, where $\psi_{m,k}$ is obtained after the m steps

$$\psi_{i,k} = \psi_{i-1,k} - \frac{\alpha_k}{\|c_i\|^2} \max\{0,\ c_i'\psi_{i-1,k} - b_i\} c_i, \qquad i = 1, \dots, m,$$

starting with $\psi_{0,k} = x_k$. Show that the algorithm can be viewed as an incremental gradient method for a suitable differentiable cost function.

(b) Implement the algorithm of (a) for two examples where $n = 2$ and $m = 100$. In the first example, the vectors c_i have the form $c_i = (\xi_i, \zeta_i)$, where ξ_i, ζ_i, as well as b_i, are chosen randomly and independently from $[-100, 100]$ according to a uniform distribution. In the second example, the vectors c_i have the form $c_i = (\xi_i, \zeta_i)$, where ξ_i, ζ_i are chosen randomly and independently within $[-10, 10]$ according to a uniform distribution, while b_i is chosen randomly and independently within $[0, 1000]$ according to a uniform distribution. Experiment with different starting points and stepsize choices, and deterministic and randomized orders of selection of the indexes i for iteration. Explain your experimental results in terms of the theoretical behavior described in Section 2.1.

2.10 (Convergence of the Incremental Gradient Method)

Consider the minimization of a cost function

$$f(x) = \sum_{i=1}^{m} f_i(x),$$

where $f_i : \Re^n \mapsto \Re$ are continuously differentiable, and let $\{x_k\}$ be a sequence generated by the incremental gradient method. Assume that for some constants L, C, D, and all $i = 1, \ldots, m$, we have

$$\big\|\nabla f_i(x) - \nabla f_i(y)\big\| \leq L\|x - y\|, \qquad \forall\, x, y \in \Re^n,$$

and

$$\big\|\nabla f_i(x)\big\| \leq C + D\big\|\nabla f(x)\big\|, \qquad \forall\, x \in \Re^n.$$

Assume also that

$$\sum_{k=0}^{\infty} \alpha_k = \infty, \qquad \sum_{k=0}^{\infty} \alpha_k^2 < \infty.$$

Show that either $f(x_k) \to -\infty$ or else $f(x_k)$ converges to a finite value and $\lim_{k\to\infty} \nabla f(x_k) = 0$. Furthermore, every limit point $\overline{x}$ of $\{x_k\}$ satisfies $\nabla f(\overline{x}) = 0$. *Abbreviated Solution*: The idea is to view the incremental gradient method as a gradient method with errors, so that the result of Exercise 2.6 can be used. For simplicity we assume that $m = 2$. The proof is similar when $m > 2$. We have

$$\psi_1 = x_k - \alpha_k \nabla f_1(x_k), \qquad x_{k+1} = \psi_1 - \alpha_k \nabla f_2(\psi_1).$$

By adding these two relations, we obtain

$$x_{k+1} = x_k + \alpha_k\big(-\nabla f(x_k) + w_k\big),$$

where

$$w_k = \nabla f_2(x_k) - \nabla f_2(\psi_1).$$

We have

$$\|w_k\| \leq L\|x_k - \psi_1\| = \alpha_k L\big\|\nabla f_1(x_k)\big\| \leq \alpha_k\big(LC + LD\big\|\nabla f(x_k)\big\|\big).$$

Thus Exercise 2.6 applies and the result follows.

2.11 (Convergence Rate of the Kaczmarz Algorithm with Random Projection [StV09])

Consider a consistent system of linear equations $c_i'x = b_i$, $i = 1, \ldots, m$, and assume for convenience that the vectors c_i have been scaled so that $\|c_i\| = 1$ for all i. A randomized version of the Kaczmarz method is given by

$$x_{k+1} = x_k - (c_{i_k}'x - b_{i_k})c_{i_k},$$

where i_k is an index randomly chosen from the set $\{1, \ldots, m\}$ with equal probabilities $1/m$, independently of previous choices. Let $P(x)$ denote the Euclidean projection of a vector $x \in \Re^n$ onto the set of solutions of the system, and let C be the matrix whose rows are $c_1, \ldots, c_m$. Show that

$$E\big\{\|x_{k+1} - P(x_{k+1})\|^2\big\} \leq \left(1 - \frac{\lambda_{min}}{m}\right) E\big\{\|x_k - P(x_k)\|^2\big\},$$

where λ_{min} is the minimum eigenvalue of the matrix $C'C$. *Hint:* Show that

$$\big\|x_{k+1} - P(x_{k+1})\big\|^2 \leq \big\|x_{k+1} - P(x_k)\big\|^2 = \big\|x_k - P(x_k)\big\|^2 - (c_{i_k}'x_{i_k} - b_{i_k})^2,$$

and take conditional expectation of both sides to show that

$$\begin{aligned} E\big\{\|x_{k+1} - P(x_{k+1})\|^2 \mid x_k\big\} &\leq \big\|x_k - P(x_k)\big\|^2 - \frac{1}{m}\|Cx_k - b\|^2 \\ &\leq \left(1 - \frac{\lambda_{min}}{m}\right)\big\|x_k - P(x_k)\big\|^2. \end{aligned}$$

2.12 (Limit Cycle of Incremental Gradient Method [Luo91])

Consider the scalar least squares problem

$$\begin{aligned} &\text{minimize} \quad \tfrac{1}{2}\big((b_1 - x)^2 + (b_2 - x)^2\big) \\ &\text{subject to} \quad x \in \Re, \end{aligned}$$

where b_1 and b_2 are given scalars, and the incremental gradient algorithm that generates x_{k+1} from x_k according to

$$x_{k+1} = \psi_k - \alpha(\psi_k - b_2),$$

where

$$\psi_k = x_k - \alpha(x_k - b_1),$$

and α is a positive stepsize. Assuming that $\alpha < 1$, show that $\{x_k\}$ and $\{\psi_k\}$ converge to limits $x(\alpha)$ and $\psi(\alpha)$, respectively. However, unless $b_1 = b_2$, $x(\alpha)$ and $\psi(\alpha)$ are neither equal to each other, nor equal to the least squares solution $x^* = (b_1 + b_2)/2$. Verify that

$$\lim_{\alpha \to 0} x(\alpha) = \lim_{\alpha \to 0} \psi(\alpha) = x^*.$$

2.13 (Convergence of Incremental Gradient Method for Linear Least Squares Problems)

Consider the linear least squares problem of minimizing

$$f(x) = \frac{1}{2}\sum_{i=1}^{m} \|z_i - C_i x\|^2$$

over $x \in \Re^n$, where the vectors z_i and the matrices C_i are given. Let x_k be the vector at the start of cycle k of the incremental gradient method that operates in cycles where components are selected according to a fixed order. Thus we have

$$x_{k+1} = x_k + \alpha_k \sum_{i=1}^{m} C_i'(z_i - C_i \psi_{i-1}),$$

where $\psi_0 = x_k$ and

$$\psi_i = \psi_{i-1} + \alpha_k C_i'(z_i - C_i \psi_{i-1}), \qquad i = 1, \ldots, m.$$

Assume that $\sum_{i=1}^{m} C_i' C_i$ is a positive definite matrix and let x^* be the optimal solution. Then:

(a) There exists $\overline{\alpha} > 0$ such that if α_k is equal to some constant $\alpha \in (0, \overline{\alpha}]$ for all k, $\{x_k\}$ converges to some vector $x(\alpha)$. Furthermore, the error $\|x_k - x(\alpha)\|$ converges to 0 linearly. In addition, we have $\lim_{\alpha \to 0} x(\alpha) = x^*$. *Hint:* Show that the mapping that produces x_{k+1} starting from x_k is a contraction mapping for α sufficiently small.

(b) If $\alpha_k > 0$ for all k, and

$$\alpha_k \to 0, \qquad \sum_{k=0}^{\infty} \alpha_k = \infty,$$

then $\{x_k\}$ converges to x^*. *Hint:* Use Prop. A.4.3 of Appendix A.

Note: The ideas of this exercise are due to [Luo91]. For a complete solution, see [BeT96], Section 3.2, or [Ber99], Section 1.5.

2.14 (Linear Convergence Rate of Incremental Gradient Method [Ber99], [NeB00])

This exercise quantifies the rate of convergence of the incremental gradient method to the "region of confusion" (cf. Fig. 2.1.11), for any order of processing the additive cost components, assuming these components are positive definite quadratic. Consider the incremental gradient method

$$x_{k+1} = x_k - \alpha \nabla f_k(x_k) \qquad k = 0, 1, \ldots,$$

where $f_0, f_1, \ldots,$ are quadratic functions with eigenvalues lying within some interval $[\gamma, \Gamma]$, where $\gamma > 0$. Suppose that for a given $\epsilon > 0$, there is a vector x^* such that

$$\big\|\nabla f_k(x^*)\big\| \le \epsilon, \qquad \forall\, k = 0, 1, \ldots.$$

Show that for all α with $0 < \alpha \leq 2/(\gamma + \Gamma)$, the generated sequence $\{x_k\}$ converges to a $2\epsilon/\gamma$-neighborhood of x^*, i.e.,

$$\limsup_{k\to\infty} \|x_k - x^*\| \leq \frac{2\epsilon}{\gamma}.$$

Moreover the rate of convergence to this neighborhood is linear, in the sense that

$$\|x_k - x^*\| > \frac{2\epsilon}{\gamma} \quad \Rightarrow \quad \|x_{k+1} - x^*\| < \left(1 - \frac{\alpha\gamma}{2}\right) \|x_k - x^*\|,$$

while

$$\|x_k - x^*\| \leq \frac{2\epsilon}{\gamma} \quad \Rightarrow \quad \|x_{k+1} - x^*\| \leq \frac{2\epsilon}{\gamma}.$$

Hint: Let $f_k(x) = \frac{1}{2}x'Q_k x - b_k' x$, where Q_k is positive definite symmetric, and write

$$x_{k+1} - x^* = (I - \alpha Q_k)(x_k - x^*) - \alpha \nabla f_k(x^*).$$

For other related convergence rate results, see [NeB00] and [Sch14a].

2.15 (Proximal Gradient Method, ℓ_1-Regularization, and the Shrinkage Operation)

The proximal gradient iteration (2.27) is well suited for problems involving a nondifferentiable function component that is convenient for a proximal iteration. This exercise considers the important case of the ℓ_1 norm. Consider the problem

$$\begin{aligned} &\text{minimize} \quad f(x) + \gamma\|x\|_1 \\ &\text{subject to } x \in \Re^n, \end{aligned}$$

where $f : \Re^n \mapsto \Re$ is a differentiable convex function, $\|\cdot\|_1$ is the ℓ_1 norm, and $\gamma > 0$. The proximal gradient iteration is given by the gradient step

$$z_k = x_k - \alpha \nabla f(x_k),$$

followed by the proximal step

$$x_{k+1} \in \arg\min_{x\in\Re^n} \left\{ \gamma\|x\|_1 + \frac{1}{2\alpha}\|x - z_k\|^2 \right\};$$

[cf. Eq. (2.28)]. Show that the proximal step can be performed separately for each coordinate x^i of x, and is given by the so-called *shrinkage operation*:

$$x_{k+1}^i = \begin{cases} z_k^i - \alpha\gamma & \text{if } z_k^i > \alpha\gamma, \\ 0 & \text{if } |z_k^i| \leq \alpha\gamma, \\ z_k^i + \alpha\gamma & \text{if } z_k^i < -\alpha\gamma, \end{cases} \qquad i = 1, \ldots, n.$$

Note: Since the shrinkage operation tends to set many coordinates x_{k+1}^i to 0, it tends to produce "sparse" iterates.

2.16 (Determining Feasibility of Nonlinear Inequalities by Exponential Smoothing, [Ber82a], p. 314, [Sch82])

Consider the problem of finding a solution of a system of inequality constraints

$$g_i(x) \leq 0, \qquad i = 1, \ldots, m,$$

where $g_i : \Re^n \mapsto \Re$ are convex functions. A smoothing method based on the exponential penalty function is to minimize instead

$$f_{c,\lambda}(x) = \frac{1}{c} \ln \sum_{i=1}^{m} \lambda_i e^{cg_i(x)},$$

where $c > 0$ is some scalar, and the scalars λ_i, $i = 1, \ldots, m$, are such that

$$\lambda_i > 0, \quad i = 1, \ldots, m, \qquad \sum_{i=1}^{m} \lambda_i = 1.$$

(a) Show that if the system is feasible (or strictly feasible) the optimal value is nonpositive (or strictly negative, respectively). If the system is infeasible, then

$$\lim_{c \to \infty} \inf_{x \in \Re^n} f_{c,\lambda}(x) = \inf_{x \in \Re^n} \max \big\{ g_1(x), \ldots, g_m(x) \big\}.$$

(b) (*Computational Exercise*) Apply the incremental gradient method and the incremental Newton method for minimizing $\sum_{i=1}^{m} e^{cg_i(x)}$ [which is equivalent to minimizing $f_{c,\lambda}(x)$ with $\lambda_i \equiv 1/m$], for the case

$$g_i(x) = c_i'x - b_i, \qquad i = 1, \ldots, m,$$

where (c_i, b_i) are randomly generated as in the two problems of Exercise 2.9(b). Experiment with different starting points and stepsize choices, and deterministic and randomized orders of selection of the indexes i for iteration.

(c) Repeat part (b) where the problem is instead to minimize $f(x) = \max_{i=1,\ldots,m} g_i(x)$ and the exponential smoothing method of Section 2.2.5 is used, possibly with the augmented Lagrangian update (2.67) of λ. Compare three methods of operation: (1) c is kept constant and λ is updated, (2) c is increased to ∞ and λ is kept constant, and (3) c is increased to ∞ and λ is updated.

3

Subgradient Methods

Contents

In this chapter we discuss subgradient methods for minimizing a real-valued convex function $f : \Re^n \mapsto \Re$ over a closed convex set X. The simplest form of a subgradient method is given by

$$x_{k+1} = P_X(x_k - \alpha_k g_k),$$

where g_k is *any* subgradient of f at x_k, α_k is a positive stepsize, and $P_X(\cdot)$ denotes Euclidean projection on the set X. Note the similarity with the gradient projection iteration of Section 2.1.2: at points x_k where f is differentiable, $\nabla f(x_k)$ is the unique subgradient, and the two iterations are identical.†

We first review in Section 3.1 the theory of subgradients of real-valued convex functions, with an emphasis on properties of algorithmic significance. We also discuss the computation of a subgradient for functions arising in duality and minimax contexts. In Section 3.2 we provide a convergence analysis of the principal form of subgradient method, with a variety of stepsize rules. In Section 3.3, we discuss variants of subgradient methods involving approximations of various kinds.

3.1 SUBGRADIENTS OF CONVEX REAL-VALUED FUNCTIONS

Given a proper convex function $f : \Re^n \mapsto (-\infty, \infty]$, we say that a vector $g \in \Re^n$ is a *subgradient* of f at a point $x \in \text{dom}(f)$ if

$$f(z) \geq f(x) + g'(z - x), \qquad \forall\, z \in \Re^n; \tag{3.1}$$

see Fig. 3.1.1. The set of all subgradients of f at $x \in \Re^n$ is called the *subdifferential* of f at x, and is denoted by $\partial f(x)$. For $x \notin \text{dom}(f)$ we use the convention $\partial f(x) = \emptyset$. Figure 3.1.2 provides some examples of subdifferentials. Note that $\partial f(x)$ *is a closed convex set*, since based on Eq. (3.1), it is the intersection of a collection of closed halfspaces (one for each $z \in \Re^n$).

It is generally true that $\partial f(x)$ is nonempty for all $x \in \text{ri}\big(\text{dom}(f)\big)$, the relative interior of the domain of f, but it is possible that $\partial f(x) = \emptyset$ at some points in the relative boundary of dom(f). The properties of subgradients of extended real-valued functions are summarized in Section

† Since the gradient projection method can be viewed as the subgradient method applied to a differentiable cost function, the analysis of the present section also applies to gradient projection. However, when f is differentiable, the stepsize α_k is often chosen so that the cost function value is reduced at each iteration, thereby allowing a more powerful analysis, based on a cost function descent approach, and improved convergence and rate of convergence results. We will postpone this analysis to Section 6.1.

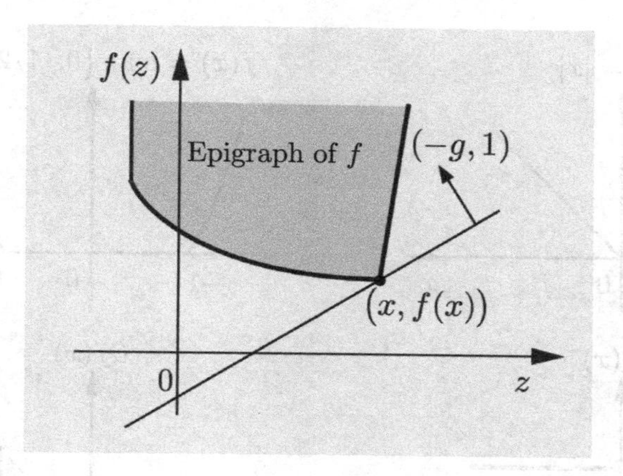

Figure 3.1.1. Illustration of the definition of a subgradient. The subgradient inequality (3.1) can be written as

$$f(z) - g'z \geq f(x) - g'x, \qquad \forall\, z \in \Re^n.$$

Thus, g is a subgradient of f at x if and only if the hyperplane in $\Re^{n+1}$ that has normal $(-g, 1)$ and passes through $\big(x, f(x)\big)$ supports the epigraph of f, as shown in the figure.

5.4 of Appendix B. When f is real-valued, however, stronger results can be shown: $\partial f(x)$ is not only closed and convex, but also nonempty and compact for all $x \in \Re^n$. Moreover the proofs of this and other related results are generally simpler than for the extended real-valued case. For this reason, we will provide an independent development of the results that we need for the case where f is real-valued (which is the primary case of interest in algorithms).

To this end, we recall the definition of the directional derivative of f at a point x in a direction d:

$$f'(x; d) = \lim_{\alpha \downarrow 0} \frac{f(x + \alpha d) - f(x)}{\alpha} \tag{3.2}$$

(cf. Section 5.4.4 of Appendix B). The ratio on the right-hand side is monotonically nonincreasing to $f'(x; d)$, as shown in Section 5.4.4 of Appendix B; also see Fig. 3.1.3.

Our first result shows some basic properties, and provides the connection between $\partial f(x)$ and $f'(x; d)$ for real-valued f. A related and more refined result is given in Prop. 5.4.8 in Appendix B for extended real-valued f. Its proof, however, is more intricate and includes some conditions that are unnecessary for the case where f is real-valued.

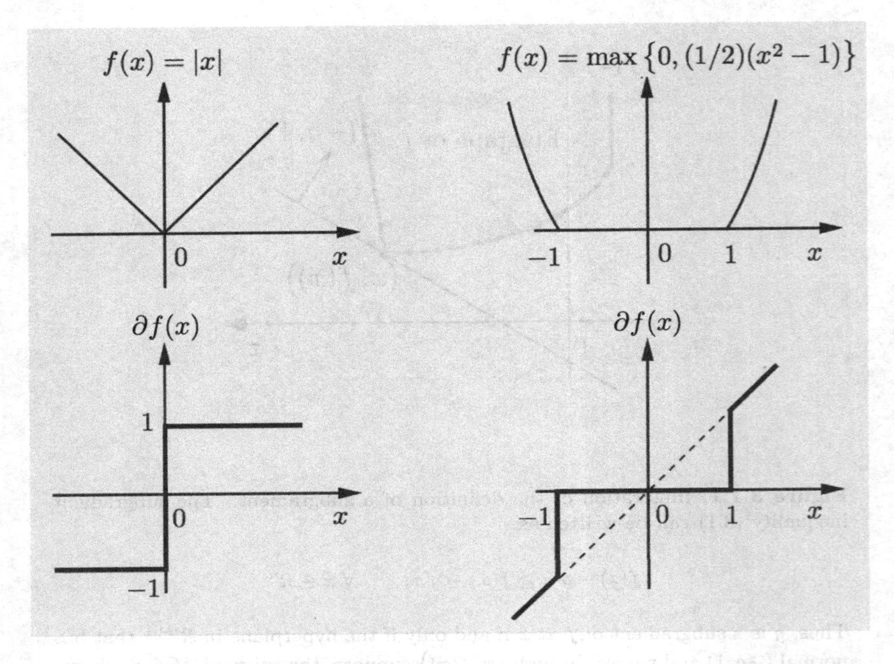

Figure 3.1.2. The subdifferentials of some real-valued scalar convex functions as a function of the argument x.

Proposition 3.1.1: (Subdifferential and Directional Derivative) Let $f : \Re^n \mapsto \Re$ be a convex function. For every $x \in \Re^n$, the following hold:

(a) The subdifferential $\partial f(x)$ is a nonempty, convex, and compact set, and we have

$$f'(x; d) = \max_{g \in \partial f(x)} g'd, \qquad \forall\, d \in \Re^n, \tag{3.3}$$

i.e., $f'(x; \cdot)$ is the support function of $\partial f(x)$. In particular, the directional derivative $f'(x; \cdot)$ is a real-valued convex function.

(b) If f is differentiable at x with gradient $\nabla f(x)$, then $\nabla f(x)$ is its unique subgradient at x, and we have $f'(x; d) = \nabla f(x)'d$.

Proof: (a) We first provide a characterization of subgradients. The subgradient inequality (3.1) is equivalent to

$$\frac{f(x + \alpha d) - f(x)}{\alpha} \geq g'd, \qquad \forall\, d \in \Re^n,\ \alpha > 0.$$

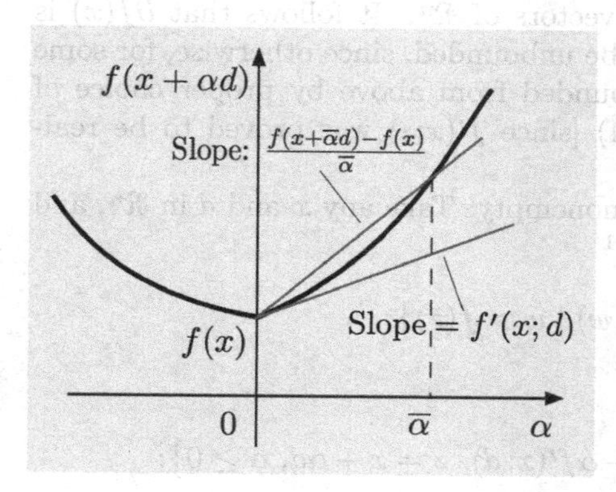

Figure 3.1.3. Illustration of the directional derivative of a convex function f. The ratio

$$\frac{f(x+\alpha d)-f(x)}{\alpha}$$

is monotonically nonincreasing and converges to $f'(x;d)$ as $\alpha \downarrow 0$.

Since the quotient on the left above decreases monotonically to $f'(x;d)$ as $\alpha \downarrow 0$, we conclude that the subgradient inequality is equivalent to $f'(x;d) \geq g'd$ for all $d \in \Re^n$:

$$g \in \partial f(x) \qquad \Longleftrightarrow \qquad f'(x;d) \geq g'd, \qquad \forall\ d \in \Re^n. \tag{3.4}$$

We next show that $f'(x;\cdot)$ is a real-valued function. For a fixed $d \in \Re^n$, consider the scalar convex function $\phi(\alpha) = f(x+\alpha d)$. From the convexity of ϕ, we have for all $\alpha > 0$ and $\beta > 0$

$$\phi(0) \leq \frac{\alpha}{\alpha+\beta}\phi(-\beta) + \frac{\beta}{\alpha+\beta}\phi(\alpha),$$

or equivalently

$$f(x) \leq \frac{\alpha}{\alpha+\beta} f(x-\beta d) + \frac{\beta}{\alpha+\beta} f(x+\alpha d).$$

This relation can be written as

$$\frac{f(x)-f(x-\beta d)}{\beta} \leq \frac{f(x+\alpha d)-f(x)}{\alpha}.$$

Noting that the quotient in the right-hand side is monotonically nonincreasing with α (cf. Fig. 3.1.3), and taking limit as $\alpha \downarrow 0$, we see that

$$\frac{f(x)-f(x-\beta d)}{\beta} \leq f'(x;d) \leq \frac{f(x+\alpha d)-f(x)}{\alpha}, \qquad \forall\ \alpha, \beta > 0.$$

Thus $f'(x;d)$ is a real number for all $x, d \in \Re^n$.

Next we show that $\partial f(x)$ is a convex and compact set. From Eq. (3.4), we see that $\partial f(x)$ is the intersection of the closed halfspaces

$$\{g \mid f'(x;d) \geq g'd\},$$

where d ranges over the nonzero vectors of $\Re^n$. It follows that $\partial f(x)$ is closed and convex. Also it cannot be unbounded, since otherwise, for some $d \in \Re^n$, $g'd$ could be made unbounded from above by proper choice of $g \in \partial f(x)$, contradicting Eq. (3.4) [since $f'(x;\cdot)$ was proved to be real-valued].

Next we show that $\partial f(x)$ is nonempty. Take any x and d in $\Re^n$, and consider the convex subset of $\Re^{n+1}$

$$C_1 = \{(z,w) \mid w > f(z)\},$$

and the half-line

$$C_2 = \{(z,w) \mid w = f(x) + \alpha f'(x;d),\ z = x + \alpha d,\ \alpha \geq 0\};$$

see Fig. 3.1.4. Since the quotient on the right in Eq. (3.2) is monotonically nonincreasing and converges to $f'(x;d)$ as $\alpha \downarrow 0$ (cf. Fig. 3.1.3), we have

$$f(x) + \alpha f'(x;d) \leq f(x + \alpha d), \qquad \forall\, \alpha \geq 0.$$

It follows that the convex sets C_1 and C_2 are disjoint. By applying the Separating Hyperplane Theorem (Prop. 1.5.2 in Appendix B), we see that there exists a nonzero vector $(\mu,\gamma) \in \Re^{n+1}$ such that

$$\gamma w + \mu' z \leq \gamma\big(f(x) + \alpha f'(x;d)\big) + \mu'(x + \alpha d), \qquad \forall\, \alpha \geq 0,\ z \in \Re^n,\ w > f(z). \tag{3.5}$$

We cannot have $\gamma > 0$ since then the left-hand side above could be made arbitrarily large by choosing w sufficiently large. Also if $\gamma = 0$, then Eq. (3.5) implies that $\mu = 0$, which is a contradiction, since (μ,γ) must be nonzero. Therefore, $\gamma < 0$ and by dividing with γ in Eq. (3.5), we obtain

$$w + (z - x)'(\mu/\gamma) \geq f(x) + \alpha f'(x;d) + \alpha(\mu/\gamma)'d,\ \forall\, \alpha \geq 0,\ z \in \Re^n,\ w > f(z). \tag{3.6}$$

By taking the limit in the above relation as $\alpha \downarrow 0$ and $w \downarrow f(z)$, we obtain

$$f(z) \geq f(x) + (-\mu/\gamma)'(z - x), \qquad \forall\, z \in \Re^n,$$

implying that $(-\mu/\gamma) \in \partial f(x)$, so $\partial f(x)$ is nonempty.

Finally, we show that Eq. (3.3) holds. We take $z = x$ and $\alpha = 1$ in Eq. (3.6), and taking the limit as $w \downarrow f(x)$, we obtain that the subgradient $(-\mu/\gamma)$ satisfies

$$(-\mu/\gamma)'d \geq f'(x;d).$$

Together with Eq. (3.4), this shows Eq. (3.3). The latter equation also implies that $f'(x;\cdot)$ is convex, being the supremum of a collection of linear functions.

(b) From the definition of directional derivative and gradient, we see that if f is differentiable at x with gradient $\nabla f(x)$, its directional derivative

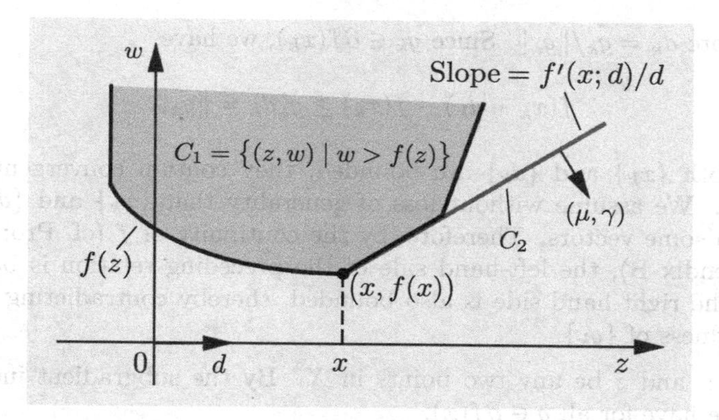

Figure 3.1.4. Illustration of the sets C_1 and C_2 used in the hyperplane separation argument of the proof of Prop. 3.1.1(a).

is $f'(x;d) = \nabla f(x)'d$. Thus, from Eq. (3.3), f has $\nabla f(x)$ as its unique subgradient at x. **Q.E.D.**

Part (b) of the preceding proposition can be used to show that *if f is convex and differentiable, it is continuously differentiable* (see Exercise 3.4). The following proposition extends the boundedness property of the subdifferential and establishes a connection with Lipschitz continuity. It is given as Prop. 5.4.2 in Appendix B, but in view of its significance for our algorithmic purposes, we restate it here and we provide a proof.

Proposition 3.1.2: (Subdifferential Boundedness and Lipschitz Continuity) Let $f : \Re^n \mapsto \Re$ be a real-valued convex function, and let X be a nonempty bounded subset of $\Re^n$.

(a) The set $\cup_{x\in X}\partial f(x)$ is nonempty and bounded.

(b) The function f is Lipschitz continuous over X,

$$\big|f(x) - f(z)\big| \leq L\,\|x - z\|, \qquad \forall\, x, z \in X,$$

where

$$L = \sup_{g\in\cup_{x\in X}\partial f(x)} \|g\|.$$

Proof: (a) Nonemptiness follows from Prop. 3.1.1(a). To prove boundedness, assume the contrary, so that there exists a sequence $\{x_k\} \subset X$, and an unbounded sequence $\{g_k\}$ with

$$g_k \in \partial f(x_k), \qquad 0 < \|g_k\| < \|g_{k+1}\|, \quad k = 0, 1, \ldots.$$

We denote $d_k = g_k/\|g_k\|$. Since $g_k \in \partial f(x_k)$, we have

$$f(x_k + d_k) - f(x_k) \geq g_k' d_k = \|g_k\|.$$

Since both $\{x_k\}$ and $\{d_k\}$ are bounded, they contain convergent subsequences. We assume without loss of generality that $\{x_k\}$ and $\{d_k\}$ converge to some vectors. Therefore, by the continuity of f (cf. Prop. 1.3.11 in Appendix B), the left-hand side of the preceding relation is bounded. Hence the right-hand side is also bounded, thereby contradicting the unboundedness of $\{g_k\}$.

(b) Let x and z be any two points in X. By the subgradient inequality (3.1), we have for all $g \in \partial f(x)$,

$$f(x) + g'(z - x) \leq f(z),$$

so that

$$f(x) - f(z) \leq \|g\| \cdot \|x - z\| \leq L\|x - z\|.$$

By exchanging the roles of x and z, we similarly obtain

$$f(z) - f(x) \leq L \|x - z\|,$$

and by combining the preceding two relations,

$$\big|f(x) - f(z)\big| \leq L \|x - z\|.$$

Q.E.D.

The next proposition provides the analog of the chain rule for subdifferentials of real-valued convex functions. The proposition is a special case of more general results that apply to extended real-valued functions (Props. 5.4.5 and 5.4.6 of Appendix B), but admits a simpler proof.

Proposition 3.1.3:

(a) *(Chain Rule)*: Let F be the composition of a convex function $h : \Re^m \mapsto \Re$ and an $m \times n$ matrix A,

$$F(x) = h(Ax), \qquad x \in \Re^n.$$

Then

$$\partial F(x) = A'\partial h(Ax) = \big\{A'g \mid g \in \partial h(Ax)\big\}, \qquad x \in \Re^n.$$

(b) *(Subdifferential of a Sum)*: Let F be the sum of convex functions $f_i : \Re^n \mapsto \Re$, $i = 1, \ldots, m$,

$$F(x) = f_1(x) + \cdots + f_m(x), \qquad x \in \Re^n.$$

Then

$$\partial F(x) = \partial f_1(x) + \cdots + \partial f_m(x), \qquad x \in \Re^n.$$

Proof: (a) Let $g \in \partial h(Ax)$. By Prop. 3.1.1(a), we have

$$g'Ad \leq h'(Ax; Ad) = F'(x; d), \qquad \forall\ d \in \Re^n,$$

where the equality holds using the definition of directional derivative. Hence

$$(A'g)'d \leq F'(x; d), \qquad \forall\ d \in \Re^n,$$

and by Prop. 3.1.1(a), $A'g \in \partial F(x)$, so that $A'\partial h(Ax) \subset \partial F(x)$.

To prove the reverse inclusion, suppose to come to a contradiction, that there exists $g \in \partial F(x)$ such that $g \notin A'\partial h(Ax)$. By Prop. 3.1.1(a), the set $\partial h(Ax)$ is compact. Since linear transformations preserve convexity and compactness, the set $A'\partial h(Ax)$ is also convex and compact, and by Prop. 1.5.3 in Appendix B, there exists a hyperplane strictly separating the singleton set $\{g\}$ from $A'\partial h(Ax)$, i.e., a vector d and a scalar c such that

$$(A'y)'d < c < g'd, \qquad \forall\ y \in \partial h(Ax).$$

From this we obtain

$$\max_{y \in \partial h(Ax)} (Ad)'y < g'd,$$

so by using the equation $h'(Ax; Ad) = F'(x; d)$ and Prop. 3.1.1(a),

$$F'(x; d) = h'(Ax; Ad) < g'd.$$

In view of Prop. 3.1.1(a) this is a contradiction.

(b) We write F as

$$F(x) = h(Ax),$$

where $h : \Re^{mn} \mapsto \Re$ is the function

$$h(x_1, \ldots, x_m) = f_1(x_1) + \cdots + f_m(x_m),$$

and A is the matrix defined by the equation $Ax = (x, \ldots, x)$. The subdifferential sum formula then follows from part (a). **Q.E.D.**

The following proposition generalizes the classical optimality condition for optimization of a differentiable convex function f over a convex set X:

$$\nabla f(x^*)'(x - x^*) \geq 0, \qquad \forall\, x \in X,$$

(cf. Prop. 1.1.8 in Appendix B). The proposition can be generalized in turn to the case where f can be extended real-valued. In this case the optimality condition requires the additional assumption that $\mathrm{ri}\big(\mathrm{dom}(f)\big) \cap \mathrm{ri}(X) \neq \emptyset$, or some polyhedral assumption on f and/or X; see Prop. 5.4.7 in Appendix B, whose proof is simple but requires a more sophisticated version of the chain rule that applies to extended real-valued functions.

Proposition 3.1.4: (Optimality Condition) A vector x minimizes a convex function $f : \Re^n \mapsto \Re$ over a convex set $X \subset \Re^n$ if and only if there exists a subgradient $g \in \partial f(x)$ such that

$$g'(z - x) \geq 0, \qquad \forall\, z \in X.$$

Proof: Suppose that $g'(z-x) \geq 0$ for some $g \in \partial f(x)$ and all $z \in X$. Then from the subgradient inequality (3.1), we have $f(z) - f(x) \geq g'(z - x)$ for all $z \in X$, so $f(z) - f(x) \geq 0$ for all $z \in X$, and x minimizes f over X.

Conversely, suppose that x minimizes f over X. Consider the set of feasible directions of X at x, i.e., the cone

$$W = \{w \neq 0 \mid x + \alpha w \in X \text{ for some } \alpha > 0\},$$

and the dual cone

$$\hat{W} = \{y \mid y'w \geq 0,\ \forall\, w \in W\}$$

(this is equal to $-W^*$, the set of all y such that $-y$ belongs to the polar cone W^*). If $\partial f(x)$ and $\hat{W}$ have a point in common, we are done, so to arrive at a contradiction, assume the opposite, i.e., $\partial f(x) \cap \hat{W} = \emptyset$. Since $\partial f(x)$ is compact [because f is real-valued and Prop. 3.1.1(a) applies] and $\hat{W}$ is closed, there exists a hyperplane strictly separating $\partial f(x)$ and $\hat{W}$ (cf. Prop. 1.5.3 in Appendix B), i.e., a vector $d \neq 0$ and a scalar c such that

$$g'd < c < y'd, \qquad \forall\, g \in \partial f(x),\ y \in \hat{W}.$$

Since $\hat{W}$ is a cone, $\inf_{y \in \hat{W}} y'd$ is either 0 or $-\infty$. The latter case is impossible, as it would violate the preceding relation. Therefore we have

$$c < 0 \leq y'd, \qquad \forall\, y \in \hat{W}, \tag{3.7}$$

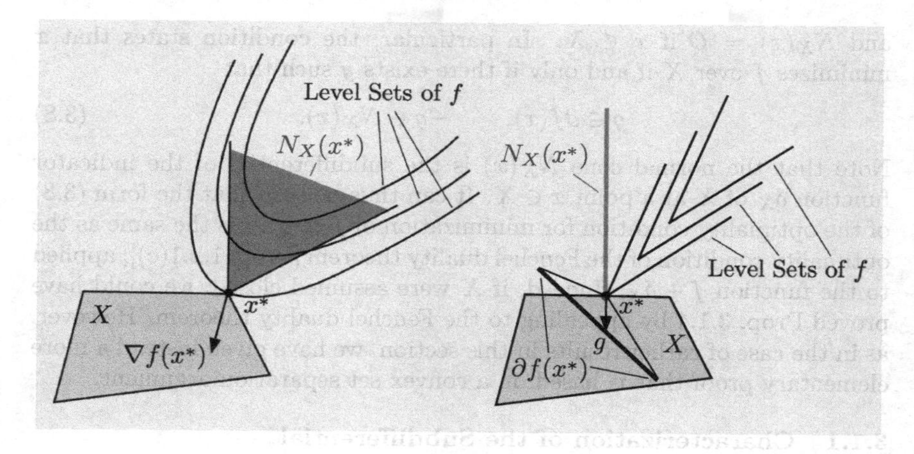

Figure 3.1.5. Illustration of the optimality condition of Prop. 3.1.4. In the figure on the left, f is differentiable and the optimality condition is

$$-\nabla f(x^*) \in N_X(x^*),$$

where $N_X(x^*)$ is the normal cone of X at x^*, which is equivalent to

$$\nabla f(x^*)'(x - x^*) \geq 0, \qquad \forall\, x \in X.$$

In the figure on the right, f is nondifferentiable, and the optimality condition is

$$-g \in N_X(x^*) \quad \text{for some } g \in \partial f(x^*).$$

which when combined with the preceding inequality, also yields

$$\max_{g \in \partial f(x)} g'd < c < 0.$$

Thus, using Prop. 3.1.1(a), we have $f'(x; d) < 0$, while from Eq. (3.7), we see that d belongs to the polar cone of W^*, which by the Polar Cone Theorem [Prop. 2.2.1(b) in Appendix B] is the closure of W. Hence there is a sequence $\{y_k\} \subset W$ that converges to d. Since $f'(x; \cdot)$ is a continuous function [being convex and real-valued by Prop. 3.1.1(a)] and $f'(x; d) < 0$, we have $f'(x; y_k) < 0$ for all k after some index, which contradicts the optimality of x. **Q.E.D.**

Figure 3.1.5 illustrates how the optimality condition of Prop. 3.1.4 is related to the *normal cone* of X at x, which is denoted by $N_X(x)$ and is defined by

$$N_X(x) = \{g \mid g'(z - x) \leq 0, \forall\, z \in X\}, \qquad x \in X,$$

and $N_X(x) = \emptyset$ if $x \notin X$. In particular, the condition states that x minimizes f over X if and only if there exists g such that

$$g \in \partial f(x), \qquad -g \in N_X(x). \tag{3.8}$$

Note that the normal cone $N_X(x)$ is the subdifferential of the indicator function δ_X of X at a point $x \in X$. It can thus be seen that the form (3.8) of the optimality condition for minimization of f over X is the same as the optimality condition of the Fenchel duality theorem [Prop. 1.2.1(c)], applied to the function $f + \delta_X$. Indeed, if X were assumed closed, we could have proved Prop. 3.1.4 by appealing to the Fenchel duality theorem. However, as in the case of earlier results in this section, we have given instead a more elementary proof that is based on a convex set separation argument.

3.1.1 Characterization of the Subdifferential

The characterization and computation of $\partial f(x)$ may not be convenient in general. It is, however, possible in some special cases. Principal among these is when

$$f(x) = \sup_{z \in Z} \phi(x, z), \tag{3.9}$$

where $x \in \Re^n$, $z \in \Re^m$, $\phi : \Re^n \times \Re^m \mapsto \Re$ is a function, Z is a compact subset of $\Re^m$, $\phi(\cdot, z)$ is convex and differentiable for each $z \in Z$, and $\nabla_x \phi(x, \cdot)$ is continuous on Z for each x. Then the form of $\partial f(x)$ is given by Danskin's Theorem [Dan67], which states that

$$\partial f(x) = \text{conv}\big\{\nabla_x \phi(x, z) \mid z \in Z(x)\big\}, \qquad x \in \Re^n, \tag{3.10}$$

where $Z(x)$ is the set of maximizing points in Eq. (3.9),

$$Z(x) = \left\{ \overline{z} \;\middle|\; \phi(x, \overline{z}) = \max_{z \in Z} \phi(x, z) \right\}.$$

The proof is somewhat long, so it is relegated to the exercises.

An important special case of Eq. (3.10) is when Z is a finite set, so f is the maximum of m differentiable convex functions $\phi_1, \ldots, \phi_m$:

$$f(x) = \max\big\{\phi_1(x), \ldots, \phi_m(x)\big\}, \qquad x \in \Re^n.$$

Then we have

$$\partial f(x) = \text{conv}\big\{\nabla \phi_i(x) \mid i \in I(x)\big\}, \tag{3.11}$$

where $I(x)$ is the set of indexes i for which the maximum is attained, i.e., $\phi_i(x) = f(x)$. Another important special case is when $\phi(\cdot, z)$ is differentiable for all $z \in Z$, and the supremum in Eq. (3.9) is attained at a unique point, so $Z(x)$ consists of a single point $z(x)$. Then f is differentiable at x and

$$\nabla f(x) = \nabla \phi\big(x, z(x)\big).$$

For various related results, see [Ber09], Examples 5.4.3, 5.4.4, 5.4.5.

Computation of Subgradients

Generally, obtaining the entire subdifferential $\partial f(x)$ at points x where f is nondifferentiable may be complicated, as indicated by Eq. (3.10): it may be difficult to determine *all* the maximizing points in Eq. (3.9). However, if we are interested in obtaining a single subgradient, as in the subgradient algorithms of the next section, the calculation is often much simpler. We illustrate this with some examples.

Example 3.1.1: (Subgradient Calculation in Minimax Problems)

Let

$$f(x) = \sup_{z \in Z} \phi(x, z), \tag{3.12}$$

where $x \in \Re^n$, $z \in \Re^m$, $\phi : \Re^n \times \Re^m \mapsto (-\infty, \infty]$ is a function, and Z is a subset of $\Re^m$. We assume that $\phi(\cdot, z)$ is convex for each $z \in Z$, so f is also convex, being the supremum of convex functions. For a fixed $x \in \text{dom}(f)$, let us assume that $z_x \in Z$ attains the supremum in Eq. (3.12), and that g_x is some subgradient of the function $\phi(\cdot, z_x)$ at x, i.e., $g_x \in \partial_x \phi(x, z_x)$. Then by using the subgradient inequality, we have for all $y \in \Re^n$,

$$f(y) = \sup_{z \in Z} \phi(y, z) \geq \phi(y, z_x) \geq \phi(x, z_x) + g_x'(y - x) = f(x) + g_x'(y - x),$$

i.e., g_x is a subgradient of f at x. Thus

$$g_x \in \partial_x \phi(x, z_x) \qquad \Rightarrow \qquad g_x \in \partial f(x).$$

This relation provides a convenient method for calculating a single subgradient of f at x with little extra computation, once a maximizer $z_x \in Z$ of $\phi(x, \cdot)$ has been found: we simply use any subgradient in $\partial_x \phi(x, z_x)$.

The next example is particularly important in the context of solving dual problems by subgradient methods. It shows that when calculating the dual function value at some point, we also obtain a subgradient with essentially no additional computation.

Example 3.1.2: (Subgradient Calculation in Dual Problems)

Consider the problem

$$\begin{aligned} &\text{minimize} \ \ f(x) \\ &\text{subject to} \ \ x \in X, \quad g(x) \leq 0, \end{aligned}$$

where $f : \Re^n \mapsto \Re$, $g : \Re^n \mapsto \Re^r$ are given functions, X is a subset of $\Re^n$. Consider the dual problem

$$\begin{aligned} &\text{maximize} \ \ \hat{q}(\mu) \\ &\text{subject to} \ \ \mu \geq 0, \end{aligned}$$

where $\hat{q}$ is the concave function

$$\hat{q}(\mu) = \inf_{x \in X} \{ f(x) + \mu' g(x) \}.$$

Thus the dual problem involves minimization of the convex function $-\hat{q}$ over $\mu \geq 0$. Note that in many cases, $\hat{q}$ is real-valued (for example when f and g are continuous, and X is compact).

For a convenient way to obtain a subgradient of $-\hat{q}$ at $\mu \in \Re^r$, suppose that x_μ minimizes the Lagrangian over $x \in X$,

$$x_\mu \in \arg\min_{x \in X} \{ f(x) + \mu' g(x) \}.$$

Then we claim that $-g(x_\mu)$ *is a subgradient of* $-\hat{q}$ *at* μ, i.e.,

$$\hat{q}(\nu) \leq \hat{q}(\mu) + (\nu - \mu)' g(x_\mu), \qquad \forall\, \nu \in \Re^r.$$

This is essentially a special case of the preceding example, and can also be verified directly by writing for all $\nu \in \Re^r$,

$$\begin{aligned} \hat{q}(\nu) &= \inf_{x \in X} \{ f(x) + \nu' g(x) \} \\ &\leq f(x_\mu) + \nu' g(x_\mu) \\ &= f(x_\mu) + \mu' g(x_\mu) + (\nu - \mu)' g(x_\mu) \\ &= \hat{q}(\mu) + (\nu - \mu)' g(x_\mu). \end{aligned}$$

Thus after computing the function value $\hat{q}(\mu)$, obtaining a single subgradient typically requires little extra calculation.

Let us finally mention another important differentiation formula that is based on an optimization operation. For any closed proper convex $f : \Re^n \mapsto (-\infty, \infty]$ and its conjugate $f^\star$, we have

$$\partial f(x) = \arg\max_{y \in \Re^n} \{ y'x - f^\star(y) \}, \qquad \forall\, x \in \Re^n,$$

$$\partial f^\star(y) = \arg\max_{x \in \Re^n} \{ x'y - f(x) \}, \qquad \forall\, y \in \Re^n.$$

This follows from the Conjugate Subgradient Theorem (see Props. 5.4.3 and 5.4.4 of Appendix B). Thus a subgradient of f at a given x can be obtained by finding a solution to a maximization problem that involves $f^\star$.

3.2 CONVERGENCE ANALYSIS OF SUBGRADIENT METHODS

In this section we consider subgradient methods for minimizing a real-valued convex function $f : \Re^n \mapsto \Re$ over a closed convex set X. In particular, we focus on methods of the form

$$x_{k+1} = P_X(x_k - \alpha_k g_k), \tag{3.13}$$

where g_k is any subgradient of f at x_k, α_k is a positive stepsize, and $P_X(\cdot)$ denotes projection on the set X (with respect to the standard Euclidean norm). An important fact here is that by projecting $(x_k - \alpha_k g_k)$ on X, we do not increase the distance to any feasible point, and hence also to any optimal solution, i.e.,

$$\big\|P_X(x_k - \alpha_k g_k) - x\big\| \le \big\|(x_k - \alpha_k g_k) - x\big\|, \qquad \forall\, x \in X.$$

This is a consequence of the following basic property of the projection.

Proposition 3.2.1: (Nonexpansiveness of the Projection) Let X be a nonempty closed convex set. We have

$$\big\|P_X(x) - P_X(y)\big\| \le \|x - y\|, \qquad \forall\, x, y \in \Re^n. \tag{3.14}$$

Proof: From the Projection Theorem (Prop. 1.1.9 in Appendix B),

$$\big(z - P_X(x)\big)'\big(x - P_X(x)\big) \le 0, \qquad \forall\, z \in X.$$

Letting $z = P_X(y)$ in this relation, we obtain

$$\big(P_X(y) - P_X(x)\big)'\big(x - P_X(x)\big) \le 0.$$

Similarly,

$$\big(P_X(x) - P_X(y)\big)'\big(y - P_X(y)\big) \le 0.$$

By adding these two inequalities, we see that

$$\big(P_X(y) - P_X(x)\big)'\big(x - P_X(x) - y + P_X(y)\big) \le 0.$$

By rearranging and by using the Schwarz inequality, we have

$$\big\|P_X(y) - P_X(x)\big\|^2 \le \big(P_X(y) - P_X(x)\big)'(y - x) \le \big\|P_X(y) - P_X(x)\big\| \cdot \|y - x\|,$$

from which the result follows. **Q.E.D.**

Another important characteristic of the subgradient method (3.13) is that the new iterate may not improve the cost for any value of the stepsize; i.e., for some k, we may have

$$f\big(P_X(x_k - \alpha g_k)\big) > f(x_k), \qquad \forall\, \alpha > 0,$$

(see Fig. 3.2.1). However, if the stepsize is small enough, the *distance of the current iterate to the optimal solution set is reduced* (this is illustrated

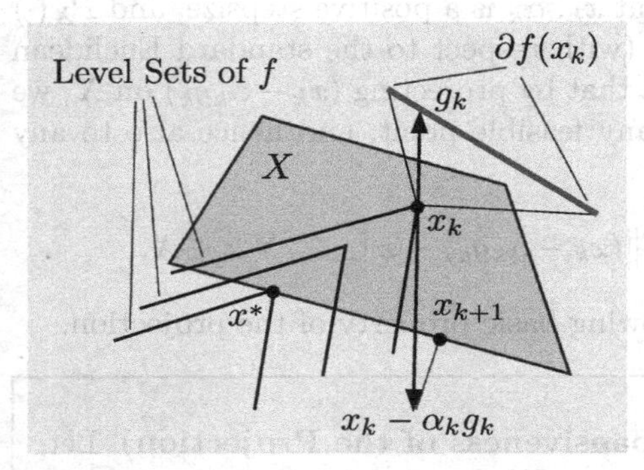

Figure 3.2.1. Illustration of how the subgradient method iterate

$$P_X(x_k - \alpha_k g_k)$$

may not improve the cost function with a particular choice of subgradient g_k, regardless of the value of the stepsize α_k.

in Fig. 3.2.2). Part (b) of the following proposition provides a formal proof of the distance reduction property and an estimate for the range of appropriate stepsizes.

Proposition 3.2.2: Let $\{x_k\}$ be the sequence generated by the subgradient method (3.13). Then, for all $y \in X$ and $k \geq 0$:

(a) We have

$$\|x_{k+1} - y\|^2 \leq \|x_k - y\|^2 - 2\alpha_k\big(f(x_k) - f(y)\big) + \alpha_k^2\|g_k\|^2.$$

(b) If $f(y) < f(x_k)$, we have

$$\|x_{k+1} - y\| < \|x_k - y\|,$$

for all stepsizes α_k such that

$$0 < \alpha_k < \frac{2\big(f(x_k) - f(y)\big)}{\|g_k\|^2}.$$

Proof: (a) Using the nonexpansion property of the projection [cf. Eq. (3.14)], we obtain for all $y \in X$ and k,

$$\begin{aligned}
\|x_{k+1} - y\|^2 &= \big\|P_X(x_k - \alpha_k g_k) - y\big\|^2 \\
&\leq \|x_k - \alpha_k g_k - y\|^2 \\
&= \|x_k - y\|^2 - 2\alpha_k g_k'(x_k - y) + \alpha_k^2\|g_k\|^2 \\
&\leq \|x_k - y\|^2 - 2\alpha_k\big(f(x_k) - f(y)\big) + \alpha_k^2\|g_k\|^2,
\end{aligned}$$

where the last step follows from the subgradient inequality.

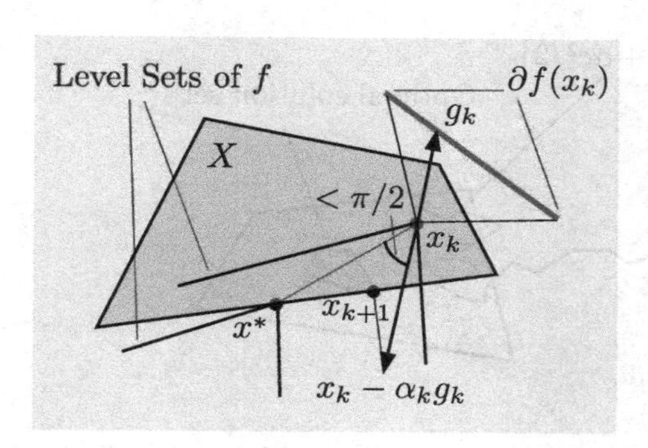

Figure 3.2.2. Illustration of how, given a nonoptimal x_k, the distance to any optimal solution x^* is reduced using a subgradient iteration with a sufficiently small stepsize. The critical fact, which follows from the definition of a subgradient, is that the angle between the negative subgradient $-g_k$ and the vector $x^* - x_k$ is less than $\pi/2$. As a result, if α_k is small enough, the vector $x_k - \alpha_k g_k$ is closer to x^* than x_k is. Through the projection on X, $P_X(x_k - \alpha_k g_k)$ gets even closer to x^*.

(b) Follows from part (a). **Q.E.D.**

Part (b) of the preceding proposition suggests the stepsize rule

$$\alpha_k = \frac{f(x_k) - f^*}{\|g_k\|^2}, \tag{3.15}$$

where f^* is the optimal value (assuming $g_k \neq 0$, otherwise x_k is optimal). This rule selects α_k to be in the middle of the range given by Prop. 3.2.2(b),

$$\left(0, \frac{2\big(f(x_k) - f(x^*)\big)}{\|g_k\|^2}\right),$$

where x^* is any optimal solution, and reduces the distance of the current iterate to x^*.

Unfortunately, however, the stepsize (3.15) requires that we know f^*, which is rare. In practice, one must either estimate f^* or use some simpler scheme for selecting a stepsize. The simplest possibility is to select α_k to be the same for all k, i.e., $\alpha_k \equiv \alpha$ for some $\alpha > 0$. Then, if the subgradients g_k are bounded, i.e., $\|g_k\| \leq c$ for some constant c and all k, Prop. 3.2.2(a) shows that for all optimal solutions x^*, we have

$$\|x_{k+1} - x^*\|^2 \leq \|x_k - x^*\|^2 - 2\alpha\big(f(x_k) - f^*\big) + \alpha^2 c^2,$$

and implies that the distance to x^* decreases if

$$0 < \alpha < \frac{2\big(f(x_k) - f^*\big)}{c^2}$$

or equivalently, if x_k is outside the level set

$$\left\{x \in X \;\middle|\; f(x) \leq f^* + \frac{\alpha c^2}{2}\right\};$$

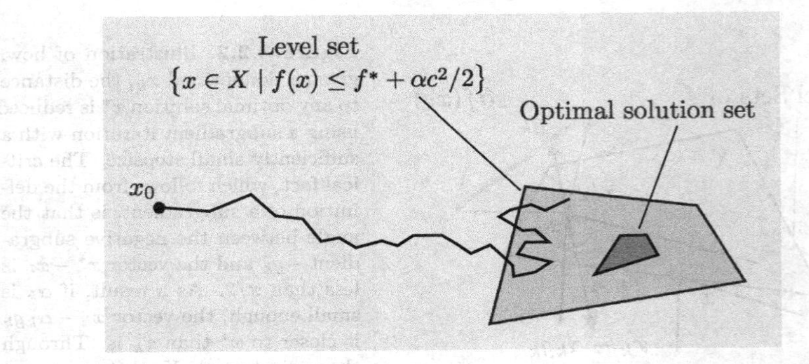

Figure 3.2.3. Illustration of a principal convergence property of the subgradient method with a constant stepsize α, assuming a bound c on the subgradient norms $\|g_k\|$. When the current iterate is outside the level set

$$\left\{ x \in X \;\middle|\; f(x) \le f^* + \frac{\alpha c^2}{2} \right\},$$

the distance to any optimal solution is reduced at the next iteration. As a result the method gets arbitrarily close to (or inside) this level set.

(see Fig. 3.2.3). Thus, if α is taken to be small enough, the convergence properties of the method are satisfactory. Since a small stepsize may result in slow initial progress, it is common to use a variant of this approach whereby we start with moderate stepsize values α_k, which are progressively reduced up to a small positive value α, using some heuristic scheme. Other possibilities for stepsize choice include a diminishing stepsize, whereby $\alpha_k \to 0$, and schemes that replace the unknown optimal value f^* in Eq. (3.15) with an estimate. We will consider these stepsize rules in what follows in the present section.

Convergence Analysis

We will now discuss the convergence of the subgradient method (3.13). Throughout our analysis, we denote by $\{x_k\}$ the corresponding generated sequence, and we denote by f^* and X^* the optimal value and optimal solution set, respectively:

$$f^* = \inf_{x \in X} f(x), \qquad X^* = \{x \in X \mid f(x) = f^*\}.$$

We will consider the subgradient method

$$x_{k+1} = P_X(x_k - \alpha_k g_k),$$

and three different types of rules for selecting the stepsize α_k:

(a) A constant stepsize.

(b) A diminishing stepsize.

(c) A dynamically chosen stepsize based on the optimal value f^* [cf. Prop. 3.2.2(b)] or a suitable estimate.

Additional stepsize rules, and related convergence and rate of convergence results can be found in [NeB00], [NeB01], and [BNO03]. For the first two stepsize rules we will assume the following:

Assumption 3.2.1: (Subgradient Boundedness) For some scalar c, we have

$$c \geq \sup\{\|g_k\| \mid k = 0, 1, \ldots\}.$$

We note that Assumption 3.2.1 is satisfied if f is polyhedral,

$$f(x) = \max_{i=1,\ldots,m} \{a_i'x + b_i\},$$

an important special case in practice. This is because the subdifferential of such a function at any point x is the convex hull of a subset of $\{a_1, \ldots, a_m\}$ [cf. Eq. (3.11)], so in this case we may use $c = \max_{i=1,\ldots,m} \|a_i\|$ in Assumption 3.2.1. Another important case where Assumption 3.2.1 is satisfied is when X is compact [see Prop. 3.1.2(a)]. More generally, Assumption 3.2.1 will hold if it can be ascertained somehow that $\{x_k\}$ is bounded.

From the point of view of analysis, the main consequence of Assumption 3.2.1 is the inequality

$$\|x_{k+1} - y\|^2 \leq \|x_k - y\|^2 - 2\alpha_k\big(f(x_k) - f(y)\big) + \alpha_k^2 c^2, \qquad \forall\, y \in X, \tag{3.16}$$

which follows from Prop. 3.2.2(a). This type of inequality allows the use of supermartingale convergence arguments (see Section A.4 in Appendix A), and lies at the heart of the convergence proofs of this section, as well as the convergence proofs of other subgradient-like methods given in the next section and Section 6.4.

Constant Stepsize

When the stepsize is constant in the subgradient method (i.e., $\alpha_k \equiv \alpha$), we cannot expect to prove convergence, in the absence of additional assumptions. As indicated in Fig. 3.2.3, we may only guarantee that asymptotically we will approach a neighborhood of the set of minima, whose size will depend on α. The following proposition quantifies the size of this neighborhood and provides an estimate on the difference between the cost value

$$f_\infty = \liminf_{k \to \infty} f(x_k)$$

that the method achieves, and the optimal cost value f^*.

Proposition 3.2.3: (Convergence within a Neighborhood) Let Assumption 3.2.1 hold, and assume that α_k is constant, $\alpha_k \equiv \alpha$.

(a) If $f^* = -\infty$, then $f_\infty = f^*$.

(b) If $f^* > -\infty$, then

$$f_\infty \le f^* + \frac{\alpha c^2}{2}.$$

Proof: We prove (a) and (b) simultaneously by contradiction. If the result does not hold, there must exist an $\epsilon > 0$ such that

$$f_\infty > f^* + \frac{\alpha c^2}{2} + 2\epsilon.$$

Let $\hat{y} \in X$ be such that

$$f_\infty \ge f(\hat{y}) + \frac{\alpha c^2}{2} + 2\epsilon,$$

and let $\bar{k}$ be large enough so that for all $k \ge \bar{k}$ we have

$$f(x_k) \ge f_\infty - \epsilon.$$

By adding the preceding two relations, we obtain for all $k \ge \bar{k}$,

$$f(x_k) - f(\hat{y}) \ge \frac{\alpha c^2}{2} + \epsilon.$$

Using Eq. (3.16) with $y = \hat{y}$, together with the above relation, we obtain for all $k \ge \bar{k}$,

$$\begin{aligned} \|x_{k+1} - \hat{y}\|^2 &\le \|x_k - \hat{y}\|^2 - 2\alpha\big(f(x_k) - f(\hat{y})\big) + \alpha^2 c^2 \\ &\le \|x_k - \hat{y}\|^2 - 2\alpha\left(\frac{\alpha c^2}{2} + \epsilon\right) + \alpha^2 c^2 \\ &= \|x_k - \hat{y}\|^2 - 2\alpha\epsilon. \end{aligned}$$

Thus we have

$$\begin{aligned} \|x_{k+1} - \hat{y}\|^2 &\le \|x_k - \hat{y}\|^2 - 2\alpha\epsilon \\ &\le \|x_{k-1} - \hat{y}\|^2 - 4\alpha\epsilon \\ &\cdots \\ &\le \|x_{\bar{k}} - \hat{y}\|^2 - 2(k + 1 - \bar{k})\alpha\epsilon, \end{aligned}$$

which cannot hold for k sufficiently large – a contradiction. **Q.E.D.**

The next proposition gives an estimate of the number of iterations needed to guarantee a level of optimality up to the threshold tolerance $\alpha c^2/2$ given in the preceding proposition. As can be expected, the number of necessary iterations depends on the distance of the initial point x_0 to the optimal solution set X^*. In the following proposition and the subsequent discussion we denote

$$d(x) = \min_{x^* \in X^*} \|x - x^*\|, \qquad x \in \Re^n.$$

Proposition 3.2.4: (Convergence Rate) Let Assumption 3.2.1 hold. Assume further that α_k is constant, $\alpha_k \equiv \alpha$, and that X^* is nonempty. Then for any positive scalar ϵ, we have

$$\min_{0 \le k \le K} f(x_k) \le f^* + \frac{\alpha c^2 + \epsilon}{2},$$

where

$$K = \left\lfloor \frac{\big(d(x_0)\big)^2}{\alpha \epsilon} \right\rfloor.$$

Proof: Assume the contrary, i.e., that for all k with $0 \le k \le K$, we have

$$f(x_k) > f^* + \frac{\alpha c^2 + \epsilon}{2}.$$

From this relation, and Eq. (3.16) with $y = x^* \in X^*$ and $\alpha_k = \alpha$, we obtain for all $x^* \in X^*$ and k with $0 \le k \le K$,

$$\begin{aligned} \|x_{k+1} - x^*\|^2 &\le \|x_k - x^*\|^2 - 2\alpha\big(f(x_k) - f^*\big) + \alpha^2 c^2 \\ &\le \|x_k - x^*\|^2 - (\alpha^2 c^2 + \alpha\epsilon) + \alpha^2 c^2 \\ &= \|x_k - x^*\|^2 - \alpha\epsilon. \end{aligned}$$

Adding the above inequalities over k for $k = 0, \ldots, K$, yields

$$0 \le \|x_{K+1} - x^*\|^2 \le \|x_0 - x^*\|^2 - (K+1)\alpha\epsilon, \qquad \forall\, x^* \in X^*.$$

Taking the minimum over $x^* \in X^*$, we obtain

$$(K+1)\alpha\epsilon \le \big(d(x_0)\big)^2,$$

which contradicts the definition of K. **Q.E.D.**

By letting $\alpha = \epsilon/c^2$, we see from the preceding proposition that we can obtain an ϵ-optimal solution in $O(1/\epsilon^2)$ iterations of the subgradient method. Equivalently, with k iterations, we can attain an optimal solution to within a $O\big(1/\sqrt{k}\big)$ cost function error. Note that the number of required iterations is independent of the dimension n of the problem.

Another interesting result is that the rate of convergence (to the appropriate neighborhood) is linear under a strong convexity type of assumption, as shown in the following proposition. Several additional convergence rate results that apply to incremental subgradient methods and other stepsize rules may be found in [NeB00].

Proposition 3.2.5: (Linear Convergence Rate) Let the assumptions of Prop. 3.2.4 hold, and assume further that for some $\gamma > 0$,

$$f(x) - f^* \geq \gamma\big(d(x)\big)^2, \qquad \forall\, x \in X, \tag{3.17}$$

and that $\alpha \leq \frac{1}{2\gamma}$. Then for all k,

$$\big(d(x_{k+1})\big)^2 \leq (1 - 2\alpha\gamma)^{k+1}\big(d(x_0)\big)^2 + \frac{\alpha c^2}{2\gamma}.$$

Proof: We let y be the projection of x_k and $\alpha_k = \alpha$ in Eq. (3.16), and strengthen the left side, and also use Eq. (3.17) to obtain

$$\begin{aligned}\big(d(x_{k+1})\big)^2 &\leq \big(d(x_k)\big)^2 - 2\alpha\big(f(x_k) - f^*\big) + \alpha^2 c^2 \\ &\leq (1 - 2\alpha\gamma)\big(d(x_k)\big)^2 + \alpha^2 c^2.\end{aligned}$$

From this we can show by induction that for all k

$$\big(d(x_{k+1})\big)^2 \leq (1 - 2\alpha\gamma)^{k+1}\big(d(x_0)\big)^2 + \alpha^2 c^2 \sum_{j=0}^{k} (1 - 2\alpha\gamma)^j,$$

and the result follows using the fact $\sum_{j=0}^{k}(1 - 2\alpha\gamma)^j \leq \frac{1}{2\alpha\gamma}$. **Q.E.D.**

The preceding proposition shows that the method converges linearly to the set of all $x \in X$ with

$$\big(d(x)\big)^2 \leq \frac{\alpha c^2}{2\gamma}.$$

The assumption (3.17) is implied by strong convexity of f, as defined in Appendix B. Moreover, it is satisfied if f is polyhedral, and X is polyhedral and compact. To see this, note that for polyhedral f and X, there exists $\beta > 0$ such that

$$f(x) - f^* \geq \beta d(x), \qquad \forall\, x \in X;$$

a proof of this is given in Prop. 5.1.6 of Chapter 5. For X compact, we have

$$\beta d(x) \geq \gamma\big(d(x)\big)^2, \qquad \forall\, x \in X,$$

for some $\gamma > 0$, and Eq. (3.17) holds.

Diminishing Stepsize

We next consider the case where the stepsize α_k diminishes to zero, but satisfies $\sum_{k=0}^{\infty} \alpha_k = \infty$. This condition is needed so that the method can "travel" infinitely far if necessary to attain convergence; otherwise, convergence to X^* may be impossible from starting points x_0 that are far from X^*, as for example in the case where $X = \Re^n$ and

$$d(x_0) > c \sum_{k=0}^{\infty} \alpha_k,$$

with c being the constant in Assumption 3.2.1. A common choice that satisfies $\alpha_k \to 0$ and $\sum_{k=0}^{\infty} \alpha_k = \infty$ is

$$\alpha_k = \frac{\beta}{k + \gamma},$$

where β and γ are some positive scalars, often determined by some preliminary experimentation.†

Proposition 3.2.6: (Convergence) Let Assumption 3.2.1 hold. If α_k satisfies

$$\lim_{k \to \infty} \alpha_k = 0, \qquad \sum_{k=0}^{\infty} \alpha_k = \infty,$$

then $f_\infty = f^*$. Moreover if

$$\sum_{k=0}^{\infty} \alpha_k^2 < \infty$$

and X^* is nonempty, then $\{x_k\}$ converges to some optimal solution.

† Larger stepsizes may also be used, provided we employ a device known as *iterate averaging*, whereby the running average $\bar{x}_k = \sum_{\ell=0}^{k} \alpha_\ell x_\ell / \sum_{\ell=0}^{k} \alpha_\ell$ of the past iterates is maintained; see Exercise 3.8.

Proof: Assume, to arrive at a contradiction, that there exists an $\epsilon > 0$ such that

$$f_\infty - 2\epsilon > f^*.$$

Then there exists a point $\hat{y} \in X$ such that

$$f_\infty - 2\epsilon > f(\hat{y}).$$

Let k_0 be large enough so that for all $k \geq k_0$, we have

$$f(x_k) \geq f_\infty - \epsilon.$$

By adding the preceding two relations, we obtain for all $k \geq k_0$,

$$f(x_k) - f(\hat{y}) > \epsilon.$$

By setting $y = \hat{y}$ in Eq. (3.16), and by using the above relation, we have for all $k \geq k_0$,

$$\|x_{k+1} - \hat{y}\|^2 \leq \|x_k - \hat{y}\|^2 - 2\alpha_k \epsilon + \alpha_k^2 c^2 = \|x_k - \hat{y}\|^2 - \alpha_k \left(2\epsilon - \alpha_k c^2\right).$$

Since $\alpha_k \to 0$, without loss of generality, we may assume that k_0 is large enough so that

$$2\epsilon - \alpha_k c^2 \geq \epsilon, \qquad \forall\ k \geq k_0.$$

Therefore for all $k \geq k_0$ we have

$$\|x_{k+1} - \hat{y}\|^2 \leq \|x_k - \hat{y}\|^2 - \alpha_k \epsilon \leq \cdots \leq \|x_{k_0} - \hat{y}\|^2 - \epsilon \sum_{j=k_0}^{k} \alpha_j,$$

which cannot hold for k sufficiently large, a contradiction showing that $f_\infty = f^*$.

Assume that X^* is nonempty. By Eq. (3.16) we have

$$\|x_{k+1} - x^*\|^2 \leq \|x_k - x^*\|^2 - 2\alpha_k\big(f(x_k) - f(x^*)\big) + \alpha_k^2 c^2, \qquad \forall\ x^* \in X^*. \tag{3.18}$$

From the convergence result of Prop. A.4.4 of Appendix A, we have that for each $x^* \in X^*$, $\|x_k - x^*\|$ converges to some real number, and hence $\{x_k\}$ is bounded. Consider a subsequence $\{x_k\}_\mathcal{K}$ such that $\lim_{k\to\infty,\, k\in\mathcal{K}} f(x_k) = f^*$, and let $\overline{x}$ be a limit point of $\{x_k\}_\mathcal{K}$. Since f is continuous, we must have $f(\overline{x}) = f^*$, so $\overline{x} \in X^*$. To prove convergence of the entire sequence to $\overline{x}$, we use $x^* = \overline{x}$ in Eq. (3.18). It then follows that $\|x_k - \overline{x}\|$ converges to a real number, which must be equal to 0 since $\overline{x}$ is a limit point of $\{x_k\}$. Thus $\overline{x}$ is the unique limit point of $\{x_k\}$.† **Q.E.D.**

---

† Note that this argument is essentially the same as the one we used to prove the Fejér Convergence Theorem (Prop. A.4.6 in Appendix A). Indeed we could have invoked that theorem for the last part of the proof.

Note that the proposition shows that $\{x_k\}$ converges to an optimal solution (assuming one exists), which is stronger than all limit points of $\{x_k\}$ being optimal. The later assertion is the type of result one typically shows for gradient methods for differentiable, possibly nonconvex problems (see e.g., nonlinear programming texts such as [Ber99], [Lue84], [NoW06]). The stronger assertion of convergence to a unique limit is possible because of the convexity of f, based on which the proof rests.

The preceding proposition can be strengthened, assuming that X^* is nonempty and α_k satisfies the slightly stronger conditions

$$\sum_{k=0}^{\infty} \alpha_k = \infty, \qquad \sum_{k=0}^{\infty} \alpha_k^2 < \infty.$$

Then convergence of $\{x_k\}$ to some optimal solution can be proved if for some scalar c, we have

$$c^2 \Big(1 + \big(d(x_k)\big)^2\Big) \geq \|g_k\|^2, \qquad \forall\, k \geq 0, \tag{3.19}$$

in place of the stronger Assumption 3.2.1 (thus covering for example the case where f is positive definite quadratic and $X = \Re^n$, which is not covered by Assumption 3.2.1). This is shown in Exercise 3.6, with essentially the same proof, after replacing Eq. (3.18) with another inequality that relies on the assumption (3.19).

Dynamic Stepsize Rules

We now discuss the stepsize rule

$$\alpha_k = \frac{f(x_k) - f^*}{\|g_k\|^2}, \qquad \forall\, k \geq 0, \tag{3.20}$$

assuming $g_k \neq 0$. This rule is motivated by Prop. 3.2.2(b) [cf. Eq. (3.15)]. Of course knowing f^* is typically unrealistic, but we will later modify the stepsize, so that f^* can be replaced by a dynamically updated estimate.

Proposition 3.2.7: (Convergence) Assume that X^* is nonempty. Then, if α_k is determined by the dynamic stepsize rule (3.20), $\{x_k\}$ converges to some optimal solution.

Proof: From Prop. 3.2.2(a) with $y = x^* \in X^*$, we have

$$\|x_{k+1} - x^*\|^2 \leq \|x_k - x^*\|^2 - 2\alpha_k\big(f(x_k) - f^*\big) + \alpha_k^2\|g_k\|^2, \quad \forall\, x^* \in X^*,\ k \geq 0.$$

By using the definition of α_k [cf. Eq. (3.20)], we obtain

$$\|x_{k+1} - x^*\|^2 \leq \|x_k - x^*\|^2 - \frac{\big(f(x_k) - f^*\big)^2}{\|g_k\|^2}, \qquad \forall\, x^* \in X^*, \quad k \geq 0.$$

This implies that $\{x_k\}$ is bounded. Furthermore, $f(x_k) \to f^*$, since otherwise we would have $\|x_{k+1} - x^*\| \leq \|x_k - x^*\| - \epsilon$ for some suitably small $\epsilon > 0$ and infinitely many k. Hence for any limit point $\overline{x}$ of $\{x_k\}$, we have $\overline{x} \in X^*$, and since the sequence $\{\|x_k - x^*\|\}$ is nonincreasing, it converges to $\|\overline{x} - x^*\|$ for every $x^* \in X^*$. If there are two distinct limit points $\tilde{x}$ and $\overline{x}$ of $\{x_k\}$, we must have $\tilde{x} \in X^*, \overline{x} \in X^*$, and $\|\tilde{x} - x^*\| = \|\overline{x} - x^*\|$ for all $x^* \in X^*$, which is possible only if $\tilde{x} = \overline{x}$. **Q.E.D.**

For most practical problems the optimal value f^* is not known. In this case we may modify the dynamic stepsize (3.20) by replacing f^* with an approximation. This leads to the stepsize rule

$$\alpha_k = \frac{f(x_k) - f_k}{\|g_k\|^2}, \qquad \forall\, k \geq 0, \tag{3.21}$$

where f_k is an estimate of f^*. One possibility is to estimate f^* by using the cost function values obtained so far, setting f_k below $\min_{0 \leq j \leq k} f(x_j)$, and adjusting f_k upwards if the algorithm appears not to be making progress. In a simple scheme proposed in [NeB01] and [BNO03], f_k is given by

$$f_k = \min_{0 \leq j \leq k} f(x_j) - \delta_k, \tag{3.22}$$

and δ_k is updated according to

$$\delta_{k+1} = \begin{cases} \theta\delta_k & \text{if } f(x_{k+1}) \leq f_k, \\ \max\{\beta\delta_k, \delta\} & \text{if } f(x_{k+1}) > f_k, \end{cases} \tag{3.23}$$

where δ, β, and θ are fixed positive constants with $\beta < 1$ and $\theta \geq 1$.

Thus in this scheme, we essentially "aspire" to reach a target level f_k that is smaller by δ_k over the best value achieved thus far [cf. Eq. (3.22)]. Whenever the target level is achieved, we increase δ_k (if $\theta > 1$) or we keep it at the same value (if $\theta = 1$). If the target level is not attained at a given iteration, δ_k is reduced up to a threshold δ. If the subgradient boundedness Assumption 3.2.1 holds, this threshold guarantees that the stepsize α_k of Eq. (3.21) is bounded away from zero, since from Eq. (3.22), we have $f(x_k) - f_k \geq \delta$ and hence $\alpha_k \geq \delta/\|g_k\|^2 \geq \delta/c^2$. As a result, the method behaves somewhat similar to the one with a constant stepsize (cf. Prop. 3.2.3), as indicated by the following proposition.

Proposition 3.2.8: (Convergence within a Neighborhood) Assume that α_k is determined by the dynamic stepsize rule (3.21) with the adjustment procedure (3.22)–(3.23). If $f^* = -\infty$, then

$$\inf_{0 \leq j} f(x_j) = f^*,$$

while if $f^* > -\infty$, then

$$\inf_{0 \leq j} f(x_j) \leq f^* + \delta.$$

Proof: Assume, to arrive at a contradiction, that

$$f^* + \delta < \inf_{0 \leq j} f(x_j). \tag{3.24}$$

Each time the target level is attained [i.e., $f(x_k) \leq f_{k-1}$], the current best function value $\min_{0 \leq j \leq k} f(x_j)$ decreases by at least δ [cf. Eqs. (3.22) and (3.23)], so in view of Eq. (3.24), the target level can be attained only a finite number of times. From Eq. (3.23) it follows that after finitely many iterations, δ_k is decreased to the threshold value δ, and remains at that value for all subsequent iterations, i.e., there is an index $\overline{k}$ such that

$$\delta_k = \delta, \qquad \forall\, k \geq \overline{k}.$$

Let us select $\overline{y} \in X$ such that $f^* \leq f(\overline{y}) \leq \inf_{0 \leq j} f(x_j) - \delta$; this is possible in view of Eq. (3.24). Then using Eq. (3.22), we have

$$f(\overline{y}) \leq \inf_{0 \leq j} f(x_j) - \delta \leq \inf_{0 \leq j \leq k} f(x_j) - \delta = f_k \leq f(x_k) - \delta, \qquad \forall\, k \geq \overline{k}. \tag{3.25}$$

Applying Prop. 3.2.2(a) with $y = \overline{y}$, together with the preceding relation, we have

$$\begin{aligned} \|x_{k+1} - \overline{y}\|^2 &\leq \|x_k - \overline{y}\|^2 - 2\alpha_k\big(f(x_k) - f(\overline{y})\big) + \alpha_k^2 \|g_k\|^2 \\ &\leq \|x_k - \overline{y}\|^2 - 2\alpha_k\big(f(x_k) - f_k\big) + \alpha_k^2 \|g_k\|^2, \qquad \forall\, k \geq \overline{k}. \end{aligned}$$

By using the definition of α_k [cf. Eq. (3.21)] and Eq. (3.25), we obtain

$$\begin{aligned} \|x_{k+1} - \overline{y}\|^2 &\leq \|x_k - \overline{y}\|^2 - 2\left(\frac{f(x_k) - f_k}{\|g_k\|}\right)^2 + \left(\frac{f(x_k) - f_k}{\|g_k\|}\right)^2 \\ &= \|x_k - \overline{y}\|^2 - \left(\frac{f(x_k) - f_k}{\|g_k\|}\right)^2 \\ &\leq \|x_k - \overline{y}\|^2 - \frac{\delta^2}{\|g_k\|^2}, \qquad \forall\, k \geq \overline{k}, \end{aligned} \tag{3.26}$$

where the last inequality follows from the right side of Eq. (3.25). Hence $\{x_k\}$ is bounded, which implies that $\{g_k\}$ is also bounded (cf. Prop. 3.1.2). Letting $\bar{c}$ be such that $\|g_k\| \leq \bar{c}$ for all k and adding Eq. (3.26) over k, we have

$$\|x_k - \overline{y}\|^2 \leq \|x_{\overline{k}} - \overline{y}\|^2 - (k - \overline{k})\frac{\delta^2}{\bar{c}^2}, \qquad \forall\, k \geq \overline{k},$$

which cannot hold for sufficiently large k – a contradiction. **Q.E.D.**

In a variation of the preceding scheme, we may use in place of the stepsize rule (3.21), one of the two rules

$$\alpha_k = \frac{f(x_k) - f_k}{\max\{\gamma,\ \|g_k\|^2\}} \quad \text{or} \quad \alpha_k = \min\left\{\gamma,\ \frac{f(x_k) - f_k}{\|g_k\|^2}\right\}, \qquad \forall\, k \geq 0,$$

where γ is a fixed positive scalar and f_k is given by the same adjustment procedure (3.22)–(3.23). This will guard against the potential practical difficulty of α_k becoming too large due to very small values of $\|g_k\|$. The result of the preceding proposition still holds with this modification (see Exercise 3.9).

We finally note that the line of convergence analysis of this section can be applied with small modifications to related methods that are based on subgradients, most notably to the ϵ-subgradient methods of the next section, and the incremental subgradient and incremental proximal methods of Section 6.4.

3.3 ϵ-SUBGRADIENT METHODS

In this section we briefly discuss subgradient-like methods that use approximate subgradients in place of subgradients. There may be several different motivations for such methods; for example, computational savings in the subgradient calculation, or exploitation of special problem structure.

Given a proper convex function $f : \Re^n \mapsto (-\infty, \infty]$ and a scalar $\epsilon > 0$, we say that a vector g is an *ϵ-subgradient* of f at a point $x \in \text{dom}(f)$ if

$$f(z) \geq f(x) + (z - x)'g - \epsilon, \qquad \forall\, z \in \Re^n. \tag{3.27}$$

The *ϵ-subdifferential* $\partial_\epsilon f(x)$ is the set of all ϵ-subgradients of f at x, and by convention, $\partial_\epsilon f(x) = \emptyset$ for $x \notin \text{dom}(f)$. It can be seen that

$$\partial_{\epsilon_1} f(x) \subset \partial_{\epsilon_2} f(x) \quad \text{if } 0 < \epsilon_1 < \epsilon_2,$$

and that

$$\cap_{\epsilon \downarrow 0} \partial_\epsilon f(x) = \partial f(x).$$

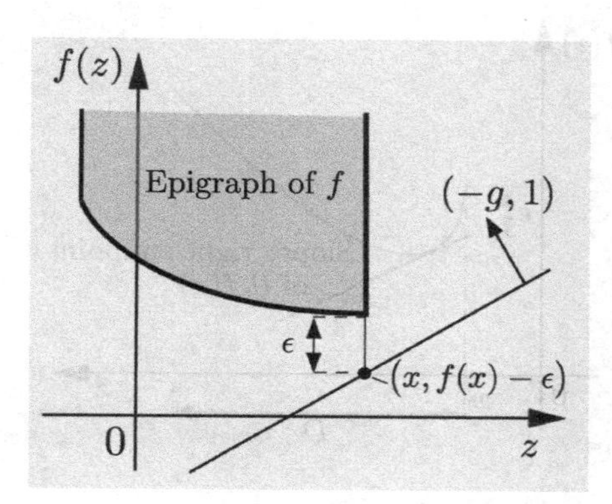

Figure 3.3.1. Illustration of an ϵ-subgradient of a convex function f. A vector g is an ϵ-subgradient at $x \in \text{dom}(f)$ if and only if there is a hyperplane with normal $(-g, 1)$, which passes through the point $\big(x, f(x) - \epsilon\big)$, and separates this point from the epigraph of f.

To interpret geometrically an ϵ-subgradient, note that the defining relation (3.27) can be written as

$$f(z) - z'g \geq \big(f(x) - \epsilon\big) - x'g, \qquad \forall\, z \in \Re^n.$$

Thus g is an ϵ-subgradient at x if and only if the epigraph of f is contained in the positive halfspace corresponding to the hyperplane in $\Re^{n+1}$ that has normal $(-g, 1)$ and passes through $\big(x, f(x) - \epsilon\big)$, as illustrated in Fig. 3.3.1.

Figure 3.3.2 illustrates the definition of the ϵ-subdifferential $\partial_\epsilon f(x)$ for the case of a one-dimensional function f. The figure indicates that if f is closed, then [in contrast with $\partial f(x)$] $\partial_\epsilon f(x)$ is nonempty at all points of $\text{dom}(f)$, including the relative boundary points of $\text{dom}(f)$. This follows by the Nonvertical Hyperplane Theorem (Prop. 1.5.8 in Appendix B). As an illustration, consider the scalar function $f(x) = |x|$. Then it is straightforward to verify that for $x \in \Re$ and $\epsilon > 0$, we have

$$\partial_\epsilon f(x) = \begin{cases} \left[-1, -1 - \frac{\epsilon}{x}\right] & \text{for } x < -\frac{\epsilon}{2}, \\ [-1, 1] & \text{for } x \in \left[-\frac{\epsilon}{2}, \frac{\epsilon}{2}\right], \\ \left[1 - \frac{\epsilon}{x}, 1\right] & \text{for } x > \frac{\epsilon}{2}. \end{cases}$$

Given the problem of minimizing a real-valued convex function $f : \Re^n \mapsto \Re$ over a closed convex set X, the *ϵ-subgradient method* is given by

$$x_{k+1} = P_X(x_k - \alpha_k g_k), \tag{3.28}$$

where g_k is an ϵ_k-subgradient of f at x_k, with ϵ_k a positive scalar, α_k is a positive stepsize, and $P_X(\cdot)$ denotes projection on X. Thus the method is the same as the subgradient method, except that ϵ-subgradients are used in place of subgradients.

The following example motivates the use of ϵ-subgradients in the context of duality and minimax problems. It shows that ϵ-subgradients may be computed more economically than subgradients, through an approximate minimization.

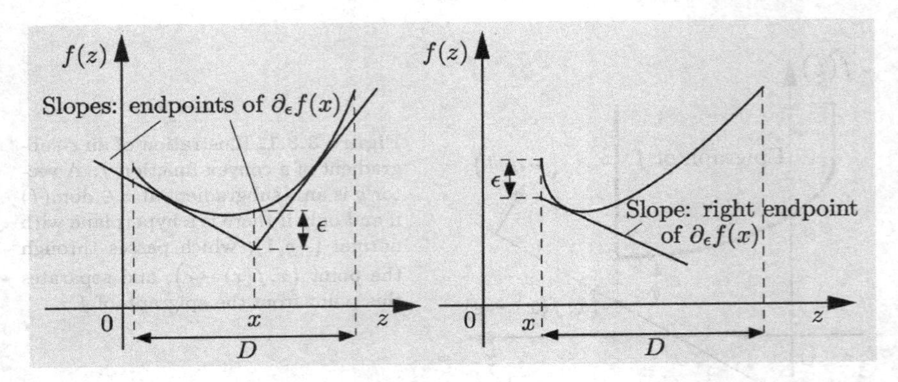

Figure 3.3.2. Illustration of the ϵ-subdifferential $\partial_\epsilon f(x)$ of a one-dimensional function $f : \Re \mapsto (-\infty, \infty]$, which is closed and convex, and has as effective domain an interval D. The ϵ-subdifferential is a nonempty interval with endpoints corresponding to the slopes indicated in the figure. At boundary points of $\text{dom}(f)$, these endpoints can be ∞ or $-\infty$ (as in the figure on the right).

Example 3.3.1: (ϵ-Subgradient Calculation in Minimax and Dual Problems)

As in Example 3.1.1, let us consider the minimization of

$$f(x) = \sup_{z \in Z} \phi(x, z), \tag{3.29}$$

where $x \in \Re^n$, $z \in \Re^m$, Z is a subset of $\Re^m$, and $\phi : \Re^n \times \Re^m \mapsto (-\infty, \infty]$ is a function such that $\phi(\cdot, z)$ is convex for each $z \in Z$. We showed in Example 3.1.1 that if we carry out exactly the maximization over z in Eq. (3.29), we can then obtain a subgradient at x. We will show with a similar argument, that if we carry out the maximization over z approximately, within ϵ, we can then obtain an ϵ-subgradient at x, which we can use in turn within an ϵ-subgradient method.

Indeed, for a fixed $x \in \text{dom}(f)$, let us assume that $z_x \in Z$ attains the supremum within $\epsilon > 0$ in Eq. (3.29), i.e.,

$$\phi(x, z_x) \geq \sup_{z \in Z} \phi(x, z) - \epsilon = f(x) - \epsilon,$$

and that g_x is some subgradient of the convex function $\phi(\cdot, z_x)$ at x, i.e., $g_x \in \partial\phi(x, z_x)$. Then, for all $y \in \Re^n$, we have using the subgradient inequality,

$$f(y) = \sup_{z \in Z} \phi(y, z) \geq \phi(y, z_x) \geq \phi(x, z_x) + g_x'(y - x) \geq f(x) - \epsilon + g_x'(y - x),$$

i.e., g_x is an ϵ-subgradient of f at x. In conclusion,

$$\phi(x, z_x) \geq \sup_{z \in Z} \phi(x, z) - \epsilon \text{ and } g_x \in \partial\phi(x, z_x) \qquad \Rightarrow \qquad g_x \in \partial_\epsilon f(x).$$

The behavior and analysis of ϵ-subgradient methods are similar to those of subgradient methods, except that *ϵ-subgradient methods generally aim to converge to the ϵ-optimal set*, where $\epsilon = \lim_{k\to\infty} \epsilon_k$, rather than the optimal set, as subgradient methods do. To get a sense of the convergence mechanism, note that there is a simple modification of the fundamental inequality of Prop. 3.2.2(a). In particular, if $\{x_k\}$ is the sequence generated by the ϵ-subgradient method, we have for all $y \in X$ and $k \geq 0$

$$\|x_{k+1} - y\|^2 \leq \|x_k - y\|^2 - 2\alpha_k\big(f(x_k) - f(y) - \epsilon_k\big) + \alpha_k^2\|g_k\|^2.$$

Using this inequality, one can essentially replicate the convergence analysis of Section 3.2, while carrying along the ϵ parameter.

As an example, consider the case where α_k and ϵ_k are constant: $\alpha_k \equiv \alpha$ for some $\alpha > 0$ and $\epsilon_k \equiv \epsilon$ for some $\epsilon > 0$. Then, if the ϵ-subgradients g_k are bounded, with $\|g_k\| \leq c$ for some constant c and all k, we obtain for all optimal solutions x^*,

$$\|x_{k+1} - x^*\|^2 \leq \|x_k - x^*\|^2 - 2\alpha\big(f(x_k) - f^* - \epsilon\big) + \alpha^2 c^2,$$

where $f^* = \inf_{x\in X} f(x)$ is the optimal value [cf. Eq. (3.16)]. This implies that the distance to all optimal x^* decreases if

$$0 < \alpha < \frac{2\big(f(x_k) - f^* - \epsilon\big)}{c^2},$$

or equivalently, if x_k is outside the level set

$$\left\{x \;\middle|\; f(x) \leq f^* + \epsilon + \frac{\alpha c^2}{2}\right\}$$

(cf. Fig. 3.2.3). With analysis similar to the one for the subgradient case, we can also show that if

$$\alpha_k \to 0, \qquad \sum_{k=0}^{\infty} \alpha_k = \infty, \qquad \epsilon_k \to \epsilon \geq 0,$$

we have

$$\liminf_{k\to\infty} f(x_k) \leq f^* + \epsilon$$

(cf. Prop. 3.2.6). There is also a related convergence result for an analog of the dynamic stepsize rule and other rules (see [NeB10]). If we have $\epsilon_k \to 0$ instead of $\epsilon_k \equiv \epsilon$, the convergence properties of the ϵ-subgradient method (3.28) are essentially the same as the ones of the ordinary subgradient method, both for a constant and for a diminishing stepsize.

3.3.1 Connection with Incremental Subgradient Methods

We discussed in Section 2.1.5 incremental variants of gradient methods, which apply to minimization over a closed convex set X of an additive cost function of the form

$$f(x) = \sum_{i=1}^{m} f_i(x),$$

where the functions $f_i : \Re^n \mapsto \Re$ are differentiable. Incremental variants of the subgradient method are also possible in the case where the f_i are nondifferentiable but convex. The idea is to sequentially take steps along the subgradients of the component functions f_i, with intermediate adjustment of x after processing each f_i. We simply use an arbitrary subgradient of f_i at a point where f_i is nondifferentiable, in place of the gradient that would be used if f_i were differentiable at that point.

Incremental methods are particularly interesting when the number of cost terms m is very large. Then a full subgradient step is very costly, and one hopes to make progress with approximate but much cheaper incremental steps. We will discuss in detail incremental subgradient methods and their combinations with other methods, such as incremental proximal methods, in Section 6.4. In this section we will discuss the most common type of incremental subgradient method, and highlight its connection with the ϵ-subgradient method.

Let us consider the minimization of $\sum_{i=1}^{m} f_i$ over $x \in X$, for the case where each f_i is a convex real-valued function. Similar to the incremental gradient methods of Section 2.1.5, we view an iteration as a cycle of m subiterations. If x_k is the vector obtained after k cycles, the vector x_{k+1} obtained after one more cycle is

$$x_{k+1} = \psi_{m,k},$$

where starting with $\psi_{0,k} = x_k$, we obtain $\psi_{m,k}$ after the m steps

$$\psi_{i,k} = P_X(\psi_{i-1,k} - \alpha_k g_{i,k}), \qquad i = 1, \ldots, m, \tag{3.30}$$

with $g_{i,k}$ being an arbitrary subgradient of f_i at $\psi_{i-1,k}$.

To see the connection with ϵ-subgradients, we first note that *if two vectors x and $\overline{x}$ are "near" each other, then subgradients at $\overline{x}$ can be viewed as ϵ-subgradients at x*, with ϵ "small." In particular, if $g \in \partial f(\overline{x})$, we have for all $z \in \Re^n$,

$$\begin{aligned} f(z) &\geq f(\overline{x}) + g'(z - \overline{x}) \\ &\geq f(x) + g'(z - x) + f(\overline{x}) - f(x) + g'(x - \overline{x}) \\ &\geq f(x) + g'(z - x) - \epsilon, \end{aligned}$$

where

$$\epsilon = \max\{0, f(x) - f(\overline{x})\} + \|g\| \cdot \|\overline{x} - x\|.$$

Thus, $g \in \partial f(\overline{x})$ implies that $g \in \partial_\epsilon f(x)$, with ϵ: small when $\overline{x}$ is near x.

We now observe from Eq. (3.30) that the ith step within a cycle of the incremental subgradient method involves the direction $g_{i,k}$, which is a subgradient of f_i at the corresponding vector $\psi_{i-1,k}$. If the stepsize α_k is small, then $\psi_{i-1,k}$ is close to the vector x_k available at the start of the cycle, and hence $g_{i,k}$ is an ϵ_i-subgradient of f_i at x_k, where ϵ_i is small. In particular, assuming for simplicity that $X = \Re^n$, we have

$$x_{k+1} = x_k - \alpha_k \sum_{i=1}^{m} g_{i,k}, \tag{3.31}$$

where $g_{i,k}$ is a subgradient of f_i at $\psi_{i-1,k}$, and hence an ϵ_i-subgradient of f_i at x_k, for an ϵ_i that is small (proportional to α_k). Thus, using also the definition of ϵ-subgradient, we have

$$\sum_{i=1}^{m} g_{i,k} \in \partial_{\epsilon_1} f_1(x_k) + \cdots + \partial_{\epsilon_m} f_m(x_k) \subset \partial_\epsilon f(x_k),$$

where $\epsilon = \epsilon_1 + \cdots + \epsilon_m$.

From this analysis it follows that the incremental subgradient iteration (3.31) can be viewed as an ϵ-subgradient iteration at x_k, the starting point of the cycle. The size of ϵ depends on the size of the stepsize α_k, as well as the function f, and we have $\epsilon \to 0$ as $\alpha_k \to 0$. As a result, when the stepsize satisfies $\alpha_k \to 0$, and $\sum_{k=0}^{\infty} \alpha_k = \infty$, the incremental subgradient method embodies a convergence mechanism similar to the one of the ordinary subgradient method, and has similar convergence properties. If α_k is kept constant, convergence to a neighborhood of the solution can be expected. These results will be established in detail later, with a somewhat different but related line of reasoning; see Section 6.4 where we will also consider methods that select the components f_i for iteration by using a randomized rather than cyclic order.

3.4 NOTES, SOURCES, AND EXERCISES

Section 3.1: Subgradients are central in the work of Fenchel [Fen51]. The original theorem by Danskin [Dan67] provides a formula for the directional derivative of the maximum of a (not necessarily convex) directionally differentiable function. When adapted to a convex function f, this formula yields Eq. (3.10) for the subdifferential of f; see Exercise 3.5. There is a more general formula, which does not require that $\phi(\cdot, z)$ is differentiable. Instead it assumes that $\phi(\cdot, z)$ is an extended real-valued closed proper convex function for each z in the compact set Z, that $\text{int}\big(\text{dom}(f)\big)$, the interior of $\text{dom}(f)$, is nonempty, and that ϕ is continuous on the set $\text{int}\big(\text{dom}(f)\big) \times Z$. Then for all $x \in \text{int}\big(\text{dom}(f)\big)$, we have

$$\partial f(x) = \text{conv}\big\{\partial \phi(x, z) \mid z \in Z_0(x)\big\},$$

where $\partial\phi(x,z)$ is the subdifferential of $\phi(\cdot,z)$ at x for any $z \in Z$; see the author's thesis [Ber71], Prop. A.22.

Another important subdifferential formula relates to the subgradients of an expected value function

$$f(x) = E\{F(x,\omega)\},$$

where ω is a random variable taking values in a set Ω, and $F(\cdot,\omega) : \Re^n \mapsto \Re$ is a real-valued convex function such that f is real-valued (note that f is easily verified to be convex). If ω takes a finite number of values with probabilities $p(\omega)$, then the formulas

$$f'(x;d) = E\{F'(x,\omega;d)\}, \qquad \partial f(x) = E\{\partial F(x,\omega)\}, \tag{3.32}$$

hold because they can be written in terms of finite sums as

$$f'(x;d) = \sum_{\omega\in\Omega} p(\omega)F'(x,\omega;d), \qquad \partial f(x) = \sum_{\omega\in\Omega} p(\omega)\partial F(x,\omega),$$

so Prop. 3.1.3(b) applies. However, the formulas (3.32) hold even in the case where Ω is uncountably infinite, with appropriate mathematical interpretation of the integral of set-valued functions $E\{\partial F(x,\omega)\}$ as the set of integrals

$$\int_{\omega\in\Omega} g(x,\omega)\,dP(\omega), \tag{3.33}$$

where $g(x,\omega) \in \partial F(x,\omega)$, $\omega \in \Omega$ (measurability issues must be addressed in this context). For a formal proof and analysis, see the author's papers [Ber72], [Ber73], which also provide a necessary and sufficient condition for f to be differentiable, even when $F(\cdot,\omega)$ is not. In this connection, it is important to note that the integration over ω in Eq. (3.33) may smooth out the nondifferentiabilities of $F(\cdot,\omega)$ if ω is a "continuous" random variable. This property can be used in turn in algorithms, including schemes that bring to bear the methodology of differentiable optimization; see e.g., Yousefian, Nedić, and Shanbhag [YNS10], [YNS12], Agarwal and Duchi [AgD11], Duchi, Bartlett, and Wainwright [DBW12], Brown and Smith [BrS13], Abernethy et al. [ALS14], and Jiang and Zhang [JiZ14].

Section 3.2: Subgradient methods were first introduced in the middle 60s by Shor; the works of Ermoliev and Poljak were also particularly influential. Description of these works can be found in many sources, including the books by Ermoliev [Erm76], Shor [Sho85], and Poljak [Pol87]. An extensive bibliography for the early period of the subject is given in the edited volume by Balinski and Wolfe [BaW75]. Some of the first papers in the Western literature on nondifferentiable optimization appeared in this volume. There are many works dealing with analysis of subgradient methods.

There are also several variations of subgradient methods that aim to accelerate the convergence of the basic method (see e.g., [CFM75], [Sho85], [Min86], [Str97], [LPS98], [Sho98], [ZLW99], [BLY14]).

The line of analysis given here is based on the joint work of the author with A. Nedić [NeB00], [NeB01], and has been used in several subsequent works, [NBB01], [NeO09a], [NeO09b], [NeB10], [Ber11], [Ned11], [WaB13a]. The book by Bertsekas, Nedić, and Ozdaglar [BNO03] contains a more extensive convergence analysis, which relates to a greater variety of stepsize rules, including dynamic rules for nonincremental and incremental subgradient methods.

Section 3.3: Methods using ϵ-subgradients (cf. Section 3.3) have been investigated by several authors, including Robinson [Rob99], Auslender and Teboulle [AuT04], and Nedić and Bertsekas [NeB10]. Subgradient methods are often implemented in approximate form, with errors in the calculation of the subgradient or the cost function value. For analysis related to such implementations, see Nedić and Bertsekas [NeB10], and Hu, Yang, and Sim [HYS15]. For some nondifferentiable problems involving sharp minima, exact convergence may be obtained despite persistent errors in the calculation of the subgradient (see Exercise 3.10 and [NeB10]). Moreover, additional algorithms based on the ϵ-subdifferential, called *ϵ-descent methods*, will be discussed in Section 6.7.

Problems of minimization of additive cost functions $f(x)=\sum_{i=1}^{m} f_i(x)$ (cf. Section 3.3.1) arise in many applications, as noted in Section 1.3. Incremental subgradient methods for such problems will be discussed in Section 6.4, and detailed references will be given in Chapter 6. Extensions of incremental subgradient methods, called *incremental constraint projection methods*, will also be considered for constraint sets of the form $X=\cap_{i=1}^{m} X_i$, where the component sets X_i are more suitable for the projection operation than X itself, and the constraint projections are done incrementally (see Section 6.4.4).

EXERCISES

3.1 (Computational Exercise)

Consider the unconstrained minimization of

$$f(x)=\sum_{i=1}^{m}\max\{0,c_i'x-b_i\},$$

where c_i are given vectors in $\Re^n$ and b_i are given scalars.

(a) Verify that a subgradient method has the form

$$x_{k+1} = x_k - \alpha_k \sum_{i=1}^{m} g_{i,k},$$

where

$$g_{i,k} = \begin{cases} c_i & \text{if } c_i' x_k > b_i, \\ 0 & \text{otherwise.} \end{cases}$$

(b) Consider an incremental subgradient method that operates in cycles as follows. At the end of cycle k, we set $x_{k+1} = \psi_{m,k}$, where $\psi_{m,k}$ is obtained after the m steps

$$\psi_{i,k} = \begin{cases} \psi_{i-1,k} - \alpha_k c_i & \text{if } c_i' \psi_{i-1,k} > b_i, \\ \psi_{i-1,k} & \text{otherwise,} \end{cases} \qquad i = 1, \ldots, m,$$

starting with $\psi_{0,k} = x_k$. Compare this method with the algorithm of (a) computationally with two examples where $n = 2$ and $m = 100$. In the first example, the vectors c_i have the form $c_i = (\xi_i, \zeta_i)$, where ξ_i, ζ_i, as well as b_i, are chosen randomly and independently from $[-100, 100]$ according to a uniform distribution. In the second example, the vectors c_i have the form $c_i = (\xi_i, \zeta_i)$, where ξ_i, ζ_i are chosen randomly and independently within $[-10, 10]$ according to a uniform distribution, while b_i is chosen randomly and independently within $[0, 1000]$ according to a uniform distribution. Experiment with different starting points and stepsize choices, and deterministic and randomized orders of selection of the indexes i for iteration. In the case of the second example, under what circumstances does the method stop after a finite number of iterations?

3.2 (Optimality Condition with Directional Derivatives)

The purpose of this exercise is to express the necessary and sufficient condition for optimality of Prop. 3.1.4 in terms of the directional derivative of the cost function. Consider the minimization of a convex function $f : \Re^n \mapsto \Re$ over a convex set $X \subset \Re^n$. For any $x \in X$, the set of feasible directions of f at x is defined to be the convex cone

$$D(x) = \big\{\alpha(\bar{x} - x) \mid \bar{x} \in X, \alpha > 0\big\}.$$

Show that a vector x minimizes f over X if and only if $x \in X$ and

$$f'(x; d) \geq 0, \qquad \forall\, d \in D(x). \tag{3.34}$$

Note: In words, this condition says that x is optimal if and only if there is no feasible descent direction of f at x. *Solution*: Let $\overline{D(x)}$ denote the closure of $D(x)$. By Prop. 3.1.4, x minimizes f over X if and only if there exists $g \in \partial f(x)$ such that

$$g'd \geq 0, \quad \forall\, d \in D(x),$$

which is equivalent to

$$g'd \geq 0, \quad \forall\, d \in \overline{D(x)}.$$

Thus, x minimizes f over X if and only if

$$\max_{g \in \partial f(x)} \min_{\|d\| \leq 1,\, d \in \overline{D(x)}} g'd \geq 0.$$

Since the minimization and maximization above are over convex and compact sets, by the Saddle Point Theorem of Prop. 5.5.3 in Appendix B, this is equivalent to

$$\min_{\|d\| \leq 1,\, d \in \overline{D(x)}} \max_{g \in \partial f(x)} g'd \geq 0,$$

or by Prop. 3.1.1(a),

$$\min_{\|d\| \leq 1,\, d \in \overline{D(x)}} f'(x; d) \geq 0.$$

This is in turn equivalent to the desired condition (3.34), since $f'(x; \cdot)$ is continuous being convex and real-valued.

3.3 (Subdifferential of an Extended Real-Valued Function)

Extended real-valued convex functions arising in algorithmic practice are often of the form

$$f(x) = \begin{cases} h(x) & \text{if } x \in X, \\ \infty & \text{if } x \notin X, \end{cases} \tag{3.35}$$

where $h : \Re^n \mapsto \Re$ is a real-valued convex function and X is a nonempty convex set. The purpose of this exercise is to show that the subdifferential of such functions admits a more favorable characterization compared to the case where h is extended real-valued.

(a) Use Props. 3.1.3 and 3.1.4 to show that the subdifferential of such a function is nonempty for all $x \in X$, and has the form

$$\partial f(x) = \partial h(x) + N_X(x), \qquad \forall\, x \in X,$$

where $N_X(x)$ is the normal cone of X at $x \in X$. *Note*: If h is convex but extended-real valued, this formula requires the assumption $\text{ri}\big(\text{dom}(h)\big) \cap \text{ri}(X) \neq \emptyset$ or some polyhedral conditions on h and X; see Prop. 5.4.6 of Appendix B. *Proof*: By the subgradient inequality (3.1), we have $g \in \partial f(x)$ if and only if x minimizes $p(z) = h(z) - g'z$ over $z \in X$, or equivalently, some subgradient of p at x [i.e., a vector in $\partial h(x) - \{g\}$, by Prop. 3.1.3] belongs to $-N_X(x)$ (cf. Prop. 3.1.4).

(b) Let $f(x) = -\sqrt{x}$ if $x \geq 0$ and $f(x) = \infty$ if $x < 0$. Verify that f is a closed convex function that cannot be written in the form (3.35) and does not have a subgradient at $x = 0$.

(c) Show the following formula for the subdifferential of the sum of functions f_i that have the form (3.35) for some h_i and X_i:

$$\partial(f_1 + \cdots + f_m)(x) = \partial h_1(x) + \cdots + \partial h_m(x) + N_{X_1 \cap \cdots \cap X_m}(x),$$

for all $x \in X_1 \cap \cdots \cap X_m$. Demonstrate by example that in this formula we cannot replace $N_{X_1 \cap \cdots \cap X_m}(x)$ by $N_{X_1}(x) + \cdots + N_{X_m}(x)$. *Proof*: Write $f_1 + \cdots + f_m = h + \delta_X$, where $h = h_1 + \cdots + h_m$ and $X = X_1 \cap \cdots \cap X_m$. For a counterexample, let $m = 2$, and X_1 and X_2 be unit spheres in the plane with centers at $(-1, 0)$ and $(1, 0)$, respectively.

3.4 (Continuity of Gradient and Directional Derivative)

The following exercise provides a basic continuity property of directional derivatives and gradients of convex functions. Let $f : \Re^n \mapsto \Re$ be a convex function, and let $\{f_k\}$ be a sequence of convex functions $f_k : \Re^n \mapsto \Re$ with the property that $\lim_{k\to\infty} f_k(x_k) = f(x)$ for every $x \in \Re^n$ and every sequence $\{x_k\}$ that converges to x. Show that for any $x \in \Re^n$ and $y \in \Re^n$, and any sequences $\{x_k\}$ and $\{y_k\}$ converging to x and y, respectively, we have

$$\limsup_{k\to\infty} f_k'(x_k; y_k) \leq f'(x; y).$$

Furthermore, if f is differentiable over $\Re^n$, then it is continuously differentiable over $\Re^n$. *Solution*: From the definition of directional derivative, it follows that for any $\epsilon > 0$, there exists an $\alpha > 0$ such that

$$\frac{f(x + \alpha y) - f(x)}{\alpha} < f'(x; y) + \epsilon.$$

Hence, using also the equation

$$f'(x; y) = \inf_{\alpha > 0} \frac{f(x + \alpha y) - f(x)}{\alpha},$$

we have for all sufficiently large k,

$$f_k'(x_k; y_k) \leq \frac{f_k(x_k + \alpha y_k) - f_k(x_k)}{\alpha} < f'(x; y) + \epsilon,$$

so by taking the limit as $k \to \infty$,

$$\limsup_{k\to\infty} f_k'(x_k; y_k) \leq f'(x; y) + \epsilon.$$

Since this is true for all $\epsilon > 0$, we obtain $\limsup_{k\to\infty} f_k'(x_k; y_k) \leq f'(x; y)$.

If f is differentiable at all $x \in \Re^n$, then by Prop. 3.1.1(b), we have $f'(x; y) = \nabla f(x)'y$ for all $x, y \in \Re^n$, and by using the part of the proposition just proved, it follows that for every sequence $\{x_k\}$ converging to x and every $y \in \Re^n$,

$$\limsup_{k\to\infty} \nabla f(x_k)'y = \limsup_{k\to\infty} f'(x_k; y) \leq f'(x; y) = \nabla f(x)'y.$$

By replacing y with $-y$ in the preceding argument, we obtain

$$-\liminf_{k\to\infty} \nabla f(x_k)'y = \limsup_{k\to\infty} \big(-\nabla f(x_k)'y\big) \leq -\nabla f(x)'y.$$

Combining the preceding two relations, we have $\nabla f(x_k)'y \to \nabla f(x)'y$ for every y, which implies that $\nabla f(x_k) \to \nabla f(x)$. Hence, $\nabla f(\cdot)$ is continuous.

3.5 (Danskin's Theorem)

Let Z be a compact subset of $\Re^m$, and let $\phi : \Re^n \times Z \mapsto \Re$ be continuous and such that $\phi(\cdot, z) : \Re^n \mapsto \Re$ is convex for each $z \in Z$.

(a) Show that the function $f : \Re^n \mapsto \Re$ given by

$$f(x) = \max_{z \in Z} \phi(x, z) \tag{3.36}$$

is convex and has directional derivative given by

$$f'(x; y) = \max_{z \in Z(x)} \phi'(x, z; y), \tag{3.37}$$

where $\phi'(x, z; y)$ is the directional derivative of the function $\phi(\cdot, z)$ at x in the direction y, and $Z(x)$ is the set of maximizing points in Eq. (3.36)

$$Z(x) = \left\{ \overline{z} \;\middle|\; \phi(x, \overline{z}) = \max_{z \in Z} \phi(x, z) \right\}.$$

Furthermore, the maximum in Eq. (3.37) is attained. In particular, if $Z(x)$ consists of a unique point $\overline{z}$ and $\phi(\cdot, \overline{z})$ is differentiable at x, then f is differentiable at x, and $\nabla f(x) = \nabla_x \phi(x, \overline{z})$, where $\nabla_x \phi(x, \overline{z})$ is the vector with components

$$\frac{\partial \phi(x, \overline{z})}{\partial x_i}, \qquad i = 1, \ldots, n.$$

(b) Show that if $\phi(\cdot, z)$ is differentiable for all $z \in Z$ and $\nabla_x \phi(x, \cdot)$ is continuous on Z for each x, then

$$\partial f(x) = \text{conv}\big\{ \nabla_x \phi(x, z) \mid z \in Z(x) \big\}, \qquad \forall\, x \in \Re^n.$$

Note: A more general version of this part, which does not assume differentiability of $\phi(\cdot, z)$ is given in in the author's thesis, [Ber71], Prop. A.22.

Solution: (a) We note that since ϕ is continuous and Z is compact, the set $Z(x)$ is nonempty by Weierstrass' Theorem and f is real-valued. For any $z \in Z(x)$, $y \in \Re^n$, and $\alpha > 0$, we use the definition of f to obtain

$$\frac{f(x + \alpha y) - f(x)}{\alpha} \geq \frac{\phi(x + \alpha y, z) - \phi(x, z)}{\alpha}.$$

Taking the limit as α decreases to zero, we obtain $f'(x; y) \geq \phi'(x, z; y)$. Since this is true for every $z \in Z(x)$, we conclude that

$$f'(x; y) \geq \sup_{z \in Z(x)} \phi'(x, z; y), \qquad \forall\, y \in \Re^n. \tag{3.38}$$

We will next prove the reverse inequality and that the supremum in the right-hand side of the above inequality is attained. To this end, we fix x, we consider a sequence $\{\alpha_k\}$ of positive scalars that converges to zero, and we let $x_k = x + \alpha_k y$. For each k, let z_k be a vector in $Z(x_k)$. Since $\{z_k\}$ belongs to the compact set Z, it has a subsequence converging to some $\overline{z} \in Z$. Without loss of generality, we assume that the entire sequence $\{z_k\}$ converges to $\overline{z}$. We have

$$\phi(x_k, z_k) \geq \phi(x_k, z), \qquad \forall\, z \in Z,$$

so by taking the limit as $k \to \infty$ and by using the continuity of ϕ, we obtain

$$\phi(x, \overline{z}) \geq \phi(x, z), \qquad \forall\, z \in Z.$$

Therefore, $\overline{z} \in Z(x)$. We now have

$$\begin{aligned} f'(x;y) &\leq \frac{f(x+\alpha_k y) - f(x)}{\alpha_k} \\ &= \frac{\phi(x+\alpha_k y, z_k) - \phi(x, \overline{z})}{\alpha_k} \\ &\leq \frac{\phi(x+\alpha_k y, z_k) - \phi(x, z_k)}{\alpha_k} \\ &\leq -\phi'(x+\alpha_k y, z_k; -y) \\ &\leq \phi'(x+\alpha_k y, z_k; y), \end{aligned} \tag{3.39}$$

where the last inequality follows from the fact $-f'(x;-y) \leq f'(x;y)$. We apply the result of Exercise 3.4 to the functions f_k defined by $f_k(\cdot) = \phi(\cdot, z_k)$, and with $x_k = x + \alpha_k y$, to obtain

$$\limsup_{k\to\infty} \phi'(x+\alpha_k y, z_k; y) \leq \phi'(x, \overline{z}; y). \tag{3.40}$$

We take the limit in inequality (3.39) as $k \to \infty$, and we use inequality (3.40) to conclude that

$$f'(x;y) \leq \phi'(x, \overline{z}; y).$$

This relation together with inequality (3.38) proves Eq. (3.37).

For the last statement of part (a), if $Z(x)$ consists of the unique point $\overline{z}$, the differentiability assumption on ϕ and Eq. (3.37) yield

$$f'(x;y) = \phi'(x, \overline{z}; y) = y'\nabla_x \phi(x, \overline{z}), \qquad \forall\, y \in \Re^n,$$

which implies that $\nabla f(x) = \nabla_x \phi(x, \overline{z})$.

(b) By part (a), we have

$$f'(x;y) = \max_{z \in Z(x)} \nabla_x \phi(x,z)'y,$$

while by Prop. 3.1.1(a),

$$f'(x;y) = \max_{z \in \partial f(x)} d'y.$$

For all $\overline{z} \in Z(x)$ and $y \in \Re^n$, we have

$$\begin{aligned} f(y) &= \max_{z \in Z} \phi(y,z) \\ &\geq \phi(y, \overline{z}) \\ &\geq \phi(x, \overline{z}) + \nabla_x \phi(x, \overline{z})'(y-x) \\ &= f(x) + \nabla_x \phi(x, \overline{z})'(y-x). \end{aligned}$$

Therefore, $\nabla_x \phi(x, \overline{z})$ is a subgradient of f at x, implying that

$$\text{conv}\{\nabla_x \phi(x, z) \mid z \in Z(x)\} \subset \partial f(x).$$

To prove the reverse inclusion, we use a hyperplane separation argument. By the continuity of $\nabla_x \phi(x, \cdot)$ and the compactness of Z, we see that $Z(x)$ is compact, and therefore also the set $\{\nabla_x \phi(x, z) \mid z \in Z(x)\}$ is compact. By Prop. 1.2.2 in Appendix B, it follows that $\text{conv}\{\nabla_x \phi(x, z) \mid z \in Z(x)\}$ is compact. If $d \in \partial f(x)$ while $d \notin \text{conv}\{\nabla_x \phi(x, z) \mid z \in Z(x)\}$, by the Strict Separation Theorem (Prop. 1.5.3 in Appendix B), there exists $y \neq 0$, and $\gamma \in \Re$, such that

$$d'y > \gamma > \nabla_x \phi(x, z)'y, \qquad \forall\, z \in Z(x).$$

Therefore, we have

$$d'y > \max_{z \in Z(x)} \nabla_x \phi(x, z)'y = f'(x; y),$$

contradicting Prop. 3.1.1(a). Therefore,

$$\partial f(x) \subset \text{conv}\{\nabla_x \phi(x, z) \mid z \in Z(x)\}$$

and the proof is complete.

3.6 (Convergence of Subgradient Method with Diminishing Stepsize Under Weaker Conditions)

This exercise shows an enhanced version of Prop. 3.2.6, whereby we assume that for some scalar c, we have

$$c^2 \left(1 + \min_{x^* \in X^*} \|x_k - x^*\|^2\right) \geq \|g_k\|^2, \qquad \forall\, k, \tag{3.41}$$

in place of the stronger Assumption 3.2.1. Assume also that X^* is nonempty and that

$$\sum_{k=0}^{\infty} \alpha_k = \infty, \qquad \sum_{k=0}^{\infty} \alpha_k^2 < \infty. \tag{3.42}$$

Show that $\{x_k\}$ converges to some optimal solution. *Abbreviated proof*: Similar to the proof of Prop. 3.2.6 [cf. Eq. (3.18)], we apply Prop. 3.2.2(a) with y equal to any $x^* \in X^*$, and then use the assumption (3.41) to obtain

$$\|x_{k+1} - x^*\|^2 \leq (1 + \alpha_k^2 c^2)\|x_k - x^*\|^2 - 2\alpha_k\big(f(x_k) - f^*\big) + \alpha_k^2 c^2. \tag{3.43}$$

In view of the assumption (3.42), the convergence result of Prop. A.4.4 of Appendix A applies, and shows that $\{x_k\}$ is bounded and that $\liminf_{k \to \infty} f(x_k) = f^*$. From this point the proof follows the one of Prop. 3.2.6.

3.7 (Convergence Rate of Subgradient Method with Dynamic Stepsize)

Consider the subgradient method $x_{k+1} = P_X(x_k - \alpha_k g_k)$ with the dynamic stepsize rule

$$\alpha_k = \frac{f(x_k) - f^*}{\|g_k\|^2}, \tag{3.44}$$

and assume that the optimal solution set X^* is nonempty. Show that:

(a) $\{x_k\}$ and $\{g_k\}$ are bounded sequences. *Proof*: Let x^* be an optimal solution. From Prop. 3.2.2(a), we have

$$\|x_{k+1} - x^*\|^2 \le \|x_k - x^*\|^2 - 2\alpha_k\big(f(x_k) - f^*\big) + \alpha_k^2\|g_k\|^2.$$

Using the stepsize form (3.44) in this relation, we obtain

$$\|x_{k+1} - x^*\|^2 \le \|x_k - x^*\|^2 - \frac{\big(f(x_k) - f^*\big)^2}{\|g_k\|^2}. \tag{3.45}$$

Therefore

$$\|x_{k+1} - x^*\| \le \|x_k - x^*\|, \quad \forall\, k,$$

implying that $\{x_k\}$ is bounded, and by Prop. 3.1.2, that $\{g_k\}$ is bounded.

(b) (*Sublinear Convergence*) We have

$$\liminf_{k\to\infty} \sqrt{k}\,\big(f(x_k) - f^*\big) = 0.$$

Proof: Assume to obtain a contradiction that there is an $\epsilon > 0$ and large enough $\bar{k}$ such that $\sqrt{k}\,\big(f(x_k) - f^*\big) \ge \epsilon$ for all $k \ge \bar{k}$. Then

$$\big(f(x_k) - f^*\big)^2 \ge \frac{\epsilon^2}{k}, \quad \forall\, k \ge \bar{k},$$

implying that

$$\sum_{k=\bar{k}}^{\infty}\big(f(x_k) - f^*\big)^2 \ge \epsilon^2 \sum_{k=\bar{k}}^{\infty}\frac{1}{k} = \infty.$$

On the other hand, by adding Eq. (3.45) over all k, and using the boundedness of $\{g_k\}$, shown in part (a), we have

$$\sum_{k=0}^{\infty}\big(f(x_k) - f^*\big)^2 < \infty,$$

a contradiction.

(c) (*Linear Convergence*) Assume that there exists a scalar $\beta > 0$ such that

$$f^* + \beta\, d(x) \le f(x), \qquad \forall\, x \in X, \tag{3.46}$$

where we denote $d(x) = \min_{x^* \in X^*} \|x - x^*\|$; this assumption, known as a *sharp minimum* condition, is satisfied in particular if f and X are polyhedral (see Prop. 5.1.6); problems where this condition holds have especially favorable properties in several convex optimization algorithmic contexts (see Exercise 3.10 and the exercises of Chapter 6). Show that for all k,

$$d(x_{k+1}) \leq \rho\, d(x_k),$$

where $\rho = \sqrt{1 - \beta^2/\gamma^2}$ and γ is any upper bound to $\|g_k\|$ with $\gamma > \beta$ [cf. part (a)]. *Proof*: From Eqs. (3.45), (3.46), we have for all k

$$\big(d(x_{k+1})\big)^2 \leq \big(d(x_k)\big)^2 - \frac{\big(f(x_k) - f^*\big)^2}{\|g_k\|^2} \leq \big(d(x_k)\big)^2 - \frac{\beta^2\big(d(x_k)\big)^2}{\|g_k\|^2}.$$

Using the fact $\sup_{k \geq 0} \|g_k\| \leq \gamma$, the desired relation follows.

3.8 (Subgradient Methods with Iterate Averaging [NJL09])

If the stepsize α_k in the subgradient method

$$x_{k+1} = P_X(x_k - \alpha_k g_k)$$

is chosen to be large (such as constant or such that the condition $\sum_{k=0}^{\infty} \alpha_k^2 < \infty$ is violated) the method may not converge. This exercise shows that by averaging the iterates of the method, we may obtain convergence with larger stepsizes. Let the optimal solution set X^* be nonempty, and assume that for some scalar c, we have

$$c \geq \sup\big\{\|g_k\| \mid k = 0, 1, \ldots\big\}, \qquad \forall\, k \geq 0,$$

(cf. Assumption 3.2.1). Assume further that α_k is chosen according to

$$\alpha_k = \frac{\theta}{c\sqrt{k+1}}, \qquad k = 0, 1, \ldots,$$

where θ is a positive constant. Show that

$$f(\bar{x}_k) - f^* \leq c\left(\frac{\min_{x^* \in X^*} \|x_0 - x^*\|^2}{2\theta} + \theta \ln(k+2)\right)\frac{1}{\sqrt{k+1}}, \qquad k = 0, 1, \ldots,$$

where $\bar{x}_k$ is the averaged iterate, generated according to

$$\bar{x}_k = \frac{\sum_{\ell=0}^{k} \alpha_\ell x_\ell}{\sum_{\ell=0}^{k} \alpha_\ell}.$$

Note: The averaging approach seems to be less sensitive to the choice of stepsize parameters. Practical variants include restarting the method with the most recent averaged iterate, and averaging over just a subset of recent iterates. A similar analysis applies to incremental and to stochastic subgradient methods. *Abbreviated proof*: Denote

$$\delta_k = \tfrac{1}{2} \min_{x^* \in X^*} \|x_k - x^*\|^2.$$

Applying Prop. 3.2.2(a) with y equal to the projection of x_k onto X^*, we obtain

$$\delta_{k+1} \le \delta_k - \alpha_k \big(f(x_k) - f^*\big) + \tfrac{1}{2}\alpha_k^2 c^2.$$

Adding this inequality from 0 to k, and using the fact $\delta_{k+1} \ge 0$,

$$\sum_{\ell=0}^{k} \alpha_\ell \big(f(x_k) - f^*\big) \le \delta_0 + \tfrac{1}{2}c^2 \textstyle\sum_{\ell=0}^{k} \alpha_\ell^2,$$

so by dividing with $\sum_{\ell=0}^{k} \alpha_\ell$,

$$\frac{\sum_{\ell=0}^{k} \alpha_\ell f(x_k)}{\sum_{\ell=0}^{k} \alpha_\ell} - f^* \le \frac{\delta_0 + \frac{1}{2}c^2 \sum_{\ell=0}^{k} \alpha_\ell^2}{\sum_{\ell=0}^{k} \alpha_\ell}.$$

The convexity of f implies that

$$f(\bar{x}_k) \le \frac{\sum_{\ell=0}^{k} \alpha_\ell f(x_k)}{\sum_{\ell=0}^{k} \alpha_\ell}.$$

Combining the preceding two relations, we obtain

$$f(\bar{x}_k) - f^* \le \frac{\delta_0 + \frac{1}{2}c^2 \sum_{\ell=0}^{k} \alpha_\ell^2}{\sum_{\ell=0}^{k} \alpha_\ell}.$$

Substituting the stepsize expression $\alpha_\ell = \theta / \big(c\sqrt{\ell + 1}\big)$, we have

$$f(\bar{x}_k) - f^* \le \frac{\delta_0 + \frac{1}{2}\theta^2 \sum_{\ell=0}^{k} \frac{1}{\ell+1}}{\frac{\theta}{c} \sum_{\ell=0}^{k} \frac{1}{\sqrt{\ell+1}}} \le \frac{\delta_0 + \theta^2 \ln(k+2)}{\frac{\theta}{c}\sqrt{k+1}},$$

which implies the result.

3.9 (Modified Dynamic Stepsize Rules)

Consider the subgradient method

$$x_{k+1} = P_X(x_k - \alpha_k g_k)$$

with the stepsize chosen according to one of the two rules

$$\alpha_k = \frac{f(x_k) - f_k}{\max\big\{\gamma, \|g_k\|^2\big\}} \quad \text{or} \quad \alpha_k = \min\left\{\gamma, \frac{f(x_k) - f_k}{\|g_k\|^2}\right\}, \qquad \forall\, k \ge 0, \quad (3.47)$$

where γ is a fixed positive scalar and f_k is given by the dynamic adjustment procedure (3.22)–(3.23). Show that the convergence result of Prop. 3.2.8 still holds. *Abbreviated Proof*: We proceed by contradiction, as in the proof of Prop. 3.2.8. From Prop. 3.2.2(a) with $y = \overline{y}$, we have for all $k \geq \overline{k}$,

$$\begin{aligned}\|x_{k+1} - \overline{y}\|^2 &\leq \|x_k - \overline{y}\|^2 - 2\alpha_k\big(f(x_k) - f(\overline{y})\big) + \alpha_k^2\|g_k\|^2 \\ &\leq \|x_k - \overline{y}\|^2 - 2\alpha_k\big(f(x_k) - f(\overline{y})\big) + \alpha_k\big(f(x_k) - f_k\big) \\ &= \|x_k - \overline{y}\|^2 - \alpha_k\big(f(x_k) - f_k\big) - 2\alpha_k\big(f_k - f(\overline{y})\big) \\ &\leq \|x_k - \overline{y}\|^2 - \alpha_k\big(f(x_k) - f_k\big).\end{aligned}$$

Hence $\{x_k\}$ is bounded, which implies that $\{g_k\}$ is also bounded (cf. Prop. 3.1.2). Let $\bar{c}$ be such that $\|g_k\| \leq \bar{c}$ for all k. Assume that α_k is chosen according to the first rule in Eq. (3.47). Then from the preceding relation we have for all $k \geq \overline{k}$,

$$\|x_{k+1} - \overline{y}\|^2 \leq \|x_k - \overline{y}\|^2 - \frac{\delta^2}{\max\{\gamma, \bar{c}^2\}}.$$

As in the proof of Prop. 3.2.8, this leads to a contradiction and the result follows. The proof is similar if α_k is chosen according to the second rule in Eq. (3.47).

3.10 (Subgradient Methods with Low Level Errors for Sharp Minima [NeB10])

Consider the problem of minimizing a convex function $f : \Re^n \to \Re$ over a closed convex set X, and assume that the optimal solution set X^* is nonempty. The purpose of this exercise is to show that under certain conditions, which are satisfied if f and X are polyhedral, the subgradient method is convergent even with "small" errors in the calculation in the subgradient. Assume that for some $\beta > 0$, we have

$$f^* + \beta d(x) \leq f(x), \qquad \forall\, x \in X, \tag{3.48}$$

where $f^* = \min_{x\in X} f(x)$ and $d(x) = \min_{x^*\in X^*} \|x - x^*\|$ [a sharp minimum condition; cf. Exercise 3.7(c)]. Consider the iteration

$$x_{k+1} = P_X\big(x_k - \alpha_k(g_k + e_k)\big),$$

where for all k, g_k is a subgradient of f at x_k, and e_k is an error such that

$$\|e_k\| \leq \epsilon, \qquad k = 0, 1, \ldots,$$

where ϵ is some positive scalar with $\epsilon < \beta$. Assume further that for some $c > 0$, we have

$$c \geq \sup\big\{\|g_k\| \mid k = 0, 1, \ldots\big\}, \qquad \forall\, k \geq 0,$$

cf. the subgradient boundedness Assumption 3.2.1.

(a) Show that if α_k is equal to some constant α for all k, then

$$\liminf_{k\to\infty} f(x_k) \leq f^* + \frac{\alpha\beta(c+\epsilon)^2}{2(\beta - \epsilon)}, \tag{3.49}$$

while if

$$\alpha_k \to 0, \qquad \sum_{k=0}^{\infty} \alpha_k = \infty,$$

then $\liminf_{k\to\infty} f(x_k) = f^*$. *Hint*: Show that

$$d(x_{k+1})^2 \le d(x_k)^2 - 2\alpha_k\big(f(x_k) - f^*\big) + 2\alpha_k\|e_k\|d(x_k) + \alpha_k^2\|\tilde{g}_k\|^2,$$

and hence

$$d(x_{k+1})^2 \le d(x_k)^2 - 2\alpha_k\frac{\beta - \|e_k\|}{\beta}\big(f(x_k) - f^*\big) + 2\alpha_k\|e_k\|d(x_k) + \alpha_k^2\|\tilde{g}_k\|^2.$$

(b) Use the scalar function $f(x) = |x|$ to show that the estimate (3.49) is tight.

3.11 (ϵ-Complementary Slackness and ϵ-Subgradients)

The purpose of this exercise (based on unpublished joint work with P. Tseng) is to show how to calculate ϵ-subgradients of the dual function of the separable problem

$$\text{minimize} \quad \sum_{i=1}^{n} f_i(x_i)$$

$$\text{subject to} \sum_{i=1}^{n} g_{ij}(x_i) \le 0, \quad j = 1,\ldots,r, \qquad \alpha_i \le x_i \le \beta_i, \quad i = 1,\ldots,n,$$

where $f_i : \Re \mapsto \Re$, $g_{ij} : \Re \mapsto \Re$ are convex functions. For an $\epsilon > 0$, we say that a pair $(\overline{x}, \overline{\mu})$ satisfies *ϵ-complementary slackness* if $\overline{\mu} \ge 0$, $\overline{x}_i \in [\alpha_i, \beta_i]$ for all i, and

$$0 \le f_i^+(\overline{x}_i) + \sum_{j=1}^{r} \overline{\mu}_j g_{ij}^+(\overline{x}_i) + \epsilon, \ \forall\, i \in I^-, \quad f_i^-(\overline{x}_i) + \sum_{j=1}^{r} \overline{\mu}_j g_{ij}^-(\overline{x}_i) - \epsilon \le 0, \ \forall\, i \in I^+,$$

where $I^- = \{i \mid x_i < \beta_i\}$, $I^+ = \{i \mid \alpha_i < x_i\}$, f_i^-, g_{ij}^- and f_i^+, g_{ij}^+ denote the left and right derivatives of f_i, g_{ij}, respectively. Show that if $(\overline{x}, \overline{\mu})$ satisfies ϵ-complementary slackness, the r-dimensional vector with jth component $\sum_{i=1}^{n} g_{ij}(\overline{x}_i)$ is an $\overline{\epsilon}$-subgradient of the dual function q at $\overline{\mu}$, where

$$\overline{\epsilon} = \epsilon \sum_{i=1}^{n} (\beta_i - \alpha_i).$$

Note: The notion of ϵ-complementary slackness, sometimes also referred to as *ϵ-optimality*, is important, among others, in network optimization algorithms, dating to the auction algorithm of [Ber79], and the related ϵ-relaxation and preflow push methods; see the books [BeT89a], [Ber98], and the references given there.

4

Polyhedral Approximation Methods

Contents

In this chapter, we discuss polyhedral approximation methods for minimizing a real-valued convex function f over a closed convex set X. Here we generate a sequence $\{x_k\}$ by solving at each k the approximate problem

$$x_{k+1} \in \arg\min_{x \in X_k} F_k(x),$$

where F_k is a polyhedral function that approximates f and X_k is a polyhedral set that approximates X (in some variants only one of f or X is approximated). The idea is that the approximate problem, thanks to its polyhedral structure, may be easier to solve than the original. The methods include mechanisms for progressively refining the approximation, thereby obtaining a solution of the original problem in the limit.

We first discuss in Sections 4.1 and 4.2 the two main approaches for polyhedral approximation: *outer linearization* (also called the *cutting plane* approach) and *inner linearization* (also called the *simplicial decomposition* approach). In Section 4.3, we show how these two approaches are intimately connected by conjugacy and duality. In Section 4.4, we generalize our framework for polyhedral approximation when the cost function is a sum of two or more convex (possibly nondifferentiable) component functions. Duality plays a central role here: each generalized polyhedral approximation algorithm has a dual where the roles of outer and inner linearization are exchanged.

Our generalized polyhedral approximation approach of Section 4.4 not only connects outer and inner linearization, but also gives rise to a diverse class of methods that can exploit a broad variety of special structures. There are two characteristics that are important in this respect:

(a) Multiple component functions can be linearized individually, based on the solution of the approximate problem. This speeds up the approximation process, and can exploit better the special structure of the cost function components.

(b) Outer and inner linearization may be simultaneously applied to different component functions. This allows additional flexibility to exploit special features of the problem at hand.

In Section 4.5, we consider various special cases of the framework of Section 4.4, some involving large-scale network flow problems, while in Section 4.6, we develop algorithmic variants that are useful when there are conic constraints. In these sections we focus on inner linearization, although there are similar (dual) algorithms based on outer linearization.

4.1 OUTER LINEARIZATION – CUTTING PLANE METHODS

Cutting plane methods are rooted in the representation of a closed convex set as the intersection of its supporting halfspaces, cf. Prop. 1.5.4 in

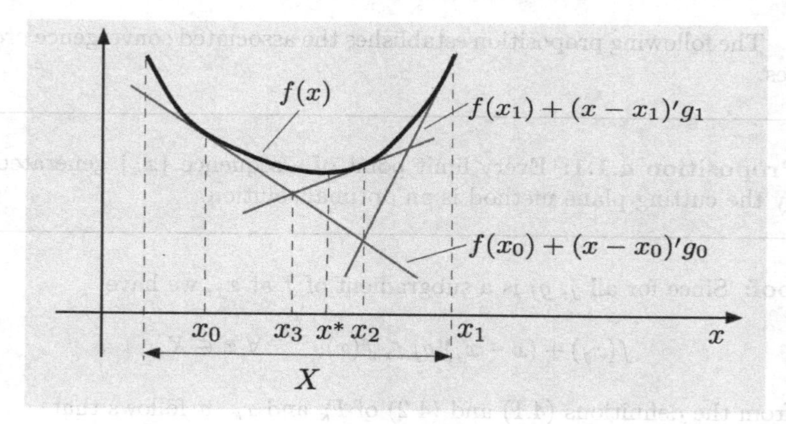

Figure 4.1.1. Illustration of the cutting plane method. With each new iterate x_k, a new hyperplane $f(x_k) + (x - x_k)'g_k$ is added to the polyhedral approximation of the cost function, where g_k is a subgradient of f at x_k.

Appendix B. The idea is to approximate either the constraint set or the epigraph of the cost function by the intersection of a limited number of halfspaces, and to gradually refine the approximation by generating additional halfspaces through the use of subgradients.

Throughout this section, we consider the problem of minimizing a convex function $f : \Re^n \mapsto \Re$ over a closed convex set X. In the simplest cutting plane method, we start with a point $x_0 \in X$ and a subgradient $g_0 \in \partial f(x_0)$. At the typical iteration we solve the approximate problem

$$\begin{aligned} &\text{minimize} \quad F_k(x) \\ &\text{subject to } x \in X, \end{aligned}$$

where f is replaced by a polyhedral approximation F_k, constructed using the points $x_0, \ldots, x_k$ generated so far, and associated subgradients $g_0, \ldots, g_k$, with $g_i \in \partial f(x_i)$ for all $i \leq k$. In particular, for $k = 0, 1, \ldots$, we define

$$F_k(x) = \max\big\{f(x_0) + (x - x_0)'g_0, \ldots, f(x_k) + (x - x_k)'g_k\big\}, \tag{4.1}$$

and compute x_{k+1} that minimizes $F_k(x)$ over $x \in X$,

$$x_{k+1} \in \arg\min_{x \in X} F_k(x); \tag{4.2}$$

see Fig. 4.1.1. We assume that the minimum of $F_k(x)$ above is attained for all k. For those k for which this is not guaranteed (as may happen in the early iterations if X is unbounded), artificial bounds may be placed on the components of x, so that the minimization will be carried out over a compact set and consequently the minimum will be attained by Weierstrass' Theorem.

The following proposition establishes the associated convergence properties.

Proposition 4.1.1: Every limit point of a sequence $\{x_k\}$ generated by the cutting plane method is an optimal solution.

Proof: Since for all j, g_j is a subgradient of f at x_j, we have

$$f(x_j) + (x - x_j)'g_j \le f(x), \qquad \forall\, x \in X,$$

so from the definitions (4.1) and (4.2) of F_k and x_k, it follows that

$$f(x_j) + (x_k - x_j)'g_j \le F_{k-1}(x_k) \le F_{k-1}(x) \le f(x), \qquad \forall\, x \in X,\ j < k. \tag{4.3}$$

Suppose that a subsequence $\{x_k\}_\mathcal{K}$ converges to $\overline{x}$. Then, since X is closed, we have $\overline{x} \in X$, and by using Eq. (4.3), we obtain for all k and all $j < k$,

$$f(x_j) + (x_k - x_j)'g_j \le F_{k-1}(x_k) \le F_{k-1}(\overline{x}) \le f(\overline{x}).$$

By taking the upper limit above as $j \to \infty$, $k \to \infty$, $j < k$, $j \in \mathcal{K}$, $k \in \mathcal{K}$, we obtain

$$\limsup_{\substack{j\to\infty,\, k\to\infty,\, j<k \\ j\in\mathcal{K},\, k\in\mathcal{K}}} \{f(x_j) + (x_k - x_j)'g_j\} \le \limsup_{k\to\infty,\, k\in\mathcal{K}} F_{k-1}(x_k) \le f(\overline{x}). \tag{4.4}$$

Since the subsequence $\{x_k\}_\mathcal{K}$ is bounded and the union of the subdifferentials of a real-valued convex function over a bounded set is bounded (cf. Prop. 3.1.2), it follows that the subgradient subsequence $\{g_j\}_\mathcal{K}$ is bounded. Moreover, we have

$$\lim_{\substack{j\to\infty,\, k\to\infty,\, j<k \\ j\in\mathcal{K},\, k\in\mathcal{K}}} (x_k - x_j) = 0,$$

so that

$$\lim_{\substack{j\to\infty,\, k\to\infty,\, j<k \\ j\in\mathcal{K},\, k\in\mathcal{K}}} (x_k - x_j)'g_j = 0. \tag{4.5}$$

Also by the continuity of f, we have

$$\lim_{j\to\infty,\, j\in\mathcal{K}} f(x_j) = f(\overline{x}). \tag{4.6}$$

Combining Eqs. (4.4)-(4.6), we obtain

$$\limsup_{k\to\infty,\, k\in\mathcal{K}} F_{k-1}(x_k) = f(\overline{x}).$$

This relation together with Eq. (4.3) yields

$$f(\overline{x}) \leq f(x), \qquad \forall\, x \in X,$$

showing that $\overline{x}$ is an optimal solution. **Q.E.D.**

In practice, it is common to use the inequalities

$$F_{k-1}(x_k) \leq f^* \leq \min_{j \leq k} f(x_j), \qquad k = 0, 1, \ldots,$$

to bound the optimal value f^* of the problem. In such a scheme, the iterations are stopped when the upper and lower bound difference

$$\min_{j \leq k} f(x_j) - F_{k-1}(x_k)$$

comes within some small tolerance.

An important special case arises when f is polyhedral of the form

$$f(x) = \max_{i \in I} \{a_i' x + b_i\}, \tag{4.7}$$

where I is a finite index set, and a_i and b_i are given vectors and scalars, respectively. Then, any vector a_{i_k} that maximizes $a_i' x_k + b_i$ over $\{a_i \mid i \in I\}$ is a subgradient of f at x_k (cf. Example 3.1.1). We assume that the cutting plane method selects such a vector at iteration k, call it a_{i_k}. We also assume that the method terminates when

$$F_{k-1}(x_k) = f(x_k).$$

Then, since $F_{k-1}(x) \leq f(x)$ for all $x \in X$ and x_k minimizes F_{k-1} over X, we see that, upon termination, x_k minimizes f over X and is therefore optimal. The following proposition shows that the method converges finitely; see also Fig. 4.1.2.

Proposition 4.1.2: Assume that the cost function f is polyhedral of the form (4.7). Then the cutting plane method, with the subgradient selection and termination rules just described, obtains an optimal solution in a finite number of iterations.

Proof: If (a_{i_k}, b_{i_k}) is equal to some pair (a_{i_j}, b_{i_j}) generated at some earlier iteration $j < k$, then

$$f(x_k) = a_{i_k}' x_k + b_{i_k} = a_{i_j}' x_k + b_{i_j} \leq F_{k-1}(x_k) \leq f(x_k),$$

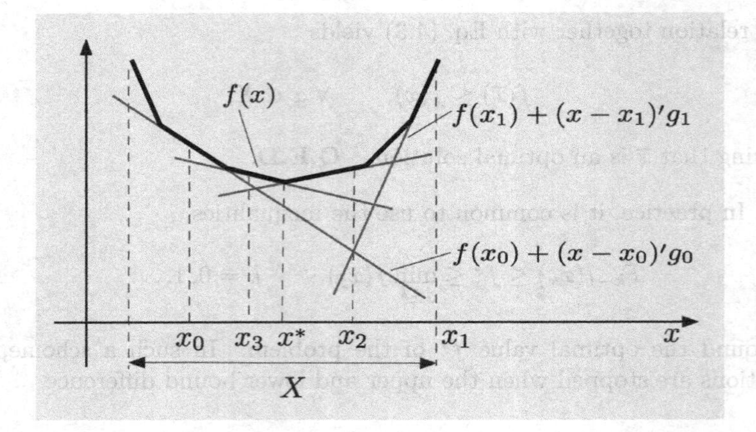

Figure 4.1.2. Illustration of the finite convergence property of the cutting plane method in the case where f is polyhedral. What happens here is that if x_k is not optimal, a new cutting plane will be added at the corresponding iteration, and there can be only a finite number of cutting planes.

where the first inequality follows since $a'_{i_j} x_k + b_{i_j}$ corresponds to one of the hyperplanes defining F_{k-1}, and the last inequality follows from the fact $F_{k-1}(x) \leq f(x)$ for all $x \in X$. Hence equality holds throughout in the preceding relation, and it follows that the method terminates if the pair (a_{i_k}, b_{i_k}) has been generated at some earlier iteration. Since the number of pairs (a_i, b_i), $i \in I$, is finite, the method must terminate finitely. **Q.E.D.**

Despite the finite convergence property shown in Prop. 4.1.2, the cutting plane method has several drawbacks:

(a) It can take large steps away from the optimum, resulting in large cost increases, even when it is close to (or even at) the optimum. For example, in Fig. 4.1.2, $f(x_1)$ is much larger than $f(x_0)$. This phenomenon is referred to as *instability*, and has another undesirable effect, namely that the current point x_k may not be a good starting point for the algorithm that minimizes the new approximate cost function $F_k(x)$ over X.

(b) The number of subgradients used in the cutting plane approximation F_k increases without bound as $k \to \infty$ leading to a potentially large and difficult optimization problem to find x_k. To remedy this, one may occasionally discard some of the cutting planes. To guarantee convergence, it is essential to do so only at times when improvement in the cost is recorded, e.g., $f(x_k) \leq \min_{j<k} f(x_j) - \delta$ for some small positive δ. Still one has to be judicious about discarding cutting planes, as some of them may reappear later.

(c) The convergence is often slow. Indeed, for challenging problems, even when f is polyhedral, one should base termination on the upper and lower bounds

$$F_{k-1}(x_k) \le \min_{x\in X} f(x) \le \min_{0\le j\le k} f(x_j),$$

rather than wait for finite termination to occur.

To overcome some of the limitations of the cutting plane method, a number of variants have been proposed, some of which are discussed in the present section. In Chapter 5, we will discuss additional methods, including *bundle methods*, which are aimed at limiting the effects of instability though combinations with the proximal algorithm.

Partial Cutting Plane Methods

In some cases the cost function has the form

$$f(x) + c(x),$$

where $f : X \mapsto \Re$ and $c : X \mapsto \Re$ are convex functions, but one of them, say c, is convenient for optimization, e.g., is quadratic. It may then be preferable to use a piecewise linear approximation of f only, while leaving c unchanged. This leads to a partial cutting plane algorithm, involving solution of the problems

$$\begin{aligned} &\text{minimize} \quad F_k(x) + c(x) \\ &\text{subject to} \quad x \in X, \end{aligned}$$

where as before

$$F_k(x) = \max\big\{f(x_0) + (x - x_0)'g_0, \ldots, f(x_k) + (x - x_k)'g_k\big\},$$

with $g_j \in \partial f(x_j)$ for all j, and x_{k+1} minimizes $F_k(x)$ over $x \in X$,

$$x_{k+1} \in \arg\min_{x\in X}\big\{F_k(x) + c(x)\big\}.$$

The convergence properties of this algorithm are similar to the ones shown earlier. In particular, if f is polyhedral, the method terminates finitely, cf. Prop. 4.1.2. The idea of partial piecewise approximation can be generalized to the case of more than two cost function components and arises also in a few other contexts to be discussed later in Sections 4.4-4.6.

Linearly Constrained Versions

Consider the case where the constraint set X is polyhedral of the form

$$X = \{x \mid c_i'x + d_i \leq 0,\, i \in I\},$$

where I is a finite set, and c_i and d_i are given vectors and scalars, respectively. Let

$$p(x) = \max_{i \in I}\{c_i'x + d_i\},$$

so the problem is to maximize $f(x)$ subject to $p(x) \leq 0$. It is then possible to consider a variation of the cutting plane method, where both functions f and p are replaced by polyhedral approximations F_k and P_k, respectively:

$$x_{k+1} \in \arg \max_{P_k(x) \leq 0} F_k(x).$$

As earlier,

$$F_k(x) = \min\{f(x_0) + (x - x_0)'g_0, \ldots, f(x_k) + (x - x_k)'g_k\},$$

with g_j being a subgradient of f at x_j. The polyhedral approximation P_k is given by

$$P_k(x) = \max_{i \in I_k}\{c_i'x + d_i\},$$

where I_k is a subset of I generated as follows: I_0 is an arbitrary subset of I, and I_k is obtained from I_{k-1} by setting $I_k = I_{k-1}$ if $p(x_k) \leq 0$, and by adding to I_{k-1} one or more of the indices $i \notin I_{k-1}$ such that $c_i'x_k + d_i > 0$ otherwise.

Note that this method applies even when f is a linear function. In this case there is no cost function approximation, i.e., $F_k = f$, just outer approximation of the constraint set, i.e., $X \subset \{x \mid P_k(x) \leq 0\}$.

The convergence properties of the method are very similar to the ones of the earlier method. In particular, propositions analogous to Props. 4.1.1 and 4.1.2 can be formulated and proved. There are also versions of this method where X is a general closed convex set, which is iteratively approximated by a polyhedral set. Variants of this type will be discussed later in Sections 4.4 and 4.5.

4.2 INNER LINEARIZATION – SIMPLICIAL DECOMPOSITION

In this section we consider an *inner approximation* approach for the problem of minimizing a convex function $f : \Re^n \mapsto \Re$ over a closed convex set X. In particular, we approximate X with the convex hull of an ever expanding finite set $X_k \subset X$ that consists of extreme points of X plus an arbitrary starting point $x_0 \in X$.† The addition of new extreme points to X_k is done in a way that guarantees a cost improvement each time we minimize f over $\text{conv}(X_k)$ (unless we are already at the optimum).

In this section we assume a *differentiable* convex cost function $f : \Re^n \mapsto \Re$ and a *bounded polyhedral constraint set* X. The method is then appealing under two conditions:

(1) Minimizing a linear function over X is much simpler than minimizing f over X. (The method makes sense only if f is nonlinear.)

(2) Minimizing f over the convex hull of a relative small number of extreme points is much simpler than minimizing f over X. The method makes sense only if X has a large number of extreme points.

Several classes of important large-scale problems, arising for example in communication and transportation networks, have structure that satisfies these conditions (see the discussion on multicommodity flows later in this section, and the end-of-chapter references).

Note that the minimization of f over the convex hull of m points $\tilde{x}_1, \ldots, \tilde{x}_m$ is a differentiable m-dimensional optimization problem over a simplex:

$$\begin{aligned} &\text{minimize} \quad \phi(\alpha_1, \ldots, \alpha_m) \overset{\text{def}}{=} f\left(\sum_{j=1}^{m} \alpha_j \tilde{x}_j\right) \\ &\text{subject to} \quad \sum_{j=1}^{m} \alpha_j = 1, \qquad \alpha_j \geq 0,\ j = 1, \ldots, m. \end{aligned} \tag{4.8}$$

The function ϕ inherits its smoothness properties from f, and in particular, if f is twice differentiable, so is ϕ, and the problem above can be solved with Newton-like versions of the two-metric projection method.

Note also that the solution of the above problem may be simplified by exploiting special structure that may be present in f. For example, if $f(x) = h(Ax)$ and computation of Ax is by far the most expensive part of computing $f(x)$, then one may compute $\tilde{y}_j = A\tilde{x}_j$ just once for each j, and write the cost function of problem (4.8) in the inexpensively computed form

$$\phi(\alpha_1, \ldots, \alpha_m) = h\left(\sum_{j=1}^{m} \alpha_j \tilde{y}_j\right).$$

This simplification applies to other inner linearization algorithms of this chapter as well, and depending on the available structure, can be exploited to solve problems where x has extraordinarily large dimension (see the subsequent discussion on multicommodity flows in this section).

---

† Extreme points and related notions of polyhedral convexity are discussed in Sections 2.1-2.4 of Appendix B.

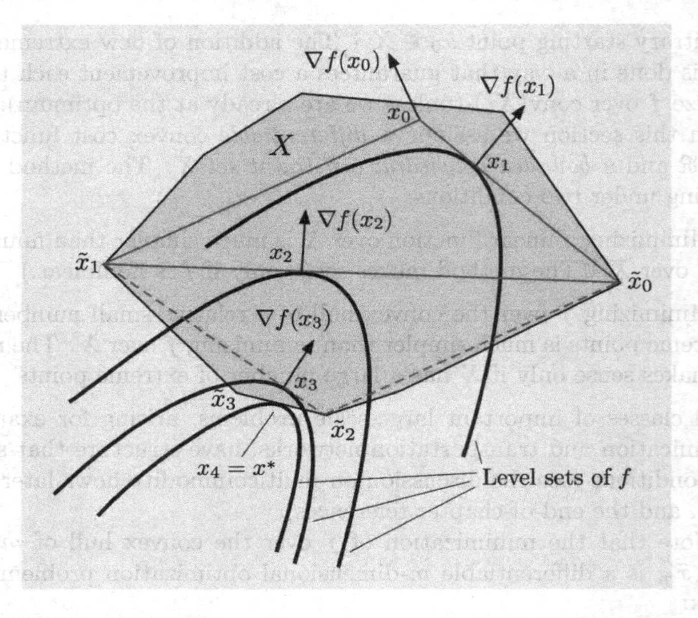

Figure 4.2.1. Successive iterates of the simplicial decomposition method. For example, the figure shows how given the initial point x_0, and the calculated extreme points $\tilde{x}_0$, $\tilde{x}_1$, we determine the next iterate x_2 as a minimizing point of f over the convex hull of $\{x_0, \tilde{x}_0, \tilde{x}_1\}$. At each iteration, a new extreme point of X is added, and after four iterations, the optimal solution is obtained.

At the typical iteration of the simplest type of inner linearization algorithm (also called *simplicial decomposition*) we have the current iterate x_k, and the finite set X_k that consists of the starting point x_0 together with a finite collection of extreme points of X (initially $X_0 = \{x_0\}$). We first generate $\tilde{x}_k$ as an extreme point of X that solves the linear program

$$\begin{aligned} &\text{minimize} \quad \nabla f(x_k)'(x - x_k) \\ &\text{subject to} \quad x \in X. \end{aligned} \tag{4.9}$$

We then add $\tilde{x}_k$ to X_k,

$$X_{k+1} = \{\tilde{x}_k\} \cup X_k,$$

and we generate x_{k+1} as an optimal solution of the problem

$$\begin{aligned} &\text{minimize} \quad f(x) \\ &\text{subject to} \quad x \in \text{conv}(X_{k+1}). \end{aligned} \tag{4.10}$$

Note that this is a problem of the form (4.8). The process is illustrated in Fig. 4.2.1.

The following proposition shows finite convergence of the method.

Proposition 4.2.1: Assume that the cost function f is convex and differentiable and the constraint set X is bounded and polyhedral. Then the simplicial decomposition method obtains an optimal solution in a finite number of iterations.

Proof: There are two possibilities for the extreme point $\tilde{x}_k$ that minimizes $\nabla f(x_k)'(x - x_k)$ over $x \in X$ [cf. problem (4.9)]:

(a) We have

$$0 \leq \nabla f(x_k)'(\tilde{x}_k - x_k) = \min_{x \in X} \nabla f(x_k)'(x - x_k),$$

in which case x_k minimizes f over X, since it satisfies the necessary and sufficient optimality condition of Prop. 1.1.8 in Appendix B.

(b) We have

$$0 > \nabla f(x_k)'(\tilde{x}_k - x_k), \tag{4.11}$$

in which case $\tilde{x}_k \notin \text{conv}(X_k)$, since x_k minimizes f over $x \in \text{conv}(X_k)$, so that $\nabla f(x_k)'(x - x_k) \geq 0$ for all $x \in \text{conv}(X_k)$.

Since case (b) cannot occur an infinite number of times ($\tilde{x}_k \notin X_k$ and X has finitely many extreme points, cf. Prop. 2.3.3 in Appendix B), case (a) must eventually occur, so the method will find a minimizer of f over X in a finite number of iterations. **Q.E.D.**

The simplicial decomposition method has been applied to several types of problems that have a suitable structure. Experience has generally been favorable and suggests that the method requires a lot fewer iterations than the cutting plane method that uses an outer approximation of the constraint set. As an indication of this, we note that if f is linear, the simplicial decomposition method terminates in a single iteration, whereas the cutting plane method may require a very large number of iterations to attain the required solution accuracy. Moreover simplicial decomposition does not exhibit the kind of instability phenomenon that is typical of the cutting plane method. In particular, once an optimal solution belongs to X_k, the method will terminate at the next iteration. By contrast, the cutting plane method, even after generating an optimal solution, it may move away from that solution.

The method is also asymptotically much faster than the conditional gradient method, which is similar and can exploit similar problem structure. Indeed the simplicial decomposition and the conditional gradient methods require solution of the same linear cost problem (4.9) to obtain $\tilde{x}_k$ at each iteration. They differ only in that the former requires minimization of f

over the convex hull of a finite number of points [cf. problem (4.8)], while the latter requires a search over the line segment $[x_k, \tilde{x}_k]$.

Variants of the Simplicial Decomposition Method

We will now discuss some variations and extensions of the simplicial decomposition method. The essence of the convergence proof of Prop. 4.2.1 is that the extreme point $\tilde{x}_k$ does not belong to X_k, unless the optimal solution has been reached. Thus it is not necessary that $\tilde{x}_k$ solves exactly the linearized problem (4.9). Instead it is sufficient that $\tilde{x}_k$ is an extreme point and that the inner product $\nabla f(x_k)'(\tilde{x}_k - x_k)$ is negative [cf. Eq. (4.11)]. This idea may be used in variants of the simplicial decomposition method whereby $\nabla f(x_k)'(x-x_k)$ is minimized inexactly over $x \in X$. Moreover, one may add multiple extreme points $\tilde{x}_k$, as long as they satisfy the condition $\nabla f(x_k)'(\tilde{x}_k - x_k) < 0$.

There are a few other variants of the method. For example to address the case where X is an unbounded polyhedral set, one may augment X with additional constraints to make it bounded (an alternative for the case where X is a cone is discussed in Section 4.6). There are extensions that allow for a nonpolyhedral constraint set, which is approximated by the convex hull of some of its extreme points in the course of the algorithm; see the discussion in Sections 4.4-4.6. Finally, one may use variants, known as *restricted simplicial decomposition* methods, which allow discarding some of the extreme points generated so far. In particular, given the minimum x_{k+1} of f over X_{k+1} [cf. problem (4.10)], we may discard from X_{k+1} all points $\tilde{x}$ such that

$$\nabla f(x_{k+1})'(\tilde{x} - x_{k+1}) > 0,$$

while possibly augmenting the constraint set with the additional constraint

$$\nabla f(x_{k+1})'(x - x_{k+1}) \leq 0. \tag{4.12}$$

The idea is that the costs of the subsequent points $x_{k+2}, x_{k+3}, \ldots$, generated by the method will all be no greater than the cost of x_{k+1}, so they will satisfy the constraint (4.12).

In fact a stronger result can be shown: any number of extreme points may be discarded, as long as $\text{conv}(X_{k+1})$ contains x_k and $\tilde{x}_k$. The proof is based on the theory of feasible direction methods (cf. Section 2.1.2), and the fact that $\tilde{x}_k - x_k$ is a descent direction for f, since if x_k is not optimal, we have

$$\nabla f(x_k)'(\tilde{x}_k - x_k) = \min_{x \in X} \nabla f(x_k)'(x - x_k) < 0,$$

so a point with improved cost can be found along the line segment connecting x_k and $\tilde{x}_k$. Indeed, the method that discards *all* previous points $x_0, \tilde{x}_0, \ldots, \tilde{x}_{k-1}$, replacing them with just x_k, is essentially the same as the conditional gradient method that was discussed in Section 2.1.2.

In Section 4.5, we will discuss additional variations and extensions of the simplicial decomposition method, where among others, we will allow f to be nondifferentiable and X to be nonpolyhedral. We will also allow the presence of additional inequality constraints, which are not approximated by linearization. Moreover, in Section 4.6, we will discuss specialized simplicial decomposition methods for conical constraints, a case of an unbounded constraint set.

Simplicial Decomposition and Multicommodity Flows

Let us now discuss an important application in network optimization, described in Example 1.4.5. As noted earlier, simplicial decomposition is well suited to problems where (a) minimizing a linear function over X is much simpler than minimizing f over X, and (b) minimizing f over the convex hull of a relative small number of extreme points is much simpler than minimizing f over X. These two conditions are eminently satisfied in multicommodity network flow problems such as the one of Example 1.4.5.

Here we have a set W of origin-destination pairs, where the origin and destination are some distinct nodes of a given directed graph. Traffic of some kind (cars, material, or packets of information, for example) enters the origins and must be routed to the corresponding destinations, while minimizing a certain cost. We denote by r_w the input traffic of $w \in W$ (a given positive scalar), and we are given a set P_w of paths (possibly all acyclic paths) that start at the origin and end at the destination of w. We wish to divide each r_w into path flows $x_p \geq 0$, $p \in P_w$, such that $\sum_{p\in P_w} x_p = r_w$. The optimization variables are the path flows x_p, $p \in P_w$, $w \in W$, and we denote by x the generic vector of path flows, $x = \{x_p \mid p \in P_w,\ w \in W\}$. The problem is

$$\begin{aligned} &\text{minimize} \quad D(x) \overset{\text{def}}{=} \sum_{(i,j)} D_{ij}(F_{ij}) \\ &\text{subject to} \quad \sum_{p\in P_w} x_p = r_w, \quad \forall\, w \in W, \\ &\qquad x_p \geq 0, \quad \forall\, p \in P_w,\ w \in W, \end{aligned}$$

where F_{ij} is the total flow that passes through arc (i,j):

$$F_{ij} = \sum_{\substack{\text{all paths } p \\ \text{containing } (i,j)}} x_p, \tag{4.13}$$

and D_{ij} is a differentiable monotonically increasing convex one-dimensional function for each arc (i,j).

Sometimes D has a more complicated form, and there may be additional constraints on the total flows F_{ij}, but we restrict ourselves to the problem above, which is the "standard" formulation. Later in Section 4.5.2, we will discuss more general versions of simplicial decomposition, which may apply to more complex multicommodity flow problems. Note that the cost function is of the form $h(Ax)$ where x is the vector of path flows x_p, and A is the matrix that maps x into the vector of arc flows F_{ij}. Calculating $F = Ax$ is far more complicated than calculating $h(F)$, so an important favorable structure noted earlier for the application of simplicial decomposition is present in multicommodity flow problems.

Another major structural characteristic of the problem relates to the linear approximation of the cost function

$$\nabla D(x_k)'(x-x_k) = \sum_{w\in W}\sum_{p\in P_w}\sum_{\{(i,j)\mid(i,j)\in p\}} \nabla D_{ij}\left(\sum_{\{p\mid(i,j)\in p\}} x_{p,k}\right)(x_p - x_{p,k})$$

at the kth iterate x_k of the simplicial decomposition method. Here $x_{p,k}$ denotes the path flow/component of x_k that goes through path p and "$(i,j) \in p$" means that (i,j) is part of path p [we use Eq. (4.13) in the preceding expression]. The key fact is that *minimizing this linear approximation over the constraint set is a shortest path problem,* which can be solved with very fast algorithms: the length of arc (i,j) is $\nabla D_{ij}\left(\sum_{\{p\mid(i,j)\in p\}} x_{p,k}\right)$, the length of path p is the sum of the lengths of the arcs on the path, and the computation of the path of minimum length over paths $p \in P_w$ can be done separately for each $w \in W$. Once the shortest path for each w is determined, the input flow r_w is placed on that shortest path, and the new extreme point $\tilde{x}_k$ is the flow vector formed by these shortest path flows.

We also note that the minimization of D over the convex hull of the extreme points forming X_{k+1} [cf. Eq. (4.10)] is a low-dimensional problem that can be conveniently solved by two-metric Newton-like methods (in practice, few extreme points are typically required). In conclusion, the multicommodity flow problem combines all the important structural elements that are necessary for the effective application of simplicial decomposition. We refer to the end-of-chapter references for further discussion, including the application of alternative algorithms.

4.3 DUALITY OF INNER AND OUTER LINEARIZATION

We have considered so far cutting plane and simplicial decomposition methods, and we will now aim to connect them via duality. To this end, we define in this section outer and inner linearizations, and we formalize their conjugacy relation and other related properties. An outer linearization of a closed proper convex function $f : \Re^n \mapsto (-\infty, \infty]$ is defined by a finite set of vectors $\{y_1, \ldots, y_\ell\}$ such that for every $j = 1, \ldots, \ell$, we have $y_j \in \partial f(x_j)$ for some $x_j \in \Re^n$. It is given by

$$F(x) = \max_{j=1,\ldots,\ell} \{f(x_j) + (x - x_j)'y_j\}, \qquad x \in \Re^n, \tag{4.14}$$

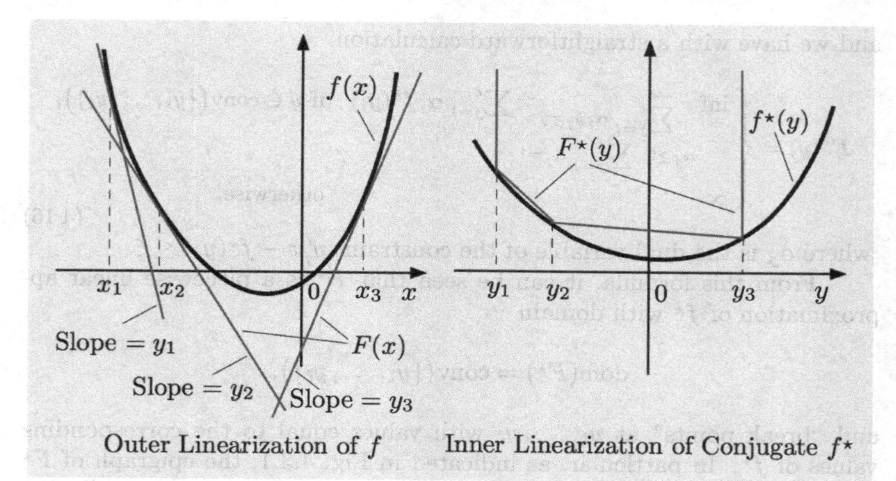

Figure 4.3.1. Illustration of the conjugate $F^\star$ of an outer linearization F of a convex function f defined by a finite set of "slopes" $y_1, \ldots, y_\ell$ and corresponding points $x_1, \ldots, x_\ell$ such that $y_j \in \partial f(x_j)$ for all $j = 1, \ldots, \ell$. It is an inner linearization of the conjugate $f^\star$ of f, a piecewise linear function whose break points are $y_1, \ldots, y_\ell$.

and it is illustrated in the left side of Fig. 4.3.1. The choices of x_j such that $y_j \in \partial f(x_j)$ may not be unique, but result in the same function $F(x)$: the epigraph of F is determined by the supporting hyperplanes to the epigraph of f with normals defined by y_j, and the points of support x_j are immaterial. In particular, the definition (4.14) can be equivalently written in terms of the conjugate $f^\star$ of f as

$$F(x) = \max_{j=1,\ldots,\ell} \{x'y_j - f^\star(y_j)\}, \tag{4.15}$$

using the relation $x_j'y_j = f(x_j) + f^\star(y_j)$, which is implied by $y_j \in \partial f(x_j)$ (the Conjugate Subgradient Theorem, Prop. 5.4.3 in Appendix B).

Note that $F(x) \leq f(x)$ for all x, so as is true for any outer approximation of f, the conjugate $F^\star$ satisfies $F^\star(y) \geq f^\star(y)$ for all y. Moreover, it can be shown that $F^\star$ is an inner linearization of the conjugate $f^\star$, as illustrated in the right side of Fig. 4.3.1. Indeed we have, using Eq. (4.15),

$$\begin{aligned} F^\star(y) &= \sup_{x \in \Re^n} \{y'x - F(x)\} \\ &= \sup_{x \in \Re^n} \left\{ y'x - \max_{j=1,\ldots,\ell} \{y_j'x - f^\star(y_j)\} \right\}, \\ &= \sup_{\substack{x \in \Re^n,\, \xi \in \Re \\ y_j'x - f^\star(y_j) \leq \xi,\ j=1,\ldots,\ell}} \{y'x - \xi\}. \end{aligned}$$

By linear programming duality, the optimal value of the linear program in (x, ξ) of the preceding equation can be replaced by the dual optimal value, and we have with a straightforward calculation

$$F^\star(y) = \begin{cases} \inf_{\substack{\sum_{j=1}^{\ell} \alpha_j y_j = y \\ \alpha_j \geq 0,\ \sum_{j=1}^{\ell} \alpha_j = 1}} \sum_{j=1}^{\ell} \alpha_j f^\star(y_j) & \text{if } y \in \text{conv}\big(\{y_1, \ldots, y_\ell\}\big), \\ \infty & \text{otherwise,} \end{cases} \tag{4.16}$$

where α_j is the dual variable of the constraint $y_j'x - f^\star(y_j) \leq \xi$.

From this formula, it can be seen that $F^\star$ is a piecewise linear approximation of $f^\star$ with domain

$$\text{dom}(F^\star) = \text{conv}\big(\{y_1, \ldots, y_\ell\}\big),$$

and "break points" at $y_1, \ldots, y_\ell$ with values equal to the corresponding values of $f^\star$. In particular, as indicated in Fig. 4.3.1, the epigraph of $F^\star$ is the convex hull of the union of the vertical halflines corresponding to $y_1, \ldots, y_\ell$:

$$\text{epi}(F^\star) = \text{conv}\Big(\cup_{j=1,\ldots,\ell} \big\{(y_j, w) \mid f^\star(y_j) \leq w\big\}\Big). \tag{4.17}$$

In what follows, by an inner linearization of a closed proper convex function $f^\star$ defined by a finite set $\{y_1, \ldots, y_\ell\}$ we will mean the function $F^\star$ given by Eq. (4.16). Note that not all sets $\{y_1, \ldots, y_\ell\}$ define conjugate pairs of outer and inner linearizations via Eqs. (4.15) and (4.16), respectively, within our framework: it is necessary that for every y_j there exists x_j such that $y_j \in \partial f(x_j)$, or equivalently that $\partial f^\star(y_j) \neq \emptyset$ for all j [which implies in particular that $y_j \in \text{dom}(f^\star)$]. By exchanging the roles of f and $f^\star$, we also obtain a dual statement, namely that for a set $\{x_1, \ldots, x_\ell\}$ to define an inner linearization of a closed proper convex function f as well as an outer linearization of its conjugate $f^\star$, it is necessary that $\partial f(x_j) \neq \emptyset$ for all j.

4.4 GENERALIZED POLYHEDRAL APPROXIMATION

We will now consider a unified framework for polyhedral approximation, which combines the cutting plane and simplicial decomposition methods. We consider the problem

$$\begin{aligned} &\text{minimize} \quad \sum_{i=1}^{m} f_i(x_i) \\ &\text{subject to} \quad x \in S, \end{aligned} \tag{4.18}$$

where

$$x \stackrel{\text{def}}{=} (x_1, \ldots, x_m),$$

is a vector in $\Re^{n_1+\cdots+n_m}$, with components $x_i \in \Re^{n_i}$, $i = 1, \ldots, m$, and

$f_i : \Re^{n_i} \mapsto (-\infty, \infty]$ is a closed proper convex function for each i,

S is a subspace of $\Re^{n_1+\cdots+n_m}$.

We refer to this as an *extended monotropic program* (EMP for short).†

A classical example of EMP is a single commodity network optimization problem, where x_i represents the (scalar) flow of an arc of a directed graph and S is the circulation subspace of the graph (see e.g., [Ber98]). Also problems involving general linear constraints and an additive extended real-valued convex cost function can be converted to EMP. In particular, the problem

$$\begin{aligned} &\text{minimize} \quad \sum_{i=1}^{m} f_i(x_i) \\ &\text{subject to} \quad Ax = b, \end{aligned} \tag{4.19}$$

where A is a given matrix and b is a given vector, is equivalent to

$$\begin{aligned} &\text{minimize} \quad \sum_{i=1}^{m} f_i(x_i) + \delta_Z(z) \\ &\text{subject to} \quad Ax - z = 0, \end{aligned}$$

where z is a vector of artificial variables, and δ_Z is the indicator function of the set $Z = \{z \mid z = b\}$. This is an EMP with constraint subspace

$$S = \{(x, z) \mid Ax - z = 0\}.$$

When all components x_i are one-dimensional and the functions f_i are linear within $\text{dom}(f_i)$, problem (4.19) reduces to a linear program. When the functions f_i are positive semidefinite quadratic within $\text{dom}(f_i)$, and $\text{dom}(f_i)$ are polyhedral, problem (4.19) reduces to a convex quadratic program.

Note that while the vectors $x_1, \ldots, x_m$ appear independently in the cost function

$$\sum_{i=1}^{m} f_i(x_i),$$

they are coupled through the subspace constraint. This allows a variety of transformations to the EMP format. For example, consider a cost function of the form

$$f(x) = F(x_1, \ldots, x_m) + \sum_{i=1}^{m} f_i(x_i),$$

† Monotropic programming, a class of problems introduced and extensively analyzed in the book [Roc84], is the special case of problem (4.18) where each component x_i is one-dimensional (i.e., $n_i = 1$). The name "monotropic" means "turning in a single direction" in Greek, and captures the characteristic monotonicity property of convex functions of a single variable such as f_i.

where F is a closed proper convex function of all the components x_i. Then, by introducing an auxiliary vector $z \in \Re^{n_1+\cdots+n_m}$, the problem of minimizing f over a subspace X can be transformed to the problem

$$\begin{aligned} &\text{minimize} \quad F(z) + \sum_{i=1}^{m} f_i(x_i) \\ &\text{subject to} \quad (x, z) \in S, \end{aligned}$$

where S is the subspace of $\Re^{2(n_1+\cdots+n_m)}$

$$S = \{(x, x) \mid x \in X\}.$$

This problem is of the form (4.18).

Another problem that can be converted to the EMP format (4.18) is

$$\begin{aligned} &\text{minimize} \quad \sum_{i=1}^{m} f_i(x) \\ &\text{subject to} \quad x \in X, \end{aligned}$$

where $f_i : \Re^n \mapsto (-\infty, \infty]$ are closed proper convex functions, and X is a subspace of $\Re^n$. This can be done by introducing m copies of x, i.e., auxiliary vectors $z_i \in \Re^n$ that are constrained to be equal, and write the problem as

$$\begin{aligned} &\text{minimize} \quad \sum_{i=1}^{m} f_i(z_i) \\ &\text{subject to} \quad (z_1, \ldots, z_m) \in S, \end{aligned}$$

where S is the subspace

$$S = \{(x, \ldots, x) \mid x \in X\}.$$

It can thus be seen that convex problems with linear constraints can generally be formulated as EMP. We will see that these problems share a powerful and symmetric duality theory, which is similar to Fenchel duality, and forms the basis for a symmetric and general framework for polyhedral approximation.

The Dual Problem

To derive the appropriate dual problem, we introduce auxiliary vectors $z_i \in \Re^{n_i}$ and we convert the EMP (4.18) to the equivalent form

$$\begin{aligned} &\text{minimize} \quad \sum_{i=1}^{m} f_i(z_i) \\ &\text{subject to} \quad z_i = x_i, \quad i = 1, \ldots, m, \qquad (x_1, \ldots, x_m) \in S. \end{aligned} \tag{4.20}$$

We then assign a multiplier vector $\lambda_i \in \Re^{n_i}$ to the constraint $z_i = x_i$, thereby obtaining the Lagrangian function

$$L(x_1, \ldots, x_m, z_1, \ldots, z_m, \lambda_1, \ldots, \lambda_m) = \sum_{i=1}^{m} \big(f_i(z_i) + \lambda_i'(x_i - z_i)\big). \quad (4.21)$$

The dual function is

$$\begin{aligned} q(\lambda) &= \inf_{(x_1,\ldots,x_m)\in S,\, z_i\in\Re^{n_i}} L(x_1, \ldots, x_m, z_1, \ldots, z_m, \lambda_1, \ldots, \lambda_m) \\ &= \inf_{(x_1,\ldots,x_m)\in S} \sum_{i=1}^{m} \lambda_i' x_i + \sum_{i=1}^{m} \inf_{z_i\in\Re^{n_i}} \big\{f_i(z_i) - \lambda_i' z_i\big\} \\ &= \begin{cases} \sum_{i=1}^{m} q_i(\lambda_i) & \text{if } (\lambda_1, \ldots, \lambda_m) \in S^\perp, \\ -\infty & \text{otherwise,} \end{cases} \end{aligned}$$

where

$$q_i(\lambda_i) = \inf_{z_i\in\Re^{n_i}} \big\{f_i(z_i) - \lambda_i' z_i\big\}, \qquad i = 1, \ldots, m,$$

and $S^\perp$ is the orthogonal subspace of S.

Note that since q_i can be written as

$$q_i(\lambda_i) = -\sup_{z_i\in\Re^{n_i}} \big\{\lambda_i' z_i - f_i(z_i)\big\},$$

it follows that $-q_i$ is the conjugate of f_i, so by Prop. 1.6.1 of Appendix B, $-q_i$ is a closed proper convex function. The dual problem is

$$\begin{aligned} &\text{maximize} \quad \sum_{i=1}^{m} q_i(\lambda_i) \\ &\text{subject to} \quad (\lambda_1, \ldots, \lambda_m) \in S^\perp, \end{aligned} \quad (4.22)$$

or with a change of sign to convert maximization to minimization,

$$\begin{aligned} &\text{minimize} \quad \sum_{i=1}^{m} f_i^\star(\lambda_i) \\ &\text{subject to} \quad (\lambda_1, \ldots, \lambda_m) \in S^\perp, \end{aligned} \quad (4.23)$$

where $f_i^\star$ is the conjugate of f_i. Thus the dual problem has the same form as the primal. Moreover, assuming that the functions f_i are closed, when the dual problem is dualized, it yields the primal, so the duality has a symmetric character, like Fenchel duality. We will discuss further the duality theory for EMP in Sections 6.7.3 and 6.7.4, using algorithmic ideas (the ϵ-descent method of Section 6.7.2).

Throughout our analysis of this section, we denote by f_{opt} and q_{opt} the optimal values of the primal and dual problems (4.18) and (4.22), respectively, and in addition to the convexity assumption on f_i made earlier, we will assume that appropriate conditions hold that guarantee the strong duality relation $f_{opt} = q_{opt}$. In Section 6.7, we will revisit EMP and we will develop conditions under which strong duality holds, using the algorithmic ideas developed there.

Since EMP in the form (4.20) can be viewed as a special case of the convex programming problem with equality constraints of Section 1.1, it is possible to obtain optimality conditions as special cases of the corresponding conditions in that section (cf. Prop. 1.1.5). In particular, it can be seen that a pair (x, λ) satisfies the Lagrangian optimality condition of Prop. 1.1.5(b), applied to the Lagrangian (4.21), if and only if x_i attains the infimum in the equation

$$q_i(\lambda_i) = \inf_{z_i \in \Re^{n_i}} \big\{ f_i(z_i) - \lambda_i' z_i \big\}, \qquad i = 1, \ldots, m.$$

Thus, by applying Prop. 1.1.5(b), we obtain the following.

Proposition 4.4.1: (EMP Optimality Conditions) There holds $-\infty < q_{opt} = f_{opt} < \infty$ and $(x_1^{opt}, \ldots, x_m^{opt}, \lambda_1^{opt}, \ldots, \lambda_m^{opt})$ are an optimal primal and dual solution pair of the EMP problem if and only if

$$(x_1^{opt}, \ldots, x_m^{opt}) \in S, \qquad (\lambda_1^{opt}, \ldots, \lambda_m^{opt}) \in S^{\perp},$$

and

$$x_i^{opt} \in \arg \min_{x_i \in \Re^{n_i}} \big\{ f_i(x_i) - x_i' \lambda_i^{opt} \big\}, \quad i = 1, \ldots, m. \tag{4.24}$$

Note that by the Conjugate Subgradient Theorem (Prop. 5.4.3 in Appendix B), the condition (4.24) of the preceding proposition is equivalent to either one of the following two subgradient conditions:

$$\lambda_i^{opt} \in \partial f_i(x_i^{opt}), \qquad x_i^{opt} \in \partial f_i^{\star}(\lambda_i^{opt}), \qquad i = 1, \ldots, m.$$

These conditions are significant, because they show that once $(x_1^{opt}, \ldots, x_m^{opt})$ or $(\lambda_1^{opt}, \ldots, \lambda_m^{opt})$ is obtained, its dual counterpart can be computed by "differentiation" of the functions f_i or $f_i^{\star}$, respectively. We will often use the equivalences of the preceding formulas in the remainder of this chapter, so we depict them in Fig. 4.4.1.

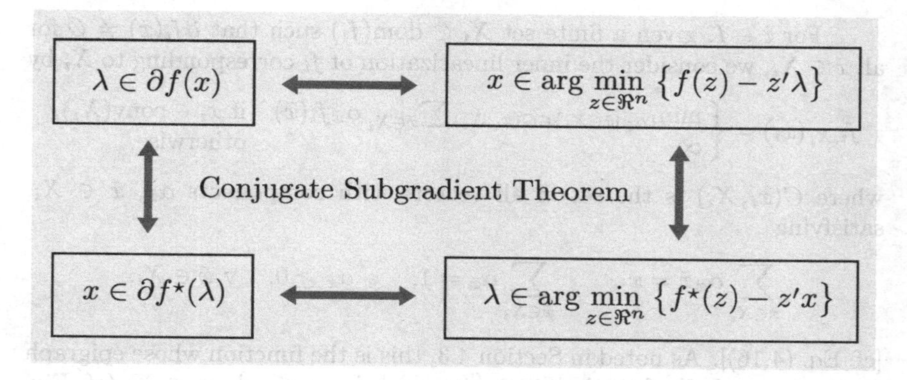

Figure 4.4.1. Equivalent "differentiation" formulas for a closed proper convex function f and its conjugate $f^\star$. All of these four relations are also equivalent to

$$\lambda' x = f(x) + f^\star(\lambda);$$

cf. the Conjugate Subgradient Theorem (Prop. 5.4.3 in Appendix B).

General Polyhedral Approximation Scheme

The EMP formalism allows a broad and elegant algorithmic framework that combines elements of the cutting plane and simplicial decomposition methods of the preceding sections. In particular, the primal and dual EMP problems (4.18) and (4.23) will be approximated, by using inner or outer linearization of some of the functions f_i and $f_i^\star$. The optimal solution of the dual approximate problem will then be used to construct more refined inner and outer linearizations.

We introduce an algorithm, referred to as the *generalized polyhedral approximation* or GPA algorithm. It uses a fixed partition of the index set $\{1,\ldots,m\}$:

$$\{1,\ldots,m\} = I \cup \underline{I} \cup \bar{I}.$$

This partition determines which of the functions f_i are outer approximated (set $\underline{I}$) and which are inner approximated (set $\bar{I}$).

For $i \in \underline{I}$, given a finite set $\Lambda_i \subset \mathrm{dom}(f_i^\star)$ such that $\partial f_i^\star(\tilde{\lambda}) \neq \varnothing$ for all $\tilde{\lambda} \in \Lambda_i$, we consider the outer linearization of f_i corresponding to Λ_i:

$$\underline{f}_{i,\Lambda_i}(x_i) = \max_{\tilde{\lambda} \in \Lambda_i} \{\tilde{\lambda}' x_i - f_i^\star(\tilde{\lambda})\},$$

or equivalently, as mentioned in Section 4.3,

$$\underline{f}_{i,\Lambda_i}(x_i) = \max_{\tilde{\lambda} \in \Lambda_i} \{f_i(x_{\tilde{\lambda}}) + \tilde{\lambda}'(x_i - x_{\tilde{\lambda}})\},$$

where for each $\tilde{\lambda} \in \Lambda_i$, $x_{\tilde{\lambda}}$ is such that $\tilde{\lambda} \in \partial f_i(x_{\tilde{\lambda}})$.

For $i \in \bar{I}$, given a finite set $X_i \subset \text{dom}(f_i)$ such that $\partial f_i(\tilde{x}) \neq \emptyset$ for all $\tilde{x} \in X_i$, we consider the inner linearization of f_i corresponding to X_i by

$$\bar{f}_{i,X_i}(x_i) = \begin{cases} \min_{\{\alpha_{\tilde{x}} \mid \tilde{x} \in X_i\} \in C(x_i,X_i)} \sum_{\tilde{x} \in X_i} \alpha_{\tilde{x}} f_i(\tilde{x}) & \text{if } x_i \in \text{conv}(X_i), \\ \infty & \text{otherwise,} \end{cases}$$

where $C(x_i, X_i)$ is the set of all vectors with components $\alpha_{\tilde{x}}$, $\tilde{x} \in X_i$, satisfying

$$\sum_{\tilde{x} \in X_i} \alpha_{\tilde{x}} \tilde{x} = x_i, \qquad \sum_{\tilde{x} \in X_i} \alpha_{\tilde{x}} = 1, \qquad \alpha_{\tilde{x}} \geq 0, \quad \forall\, \tilde{x} \in X_i,$$

[cf. Eq. (4.16)]. As noted in Section 4.3, this is the function whose epigraph is the convex hull of the halflines $\{(x_i, w) \mid f_i(x_i) \leq w\}$, $x_i \in X_i$ (cf. Fig. 4.3.1).

We assume that at least one of the sets $\underline{I}$ and $\bar{I}$ is nonempty. At the start of the typical iteration, we have for each $i \in \underline{I}$, a finite subset $\Lambda_i \subset \text{dom}(f_i^\star)$, and for each $i \in \bar{I}$, a finite subset $X_i \subset \text{dom}(f_i)$. The iteration is as follows:

Typical Iteration of GPA Algorithm:

Step 1: (Approximate Problem Solution) Find a primal-dual optimal solution pair $(\hat{x}, \hat{\lambda}) = (\hat{x}_1, \hat{\lambda}_1, \ldots, \hat{x}_m, \hat{\lambda}_m)$ of the EMP

$$\begin{aligned} &\text{minimize} \quad \sum_{i \in I} f_i(x_i) + \sum_{i \in \underline{I}} \underline{f}_{i,\Lambda_i}(x_i) + \sum_{i \in \bar{I}} \bar{f}_{i,X_i}(x_i) \\ &\text{subject to} \quad (x_1, \ldots, x_m) \in S, \end{aligned} \tag{4.25}$$

where $\underline{f}_{i,\Lambda_i}$ and $\bar{f}_{i,X_i}$ are the outer and inner linearizations of f_i corresponding to Λ_i and X_i, respectively.

Step 2: (Test for Termination and Enlargement) Enlarge the sets Λ_i and X_i as follows (see Fig. 4.4.2):

(a) For $i \in \underline{I}$, we add any subgradient $\tilde{\lambda}_i \in \partial f_i(\hat{x}_i)$ to Λ_i.

(b) For $i \in \bar{I}$, we add any subgradient $\tilde{x}_i \in \partial f_i^\star(\hat{\lambda}_i)$ to X_i.

If there is no strict enlargement, i.e., for all $i \in \underline{I}$ we have $\tilde{\lambda}_i \in \Lambda_i$, and for all $i \in \bar{I}$ we have $\tilde{x}_i \in X_i$, the algorithm terminates. Otherwise, we proceed to the next iteration, using the enlarged sets Λ_i and X_i.

We will show in a subsequent proposition that if the algorithm terminates, the current vector $(\hat{x}, \hat{\lambda})$ is a primal and dual optimal solution pair. If there is strict enlargement and the algorithm does not terminate, we proceed to the next iteration, using the enlarged sets Λ_i and X_i.

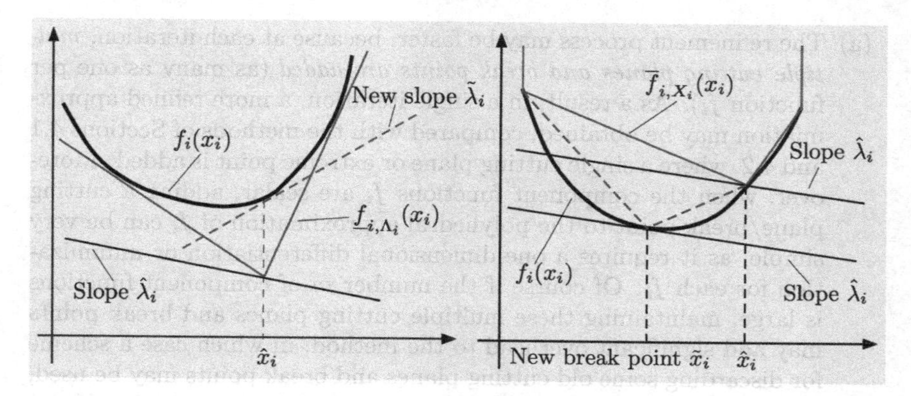

Figure 4.4.2. Illustration of the enlargement step in the GPA algorithm, after we obtain a primal-dual optimal solution pair

$$(\hat{x}, \hat{\lambda}) = (\hat{x}_1, \hat{\lambda}_1, \ldots, \hat{x}_m, \hat{\lambda}_m).$$

Note that in the figure on the right, we use the fact

$$\tilde{x}_i \in \partial f_i^\star(\hat{\lambda}_i) \quad \Longleftrightarrow \quad \hat{\lambda}_i \in \partial f_i(\tilde{x}_i)$$

(cf. Fig. 4.4.1 – the Conjugate Subgradient Theorem, Prop. 5.4.3 in Appendix B). The enlargement step on the left (finding $\tilde{\lambda}_i$) is also equivalent to $\tilde{\lambda}_i$ satisfying $\hat{x}_i \in \partial f_i^\star(\tilde{\lambda}_i)$, or equivalently, solving the optimization problem

$$\begin{array}{ll} \text{maximize} & \left\{\lambda_i'\hat{x}_i - f_i^\star(\lambda_i)\right\} \\ \text{subject to} & \lambda_i \in \Re^{n_i}. \end{array}$$

The enlargement step on the right (finding $\tilde{x}_i$) is also equivalent to solving the optimization problem

$$\begin{array}{ll} \text{maximize} & \left\{\hat{\lambda}_i' x_i - f_i(x_i)\right\} \\ \text{subject to} & x_i \in \Re^{n_i}; \end{array}$$

cf. Fig. 4.4.1.

Note that we implicitly assume that at each iteration, there exists a primal and dual optimal solution pair of problem (4.25). Furthermore, we assume that the enlargement step can be carried out, i.e., that $\partial f_i(\hat{x}_i) \neq \emptyset$ for all $i \in \underline{I}$ and $\partial f_i^\star(\hat{\lambda}_i) \neq \emptyset$ for all $i \in \bar{I}$. Sufficient assumptions may need to be imposed on the problem to guarantee that this is so.

There are two potential advantages of the GPA algorithm over the earlier cutting plane and simplicial decomposition methods of Sections 4.1 and 4.2, depending on the problem's structure:

(a) The refinement process may be faster, because at each iteration, *multiple cutting planes and break points are added* (as many as one per function f_i). As a result, in a single iteration, a more refined approximation may be obtained, compared with the methods of Sections 4.1 and 4.2, where a single cutting plane or extreme point is added. Moreover, when the component functions f_i are scalar, adding a cutting plane/break point to the polyhedral approximation of f_i can be very simple, as it requires a one-dimensional differentiation or minimization for each f_i. Of course if the number m of component functions is large, maintaining these multiple cutting planes and break points may add significant overhead to the method, in which case a scheme for discarding some old cutting planes and break points may be used, similar to the case of the restricted simplicial decomposition scheme.

(b) The approximation process may *preserve some of the special structure of the cost function and/or the constraint set*. For example if the component functions f_i are scalar, or have partially overlapping dependences, such as for example,

$$f(x_1,\ldots,x_m) = f_1(x_1,x_2) + f_2(x_2,x_3) + \cdots + f_{m-1}(x_{m-1},x_m) + f_m(x_m),$$

the minimization of f by the cutting plane method of Section 4.1 leads to general/unstructured linear programming problems. By contrast, using separate outer approximation of the components functions leads to linear programs with special structure, which can be solved efficiently by specialized methods, such as network flow algorithms, or interior point algorithms that can exploit the sparsity structure of the problem.

Generally, in specially structured problems, the preceding two advantages can be of decisive importance.

Note two prerequisites for the GPA algorithm to be effective:

(1) The (partially) linearized problem (4.25) must be easier to solve than the original problem (4.18). For example, problem (4.25) may be a linear program, while the original may be nonlinear (cf. the cutting plane method of Section 4.1); or it may effectively have much smaller dimension than the original (cf. the simplicial decomposition method of Section 4.2).

(2) Finding the enlargement vectors ($\tilde{\lambda}_i$ for $i \in \underline{I}$, and $\tilde{x}_i$ for $i \in \bar{I}$) must not be too difficult. This can be done by the differentiation $\tilde{\lambda}_i \in \partial f_i(\hat{x}_i)$ for $i \in \underline{I}$, and $\tilde{x}_i \in \partial f_i^\star(\hat{\lambda}_i)$ or $i \in \bar{I}$. Alternatively, if this is not convenient for some of the functions (e.g., because some of the f_i or the $f_i^\star$ are not available in closed form), one may calculate $\tilde{\lambda}_i$ and/or $\tilde{x}_i$ via the relations

$$\hat{x}_i \in \partial f_i^\star(\tilde{\lambda}_i), \qquad \hat{\lambda}_i \in \partial f_i(\tilde{x}_i);$$

(cf. Fig. 4.4.1 – the Conjugate Subgradient Theorem, Prop. 5.4.3 in Appendix B). This involves solving optimization problems. For example, finding $\tilde{x}_i$ such that $\hat{\lambda}_i \in \partial f_i(\tilde{x}_i)$ for $i \in \bar{I}$ is equivalent to solving the problem

$$\begin{aligned} &\text{maximize} \quad \{\hat{\lambda}_i' x_i - f_i(x_i)\} \\ &\text{subject to} \quad x_i \in \Re^{n_i}, \end{aligned}$$

and may be nontrivial (cf. Fig. 4.4.2).

Generally, the facility of solving the linearized problem (4.25) and carrying out the subsequent enlargement step may guide the choice of functions that are inner or outer linearized. If x_i is one-dimensional, as is often true in separable-type problems, the enlargement step is typically quite easy.

We finally note that the symmetric duality of the EMP can be exploited in the implementation of the GPA algorithm. In particular, the algorithm may be applied to the dual problem of problem (4.18):

$$\begin{aligned} &\text{minimize} \quad \sum_{i=1}^{m} f_i^\star(\lambda_i) \\ &\text{subject to} \quad (\lambda_1, \ldots, \lambda_m) \in S^\perp, \end{aligned} \tag{4.26}$$

where $f_i^\star$ is the conjugate of f_i. Then the inner (or outer) linearized index set $\bar{I}$ of the primal becomes the outer (or inner, respectively) linearized index set of the dual. At each iteration, the algorithm solves the approximate dual EMP,

$$\begin{aligned} &\text{minimize} \quad \sum_{i \in I} f_i^\star(\lambda_i) + \sum_{i \in \underline{I}} \bar{f}_{i,\Lambda_i}^\star(\lambda_i) + \sum_{i \in \bar{I}} \underline{f}_{i,X_i}^\star(\lambda_i) \\ &\text{subject to} \quad (\lambda_1, \ldots, \lambda_m) \in S^\perp, \end{aligned} \tag{4.27}$$

which is simply the dual of the approximate primal EMP (4.25) [since the outer (or inner) linearization of $f_i^\star$ is the conjugate of the inner (or respectively, outer) linearization of f_i]. Thus the algorithm produces mathematically identical results when applied to the primal or the dual EMP. The choice of whether to apply the algorithm in its primal or its dual form is simply a matter of whether calculations with f_i or with their conjugates $f_i^\star$ are more or less convenient. In fact, when the algorithm makes use of both the primal solution $\hat{x}$ and the dual solution $\hat{\lambda}$ in the enlargement step, the question of whether the starting point is the primal or the dual EMP becomes moot: it is best to view the algorithm as applied to the pair of primal and dual EMP, without designating which is primal and which is dual.

Termination and Convergence

Now let us discuss the validity of the GPA algorithm. To this end, we will use two basic properties of outer approximations. The first is that for any closed proper convex functions f and $\underline{f}$, and vector $x \in \text{dom}(f)$, we have

$$\underline{f} \leq f, \quad \underline{f}(x) = f(x) \qquad \Longrightarrow \qquad \partial f(x) \subset \partial \underline{f}(x). \tag{4.28}$$

To see this, use the subgradient inequality to write for any $g \in \partial \underline{f}(x)$,

$$f(z) \geq \underline{f}(z) \geq \underline{f}(x) + g'(z-x) = f(x) + g'(z-x), \qquad z \in \Re^n,$$

which implies that $g \in \partial f(x)$. The second property is that for any outer linearization $\underline{f}_\Lambda$ of f, we have

$$\tilde{\lambda} \in \Lambda, \quad \tilde{\lambda} \in \partial f(x) \qquad \Longrightarrow \qquad \underline{f}_\Lambda(x) = f(x). \tag{4.29}$$

To see this, consider vectors x_λ such that $\lambda \in \partial f(x_\lambda)$, $\lambda \in \Lambda$, and write

$$\underline{f}_\Lambda(x) = \max_{\lambda \in \Lambda} \{ f(x_\lambda) + \lambda'(x - x_\lambda) \} \geq f(x_{\tilde{\lambda}}) + \tilde{\lambda}'(x - x_{\tilde{\lambda}}) \geq f(x),$$

where the second inequality follows from $\tilde{\lambda} \in \partial f(x_{\tilde{\lambda}})$. Since we also have $f \geq \underline{f}_\Lambda$, we obtain $f_\Lambda(x) = f(x)$. We first show the optimality of the primal and dual solution pair obtained upon termination of the algorithm.

Proposition 4.4.2: (Optimality at Termination) If the GPA algorithm terminates at some iteration, the corresponding primal and dual solutions, $(\hat{x}_1, \ldots, \hat{x}_m)$ and $(\hat{\lambda}_1, \ldots, \hat{\lambda}_m)$, form a primal and dual optimal solution pair of the EMP problem.

Proof: From Prop. 4.4.1 and the definition of $(\hat{x}_1, \ldots, \hat{x}_m)$ and $(\hat{\lambda}_1, \ldots, \hat{\lambda}_m)$ as a primal and dual optimal solution pair of the approximate problem (4.25), we have

$$(\hat{x}_1, \ldots, \hat{x}_m) \in S, \qquad (\hat{\lambda}_1, \ldots, \hat{\lambda}_m) \in S^\perp.$$

We will show that upon termination, we have for all i

$$\hat{\lambda}_i \in \partial f_i(\hat{x}_i), \tag{4.30}$$

which by Prop. 4.4.1 implies the desired conclusion.

Since $(\hat{x}_1, \ldots, \hat{x}_m)$ and $(\hat{\lambda}_1, \ldots, \hat{\lambda}_m)$ are a primal and dual optimal solution pair of problem (4.25), Eq. (4.30) holds for all $i \notin \underline{I} \cup \bar{I}$ (cf. Prop.

4.4.1). We will complete the proof by showing that it holds for all $i \in \underline{I}$ (the proof for $i \in \bar{I}$ follows by a dual argument).

Indeed, let us fix $i \in \underline{I}$ and let $\tilde{\lambda}_i \in \partial f_i(\hat{x}_i)$ be the vector generated by the enlargement step upon termination. We must have $\tilde{\lambda}_i \in \Lambda_i$, since there is no strict enlargement when termination occurs. Since $\underline{f}_{i,\Lambda_i}$ is an outer linearization of f_i, by Eq. (4.29), the fact $\tilde{\lambda}_i \in \Lambda_i, \tilde{\lambda}_i \in \partial f_i(\hat{x}_i)$ implies that

$$\underline{f}_{i,\Lambda_i}(\hat{x}_i) = f_i(\hat{x}_i),$$

which in turn implies by Eq. (4.28) that

$$\partial \underline{f}_{i,\Lambda_i}(\hat{x}_i) \subset \partial f_i(\hat{x}_i).$$

By Prop. 4.4.1, we also have $\hat{\lambda}_i \in \partial \underline{f}_{i,\Lambda_i}(\hat{x}_i)$, so $\hat{\lambda}_i \in \partial f_i(\hat{x}_i)$. **Q.E.D.**

As in Sections 4.1 and 4.2, convergence can be easily established in the case where the functions f_i, $i \in \underline{I} \cup \bar{I}$, are polyhedral, assuming that care is taken to ensure that the corresponding enlargement vectors $\tilde{\lambda}_i$ are chosen from a finite set of extreme points. In particular, assume that:

(a) All outer linearized functions f_i are real-valued and polyhedral, and for all inner linearized functions f_i, the conjugates $f_i^\star$ are real-valued and polyhedral.

(b) The vectors $\tilde{\lambda}_i$ and $\tilde{x}_i$ added to the polyhedral approximations are elements of the finite representations of the corresponding $f_i^\star$ and f_i.

Then at each iteration there are two possibilities: either $(\hat{x}, \hat{\lambda})$ is an optimal primal-dual pair for the original problem and the algorithm terminates, or the approximation of one of the f_i, $i \in \underline{I} \cup \bar{I}$, will be refined/improved. Since there can be only a finite number of refinements, convergence in a finite number of iterations follows.

Other convergence results are possible, extending some of the analysis of Sections 4.1 and 4.2. In particular, let $(\hat{x}^k, \hat{\lambda}^k)$ be the primal and dual pair generated at iteration k, and let

$$\tilde{\lambda}_i^k \in \partial f_i(\hat{x}_i^k), \quad i \in \underline{I}, \quad \tilde{x}_i^k \in \partial f_i^\star(\hat{\lambda}_i^k), \quad \in \bar{I},$$

be the vectors used for the corresponding enlargements. If the set $\bar{I}$ is empty (no inner approximation) and the sequence $\{\tilde{\lambda}_i^k\}$ is bounded for every $i \in \underline{I}$, then we can easily show that every limit point of $\{\hat{x}^k\}$ is primal optimal. To see this, note that for all k, $\ell \leq k-1$, and $(x_1, \ldots, x_m) \in S$, we have

$$\begin{aligned}\sum_{i \notin \underline{I}} f_i(\hat{x}_i^k) + \sum_{i \in \underline{I}} \big(f_i(\hat{x}_i^\ell) + (\hat{x}_i^k - \hat{x}_i^\ell)'\tilde{\lambda}_i^\ell\big) &\leq \sum_{i \notin \underline{I}} f_i(\hat{x}_i^k) + \sum_{i \in \underline{I}} \underline{f}_{i,\Lambda_i^{k-1}}(\hat{x}_i^k) \\ &\leq \sum_{i=1}^m f_i(x_i).\end{aligned}$$

Let $\{\hat{x}^k\}_{\mathcal{K}}$ be a subsequence converging to a vector $\bar{x}$. By taking the limit as $\ell \to \infty$, $k \in \mathcal{K}$, $\ell \in \mathcal{K}$, $\ell < k$, and using the closedness of f_i, we obtain

$$\sum_{i=1}^{m} f_i(\bar{x}_i) \leq \liminf_{k\to\infty,\, k\in\mathcal{K}} \sum_{i\notin \underline{I}} f_i(\hat{x}_i^k) + \liminf_{\ell\to\infty,\, \ell\in\mathcal{K}} \sum_{i\in \underline{I}} f_i(\hat{x}_i^\ell) \leq \sum_{i=1}^{m} f_i(x_i)$$

for all $(x_1, \ldots, x_m) \in S$. It follows that $\bar{x}$ is primal optimal, i.e., every limit point of $\{\hat{x}^k\}$ is optimal. The preceding convergence argument also goes through even if the sequences $\{\tilde{\lambda}_i^k\}$ are not assumed bounded, as long as the limit points $\bar{x}_i$ belong to the relative interior of the corresponding functions f_i (this follows from the subgradient decomposition result of Prop. 5.4.1 in Appendix B).

Exchanging the roles of primal and dual, we similarly obtain a convergence result for the case where $\underline{I}$ is empty (no outer linearization): assuming that the sequence $\{\tilde{x}_i^k\}$ is bounded for every $i \in \bar{I}$, every limit point of $\{\hat{\lambda}^k\}$ is dual optimal.

We finally state a more general convergence result from [BeY11], which applies to the mixed case where we simultaneously use outer and inner approximation (both $\bar{I}$ and $\underline{I}$ are nonempty). The proof is more complicated than the preceding ones, and we refer to [BeY11] for the corresponding analysis.

Proposition 4.4.3: (Convergence of the GPA Algorithm) Consider the GPA algorithm for the EMP problem, assuming the strong duality relation $-\infty < q_{opt} = f_{opt} < \infty$. Let $(\hat{x}^k, \hat{\lambda}^k)$ be a primal and dual optimal solution pair of the approximate problem at the kth iteration, and let $\tilde{\lambda}_i^k, i \in \underline{I}$ and $\tilde{x}_i^k, i \in \bar{I}$ be the vectors generated at the corresponding enlargement step. Suppose that there exist convergent subsequences $\{\hat{x}_i^k\}_{\mathcal{K}}, i \in \underline{I}$, $\{\hat{\lambda}_i^k\}_{\mathcal{K}}, i \in \bar{I}$, such that the sequences $\{\tilde{\lambda}_i^k\}_{\mathcal{K}}, i \in \underline{I}$, $\{\tilde{x}_i^k\}_{\mathcal{K}}, i \in \bar{I}$, are bounded. Then:

(a) Any limit point of the sequence $\{(\hat{x}^k, \hat{\lambda}^k)\}_{\mathcal{K}}$ is a primal and dual optimal solution pair of the original problem.

(b) The sequence of optimal values of the approximate problems converges to the optimal value f_{opt}.

Application to Network Optimization and Monotropic Programming

Let us consider a directed graph with set of nodes $\mathcal{N}$ and set of arcs $\mathcal{A}$. A classical network optimization problem is to minimize a cost function

$$\sum_{a\in\mathcal{A}} f_a(x_a),$$

where f_a is a scalar closed proper convex function, and x_a is the flow of arc $a \in \mathcal{A}$. The minimization is over all flow vectors $x = \{x_a \mid a \in \mathcal{A}\}$ that belong to the circulation subspace S of the graph (at each node, the sum of all incoming arc flows is equal to the sum of all outgoing arc flows).

The GPA algorithm that uses inner linearization of all the functions f_a that are nonlinear is particularly attractive for this problem, because of the favorable structure of the corresponding approximate EMP:

$$\begin{aligned} &\text{minimize} \quad \sum_{a \in \mathcal{A}} \bar{f}_{a,X_a}(x_a) \\ &\text{subject to} \quad x \in S, \end{aligned}$$

where for each arc a, $\bar{f}_{a,X_a}$ is the inner approximation of f_a, corresponding to a finite set of break points $X_a \subset \text{dom}(f_a)$. By suitably introducing multiple arcs in place of each arc, we can recast this problem as a linear minimum cost network flow problem that can be solved using very fast polynomial algorithms. These algorithms, simultaneously with an optimal primal (flow) vector, yield a dual optimal (price differential) vector (see e.g., [Ber98], Chapters 5-7). Furthermore, because the functions f_a are scalar, the enlargement step is very simple.

Some of the preceding advantages of the GPA algorithm with inner linearization carry over to monotropic programming problems ($n_i = 1$ for all i), the key idea being the simplicity of the enlargement step. Furthermore, there are effective algorithms for solving the associated approximate primal and dual EMP, such as out-of-kilter methods [Roc84], [Tse01b], and ϵ-relaxation methods [Ber98], [TsB00].

4.5 GENERALIZED SIMPLICIAL DECOMPOSITION

In this section we will aim to highlight some of the applications and the fine points of the general algorithm of the preceding section. As vehicle we will use the simplicial decomposition approach, and the problem

$$\begin{aligned} &\text{minimize} \quad f(x) + c(x) \\ &\text{subject to} \quad x \in \Re^n, \end{aligned} \tag{4.31}$$

where $f : \Re^n \mapsto (-\infty, \infty]$ and $c : \Re^n \mapsto (-\infty, \infty]$ are closed proper convex functions. This is the Fenchel duality context, and it contains as a special case the problem to which the ordinary simplicial decomposition method of Section 4.2 applies (where f is differentiable, and c is the indicator function of a bounded polyhedral set). Here we will mainly focus on the case where f is nondifferentiable and possibly extended real-valued.

We apply the polyhedral approximation scheme of the preceding section to the equivalent EMP

$$\begin{aligned} &\text{minimize} \quad f_1(x_1) + f_2(x_2) \\ &\text{subject to} \quad (x_1, x_2) \in S, \end{aligned}$$

where

$$f_1(x_1) = f(x_1), \qquad f_2(x_2) = c(x_2), \qquad S = \{(x_1, x_2) \mid x_1 = x_2\}.$$

Note that the orthogonal subspace has the form

$$S^\perp = \{(\lambda_1, \lambda_2) \mid \lambda_1 = -\lambda_2\} = \{(\lambda, -\lambda) \mid \lambda \in \Re^n\}.$$

Optimal primal and dual solutions of this EMP problem are of the form (x^{opt}, x^{opt}) and $(\lambda^{opt}, -\lambda^{opt})$, with

$$\lambda^{opt} \in \partial f(x^{opt}), \qquad -\lambda^{opt} \in \partial c(x^{opt}),$$

consistently with the optimality conditions of Prop. 4.4.1. A pair of such optimal solutions (x^{opt}, λ^{opt}) satisfies the necessary and sufficient optimality conditions of the Fenchel Duality Theorem [Prop. 1.2.1(c)] for the original problem.

In one possible polyhedral approximation method, at the typical iteration f_2 is replaced by an inner linearization involving a set of break points X, while f_1 is left unchanged. At the end of the iteration, if $(\hat{\lambda}, -\hat{\lambda})$ is a dual optimal solution, X is enlarged to include a vector $\tilde{x}$ such that $-\hat{\lambda} \in \partial f_2(\tilde{x})$ (cf. Section 4.4). We will now transcribe this method in the notation of problem (4.31).

We start with some finite set $X_0 \subset \text{dom}(c)$. At the typical iteration, given a finite set $X_k \subset \text{dom}(c)$, we use the following three steps to compute vectors $x_k, \tilde{x}_k$, and the enlarged set $X_{k+1} = X_k \cup \{\tilde{x}_k\}$ to start the next iteration:

Typical Iteration of Generalized Simplicial Decomposition Algorithm to Minimize $f + c$

(1) We obtain

$$x_k \in \arg\min_{x \in \Re^n} \{f(x) + C_k(x)\}, \tag{4.32}$$

where C_k is the polyhedral/inner linearization function whose epigraph is the convex hull of the finite collection of halflines $\{(\tilde{x}, w) \mid c(\tilde{x}) \le w\}$, $\tilde{x} \in X_k$.

(2) We obtain λ_k such that

$$\lambda_k \in \partial f(x_k), \qquad -\lambda_k \in \partial C_k(x_k). \tag{4.33}$$

(3) We obtain $\tilde{x}_k$ such that

$$-\lambda_k \in \partial c(\tilde{x}_k), \tag{4.34}$$

and form

$$X_{k+1} = X_k \cup \{\tilde{x}_k\}.$$

As in the case of the GPA algorithm, we assume that f and c are such that the steps (1)-(3) above can be carried out. In particular, the existence of the subgradient λ_k in step (2) is guaranteed by the optimality conditions of Prop. 5.4.7 in Appendix B, applied to the minimization in Eq. (4.32), under the appropriate relative interior conditions.

Note that step (3) is equivalent to finding

$$\tilde{x}_k \in \arg\min_{x\in\Re^n} \{\lambda_k' x + c(x)\}, \tag{4.35}$$

and that this is a linear or quadratic program in important special cases where c is polyhedral or quadratic, respectively. Note also that the approximate problem (4.32) is a linearized version of the original problem (4.31), where c is replaced by $C_k(x)$, which is an inner linearization of c. More specifically, if $X_k = \{\tilde{x}_j \mid j \in J_k\}$, where J_k is a finite index set, C_k is given by

$$C_k(x) = \begin{cases} \inf_{\substack{\sum_{j\in J_k} \alpha_j \tilde{x}_j = x \\ \alpha_j \geq 0, \sum_{j\in J_k} \alpha_j = 1}} \sum_{j\in J_k} \alpha_j c(\tilde{x}_j) & \text{if } x \in \text{conv}(X_k), \\ \infty & \text{if } x \notin \text{conv}(X_k), \end{cases}$$

so the minimization (4.32) involves in effect the variables α_j, $j \in J_k$, and is equivalent to

$$\begin{aligned} &\text{minimize} \quad f\left(\sum_{j\in J_k} \alpha_j \tilde{x}_j\right) + \sum_{j\in J_k} \alpha_j c(\tilde{x}_j) \\ &\text{subject to} \quad \sum_{j\in J_k} \alpha_j = 1, \quad \alpha_j \geq 0, \ j \in J_k. \end{aligned} \tag{4.36}$$

The dimension of this problem is the cardinality of J_k, which can be quite small relative to the dimension of the original problem.

Dual/Cutting Plane Implementation

Let us also provide a dual implementation, which is an equivalent outer linearization/cutting plane-type of method. The Fenchel dual of the minimization of $f + c$ [cf. Eq. (4.31)] is

$$\begin{aligned} &\text{minimize} \quad f^\star(\lambda) + c^\star(-\lambda) \\ &\text{subject to} \quad \lambda \in \Re^n, \end{aligned}$$

where $f^\star$ and $c^\star$ are the conjugates of f and c, respectively. According to the theory of the preceding section, the generalized simplicial decomposition algorithm (4.32)-(4.34) can alternatively be implemented by replacing

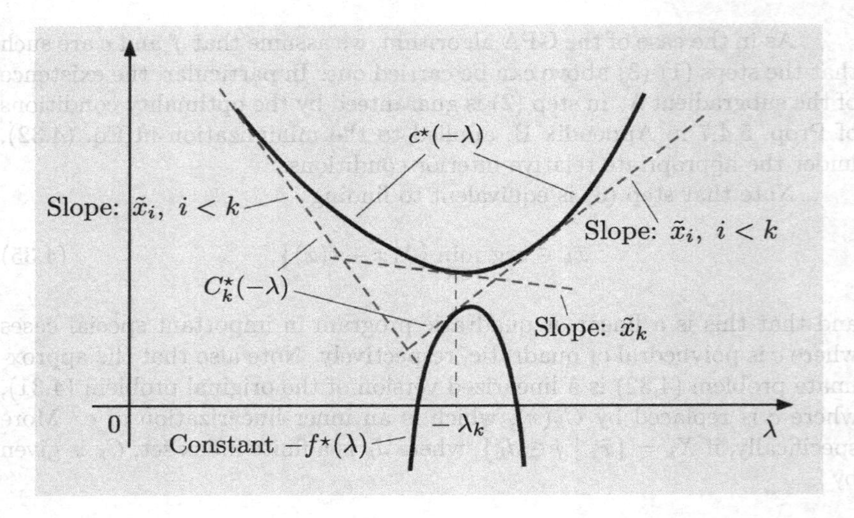

Figure 4.5.1. Illustration of the cutting plane implementation of the generalized simplicial decomposition method for minimizing the sum of two closed proper convex functions f and c [cf. Eq. (4.31)]. The ordinary cutting plane method, described in the beginning of Section 4.1, is obtained as the special case where $f^\star(x) \equiv 0$. In this case, f is the indicator function of the set consisting of just the origin, and the primal problem is to evaluate $c(0)$ (the optimal value of $c^\star$).

$c^\star$ by a piecewise linear/cutting plane outer linearization, while leaving $f^\star$ unchanged, i.e., by solving at iteration k the problem

$$\begin{aligned} &\text{minimize} \quad f^\star(\lambda) + C_k^\star(-\lambda) \\ &\text{subject to} \quad \lambda \in \Re^n, \end{aligned} \tag{4.37}$$

where $C_k^\star$ is an outer linearization of $c^\star$ (the conjugate of C_k). This problem is the (Fenchel) dual of problem (4.32) [or equivalently, the low-dimensional problem (4.36)].

Note that solutions of problem (4.37) are the subgradients λ_k satisfying $\lambda_k \in \partial f(x_k)$ and $-\lambda_k \in \partial C_k(x_k)$, where x_k is the solution of the problem (4.32) [cf. Eq. (4.33)], while the associated subgradient of $c^\star$ at $-\lambda_k$ is the vector $\tilde{x}_k$ generated by Eq. (4.34), as shown in Fig. 4.5.1. In fact, the function $C_k^\star$ has the form

$$C_k^\star(-\lambda) = \max_{j \in J_k} \big\{ c(-\lambda_j) - \tilde{x}_j'(\lambda - \lambda_j) \big\},$$

where λ_j and $\tilde{x}_j$ are vectors that can be obtained either by using the generalized simplicial decomposition method (4.32)-(4.34), or by using its dual, the cutting plane method based on solving the outer approximation problems (4.37). The ordinary cutting plane method, described in the beginning of Section 4.1, is obtained as the special case where $f^\star(\lambda) \equiv 0$ [or equivalently, $f(x) = \infty$ if $x \neq 0$, and $f(0) = 0$].

Whether the primal or the dual implementation is preferable depends on the structure of the functions f and c. When f (and hence also $f^\star$) is not polyhedral, the dual implementation may not be attractive, because it requires the n-dimensional nonlinear optimization (4.37) at each iteration, as opposed to the typically low-dimensional optimization (4.32) or (4.36). When f is polyhedral, both methods require the solution of linear programs for the enlargement step, and the choice between them may depend on whether it is more convenient to work with c rather than $c^\star$.

4.5.1 Differentiable Cost Case

Let us first consider the favorable case of the generalized simplicial decomposition algorithm (4.32)-(4.35), where f is differentiable and c is polyhedral with bounded effective domain. Then the method is essentially equivalent to the simple version of the simplicial decomposition method of Section 4.2. In particular:

(a) When c is the indicator function of a bounded polyhedral set X, and $X_0 = \{x_0\}$, the method reduces to the earlier simplicial decomposition method (4.9)-(4.10). Indeed, step (1) corresponds to the minimization (4.10), step (2) simply yields $\lambda_k = \nabla f(x_k)$, and step (3), as implemented by Eq. (4.35), corresponds to solution of the linear program (4.9) that generates a new extreme point.

(b) When c is a general polyhedral function, the method can be viewed as essentially the special case of the earlier simplicial decomposition method (4.9)-(4.10) applied to the problem of minimizing $f(x) + w$ subject to $x \in X$ and $(x, w) \in \text{epi}(c)$ [the only difference is that epi(c) is not bounded, but this is inconsequential if we assume that dom(c) is bounded, or more generally that the problem (4.32) has a solution]. In this case, the method terminates finitely, assuming that the vectors $(\tilde{x}_k, c(\tilde{x}_k))$ obtained by solving the linear program (4.35) are extreme points of epi(c) (cf. Prop. 4.2.1).

For the more general case where f is differentiable and c is a (nonpolyhedral) convex function, the method is illustrated in Fig. 4.5.2. The existence of a solution x_k to problem (4.32) [or equivalently (4.36)] is guaranteed by the compactness of conv(X_k) and Weierstrass' Theorem, while step (2) yields $\lambda_k = \nabla f(x_k)$. The existence of a solution to problem (4.35) must be guaranteed by some assumption such as for example compactness of the effective domain of c.

4.5.2 Nondifferentiable Cost and Side Constraints

Let us now consider the problem of minimizing $f + c$ [cf. Eq. (4.31)] for the more complicated case where f is extended real-valued and nondiffer-

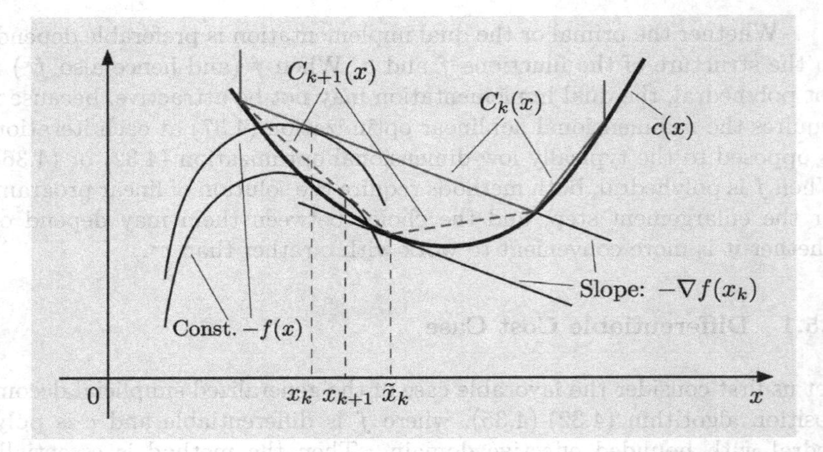

Figure 4.5.2. Illustration of successive iterates of the generalized simplicial decomposition method in the case where f is differentiable and c is a general convex function. Given the inner linearization C_k of c, we minimize $f + C_k$ to obtain x_k (graphically, we move the graph of $-f$ vertically until it touches the graph of C_k). We then compute $\tilde{x}_k$ as a point at which $-\nabla f(x_k)$ is a subgradient of c, and we use it to form the improved inner linearization C_{k+1} of c. Finally, we minimize $f + C_{k+1}$ to obtain x_{k+1} (graphically, we move the graph of $-f$ vertically until it touches the graph of C_{k+1}).

entiable. Then assuming that

$$\mathrm{ri}\big(\mathrm{dom}(f)\big) \cap \mathrm{conv}(X_0) \neq \emptyset,$$

the existence of the subgradient λ_k of Eq. (4.33) is guaranteed by the optimality conditions of Prop. 5.4.7 in Appendix B, and the existence of a solution x_k to problem (4.32) is guaranteed by Weierstrass' Theorem, since the effective domain of C_k is compact.

When c is the indicator function of a polyhedral set X, the condition of step (2) becomes

$$\lambda_k'(\tilde{x} - x_k) \geq 0, \qquad \forall\, \tilde{x} \in \mathrm{conv}(X_k), \tag{4.38}$$

i.e., $-\lambda_k$ is in the normal cone of $\mathrm{conv}(X_k)$ at x_k. The method is illustrated for this case in Fig. 4.5.3. It terminates finitely, assuming that the vector $\tilde{x}_k$ obtained by solving the linear program (4.35) is an extreme point of X. The reason is that in view of Eq. (4.38), the vector $\tilde{x}_k$ does not belong to X_k (unless x_k is optimal), so X_{k+1} is a strict enlargement of X_k.

Let us now address the calculation of a subgradient $\lambda_k \in \partial f(x_k)$ such that $-\lambda_k \in \partial C_k(x_k)$ [cf. Eq. (4.33)]. This may be a difficult problem as it may require knowledge of $\partial f(x_k)$ as well as $\partial C_k(x_k)$. However, in important special cases, λ_k may be obtained simply as a byproduct of the minimization

$$x_k \in \arg\min_{x \in \Re^n} \big\{ f(x) + C_k(x) \big\}, \tag{4.39}$$

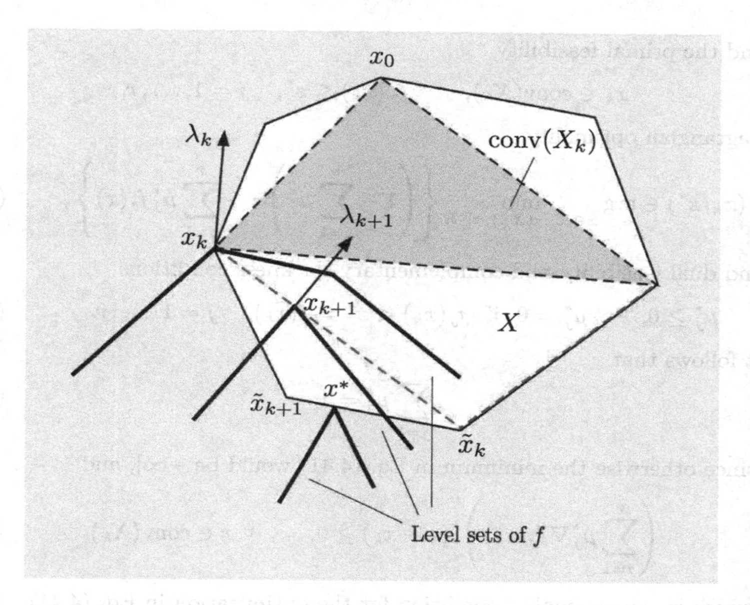

Figure 4.5.3. Illustration of the generalized simplicial decomposition method for the case where f is nondifferentiable and c is the indicator function of a polyhedral set X. For each k, we compute a subgradient $\lambda_k \in \partial f(x_k)$ such that $-\lambda_k$ lies in the normal cone of conv(X_k) at x_k, and we use it to generate a new extreme point $\tilde{x}_k$ of X using the relation

$$-\lambda_k \in \partial c(\tilde{x}_k),$$

cf. Step (3) of the algorithm.

[cf. Eq. (4.32)]. This is illustrated in the following examples.

Example 4.5.1 (Minimax Problems)

Consider the minimization of $f + c$ for the case where c is the indicator of a closed convex set X, and

$$f(x) = \max\big\{f_1(x), \ldots, f_r(x)\big\},$$

where $f_1, \ldots, f_r : \Re^n \mapsto \Re$ are differentiable functions. Then the minimization (4.39) takes the form

$$\begin{aligned} &\text{minimize} \ \ z \\ &\text{subject to} \ \ f_j(x) \le z, \ \ j = 1, \ldots, r, \quad x \in \text{conv}(X_k), \end{aligned} \tag{4.40}$$

where conv(X_k) is a polyhedral inner approximation to X. According to the optimality conditions of Prop. 1.1.3, the optimal solution (x_k, z^*) together with dual optimal variables μ_j^*, satisfy

$$z^* = f(x_k) = \max\big\{f_1(x_k), \ldots, f_r(x_k)\big\},$$

and the primal feasibility

$$x_k \in \mathrm{conv}(X_k), \qquad f_j(x_k) \le z^*, \quad j = 1, \ldots, r,$$

Lagrangian optimality

$$(x_k, z^*) \in \arg \min_{x \in \mathrm{conv}(X_k),\, z \in \Re} \left\{ \left(1 - \sum_{j=1}^r \mu_j^*\right) z + \sum_{j=1}^r \mu_j^* f_j(x) \right\}, \tag{4.41}$$

and dual feasibility and complementary slackness conditions

$$\mu_j^* \ge 0, \qquad \mu_j^* = 0 \ \text{ if } \ f_j(x_k) < z^* = f(x_k), \quad j = 1, \ldots, r. \tag{4.42}$$

It follows that

$$\sum_{j=1}^r \mu_j^* = 1, \tag{4.43}$$

[since otherwise the minimum in Eq. (4.41) would be $-\infty$], and

$$\left(\sum_{j=1}^r \mu_j^* \nabla f_j(x_k) \right)' (x - x_k) \ge 0, \qquad \forall\, x \in \mathrm{conv}(X_k), \tag{4.44}$$

[which is the optimality condition for the optimization in Eq. (4.41)]. Using Eqs. (4.42) and (4.43), it can be shown that the vector

$$\lambda_k = \sum_{j=1}^r \mu_j^* \nabla f_j(x_k) \tag{4.45}$$

is a subgradient of f at x_k (cf. Danskin's Theorem, Section 3.1). Furthermore, using Eq. (4.44), it follows that $-\lambda_k$ is in the normal cone of $\mathrm{conv}(X_k)$ at x_k.

In conclusion, λ_k as given by Eq. (4.45), is a suitable subgradient for determining a new extreme point $\tilde{x}_k$ to refine the inner approximation of the constraint set X via solution of the problem

$$\tilde{x}_k \in \arg \min_{x \in X} \lambda_k' x; \tag{4.46}$$

cf. Eq. (4.35). To obtain λ_k, we need to:

(a) Compute the primal approximation x_k as the solution of problem (4.40).

(b) Simultaneously compute the dual variables/multipliers μ_j^* of this problem.

(c) Use Eq. (4.45).

Note an important advantage of this method over potential competitors: it involves solution of often simple programs (linear programs if X is polyhedral) of the form (4.46) to generate new extreme points of X, and solution of nonlinear programs of the form (4.40), which are low-dimensional [their dimension is equal to the number of extreme points of X_k; cf. Eq. (4.36)]. When each f_j is twice differentiable, the latter programs can be solved by fast Newton-like methods, such as sequential quadratic programming (see e.g., [Ber82a], [Ber99], [NoW06]).

Example 4.5.2 (Minimax Problems with Side Constraints)

Consider a more general version of the preceding example, where there are additional inequality constraints defining the domain of f. This is the problem of minimizing $f + c$ where c is the indicator of a closed convex set X, and f is of the form

$$f(x) = \begin{cases} \max\{f_1(x), \ldots, f_r(x)\} & \text{if } g_i(x) \leq 0,\ i = 1, \ldots, p, \\ \infty & \text{otherwise,} \end{cases} \tag{4.47}$$

with f_j and g_i being convex differentiable functions. Applications of this type include multicommodity flow problems with "side constraints" [the inequalities $g_i(x) \leq 0$, which are separate from the network flow constraints that comprise the set X; cf. the discussion of Section 4.2].

Similarly, to calculate λ_k, we introduce dual variables $\nu_i^* \geq 0$ for the constraints $g_i(x) \leq 0$, and we write the Lagrangian optimality and complementary slackness conditions. Then Eq. (4.44) takes the form

$$\left(\sum_{j=1}^{r} \mu_j^* \nabla f_j(x_k) + \sum_{i=1}^{p} \nu_i^* \nabla g_i(x_k)\right)'(x - x_k) \geq 0, \qquad \forall\, x \in \text{conv}(X_k).$$

Similar to Eq. (4.45), it can be shown that the vector

$$\lambda_k = \sum_{j=1}^{r} \mu_j^* \nabla f_j(x_k) + \sum_{i=1}^{p} \nu_i^* \nabla g_i(x_k)$$

is a subgradient of f at x_k, while we have $-\lambda_k \in \partial C_k(x_k)$ as required by Eq. (4.33). Once λ_k is computed simultaneously with x_k, then as in the preceding example, the enlargement of the inner approximation of X involves the addition of $\tilde{x}_k$, which is obtained by solving the problem (4.46).

The preceding two examples involve the simple nondifferentiable function $\max\{f_1(x), \ldots, f_r(x)\}$, where f_i are differentiable. However, the ideas for the calculation of extreme points of X and the associated process of enlargement via dual variables apply more broadly. In particular, we may treat similarly problems involving other types of nondifferentiable or extended-real valued cost function f, by converting them to constrained minimization problems. Then by using the dual variables of the approximate inner-linearized problems, we may compute a vector λ_k such that $\lambda_k \in \partial f(x_k)$ and $-\lambda_k \in \partial C_k(x_k)$, which can in turn be used for enlargement of X_k.

4.6 POLYHEDRAL APPROXIMATION FOR CONIC PROGRAMMING

In this section we will aim to extend the range of applications of the generalized polyhedral approximation framework for EMP by introducing additional conic constraint sets. Our motivation is that the framework of the

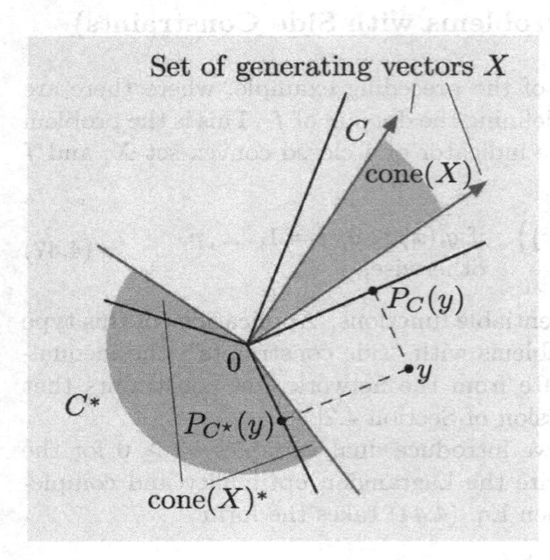

Figure 4.6.1. Illustration of cone(X), the cone generated by a subset X of a cone C, as an inner linearization of C. The polar cone$(X)^\star$ is an outer linearization of the polar cone

$$C^\star = \{y \mid y'x \leq 0, \forall\, x \in C\}.$$

preceding two sections is not well-suited for the case where some of the component functions of the cost are indicator functions of unbounded sets such as cones. There are two main reasons for this:

(1) The enlargement procedure of the GPA algorithm may not be implementable by optimization, as in Fig. 4.4.2, because this optimization may not have a solution. This may be true in particular if the function involved is the indicator function of an unbounded set.

(2) The inner linearization procedure of the GPA algorithm approximates an unbounded set by the convex hull of a finite number of points, which is a compact set. It would appear that an unbounded polyhedral set may provide a more effective approximation.

Motivated by these concerns, we extend the generalized polyhedral approximation approach of Section 4.4 so that it applies to the problem of minimizing the sum $\sum_{i=1}^{m} f_i(x_i)$ of convex extended real-valued functions f_i, subject to $(x_1, \ldots, x_m)$ being in the intersection of a given subspace and the Cartesian product of closed convex cones. To this end we first discuss an alternative method for linearization of a cone, which allows enlargements using directions of recession rather than points.

In particular, given a closed convex cone C and a finite subset $X \subset C$, we view cone(X), the cone generated by X (see Section 1.2 in Appendix B), as an inner linearization of C. Its polar, denoted cone$(X)^\star$, is an outer linearization of the polar $C^\star$ (see Fig. 4.6.1). This type of linearization has a twofold advantage: a cone is approximated by a cone (rather than by a compact set), and outer and inner linearizations yield convex functions of the same type as the original (indicator functions of cones).

As a first step in our analysis, we introduce some duality concepts relating to cones. We say that (x, λ) is a *dual pair with respect to the* *closed convex cones C and $C^\star$* if

$$x = P_C(x+\lambda) \qquad \text{and} \qquad \lambda = P_{C^\star}(x+\lambda),$$

where $P_C(y)$ and $P_{C^\star}(y)$ denote projection of a vector y onto C and $C^\star$, respectively. We also say that (x,λ) *is a dual pair representation of a vector* y if $y = x+\lambda$ and (x,λ) is a dual pair with respect to C and $C^\star$. The following proposition shows that $\big(P_C(y), P_{C^\star}(y)\big)$ is the unique dual pair representation of y, and provides a related characterization; see Fig. 4.6.1.

Proposition 4.6.1: (Cone Decomposition Theorem) Let C be a nonempty closed convex cone in $\Re^n$ and $C^\star$ be its polar cone.

(a) Any vector $y \in \Re^n$, has a unique dual pair representation, the pair $\big(P_C(y), P_{C^\star}(y)\big)$.

(b) The following conditions are equivalent:

(i) (x,λ) is a dual pair with respect to C and $C^\star$.

(ii) $x \in C$, $\lambda \in C^\star$, and $x \perp \lambda$.

Proof: (a) We denote $\xi = y - P_C(y)$, and we will show that $\xi = P_{C^\star}(y)$. This will prove that $\big(P_C(y), P_{C^\star}(y)\big)$ is a dual pair representation of y, which must be unique, since by the definition of dual pair, a vector y can have at most one dual pair representation, the pair $\big(P_C(y), P_{C^\star}(y)\big)$. Indeed, by the Projection Theorem (Prop. 1.1.9 in Appendix B), we have

$$\xi'\big(z - P_C(y)\big) \le 0, \qquad \forall\, z \in C. \tag{4.48}$$

Since C is a cone, we have $(1/2)P_C(y) \in C$ and $2P_C(y) \in C$, so by taking $z = (1/2)P_C(y)$ and $z = 2P_C(y)$ in Eq. (4.48), it follows that

$$\xi' P_C(y) = 0. \tag{4.49}$$

By combining Eqs. (4.48) and (4.49), we obtain $\xi' z \le 0$ for all $z \in C$, implying that $\xi \in C^\star$. Moreover, since $P_C(y) \in C$, we have

$$(y-\xi)'(z-\xi) = P_C(y)'(z-\xi) = P_C(y)'z \le 0, \qquad \forall\, z \in C^\star,$$

where the second equality follows from Eq. (4.49). Thus ξ satisfies the necessary and sufficient condition for being the projection $P_{C^\star}(y)$.

(b) Suppose that property (i) holds, i.e., x and λ are the projections of $x+\lambda$ on C and $C^\star$, respectively. Then we have, using also the Projection Theorem,

$$x \in C, \qquad \lambda \in C^\star, \qquad \big((x+\lambda) - x\big)'x = 0,$$

or

$$x \in C, \qquad \lambda \in C^\star, \qquad \lambda' x = 0,$$

which is property (ii).

Conversely, suppose that property (ii) holds. Then, since $\lambda \in C^\star$, we have $\lambda' z \leq 0$ for all $z \in C$, and hence

$$\big((x+\lambda) - x\big)'(z - x) = \lambda'(z - x) = \lambda' z \leq 0, \qquad \forall\, z \in C,$$

where the second equality follows from the fact $x \perp \lambda$. Thus x satisfies the necessary and sufficient condition for being the projection $P_C(x+\lambda)$. By a symmetric argument, it follows that λ is the projection $P_{C^\star}(x+\lambda)$. **Q.E.D.**

Duality and Optimality Conditions

We now introduce a version of the EMP problem of Section 4.4, generalized to include cone constraints. It is given by

$$\begin{aligned} &\text{minimize} \quad \sum_{i=1}^{m} f_i(x_i) + \sum_{i=m+1}^{r} \delta(x_i \mid C_i) \\ &\text{subject to} \quad (x_1, \ldots, x_r) \in S, \end{aligned} \tag{4.50}$$

where $(x_1, \ldots, x_r)$ is a vector in $\Re^{n_1+\cdots+n_r}$, with components $x_i \in \Re^{n_i}$, $i = 1, \ldots, r$, and

$f_i : \Re^{n_i} \mapsto (-\infty, \infty]$ is a closed proper convex function for each i,

S is a subspace of $\Re^{n_1+\cdots+n_r}$,

$C_i \subset \Re^{n_i}$, $i = m+1, \ldots, r$, is a closed convex cone, and $\delta(x_i \mid C_i)$ denotes the indicator function of C_i.

Interesting special cases are the conic programming problems of Section 1.2, as well as several other problems described in the exercises. Included, in particular, are problems where the cost function involves some positively homogeneous additive components, whose epigraphs are cones, such as norms and support functions of sets. Such cost function components may be expressed in terms of conical constraints.

Note that the conjugate of $\delta(\cdot \mid C_i)$ is $\delta(\cdot \mid C_i^\star)$, the indicator function of the polar cone $C_i^\star$. Thus, according to the EMP duality theory of Section 4.4, the dual problem is

$$\begin{aligned} &\text{minimize} \quad \sum_{i=1}^{m} f_i^\star(\lambda_i) + \sum_{i=m+1}^{r} \delta(\lambda_i \mid C_i^\star) \\ &\text{subject to} \quad (\lambda_1, \ldots, \lambda_r) \in S^\perp, \end{aligned} \tag{4.51}$$

and has the same form as the primal problem (4.50). Furthermore, since f_i is assumed closed proper and convex, and C_i is assumed closed convex, the conjugate of $f_i^\star$ is $f_i^{\star\star} = f_i$ and the polar cone of $C_i^\star$ is $(C_i^\star)^\star = C$. Thus when the dual problem is dualized, it yields the primal problem, similar to the EMP problem of Section 4.4.

Let us denote by f_{opt} and q_{opt} the optimal primal and dual values. According to Prop. 1.1.5, (x^{opt}, λ^{opt}) form an optimal primal and dual solution pair if and only if they satisfy the standard primal feasibility, dual feasibility, and Lagrangian optimality conditions. By working out these conditions similar to Section 4.4, we obtain the following proposition, which parallels Prop. 4.4.1.

Proposition 4.6.2: (Optimality Conditions) We have $-\infty < q_{opt} = f_{opt} < \infty$, and $x^{opt} = (x_1^{opt}, \ldots, x_r^{opt})$ and $\lambda^{opt} = (\lambda_1^{opt}, \ldots, \lambda_r^{opt})$ are optimal primal and dual solutions, respectively, of problems (4.50) and (4.51) if and only if

$$(x_1^{opt}, \ldots, x_r^{opt}) \in S, \qquad (\lambda_1^{opt}, \ldots, \lambda_r^{opt}) \in S^\perp, \tag{4.52}$$

$$x_i^{opt} \in \arg \min_{x_i \in \Re^n} \big\{ f_i(x_i) - x_i' \lambda_i^{opt} \big\}, \qquad i = 1, \ldots, m, \tag{4.53}$$

$(x_i^{opt}, \lambda_i^{opt})$ is a dual pair with respect to C_i and $C_i^\star$, $i = m+1, \ldots, r$. (4.54)

Note that by the Conjugate Subgradient Theorem (Prop. 5.4.3 in Appendix B), the condition (4.53) of the preceding proposition is equivalent to either one of the following two subgradient conditions

$$\lambda_i^{opt} \in \partial f_i(x_i^{opt}), \qquad x_i^{opt} \in \partial f_i^\star(\lambda_i^{opt}), \qquad i = 1, \ldots, m;$$

(cf. Fig. 4.4.1). Thus the optimality conditions are fully symmetric, consistently with the symmetric form of the primal and dual problems (4.50) and (4.51).

Generalized Simplicial Decomposition for Conical Constraints

We will now describe an algorithm, whereby problem (4.50) is approximated by using inner linearization of some of the functions f_i and of all the cones C_i. The optimal primal and dual solution pair of the approximate problem is then used to construct more refined inner linearizations. For simplicity, we are focusing on the pure simplicial decomposition approach (and by duality on the pure cutting plane approach). It is straightforward to extend our algorithms to the mixed case, where some of the component functions f_i and/or cones C_i are inner linearized while others are outer linearized.

We introduce a fixed subset $\bar{I} \subset \{1, \ldots, m\}$, which corresponds to functions f_i that are inner linearized. For notational convenience, we denote by I the complement of $\bar{I}$ in $\{1, \ldots, m\}$:

$$\{1, \ldots, m\} = I \cup \bar{I},$$

and we also denote†

$$I_c = \{m+1, \ldots, r\}.$$

At the typical iteration of the algorithm, we have for each $i \in \bar{I}$, a finite set X_i such that $\partial f_i(x_i) \neq \emptyset$ for all $x_i \in X_i$, and for each $i \in I_c$ a finite set $X_i \subset C_i$. The iteration is as follows.

Typical Iteration of Simplicial Decomposition for Conic Constraints

Step 1: (Approximate Problem Solution) Find a primal and dual optimal solution pair

$$(\hat{x}, \hat{\lambda}) = (\hat{x}_1, \ldots, \hat{x}_r, \hat{\lambda}_1, \ldots, \hat{\lambda}_r)$$

of the problem

$$\begin{aligned} &\text{minimize} \quad \sum_{i \in I} f_i(x_i) + \sum_{i \in \bar{I}} \bar{f}_{i,X_i}(x_i) + \sum_{i \in I_c} \delta\big(x_i \mid \text{cone}(X_i)\big) \\ &\text{subject to} \quad (x_1, \ldots, x_r) \in S, \end{aligned} \tag{4.55}$$

where $\bar{f}_{i,X_i}$ are the inner linearizations of f_i corresponding to X_i, $i \in \bar{I}$.

Step 2: (Test for Termination and Enlargement) Enlarge the sets X_i as follows (see Fig. 4.6.2):

(a) For $i \in \bar{I}$, we add any subgradient $\tilde{x}_i \in \partial f_i^\star(\hat{\lambda}_i)$ to X_i.

(b) For $i \in I_c$, we add the projection $\tilde{x}_i = P_{C_i}(\hat{\lambda}_i)$ to X_i.

If there is no strict enlargement for all $i \in \bar{I}$, i.e., we have $\tilde{x}_i \in X_i$, and moreover $\tilde{x}_i = 0$ for all $i \in I_c$, the algorithm terminates. Otherwise, we proceed to the next iteration, using the enlarged sets X_i.

---

† We allow $\bar{I}$ to be empty, in which case none of the functions f_i is inner linearized. Then the portions of the subsequent algorithmic descriptions and analysis that refer to the functions f_i with $i \in \bar{I}$ should be simply omitted. Also, there is no loss of generality in using $I_c = \{m+1, \ldots, r\}$, since the indicator functions of the cones that are not linearized, may be included within the set of functions f_i, $i \in I$.

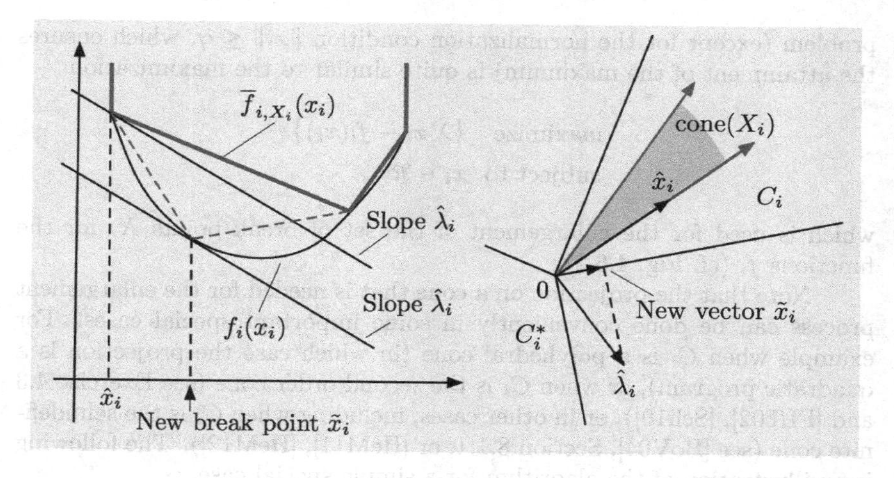

Figure 4.6.2. Illustration of the enlargement step of the algorithm, after we obtain a primal and dual optimal solution pair $(\hat{x}_1, \ldots, \hat{x}_r, \hat{\lambda}_1, \ldots, \hat{\lambda}_r)$. The enlargement step on the left [finding $\tilde{x}_i$ with $\tilde{x}_i \in \partial f_i^\star(\hat{\lambda}_i)$ for $i \in \bar{I}$] is also equivalent to finding $\tilde{x}_i$ satisfying $\hat{\lambda}_i \in \partial f_i(\tilde{x}_i)$, or equivalently, solving the optimization problem

$$\begin{aligned} &\text{maximize} \quad \left\{\hat{\lambda}_i' x_i - f_i(x_i)\right\} \\ &\text{subject to} \quad x_i \in \Re^{n_i}. \end{aligned}$$

The enlargement step on the right, for $i \in I_c$, is to add to X_i the vector $\tilde{x}_i = P_{C_i}(\hat{\lambda}_i)$, the projection on C_i of $\hat{\lambda}_i$.

The enlargement process in the preceding iteration is illustrated in Fig. 4.6.2. Note that we implicitly assume that at each iteration, there exists a primal and dual optimal solution pair of problem (4.55). The algorithm for finding such a pair is left unspecified. Furthermore, we assume that the enlargement step can be carried out, i.e., that $\partial f_i^\star(\hat{\lambda}_i) \neq \emptyset$ for all $i \in \bar{I}$. Sufficient assumptions may need to be imposed on the problem to guarantee that this is so.

The enlargement steps for f_i (left side of Fig. 4.6.2) and for C_i (right side of Fig. 4.6.2) are quite related. Indeed, it can be verified that the projection $\tilde{x}_i = P_{C_i}(\hat{\lambda}_i)$ can be obtained as a positive multiple of the solution of the problem

$$\begin{aligned} &\text{maximize} \quad \left\{\hat{\lambda}_i' x_i - \delta(x_i \mid C_i)\right\} \\ &\text{subject to} \quad \|x_i\| \le \gamma, \end{aligned}$$

where γ is any positive scalar, and $\|\cdot\|$ denotes the Euclidean norm.† This

† To see this, write the problem as

$$\begin{aligned} &\text{minimize} \quad -\hat{\lambda}_i' x_i \\ &\text{subject to} \quad x_i \in C_i, \ \|x_i\|^2 \le \gamma^2, \end{aligned}$$

problem (except for the normalization condition $\|x_i\| \le \gamma$, which ensures the attainment of the maximum) is quite similar to the maximization

$$\begin{aligned} &\text{maximize} \quad \{\hat{\lambda}_i' x_i - f_i(x_i)\} \\ &\text{subject to} \quad x_i \in \Re^{n_i}, \end{aligned}$$

which is used for the enlargement of the set of break points X_i for the functions f_i (cf. Fig. 4.6.2).

Note that the projection on a cone that is needed for the enlargement process can be done conveniently in some important special cases. For example when C_i is a polyhedral cone (in which case the projection is a quadratic program), or when C_i is the second order cone (see Exercise 4.3 and [FLT02], [Sch10]), or in other cases, including when C_i is the semidefinite cone (see [BoV04], Section 8.1.1, or [HeM11], [HeM12]). The following is an illustration of the algorithm for a simple special case.

Example 4.6.1 (Minimization Over a Cone)

Consider the problem

$$\begin{aligned} &\text{minimize} \quad f(x) \\ &\text{subject to} \quad x \in C, \end{aligned} \tag{4.56}$$

where $f : \Re^n \mapsto (-\infty, \infty]$ is a closed proper convex function and C is a closed convex cone. We reformulate this problem into our basic form (4.50) as

$$\begin{aligned} &\text{minimize} \quad f(x_1) + \delta(x_2 \mid C) \\ &\text{subject to} \quad (x_1, x_2) \in S \overset{\text{def}}{=} \{(x_1, x_2) \mid x_1 = x_2\}. \end{aligned} \tag{4.57}$$

Primal and dual optimal solutions have the form (x^*, x^*) and $(\lambda^*, -\lambda^*)$, respectively, since

$$S^\perp = \{(\lambda_1, \lambda_2) \mid \lambda_1 + \lambda_2 = 0\}.$$

By transcribing our algorithm to this special case, we see that $(\hat{x}^k, \hat{x}^k)$ and $(\hat{\lambda}^k, -\hat{\lambda}^k)$ are optimal primal and dual solutions of the corresponding approximate problem of the algorithm if and only if

$$\hat{x}^k \in \arg \min_{x \in \text{cone}(X^k)} f(x),$$

and

$$\hat{\lambda}^k \in \partial f(\hat{x}^k), \qquad -\hat{\lambda}^k \in N_{\text{cone}(X^k)}(\hat{x}^k), \tag{4.58}$$

(cf. Prop. 4.6.2). Once $\hat{\lambda}^k$ is found, X^k is enlarged by adding $\tilde{x}^k$, the projection of $-\hat{\lambda}^k$ onto C. This construction illustrated in Fig. 4.6.3.

---

introduce a dual variable μ for the constraint $\|x_i\|^2 \le \gamma^2$, and show that if $\hat{\lambda}_i \notin C_i^\star$, then the optimal solution is $\tilde{x}_i = (1/2\mu) P_{C_i}(\hat{\lambda}_i)$.

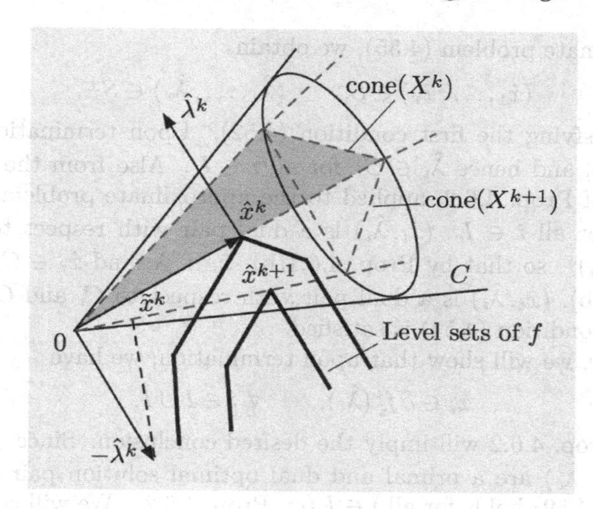

Figure 4.6.3. Illustration of the generalized simplicial decomposition method for minimizing a closed proper convex function f over a cone C (cf. Example 4.6.1). For each k, given the subset $X^k \subset C$, we find a minimum $\hat{x}^k$ of f over cone(X^k), we compute a subgradient $\hat{\lambda}^k \in \partial f(\hat{x}^k)$ such that $-\hat{\lambda}^k$ lies in the normal cone of cone(X^k) at $\hat{x}^k$ [cf. Eq. (4.58)], and we enlarge X^k with $\tilde{x}^k$, the projection of $-\hat{\lambda}^k$ onto C.

When C is a cone generated by a finite set of directions X, there is an interesting variant of the algorithm: we may represent $\tilde{x}^k$ as a positive combination of vectors in X, and simultaneously add all of these vectors to X^k in place of $\tilde{x}^k$. Because X is finite, it can be seen that this version of the algorithm terminates finitely. An example of such a possibility arises in ℓ_1-regularization, where C is the epigraph of the ℓ_1 norm (see Exercise 4.6).

Convergence Analysis

We will now discuss the convergence properties of the algorithm. We first show that if it terminates, it does so at an optimal solution.

Proposition 4.6.3: (Optimality at Termination) If the algorithm of this section terminates at some iteration, the corresponding primal and dual solutions, $(\hat{x}_1, \ldots, \hat{x}_r)$ and $(\hat{\lambda}_1, \ldots, \hat{\lambda}_r)$, form a primal and dual optimal solution pair of problem (4.50).

Proof: We will verify that upon termination, the three conditions of Prop. 4.6.2 are satisfied for the original problem (4.50). From the definition of $(\hat{x}_1, \ldots, \hat{x}_r)$ and $(\hat{\lambda}_1, \ldots, \hat{\lambda}_r)$ as a primal and dual optimal solution pair of the approximate problem (4.55), we obtain

$$(\hat{x}_1, \ldots, \hat{x}_r) \in S, \qquad (\hat{\lambda}_1, \ldots, \hat{\lambda}_r) \in S^\perp,$$

thereby satisfying the first condition (4.52). Upon termination we have $P_{C_i}(\hat{\lambda}_i) = 0$, and hence $\hat{\lambda}_i \in C_i^*$ for all $i \in I_c$. Also from the optimality conditions of Prop. 4.6.2, applied to the approximate problem (4.55), we have that for all $i \in I_c$, $(\hat{x}_i, \hat{\lambda}_i)$ is a dual pair with respect to cone(X_i) and cone$(X_i)^*$, so that by Prop. 4.6.1(b), $\hat{x}_i \perp \hat{\lambda}_i$ and $\hat{x}_i \in C_i$. Thus by Prop. 4.6.1(b), $(\hat{x}_i, \hat{\lambda}_i)$ is a dual pair with respect to C_i and C_i^*, and the optimality condition (4.54) is satisfied.

Finally, we will show that upon termination, we have

$$\hat{x}_i \in \partial f_i^\star(\hat{\lambda}_i), \qquad \forall\, i \in I \cup \bar{I}, \tag{4.59}$$

which by Prop. 4.6.2 will imply the desired conclusion. Since $(\hat{x}_1, \ldots, \hat{x}_r)$ and $(\hat{\lambda}_1, \ldots, \hat{\lambda}_r)$ are a primal and dual optimal solution pair of problem (4.55), Eq. (4.59) holds for all $i \in I$ (cf. Prop. 4.6.2). We will complete the proof by showing that it holds for all $i \in \bar{I}$.

Indeed, let us fix $i \in \bar{I}$ and let $\tilde{x}_i \in \partial f_i^\star(\hat{\lambda}_i)$ be the vector generated by the enlargement step upon termination, so that $\tilde{x}_i \in X_i$. Since $\bar{f}_{i,X_i}$ is an inner linearization of f_i, it follows that $\overline{f}^\star_{i,X_i}$ is an outer linearization of $f_i^\star$ of the form

$$\overline{f}^\star_{i,X_i}(\lambda) = \max_{x \in X_i} \big\{ f_i^\star(\lambda_x) + (\lambda - \lambda_x)'x \big\}. \tag{4.60}$$

where the vectors λ_x can be any vectors such that $x \in \partial f_i^\star(\lambda_x)$. Therefore, the relations $\tilde{x}_i \in X_i$ and $\tilde{x}_i \in \partial f_i^\star(\hat{\lambda}_i)$ imply that

$$\overline{f}^\star_{i,X_i}(\hat{\lambda}_i) = f_i^\star(\hat{\lambda}_i),$$

which by Eq. (4.28), shows that

$$\partial \overline{f}^\star_{i,X_i}(\hat{\lambda}_i) \subset \partial f_i^\star(\hat{\lambda}_i).$$

By Eq. (4.53), we also have $\hat{x}_i \in \partial \overline{f}^\star_{i,X_i}(\hat{\lambda}_i)$, so $\hat{x}_i \in \partial f_i^\star(\hat{\lambda}_i)$. Thus Eq. (4.59) is shown for $i \in \bar{I}$, and all the optimality conditions of Prop. 4.6.2 are satisfied for the original problem (4.50). **Q.E.D.**

The next proposition is a convergence result that is similar to the one we showed in Section 4.4 for the case of pure outer linearization.

Proposition 4.6.4: (Convergence) Consider the algorithm of this section, under the strong duality condition $-\infty < q_{opt} = f_{opt} < \infty$. Let $(\hat{x}^k, \hat{\lambda}^k)$ be the primal and dual optimal solution pair of the approximate problem (4.55), generated at the kth iteration, and let $\tilde{x}_i^k$, $i \in \bar{I}$, be the vectors generated at the corresponding enlargement step. Consider a subsequence $\{\hat{\lambda}^k\}_\mathcal{K}$ that converges to a vector $\hat{\lambda}$. Then:

(a) $\hat{\lambda}_i \in C_i^*$ for all $i \in I_c$.

(b) If the subsequences $\{\tilde{x}_i^k\}_{\mathcal{K}}$, $i \in \bar{I}$, are bounded, $\hat{\lambda}$ is dual optimal, and the optimal value of the inner approximation problem (4.55) converges monotonically from above to f^{opt}, while the optimal value of the dual problem of (4.55) converges monotonically from below to $-f^{opt}$.

Proof: (a) Let us fix $i \in I_c$. Since $\tilde{x}_i^k = P_{C_i}(\hat{\lambda}_i^k)$, the subsequence $\{\tilde{x}_i^k\}_{\mathcal{K}}$ converges to $\tilde{x}_i = P_{C_i}(\hat{\lambda}_i)$. We will show that $\tilde{x}_i = 0$, which implies that $\hat{\lambda}_i \in C_i^*$.

Denote $X_i^\infty = \cup_{k=0}^\infty X_i^k$. Since $\hat{\lambda}_i^k \in \text{cone}(X_i^k)^*$, we have $x_i'\hat{\lambda}_i^k \leq 0$ for all $x_i \in X_i^k$, so that $x_i'\hat{\lambda}_i \leq 0$ for all $x_i \in X_i^\infty$. Since $\tilde{x}_i$ belongs to the closure of X_i^∞, it follows that $\tilde{x}_i'\hat{\lambda}_i \leq 0$. On the other hand, since $\tilde{x}_i = P_{C_i}(\hat{\lambda}_i)$, from Prop. 4.6.1(b) we have $\tilde{x}_i'(\hat{\lambda}_i - \tilde{x}_i) = 0$, which together with $\tilde{x}_i'\hat{\lambda}_i \leq 0$, implies that $\|\tilde{x}_i\|^2 \leq 0$, or $\tilde{x}_i = 0$.

(b) From the definition of $\overline{f}_{i,X_i^k}^\star$ [cf. Eq. (4.60)], we have for all $i \in \bar{I}$ and $k, \ell \in \mathcal{K}$ with $\ell < k$,

$$f_i^\star(\hat{\lambda}_i^\ell) + (\hat{\lambda}_i^k - \hat{\lambda}_i^\ell)'\tilde{x}_i^\ell \leq \overline{f}_{i,X_i^k}^\star(\hat{\lambda}_i^k).$$

Using this relation and the optimality of $\hat{\lambda}^k$ for the kth approximate dual problem to write for all $k, \ell \in \mathcal{K}$ with $\ell < k$

$$\begin{aligned}\sum_{i\in I} f_i^\star(\hat{\lambda}_i^k) + \sum_{i\in\bar{I}}\big(f_i^\star(\hat{\lambda}_i^\ell) + (\hat{\lambda}_i^k - \hat{\lambda}_i^\ell)'\tilde{x}_i^\ell\big) &\leq \sum_{i\in I} f_i^\star(\hat{\lambda}_i^k) + \sum_{i\in\bar{I}} \overline{f}_{i,X_i^k}^\star(\hat{\lambda}_i^k)\\ &\leq \sum_{i\in I} f_i^\star(\lambda_i) + \sum_{i\in\bar{I}} \overline{f}_{i,X_i^k}^\star(\lambda_i),\end{aligned}$$

for all $(\lambda_1, \ldots, \lambda_m)$ such that there exist $\lambda_i \in \text{cone}(X_i^k)^*$, $i \in I_c$, with $(\lambda_1, \ldots, \lambda_r) \in S$. Since $C_i^* \subset \text{cone}(X_i^k)^*$, it follows that

$$\begin{aligned}\sum_{i\in I} f_i^\star(\hat{\lambda}_i^k) + \sum_{i\in\bar{I}}\big(f_i^\star(\hat{\lambda}_i^\ell) + (\hat{\lambda}_i^k - \hat{\lambda}_i^\ell)'\tilde{x}_i^\ell\big) &\leq \sum_{i\in I} f_i^\star(\lambda_i) + \sum_{i\in\bar{I}} \overline{f}_{i,X_i^k}^\star(\lambda_i)\\ &\leq \sum_{i=1}^m f_i^\star(\lambda_i),\end{aligned} \tag{4.61}$$

for all $(\lambda_1, \ldots, \lambda_m)$ such that there exist $\lambda_i \in C_i^*$, $i \in I_c$, with $(\lambda_1, \ldots, \lambda_r) \in S$, where the last inequality holds since $\overline{f}_{i,X_i^k}^\star$ is an outer linearization of $f_i^\star$.

By taking limit inferior in Eq. (4.61), as $k, \ell \to \infty$ with $k, \ell \in \mathcal{K}$, and by using the lower semicontinuity of $f_i^\star$, which implies that

$$f_i^\star(\hat{\lambda}_i) \le \liminf_{\ell \to \infty,\, \ell \in \mathcal{K}} f_i^\star(\hat{\lambda}_i^\ell), \qquad i \in I_c,$$

we obtain

$$\sum_{i=1}^m f_i^\star(\hat{\lambda}_i) \le \sum_{i=1}^m f_i^\star(\lambda_i) \tag{4.62}$$

for all $(\lambda_1, \ldots, \lambda_m)$ such that there exist $\lambda_i \in C_i^*$, $i \in I_c$, with $(\lambda_1, \ldots, \lambda_r) \in S$. We have $\hat{\lambda} \in S$ and $\hat{\lambda}_i \in C_i^*$ for all $i \in I_c$, from part (a). Thus Eq. (4.62) implies that $\hat{\lambda}$ is dual optimal. The sequence of optimal values of the dual approximation problem [the dual of problem (4.55)] is monotonically nondecreasing (since the outer approximation is monotonically refined) and converges to $-f^{opt}$ since $\hat{\lambda}$ is dual optimal. This sequence is the opposite of the sequence of optimal values of the primal approximation problem (4.55), so the latter sequence is monotonically nonincreasing and converges to f^{opt}. **Q.E.D.**

As in Prop. 4.4.3 (cf. the GPA algorithm), the preceding proposition leaves open the question whether there exists a convergent subsequence $\{\hat{\lambda}^k\}_{\mathcal{K}}$, and whether the corresponding subsequences $\{\tilde{x}_i^k\}_{\mathcal{K}}$, $i \in \bar{I}$, are bounded. This must be verified separately, for the problem at hand.

4.7 NOTES, SOURCES, AND EXERCISES

Section 4.1: Cutting plane methods were introduced by Cheney and Goldstein [ChG59], and by Kelley [Kel60]. For analysis of related methods, see Ruszczynski [Rus86], Mifflin [Mif96], Burke and Qian [BuQ98], Mifflin, Sun, and Qi [MSQ98], and Bonnans et al. [BGL09].

Section 4.2: The simplicial decomposition method was introduced as an improvement of the conditional gradient method by Holloway [Hol74]; see also Hohenbalken [Hoh77], Pang and Yu [PaY84], Hearn, Lawphongpanich, and Ventura [HLV87], Ventura and Hearn [VeH93], and Patriksson [Pat01]. The method was also independently proposed in the context of multicommodity flow problems by Cantor and Gerla [CaG74]. Some of these references describe applications to communication and transportation networks; see also the surveys by Florian and Hearn [FlH95], Patriksson [Pat04], the nonlinear programming textbook [Ber99] (Examples 2.1.3 and 2.1.4), and the discussion of the application of gradient projection methods in [BeG83], [BeG92]. Simplicial decomposition in a dual setting for problems with a large number of constraints (Exercise 4.4), was proposed by Huizhen Yu, and was developed in the context of some large-scale parameter estimation/machine learning problems in the papers [YuR07] and [YBR08].

Section 4.3: The duality relation between outer and inner linearization has been known for a long time, particularly in the context of the Dantzig-Wolfe decomposition algorithm [DaW60], which is a cutting plane/simplicial decomposition algorithm applied to separable problems (see textbooks such as [Las70], [BeT97], [Ber99] for descriptions and analysis). Our development of the conjugacy-based form of this duality follows the paper by Bertsekas and Yu [BeY11].

Section 4.4: The generalized polyhedral approximation algorithm is due to Bertsekas and Yu [BeY11], which contains a detailed convergence analysis. Extended monotropic programming and its duality theory were developed in the author's paper [Ber10a], and will be discussed in greater detail in Section 6.7.

Section 4.5: The generalized simplicial decomposition material of this section follows the paper [BeY11]. Note that there is no known extension of the conditional gradient method of Section 2.1 that works with a nondifferentiable cost. Another method for minimizing a nondifferentiable convex function over a polyhedral set, based on concepts of ergodic sequences of subgradients and a conditional subgradient method, is given by Larsson, Patriksson, and Stromberg (see [Str97], [LPS98]).

Section 4.6: The simplicial decomposition algorithm with conical approximations is new and was developed as the book was being written.

EXERCISES

4.1 (Computational Exercise)

Consider using the cutting plane method for finding a solution of a system of inequality constraints $g_i(x) \leq 0$, $i = 1, \ldots, m$, where $g_i : \Re^n \mapsto \Re$ are convex functions. Formulate this as a problem of unconstrained minimization of the convex function

$$f(x) = \max_{i=1,\ldots,m} g_i(x).$$

(a) State the cutting plane method, making sure that the method is well-defined.

(b) Implement the method of part (a) for the case where $g_i(x) = c_i'x - b_i$, $n = 2$, and $m = 100$. The vectors c_i have the form $c_i = (\xi_i, \zeta_i)$, where ξ_i, ζ_i are chosen randomly and independently within $[-1, 1]$ according to a uniform distribution, while b_i is chosen randomly and independently within $[0, 1]$ according to a uniform distribution. Does the method converge in a finite number of iterations? Is the problem solved after a finite number of iterations? How can you monitor the progress of the method towards optimality using upper and lower bounds?

4.2

Consider the conic programming problem (4.50) of Section 4.6,

$$\begin{aligned}&\text{minimize} \quad \sum_{i=1}^{m} f_i(x_i) + \sum_{i=m+1}^{r} \delta(x_i \mid C_i)\\ &\text{subject to} \quad (x_1, \ldots, x_r) \in S.\end{aligned}$$

Verify that an appropriate dual problem is the one of Eq. (4.51):

$$\begin{aligned}&\text{minimize} \quad \sum_{i=1}^{m} f_i^\star(\lambda_i) + \sum_{i=m+1}^{r} \delta(\lambda_i \mid C_i^*)\\ &\text{subject to} \quad (\lambda_1, \ldots, \lambda_r) \in S^\perp.\end{aligned}$$

Verify also the optimality conditions of Prop. 4.6.2.

4.3 (Projection on the Second Order Cone)

Consider the second order cone in $\Re^n$:

$$C = \left\{ (x_1, \ldots, x_n) \;\middle|\; x_n \geq \sqrt{x_1^2 + \cdots + x_{n-1}^2} \right\},$$

and the problem of Euclidean projection of a given vector $\bar{x} = (\bar{x}_1, \ldots, \bar{x}_n)$ onto C. Let $\bar{z} \in \Re^{n-1}$ be the vector $\bar{z} = (\bar{x}_1, \ldots, \bar{x}_{n-1})$. Show that the projection, denoted $\hat{x}$, is given by

$$\hat{x} = \begin{cases} \bar{x} & \text{if } \|\bar{z}\| \leq \bar{x}_n, \\ \frac{\|\bar{z}\| + \bar{x}_n}{2} \left(\frac{\bar{z}}{\|\bar{z}\|}, 1 \right) & \text{if } \|\bar{z}\| > \bar{x}_n, \ \|\bar{z}\| + \bar{x}_n > 0, \\ 0 & \text{if } \|\bar{z}\| > \bar{x}_n, \ \|\bar{z}\| + \bar{x}_n \leq 0. \end{cases}$$

Note: For a derivation of this formula, together with derivations of projection formulas for other cones, see [Sch10].

4.4 (Dual Conic Simplicial Decomposition and Primal Constraint Aggregation)

In this exercise the simplicial decomposition approach is applied to the dual of a constrained optimization problem, using the conic approximation framework of Section 4.6. Consider the problem

$$\begin{aligned}&\text{minimize} \quad f(x)\\ &\text{subject to} \quad Ax \leq 0, \qquad x \in X,\end{aligned}$$

where $f : \Re^n \mapsto \Re$ is a convex function, X is a convex set, and A is an $m \times n$ matrix.

(a) Derive a dual problem of the form

$$\begin{aligned} &\text{maximize} \quad h(\xi) \\ &\text{subject to} \quad \xi \in C, \end{aligned}$$

where

$$h(\xi) = \inf_{x \in X} \{f(x) + \xi' x\},$$

and C is the cone $\{A'\mu \mid \mu \geq 0\}$, the polar of the cone $\{x \mid Ax \leq 0\}$.

(b) Suppose that the cone C of the dual problem of (a) is approximated by a polyhedral cone of the form

$$\overline{C} = \text{cone}(\overline{\xi}_1, \ldots, \overline{\xi}_m),$$

where $\overline{\xi}_1, \ldots, \overline{\xi}_m$ are m vectors from C. Show that the resulting approximate problem is dual to the problem

$$\begin{aligned} &\text{minimize} \quad f(x) \\ &\text{subject to} \quad \overline{\mu}_i' Ax \leq 0, \quad i = 1, \ldots, m, \quad x \in X, \end{aligned}$$

where $\overline{\mu}_i$ satisfies $\overline{\mu}_i \geq 0$ and $\overline{\xi}_i = A'\overline{\mu}_i$. Show also that the constraint set of this approximate problem is an outer linearization of the original, and interpret the constraints $\overline{\mu}_i' Ax \leq 0$ as aggregate inequality constraints, (i.e., nonnegative combinations of constraints).

(c) Explain why the duality of parts (a) and (b) is a special case of the conic approximation duality framework of Section 4.6.

(d) Generalize the analysis of parts (a) and (b) for the case where the constraint $Ax \leq 0$ is replaced by $Ax \leq b$. *Hint*: Derive a dual problem of the form

$$\begin{aligned} &\text{maximize} \quad h(\xi) - \zeta \\ &\text{subject to} \quad (\xi, \zeta) \in C, \end{aligned}$$

where $h(\xi) = \inf_{x \in X} \{f(x) + \xi' x\}$ and C is the cone $\{(A'\mu, b'\mu) \mid \mu \geq 0\}$.

4.5 (Conic Simplicial Decomposition with Vector Sum Constraints)

The algorithms and analysis of Section 4.6 apply to cases where the constraint set involves the intersection of compact sets and cones, which can be inner linearized separately (the compact set constraints can be represented as indicator functions via the functions f_i). This exercise deals with the related case where the constraints are vector sums of compact sets and cones, which again can be linearized separately. Describe how the algorithm of Section 4.6 can be applied to the problem

$$\begin{aligned} &\text{minimize} \quad f(x) \\ &\text{subject to} \quad x \in X + C, \end{aligned}$$

where X is a compact set and C is a closed convex cone. *Hint*: Write the problem as

$$\begin{aligned} &\text{minimize} \quad f(x_1) + \delta(x_2|X) + \delta(x_3|C) \\ &\text{subject to} \quad x_1 = x_2 + x_3, \end{aligned}$$

which is of the form (4.50) with

$$S = \{(x_1, x_2, x_3) \mid x_1 = x_2 + x_3\}.$$

4.6 (Conic Polyhedral Approximations of Positively Homogeneous Functions – ℓ_1 Regularization)

We recall that an extended real-valued function is said to be positively homogeneous if its epigraph is a cone (see Section 1.6 of Appendix B); examples of such functions include norms and, more generally, support functions of sets. Consider the minimization of a sum $f + h$ of closed proper convex functions such that h is positively homogeneous.

(a) Show that this problem is equivalent to

$$\begin{aligned} &\text{minimize} \quad f(x) + w \\ &\text{subject to} \quad (x, w) \in \text{epi}(h), \end{aligned}$$

and describe how the conic polyhedral approximation algorithm of Section 4.6 (cf. Example 4.6.1) can be applied.

(b) (*Conical Approximation of the ℓ_1 Norm*) Consider the problem

$$\begin{aligned} &\text{minimize} \quad f(x) + \|x\|_1 \\ &\text{subject to} \quad x \in \Re^n, \end{aligned}$$

where $f : \Re^n \mapsto \Re$ is a convex function, and the equivalent problem

$$\begin{aligned} &\text{minimize} \quad f(x) + w \\ &\text{subject to} \quad (x, w) \in C, \end{aligned}$$

where $C \subset \Re^{n+1}$ is the cone $C = \{(x, w) \mid \|x\|_1 \le w\}$. Describe how the algorithm of Example 4.6.1 can be applied. Discuss a finitely terminating variant, where the cone C is approximated with a cone generated exclusively by coordinate directions of the form $(e_i, 1)$, where $e_i = (0, \ldots, 0, 1, 0, \ldots, 0)$, with the 1 in the ith position.

5

Proximal Algorithms

Contents

In this chapter, we continue our discussion of iterative approximation methods for minimizing a convex function f. In particular the generated sequence $\{x_k\}$ is obtained by solving at each k an approximate problem,

$$x_{k+1} \in \arg\min_{x \in \Re^n} F_k(x),$$

where F_k is a function that approximates f. However, unlike the preceding chapter, F_k is not polyhedral. Instead F_k is obtained by adding to f a quadratic regularization term centered at the current iterate x_k, and weighted by a positive scalar parameter c_k. This is a fundamental algorithm, with broad extensions, which can also be combined with the subgradient and polyhedral approximation approaches of Chapters 3 and 4, as well as the incremental approach of Section 3.3.

We develop the basic theory of the method in Section 5.1, setting the stage for related methods, which are discussed in subsequent sections in this chapter and in Chapter 6. In Section 5.2, we consider a dual version of the algorithm, which among others yields the popular augmented Lagrangian method for constrained optimization. In Section 5.3, we discuss variations of the algorithm, which include combinations with the polyhedral approximation methods of Chapter 4. In Section 5.4, we develop another type of augmented Lagrangian method, the alternating direction method of multipliers, which is well suited for the special structure of several types of large-scale problems. In Chapter 6, we will revisit the proximal algorithm in the context of combinations with gradient and subgradient methods, and develop generalizations where the regularization term is not quadratic.

5.1 BASIC THEORY OF PROXIMAL ALGORITHMS

In this section we consider the minimization of a closed proper convex function $f : \Re^n \mapsto (-\infty, \infty]$ using an approximation approach whereby we modify f by adding a regularization term. In particular, we consider the algorithm

$$x_{k+1} \in \arg\min_{x \in \Re^n} \left\{ f(x) + \frac{1}{2c_k}\|x - x_k\|^2 \right\}, \tag{5.1}$$

where x_0 is an arbitrary starting point and c_k is a positive scalar parameter; see Fig. 5.1.1. This is the *proximal algorithm* (also known as the *proximal minimization algorithm* or the *proximal point algorithm*).

The degree of regularization is controlled by the parameter c_k. For small values of c_k, x_{k+1} tends to stay close to x_k, albeit at the expense of slower convergence. The convergence mechanism is illustrated in Fig. 5.1.2. Note that the quadratic term $\|x - x_k\|^2$ makes the function that is minimized at each iteration strictly convex with compact level sets. This guarantees, among others, that x_{k+1} is well-defined as the unique minimum

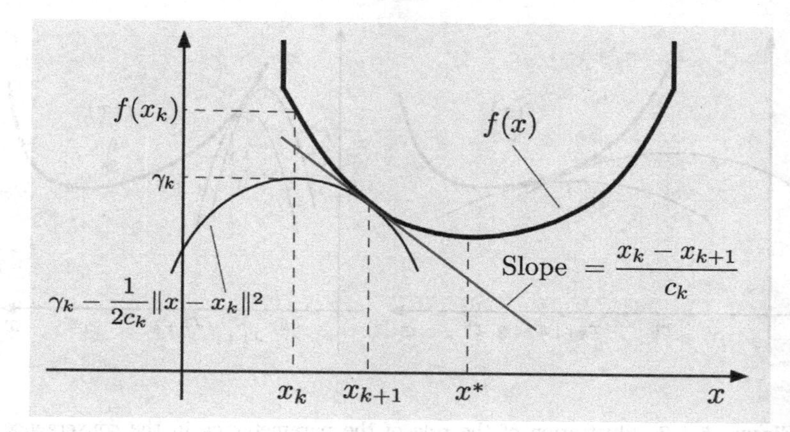

Figure 5.1.1. Geometric view of the proximal algorithm (5.1). The minimum of

$$f(x) + \frac{1}{2c_k}\|x - x_k\|^2$$

is attained at a unique point x_{k+1} as shown. In this figure, γ_k is the scalar by which the graph of the quadratic $-\frac{1}{2c_k}\|x - x_k\|^2$ must be raised so that it just touches the graph of f. The slope shown in the figure,

$$\frac{x_k - x_{k+1}}{c_k},$$

is the common subgradient of $f(x)$ and $-\frac{1}{2c_k}\|x - x_k\|^2$ at the minimizing point x_{k+1}, cf. the Fenchel Duality Theorem (Prop. 1.2.1).

in Eq. (5.1) [cf. Prop. 3.1.1 and Prop. 3.2.1 in Appendix B; also the broader discussion of existence of minima in Chapter 3 of [Ber09]].

Evidently, the algorithm is useful only for problems that can benefit from regularization. It turns out, however, that many interesting problems fall in this category, and often in unexpected and diverse ways. In particular, as we will see in this and the next chapter, the creative application of the proximal algorithm and its variations, together with duality ideas, can allow the elimination of constraints and nondifferentiabilities, the stabilization of the linear approximation methods of Chapter 4, and the effective exploitation of special problem structures.

5.1.1 Convergence

The proximal algorithm has excellent convergence properties, which we develop in this section. We first derive some preliminary results in the following two propositions.

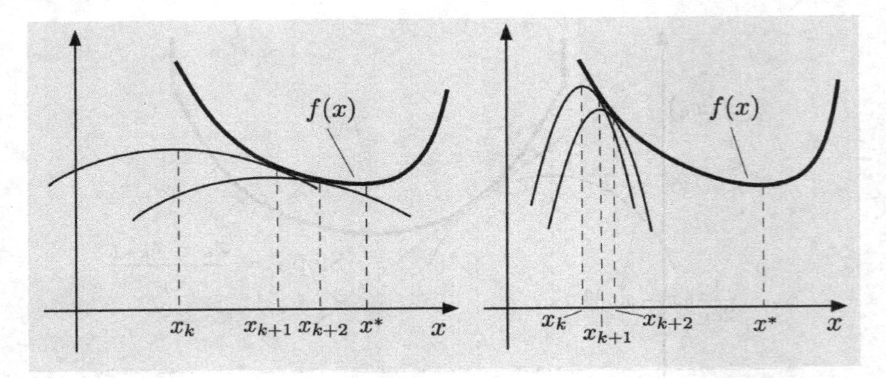

Figure 5.1.2. Illustration of the role of the parameter c_k in the convergence process of the proximal algorithm. In the figure on the left, c_k is large, the graph of the quadratic term is "blunt," and the method makes fast progress toward the optimal solution. In the figure on the right, c_k is small, the graph of the quadratic term is "pointed," and the method makes slow progress.

Proposition 5.1.1: If x_k and x_{k+1} are two successive iterates of the proximal algorithm (5.1), we have

$$\frac{x_k - x_{k+1}}{c_k} \in \partial f(x_{k+1}). \tag{5.2}$$

Proof: Since the function

$$f(x) + \frac{1}{2c_k}\|x - x_k\|^2$$

is minimized at x_{k+1}, the origin must belong to its subdifferential at x_{k+1}, which is equal to

$$\partial f(x_{k+1}) + \frac{x_{k+1} - x_k}{c_k},$$

(cf. Prop. 5.4.6 in Appendix B, which applies because its relative interior condition is satisfied since the quadratic term is real-valued). The desired relation (5.2) holds if and only if the origin belongs to the above set. **Q.E.D.**

The preceding proposition may be visualized from Fig. 5.1.1. An interesting observation is that the move from x_k to x_{k+1} is "nearly" a subgradient step [it would be a subgradient step if $\partial f(x_{k+1})$ were replaced by $\partial f(x_k)$ in Eq. (5.2)]. This fact will provide motivation later for combinations of the proximal algorithm with the subgradient method (see Section 6.4).

Generally, starting from any nonoptimal point x_k, the cost function value is reduced at each iteration, since from the minimization in the algorithm's definition [cf. Eq. (5.1)], by setting $x = x_k$, we have

$$f(x_{k+1}) + \frac{1}{2c_k}\|x_{k+1} - x_k\|^2 \leq f(x_k).$$

The following proposition provides an inequality, which among others shows that the iterate distance to any optimal solution is also reduced. This inequality resembles (but is more favorable than) the fundamental inequality of Prop. 3.2.2(a) for the subgradient method.

Proposition 5.1.2: (Three-Term Inequality) Consider a closed proper convex function $f : \Re^n \mapsto (-\infty, \infty]$, and for any $x_k \in \Re^n$ and $c_k > 0$, the proximal algorithm (5.1). Then for all $y \in \Re^n$, we have

$$\|x_{k+1} - y\|^2 \leq \|x_k - y\|^2 - 2c_k\big(f(x_{k+1}) - f(y)\big) - \|x_k - x_{k+1}\|^2. \quad (5.3)$$

Proof: We have

$$\begin{aligned}\|x_k - y\|^2 &= \|x_k - x_{k+1} + x_{k+1} - y\|^2 \\ &= \|x_k - x_{k+1}\|^2 + 2(x_k - x_{k+1})'(x_{k+1} - y) + \|x_{k+1} - y\|^2.\end{aligned}$$

Using Eq. (5.2) and the definition of subgradient, we obtain

$$\frac{1}{c_k}(x_k - x_{k+1})'(x_{k+1} - y) \geq f(x_{k+1}) - f(y).$$

By multiplying this relation with $2c_k$ and adding it to the preceding relation, the result follows. **Q.E.D.**

Let us denote by f^* the optimal value

$$f^* = \inf_{x \in \Re^n} f(x),$$

(which may be $-\infty$) and by X^* the set of minima of f (which may be empty),

$$X^* = \arg\min_{x \in \Re^n} f(x).$$

The following is the basic convergence result for the proximal algorithm.

Proposition 5.1.3: (Convergence) Let $\{x_k\}$ be a sequence generated by the proximal algorithm (5.1). Then, if $\sum_{k=0}^{\infty} c_k = \infty$, we have

$$f(x_k) \downarrow f^*,$$

and if X^* is nonempty, $\{x_k\}$ converges to some point in X^*.

Proof: We first note that since x_{k+1} minimizes $f(x)+\frac{1}{2c_k}\|x-x_k\|^2$, we have by setting $x=x_k$,

$$f(x_{k+1})+\frac{1}{2c_k}\|x_{k+1}-x_k\|^2 \le f(x_k), \qquad \forall\, k.$$

It follows that $\{f(x_k)\}$ is monotonically nonincreasing. Hence $f(x_k)\downarrow f_\infty$, where f_∞ is either a scalar or $-\infty$, and satisfies $f_\infty \ge f^*$.

From Eq. (5.3), we have for all $y\in\Re^n$,

$$\|x_{k+1}-y\|^2 \le \|x_k-y\|^2 - 2c_k\big(f(x_{k+1})-f(y)\big). \tag{5.4}$$

By adding this inequality over $k=0,\ldots,N$, we obtain

$$\|x_{N+1}-y\|^2 + 2\sum_{k=0}^{N} c_k\big(f(x_{k+1})-f(y)\big) \le \|x_0-y\|^2, \quad \forall\, y\in\Re^n,\ N\ge 0,$$

so that

$$2\sum_{k=0}^{N} c_k\big(f(x_{k+1})-f(y)\big) \le \|x_0-y\|^2, \qquad \forall\, y\in\Re^n,\ N\ge 0.$$

Taking the limit as $N\to\infty$, we have

$$2\sum_{k=0}^{\infty} c_k\big(f(x_{k+1})-f(y)\big) \le \|x_0-y\|^2, \qquad \forall\, y\in\Re^n. \tag{5.5}$$

Assume to arrive at a contradiction that $f_\infty > f^*$, and let $\hat y$ be such that

$$f_\infty > f(\hat y) > f^*.$$

Since $\{f(x_k)\}$ is monotonically nonincreasing, we have

$$f(x_{k+1}) - f(\hat y) \ge f_\infty - f(\hat y) > 0.$$

Then in view of the assumption $\sum_{k=0}^{\infty} c_k=\infty$, Eq. (5.5), with $y=\hat y$, leads to a contradiction. Thus $f_\infty=f^*$.

Consider now the case where X^* is nonempty, and let x^* be any point in X^*. Applying Eq. (5.4) with $y=x^*$, we have

$$\|x_{k+1}-x^*\|^2 \le \|x_k-x^*\|^2 - 2c_k\big(f(x_{k+1})-f(x^*)\big), \qquad k=0,1,\ldots. \tag{5.6}$$

From this relation it follows that $\|x_k-x^*\|^2$ is monotonically nonincreasing, so $\{x_k\}$ is bounded. If $\overline{x}$ is a limit point of $\{x_k\}$, we have

$$f(\overline{x}) \le \liminf_{k\to\infty,\, k\in\mathcal{K}} f(x_k) = f^*$$

for any subsequence $\{x_k\}_{\mathcal{K}}\to\overline{x}$, since $\{f(x_k)\}$ monotonically decreases to f^* and f is closed. Hence $\overline{x}$ must belong to X^*. Finally, by Eq. (5.6), the distance of x_k to every $x^*\in X^*$ is monotonically nonincreasing, so $\{x_k\}$ must converge to a unique point in X^*. **Q.E.D.**

Note some remarkable properties from the preceding proposition. Convergence to the optimal value is obtained even if X^* is empty or $f^*=-\infty$. Moreover, when X^* is nonempty, convergence to a single point of X^* occurs.

5.1.2 Rate of Convergence

The following proposition describes how the convergence rate of the proximal algorithm depends on the magnitude of c_k and on the order of growth of f near the optimal solution set (see also Fig. 5.1.3).

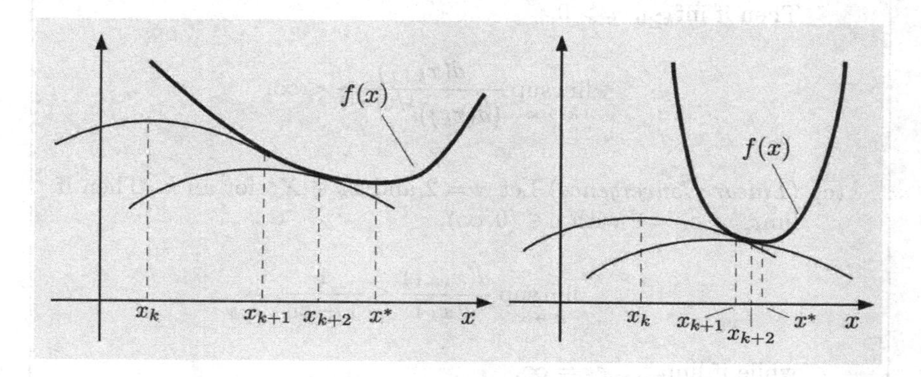

Figure 5.1.3. Illustration of the convergence rate of the proximal algorithm and the effect of the growth properties of f near the optimal solution set. In the figure on the left, f grows slowly and the convergence is slow. In the figure on the right, f grows fast and the convergence is fast.

Proposition 5.1.4: (Rate of Convergence) Assume that X^* is nonempty and that for some scalars $\beta > 0$, $\delta > 0$, and $\gamma \geq 1$, we have

$$f^* + \beta\big(d(x)\big)^\gamma \leq f(x), \qquad \forall\, x \in \Re^n \text{ with } d(x) \leq \delta, \tag{5.7}$$

where

$$d(x) = \min_{x^* \in X^*} \|x - x^*\|.$$

Let also

$$\sum_{k=0}^{\infty} c_k = \infty,$$

so that the sequence $\{x_k\}$ generated by the proximal algorithm (5.1) converges to some point in X^* by Prop. 5.1.3. Then:

(a) For all k sufficiently large, we have

$$d(x_{k+1}) + \beta c_k \big(d(x_{k+1})\big)^{\gamma-1} \leq d(x_k), \tag{5.8}$$

if $\gamma > 1$, and

$$d(x_{k+1}) + \beta c_k \leq d(x_k), \tag{5.9}$$

if $\gamma = 1$ and $x_{k+1} \notin X^*$.

(b) (*Superlinear Convergence*) Let $1 < \gamma < 2$ and $x_k \notin X^*$ for all k. Then if $\inf_{k \geq 0} c_k > 0$,

$$\limsup_{k \to \infty} \frac{d(x_{k+1})}{\big(d(x_k)\big)^{1/(\gamma-1)}} < \infty.$$

(c) (*Linear Convergence*) Let $\gamma = 2$ and $x_k \notin X^*$ for all k. Then if $\lim_{k \to \infty} c_k = \bar{c}$ with $\bar{c} \in (0, \infty)$,

$$\limsup_{k \to \infty} \frac{d(x_{k+1})}{d(x_k)} \leq \frac{1}{1 + \beta \bar{c}},$$

while if $\lim_{k \to \infty} c_k = \infty$,

$$\lim_{k \to \infty} \frac{d(x_{k+1})}{d(x_k)} = 0.$$

(d) (*Sublinear Convergence*) Let $\gamma > 2$. Then

$$\limsup_{k \to \infty} \frac{d(x_{k+1})}{d(x_k)^{2/\gamma}} < \infty.$$

Proof: (a) The proof uses an argument that can be visualized from Fig. 5.1.4. Since the conclusion clearly holds when $x_{k+1} \in X^*$, we assume that $x_{k+1} \notin X^*$ and we denote by $\hat{x}_{k+1}$ and $\hat{x}_k$ the projections of x_{k+1} and x_k on X^*, respectively. From the subgradient relation (5.2), we have

$$f(x_{k+1}) + \frac{1}{c_k}(x_k - x_{k+1})'(\hat{x}_{k+1} - x_{k+1}) \leq f(\hat{x}_{k+1}) = f^*.$$

Using the hypothesis, $\{x_k\}$ converges to some point in X^*, so it follows from Eq. (5.7) that

$$f^* + \beta\big(d(x_{k+1})\big)^\gamma \leq f(x_{k+1}),$$

for k sufficiently large. Adding the preceding two relations, we obtain

$$\beta c_k \big(d(x_{k+1})\big)^\gamma \leq (x_{k+1} - \hat{x}_{k+1})'(x_k - x_{k+1}), \tag{5.10}$$

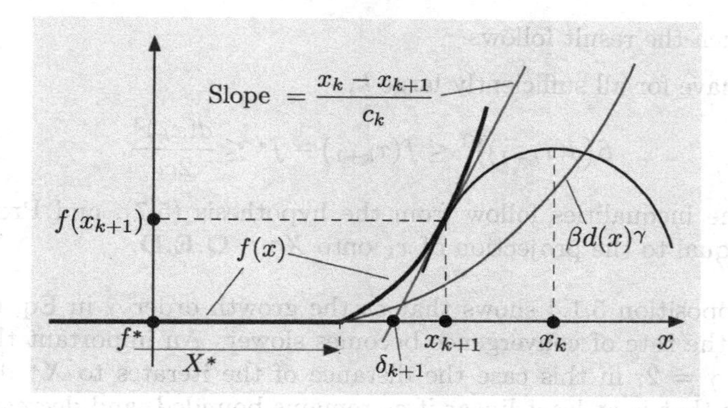

Figure 5.1.4. Visualization of the estimate $d(x_{k+1})+\beta c_k\big(d(x_{k+1})\big)^{\gamma-1} \le d(x_k)$, cf. Eq. (5.8), in one dimension. Using the hypothesis (5.7), and the triangle geometry indicated, we have

$$\begin{aligned}\beta\big(d(x_{k+1})\big)^\gamma &\le f(x_{k+1}) - f^* \\ &= \frac{x_k - x_{k+1}}{c_k}\cdot(x_{k+1}-\delta_{k+1}) \\ &\le \frac{d(x_k)-d(x_{k+1})}{c_k}\cdot d(x_{k+1}),\end{aligned}$$

where δ_{k+1} is the scalar shown in the figure. Canceling $d(x_{k+1})$ from both sides, we obtain Eq. (5.8).

for k sufficiently large. We also write the identity

$$\|x_{k+1}-\hat{x}_{k+1}\|^2-(x_{k+1}-\hat{x}_{k+1})'(x_{k+1}-\hat{x}_k) = (x_{k+1}-\hat{x}_{k+1})'(\hat{x}_k-\hat{x}_{k+1}),$$

and note that since $\hat{x}_{k+1}$ is the projection of x_{k+1} on X^* and $\hat{x}_k \in X^*$, by the Projection Theorem the above expression is nonpositive. We thus have

$$\|x_{k+1}-\hat{x}_{k+1}\|^2 \le (x_{k+1}-\hat{x}_{k+1})'(x_{k+1}-\hat{x}_k),$$

which by adding to Eq. (5.10) and using the Schwarz inequality, yields

$$\begin{aligned}\|x_{k+1}-\hat{x}_{k+1}\|^2+\beta c_k\big(d(x_{k+1})\big)^\gamma &\le (x_{k+1}-\hat{x}_{k+1})'(x_k-\hat{x}_k) \\ &\le \|x_{k+1}-\hat{x}_{k+1}\|\,\|x_k-\hat{x}_k\|.\end{aligned}$$

Dividing with $\|x_{k+1}-\hat{x}_{k+1}\|$ (which is nonzero since we assumed that $x_{k+1}\notin X^*$), Eqs. (5.8) and (5.9) follow.

(b) Since $\gamma < 2$, $d(x_{k+1})$ is dominated by $\beta c_k\big(d(x_{k+1})\big)^{\gamma-1}$ when $d(x_{k+1})$ is sufficiently small, and the desired relation follows from Eq. (5.8).

(c) For $\gamma = 2$, Eq. (5.8) becomes

$$(1+\beta c_k)d(x_{k+1}) \le d(x_k),$$

from which the result follows.

(d) We have for all sufficiently large k,

$$\beta\big(d(x_{k+1})\big)^{\gamma} \leq f(x_{k+1}) - f^* \leq \frac{d(x_k)^2}{2c_k},$$

where the inequalities follow from the hypothesis (5.7), and Prop. 5.1.2 with y equal to the projection of x_k onto X^*. **Q.E.D.**

Proposition 5.1.4 shows that as the growth order γ in Eq. (5.7) increases, the rate of convergence becomes slower. An important threshold value is $\gamma = 2$; in this case the distance of the iterates to X^* decreases at a rate that is at least linear if c_k remains bounded, and decreases even faster (superlinearly) if $c_k \to \infty$. Generally, the convergence is accelerated if c_k is increased with k, rather than kept constant; this is illustrated most clearly when $\gamma = 2$ [cf. Prop. 5.1.4(c)]. When $1 < \gamma < 2$, the convergence rate is faster than linear (superlinear) [cf. Prop. 5.1.4(b)]. When $\gamma > 2$, the convergence rate is generally slower than when $\gamma = 2$, and examples show that $d(x_k)$ may converge to 0 sublinearly, i.e., slower than any geometric progression [cf. Prop. 5.1.4(d)].

The threshold value of $\gamma = 2$ for linear convergence is related to the quadratic growth property of the regularization term. A generalized version of the proposition, with similar proof, is possible for proximal algorithms that use nonquadratic regularization functions (see [KoB76], and [Ber82a], Section 3.5, and also Example 6.6.5 in Section 6.6). In this context, the threshold value for linear convergence is related to the order of growth of the regularization function.

When $\gamma = 1$, f is said to have a *sharp minimum*, a favorable condition that we encountered in Chapter 3. Then the proximal algorithm converges finitely. This is shown in the following proposition (see also Fig. 5.1.5).

Proposition 5.1.5: (Finite Convergence) Assume that the set of minima X^* of f is nonempty and that there exists a scalar $\beta > 0$ such that

$$f^* + \beta d(x) \leq f(x), \qquad \forall\, x \in \Re^n, \tag{5.11}$$

where $d(x) = \min_{x^* \in X^*} \|x - x^*\|$. Then if $\sum_{k=0}^{\infty} c_k = \infty$, the proximal algorithm (5.1) converges to X^* finitely (i.e., there exists $\overline{k} > 0$ such that $x_k \in X^*$ for all $k \geq \overline{k}$). Furthermore, if $c_0 \geq d(x_0)/\beta$, the algorithm converges in a single iteration (i.e., $x_1 \in X^*$).

Proof: The assumption (5.7) of Prop. 5.1.4 holds with $\gamma = 1$ and all $\delta > 0$, so Eq. (5.9) yields

$$d(x_{k+1}) + \beta c_k \leq d(x_k), \qquad \text{if } x_{k+1} \notin X^*. \tag{5.12}$$

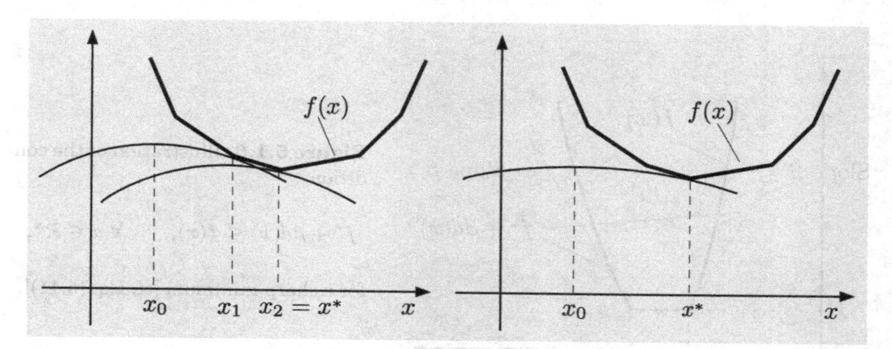

Figure 5.1.5. Finite convergence of the proximal algorithm for the case of a sharp minimum, when $f(x)$ grows at a linear rate near the optimal solution set (e.g., when f is polyhedral). In the figure on the right, convergence occurs in a single iteration for sufficiently large c_0.

If $\sum_{k=0}^{\infty} c_k = \infty$ and $x_k \notin X^*$ for all k, by adding Eq. (5.12) over all k, we obtain a contradiction. Hence we must have $x_k \in X^*$ for k sufficiently large. Also if $c_0 \geq d(x_0)/\beta$, Eq. (5.12) cannot hold with $k = 0$, so we must have $x_1 \in X^*$. **Q.E.D.**

It is also possible to prove the one-step convergence property of Prop. 5.1.5 with a simpler argument that does not rely on Prop. 5.1.4 and Eq. (5.9). Indeed, assume that $x_0 \notin X^*$, let $\hat{x}_0$ be the projection of x_0 on X^*, and consider the function

$$\tilde{f}(x) = f^* + \beta d(x) + \frac{1}{2c_0}\|x - x_0\|^2. \tag{5.13}$$

Its subdifferential at $\hat{x}_0$ is given by the sum formula of Prop. 3.1.3(b):

$$\begin{aligned}\partial \tilde{f}(\hat{x}_0) &= \left\{ \beta\xi \frac{x_0 - \hat{x}_0}{\|x_0 - \hat{x}_0\|} + \frac{1}{c_0}(\hat{x}_0 - x_0) \;\middle|\; \xi \in [0,1] \right\} \\ &= \left\{ \left(\frac{\beta\xi}{d(x_0)} - \frac{1}{c_0} \right)(x_0 - \hat{x}_0) \;\middle|\; \xi \in [0,1] \right\}.\end{aligned}$$

Therefore, if $c_0 \geq d(x_0)/\beta$, then $0 \in \partial \tilde{f}(\hat{x}_0)$, so that $\hat{x}_0$ minimizes $\tilde{f}(x)$. Since from Eqs. (5.11) and (5.13), we have

$$\tilde{f}(x) \leq f(x) + \frac{1}{2c_0}\|x - x_0\|^2, \qquad \forall\, x \in \Re^n,$$

with equality when $x = \hat{x}_0$, it follows that $\hat{x}_0$ minimizes

$$f(x) + \frac{1}{2c_0}\|x - x_0\|^2$$

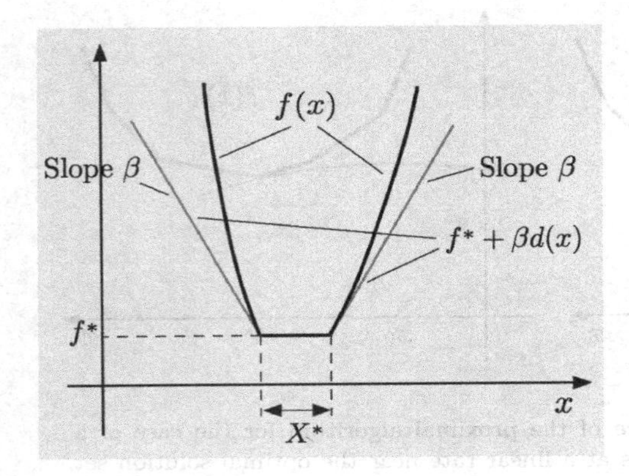

Figure 5.1.6. Illustration of the condition

$$f^* + \beta d(x) \le f(x), \qquad \forall\, x \in \Re^n,$$

for a sharp minimum [cf. Eq. (5.11)].

over $x \in X$. Thus $\hat{x}_0$ is equal to the first iterate x_1 of the proximal algorithm.

The growth condition (5.11) is illustrated in Fig. 5.1.6. The following proposition shows that the condition holds when f is a polyhedral function and X^* is nonempty.

Proposition 5.1.6: (Sharp Minimum Condition for Polyhedral Functions) Let $f : \Re^n \mapsto (-\infty, \infty]$ be a polyhedral function, and assume that X^*, the set of minima of f, is nonempty. Then there exists a scalar $\beta > 0$ such that

$$f^* + \beta d(x) \le f(x), \qquad \forall\, x \notin X^*,$$

where $d(x) = \min_{x^* \in X^*} \|x - x^*\|$.

Proof: We assume first that f is linear within $\text{dom}(f)$, and then generalize. Then, there exists $a \in \Re^n$ such that for all $x, \hat{x} \in \text{dom}(f)$, we have

$$f(x) - f(\hat{x}) = a'(x - \hat{x}).$$

For any $x \in X^*$, let S_x be the cone of vectors d that are in the normal cone $N_{X^*}(x)$ of X^* at x, and are also feasible directions in the sense that $x + \alpha d \in \text{dom}(f)$ for a small enough $\alpha > 0$. Since X^* and $\text{dom}(f)$ are polyhedral sets, there exist only a finite number of possible cones S_x as x ranges over X^*. Thus, there is a finite set of nonzero vectors $\{c_j \mid j \in J\}$, such that for any $x \in X^*$, S_x is either equal to $\{0\}$, or is the cone generated by a subset $\{c_j \mid j \in J_x\}$, where $J = \cup_{x \in X^*} J_x$. In addition, for all $x \in X^*$ and $d \in S_x$ with $\|d\| = 1$, we have

$$d = \sum_{j \in J_x} \gamma_j c_j,$$

for some scalars $\gamma_j \geq 0$ with $\sum_{j \in J_x} \gamma_j \geq \overline{\gamma}$, where $\overline{\gamma} = 1/\max_{j \in J} \|c_j\|$. Also we can show that for all $j \in J$, we have $a'c_j > 0$, by using the fact $c_j \in S_x$ for some $x \in X^*$.

For $x \in \text{dom}(f)$ with $x \notin X^*$, let $\hat{x}$ be the projection of x on X^*. Then the vector $x - \hat{x}$ belongs to $S_{\hat{x}}$, and we have

$$f(x) - f(\hat{x}) = a'(x - \hat{x}) = \|x - \hat{x}\| \frac{a'(x - \hat{x})}{\|x - \hat{x}\|} \geq \beta \|x - \hat{x}\|,$$

where $\beta = \overline{\gamma} \min_{j \in J} a'c_j$. Since J is finite, we have $\beta > 0$, and this implies the desired result for the case where f is linear within $\text{dom}(f)$.

Assume now that f is of the form

$$f(x) = \max_{i \in I}\{a_i'x + b_i\}, \qquad \forall\, x \in \text{dom}(f),$$

where I is a finite set, and a_i and b_i are some vectors and scalars, respectively. Let

$$Y = \big\{(x, z) \mid z \geq f(x),\, x \in \text{dom}(f)\big\},$$

and consider the function

$$g(x, z) = \begin{cases} z & \text{if } (x, z) \in Y, \\ \infty & \text{otherwise.} \end{cases}$$

Note that g is polyhedral and linear within $\text{dom}(g)$. Moreover, its set of minima is

$$Y^* = \big\{(x, z) \mid x \in X^*,\, z = f^*\big\},$$

and its minimal value is f^*.

Applying the result already shown to the function g, we have for some $\beta > 0$

$$f^* + \beta \hat{d}(x, z) \leq g(x, z), \qquad \forall\, (x, z) \notin Y^*,$$

where

$$\hat{d}(x, z) = \min_{(x^*, z^*) \in Y^*} \big(\|x - x^*\|^2 + |z - z^*|^2\big)^{1/2} = \min_{x^* \in X^*} \big(\|x - x^*\|^2 + |z - f^*|^2\big)^{1/2}.$$

Since

$$\hat{d}(x, z) \geq \min_{x \in X^*} \|x - x^*\| = d(x),$$

we have

$$f^* + \beta d(x) \leq g(x, z), \qquad \forall\, (x, z) \notin Y^*,$$

and by taking the infimum of the right-hand side over z for any fixed x,

$$f^* + \beta d(x) \leq f(x), \qquad \forall\, x \notin X^*.$$

Q.E.D.

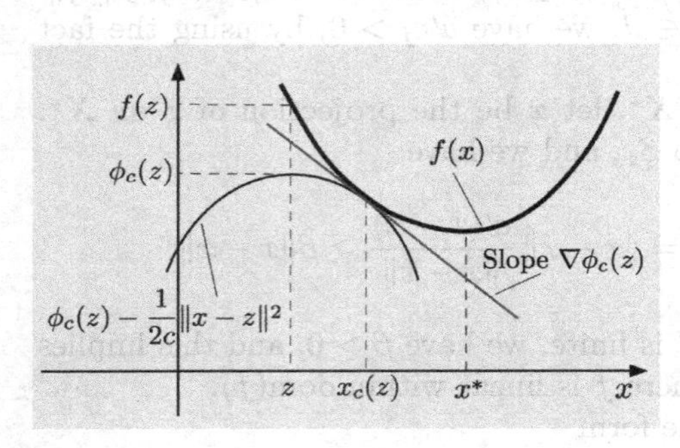

Figure 5.1.7. Illustration of the function

$$\phi_c(z) = \inf_{x\in\Re^n}\left\{f(x) + \frac{1}{2c}\|x - z\|^2\right\}.$$

We have $\phi_c(z) \leq f(z)$ for all $z \in \Re^n$, and at the set of minima of f, ϕ_c coincides with f. We also have

$$\nabla\phi_c(z) = \frac{z - x_c(z)}{c};$$

cf. Prop. 5.1.7.

From the preceding discussion and graphical illustrations, it can be seen that the rate of convergence of the proximal algorithm is improved by choosing large values of c. However, the corresponding regularization effect is reduced as c is increased, and this may adversely affect the proximal minimizations. In practice, it is often suggested to start with a moderate value of c, and gradually increase this value in subsequent proximal minimizations. How fast c can increase depends on the method used to solve the corresponding proximal minimization problems. If a fast Newton-like method is used, a fast rate of increase of c (say by a factor 5-10) may be possible, resulting in very few proximal minimizations. If instead a relatively slow first order method is used, it may be best to keep c constant at a moderate value, which is usually determined by trial and error.

5.1.3 Gradient Interpretation

An interesting interpretation of the proximal iteration is obtained by considering the function

$$\phi_c(z) = \inf_{x\in\Re^n}\left\{f(x) + \frac{1}{2c}\|x - z\|^2\right\} \tag{5.14}$$

for a fixed positive value of c. It can be seen that

$$\inf_{x\in\Re^n} f(x) \leq \phi_c(z) \leq f(z), \qquad \forall\, z \in \Re^n,$$

from which it follows that the set of minima of f and ϕ_c coincide (this is also evident from the geometric view of the proximal minimization given in Fig. 5.1.7). The following proposition shows that ϕ_c is a convex differentiable function, and derives its gradient.

Proposition 5.1.7: The function ϕ_c of Eq. (5.14) is convex and differentiable, and we have

$$\nabla \phi_c(z) = \frac{z - x_c(z)}{c} \qquad \forall\, z \in \Re^n, \tag{5.15}$$

where $x_c(z)$ is the unique minimizer in Eq. (5.14). Moreover

$$\nabla \phi_c(z) \in \partial f\big(x_c(z)\big), \qquad \forall\, z \in \Re^n.$$

Proof: We first note that ϕ_c is convex, since it is obtained by partial minimization of $f(x) + \frac{1}{2c}\|x - z\|^2$, which is convex as a function of (x, z) (cf. Prop. 3.3.1 in Appendix B). Furthermore, ϕ_c is real-valued, since the infimum in Eq. (5.14) is attained.

Let us fix z, and for notational simplicity, denote $\overline{z} = x_c(z)$. To show that ϕ_c is differentiable with the given form of gradient, we note that by the optimality condition of Prop. 3.1.4, we have $v \in \partial \phi_c(z)$, or equivalently $0 \in \partial \phi_c(z) - v$, if and only if z attains the minimum over $y \in \Re^n$ of

$$\phi_c(y) - v'y = \inf_{x \in \Re^n} \left\{ f(x) + \frac{1}{2c}\|x - y\|^2 \right\} - v'y.$$

Equivalently, $v \in \partial \phi_c(z)$ if and only if $(\overline{z}, z)$ attains the minimum over $(x, y) \in \Re^{2n}$ of the function

$$F(x, y) = f(x) + \frac{1}{2c}\|x - y\|^2 - v'y,$$

which is equivalent to $(0, 0) \in \partial F(\overline{z}, z)$, or

$$0 \in \partial f(\overline{z}) + \frac{\overline{z} - z}{c}, \qquad v = \frac{z - \overline{z}}{c}. \tag{5.16}$$

[This last step is obtained by viewing F as the sum of the function f and the differentiable function

$$\frac{1}{2c}\|x - y\|^2 - v'y,$$

and by writing

$$\partial F(x, y) = \big\{(g, 0) \mid g \in \partial f(x)\big\} + \left\{ \frac{x - y}{c}, \frac{y - x}{c} - v \right\};$$

cf. Prop. 5.4.6 in Appendix B.] The right side of Eq. (5.16) uniquely defines v, so that v is the unique subgradient of ϕ_c at z, and it has the form

$v = (z - \overline{z})/c$, as required by Eq. (5.15). From the left side of Eq. (5.16), we also see that $v = \nabla \phi_c(z) \in \partial f\big(x_c(z)\big)$. **Q.E.D.**

Using the gradient formula (5.15), we see that the proximal iteration can be written as

$$x_{k+1} = x_k - c_k \nabla \phi_{c_k}(x_k), \tag{5.17}$$

so *it is a gradient iteration for minimizing* ϕ_{c_k} with stepsize equal to c_k. This interpretation provides insight into the working mechanism of the algorithm and has formed the basis for various acceleration schemes, based on gradient and Newton-like schemes, particularly in connection with the augmented Lagrangian method, to be discussed in Section 5.2.1. In this connection, we will show in the next subsection that a stepsize as large as $2c_k$ can be used in place of c_k in Eq. (5.17). Moreover, the use of extrapolation schemes to modify the stepsize c_k has been shown to be beneficial in the constrained optimization context of the augmented Lagrangian method (see [Ber82a], Section 2.3.1).

5.1.4 Fixed Point Interpretation, Overrelaxation, and Generalization

We will now discuss the connection of the proximal algorithm with iterations for finding fixed points of mappings that are nonexpansive with respect to the Euclidean norm. As a first step, we will view the problem of minimizing f as a fixed point problem involving a special type of mapping.

For a scalar $c > 0$ and a closed proper convex function $f : \Re^n \mapsto (-\infty, \infty]$, let us consider the (single-valued) mapping $P_{c,f} : \Re^n \mapsto \Re^n$ given by

$$P_{c,f}(z) = \arg \min_{x \in \Re^n} \left\{ f(x) + \frac{1}{2c}\|x - z\|^2 \right\}, \qquad z \in \Re^n, \tag{5.18}$$

which is known as the *proximal operator* corresponding to c and f. The set of fixed points of $P_{c,f}$ coincides with the set of minima of f, and the proximal algorithm, written as

$$x_{k+1} = P_{c_k,f}(x_k),$$

may be viewed as a fixed point iteration. This alternative view leads to useful insights and some important generalizations.

The key idea is based on the mapping $N_{c,f} : \Re^n \mapsto \Re^n$ given by

$$N_{c,f}(z) = 2P_{c,f}(z) - z, \qquad z \in \Re^n. \tag{5.19}$$

We can visualize this mapping by writing

$$P_{c,f}(z) = \frac{N_{c,f}(z) + z}{2},$$

so $P_{c,f}(z)$ *is the midpoint of the line segment connecting* $N_{c,f}(z)$ *and* z. For this reason, $N_{c,f}$ is called the *reflection operator*. Some interesting facts here are that:

(a) The set of fixed points of $N_{c,f}$ is equal to the set of fixed points of $P_{c,f}$ and hence the set of minima of f. Moreover, as we will show shortly, the mapping $N_{c,f}$ is nonexpansive, i.e.,

$$\big\|N_{c,f}(z_1) - N_{c,f}(z_2)\big\| \leq \|z_1 - z_2\|, \qquad \forall\, z_1, z_2 \in \Re^n.$$

Thus for any x, $N_{c,f}(x)$ is at least as close to the set of minima of f as x.

(b) The interpolated iteration

$$x_{k+1} = (1 - \alpha_k)x_k + \alpha_k N_{c,f}(x_k), \tag{5.20}$$

where the interpolation parameter α_k satisfies $\alpha_k \in [\epsilon, 1 - \epsilon]$ for some scalar $\epsilon > 0$ and all k, converges to a fixed point of $N_{c,f}$, provided $N_{c,f}$ has at least one fixed point (this is a consequence of a classical result on the convergence of interpolated nonexpansive iterations, to be stated shortly).

(c) The preceding interpolated iteration (5.20), in view of the definition of $N_{c,f}$ [cf. Eq. (5.19)], can be written as

$$x_{k+1} = (1 - 2\alpha_k)x_k + 2\alpha_k P_{c,f}(x_k), \tag{5.21}$$

and as a special case, for $\alpha_k \equiv 1/2$, yields the proximal algorithm $x_{k+1} = P_{c,f}(x_k)$. We thus obtain a generalized form of the proximal algorithm, which depending on the parameter α_k, provides for extrapolation (when $1/2 < \alpha_k < 1$) or interpolation (when $0 < \alpha_k < 1/2$).

We will now prove the facts just stated in the following two propositions. To this end, we note that for any $z \in \Re^n$, the proximal iterate $P_{c,f}(z)$ is uniquely defined, and we have

$$\overline{z} = P_{c,f}(z) \quad \Rightarrow \quad z = \overline{z} + cv \text{ for some } v \in \partial f(\overline{z}), \tag{5.22}$$

since the right side above is the necessary condition for optimality of $\overline{z}$ in the proximal minimization (5.18) that defines $P_{c,f}(z)$. Moreover the converse also holds,

$$z = \overline{z} + cv \text{ for some } v \in \partial f(\overline{z}) \text{ and } \overline{z} \in \Re^n \quad \Rightarrow \quad \overline{z} = P_{c,f}(z), \tag{5.23}$$

since the left side above is the sufficiency condition for $\overline{z}$ to be (uniquely) optimal in the proximal minimization. An equivalent way to state the two

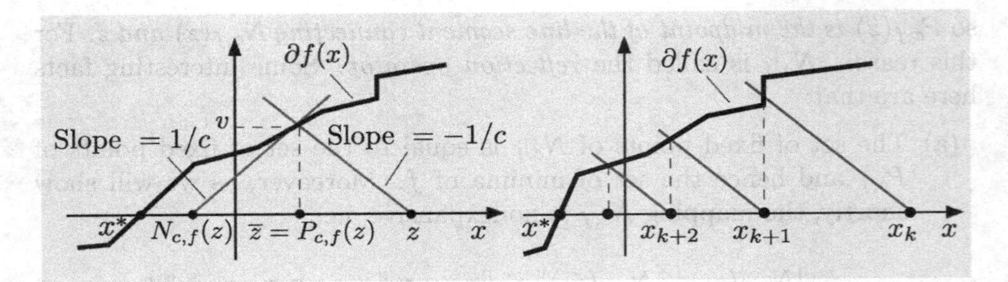

Figure 5.1.8. The figure on the left provides a graphical interpretation of the proximal iteration at a vector z for a one-dimensional problem. The line that passes through z and has slope $-1/c$ intercepts the graph of the (monotone) subdifferential mapping $\partial f(x)$ at a unique point v, which corresponds to $\overline{z}$, the unique vector $P_{c,f}(z)$ produced by the proximal iteration [cf. Eqs. (5.22)-(5.24)]. The figure on the left also illustrates the reflection operator $N_{c,f}(z) = 2P_{c,f}(z) - z$. The iterate $P_{c,f}(z)$ lies at the midpoint between z and $N_{c,f}(z)$ [cf. Eq. (5.25)]. Note that all points between z and $N_{c,f}(z)$ are at least as close to x^* as z. The figure on the right illustrates the proximal iteration $x_{k+1} = P_{c,f}(x_k)$.

relations (5.22) and (5.23) is that *any vector* $z \in \Re^n$ *can be written in exactly one way as*

$$z = \overline{z} + cv \qquad \text{where } \overline{z} \in \Re^n,\ v \in \partial f(\overline{z}), \tag{5.24}$$

and moreover the vector $\overline{z}$ *is equal to* $P_{c,f}(z)$,

$$\overline{z} = P_{c,f}(z).$$

Using Eq. (5.19), we also obtain a corresponding formula for $N_{c,f}$:

$$N_{c,f}(z) = 2P_{c,f}(z) - z = 2\overline{z} - (\overline{z} + cv) = \overline{z} - cv. \tag{5.25}$$

Figure 5.1.8 illustrates the preceding relations and provides a graphical interpretation of the proximal algorithm. The following proposition verifies the nonexpansiveness property of $N_{c,f}$.

Proposition 5.1.8: For any $c > 0$ and closed proper convex function $f : \Re^n \mapsto (-\infty, \infty]$, the mapping

$$N_{c,f}(z) = 2P_{c,f}(z) - z$$

[cf. Eqs. (5.18) and (5.19)] is nonexpansive, i.e.,

$$\big\| N_{c,f}(z_1) - N_{c,f}(z_2) \big\| \le \|z_1 - z_2\|, \qquad \forall\ z_1, z_2 \in \Re^n.$$

Moreover, any interpolated mapping $(1-\alpha)z + \alpha N_{c,f}(z)$, $\alpha \in [0,1]$ (including the proximal operator $P_{c,f}$, which corresponds to $\alpha = 1/2$) is nonexpansive.

Proof: Consider any $z_1, z_2 \in \Re^n$, and express them as

$$z_1 = \overline{z}_1 + cv_1, \qquad z_2 = \overline{z}_2 + cv_2,$$

with

$$\overline{z}_1 = P_{c,f}(z_1), \quad v_1 \in \partial f(\overline{z}_1), \qquad \overline{z}_2 = P_{c,f}(z_2), \quad v_2 \in \partial f(\overline{z}_2),$$

cf. Eq. (5.24). Then we have

$$\begin{aligned} \|z_1 - z_2\|^2 &= \big\|(\overline{z}_1 + cv_1) - (\overline{z}_2 + cv_2)\big\|^2 \\ &= \|\overline{z}_1 - \overline{z}_2\|^2 + 2c(\overline{z}_1 - \overline{z}_2)'(v_1 - v_2) + c^2\|v_1 - v_2\|^2. \end{aligned} \tag{5.26}$$

Also, from Eq. (5.25),

$$N_{c,f}(z_1) = \overline{z}_1 - cv_1, \qquad N_{c,f}(z_2) = \overline{z}_2 - cv_2,$$

and it follows that

$$\begin{aligned} \big\|N_{c,f}(z_1) - N_{c,f}(z_2)\big\|^2 &= \big\|(\overline{z}_1 - cv_1) - (\overline{z}_2 - cv_2)\big\|^2 \\ &= \|\overline{z}_1 - \overline{z}_2\|^2 - 2c(\overline{z}_1 - \overline{z}_2)'(v_1 - v_2) + c^2\|v_1 - v_2\|^2. \end{aligned} \tag{5.27}$$

By subtracting Eq. (5.26) from Eq. (5.27), we obtain

$$\big\|N_{c,f}(z_1) - N_{c,f}(z_2)\big\|^2 = \|z_1 - z_2\|^2 - 4c(\overline{z}_1 - \overline{z}_2)'(v_1 - v_2).$$

The nonexpansiveness of $N_{c,f}$ will follow if we can show that the inner product in the right-hand side is nonnegative. Indeed this is obtained by using the definition of subgradients to write

$$f(\overline{z}_2) \geq f(\overline{z}_1) + (\overline{z}_2 - \overline{z}_1)'v_1, \qquad f(\overline{z}_1) \geq f(\overline{z}_2) + (\overline{z}_1 - \overline{z}_2)'v_2,$$

so by adding these two relations, we have

$$(\overline{z}_1 - \overline{z}_2)'(v_1 - v_2) \geq 0, \tag{5.28}$$

and the result follows. Finally the nonexpansiveness of $N_{c,f}$ clearly implies the nonexpansiveness of the interpolated mapping. **Q.E.D.**

We will now use the Krasnosel'skii-Mann Theorem, which shows that fixed points of nonexpansive mappings can be found by an interpolated iteration. The theorem is proved and intuitively explained in Appendix A (Prop. A.4.2). For convenience, we reproduce its statement here.

Proposition 5.1.9: (Krasnosel'skii-Mann Theorem for Nonexpansive Iterations) Consider a mapping $T : \Re^n \mapsto \Re^n$ that is nonexpansive with respect to the Euclidean norm $\|\cdot\|$, i.e.,

$$\big\|T(x) - T(y)\big\| \leq \|x - y\|, \qquad \forall\, x, y \in \Re^n,$$

and has at least one fixed point. Then the iteration

$$x_{k+1} = (1 - \alpha_k)x_k + \alpha_k T(x_k), \tag{5.29}$$

where $\alpha_k \in [0, 1]$ for all k and $\sum_{k=0}^{\infty} \alpha_k(1 - \alpha_k) = \infty$, converges to a fixed point of T, starting from any $x_0 \in \Re^n$.

By applying the preceding theorem with $T = N_{c,f}$, we obtain the following.

Proposition 5.1.10: (Stepsize Relaxation in the Proximal Algorithm) The iteration

$$x_{k+1} = x_k + \gamma_k\big(P_{c,f}(x_k) - x_k\big), \tag{5.30}$$

where $\gamma_k \in [\epsilon, 2 - \epsilon]$ for some scalar $\epsilon > 0$ and all k, converges to a minimum of f, assuming at least one minimum exists.

Proof: Using the definition

$$N_{c,f}(x_k) = 2P_{c,f}(x_k) - x_k,$$

the iteration (5.30) is equivalent to

$$x_{k+1} = (1 - \alpha_k)x_k + \alpha_k N_{c,f}(x_k),$$

with $\gamma_k = 2\alpha_k$. Since the fixed points of $N_{c,f}$ are the minima of f, the result follows from Prop. 5.1.9 with $T = N_{c,f}$. **Q.E.D.**

In iteration (5.30) the parameter c is constant, but an extension is possible, whereby convergence can be shown for the case of the version with variable c_k:

$$x_{k+1} = x_k + \gamma_k\big(P_{c_k,f}(x_k) - x_k\big), \tag{5.31}$$

provided that $\inf_{k\geq 0} c_k > 0$ [see [Ber75d] for the present minimization context, and [EcB92] for a more general context]. This is based on the fact that the set of fixed points of $N_{c,f}$ does not depend on c as long as $c > 0$. Note that for $\gamma_k \equiv 1$, we obtain the proximal algorithm $x_{k+1} = P_{c_k,f}(x_k)$.

Another interesting fact is that the iteration (5.31) can also be written as a gradient iteration

$$x_{k+1} = x_k - \gamma_k c_k \nabla \phi_{c_k}(x_k),$$

where

$$\phi_c(z) = \inf_{x \in \Re^n} \left\{ f(x) + \frac{1}{2c}\|x - z\|^2 \right\},$$

[cf. Eq. (5.14)], based on the fact

$$\nabla \phi_{c_k}(x_k) = \frac{x_k - P_{c_k,f}(x_k)}{c_k},$$

[cf. Eq. (5.17)]. Since the performance of gradient methods is often improved by intelligent stepsize choice, this motivates stepsize selection schemes that are aimed at acceleration of convergence.

Indeed, it turns out that with extrapolation along the interval connecting x_k and $P_{c,f}(x_k)$, we can always obtain points that are closer to the set of optimal solutions X^* than $P_{c,f}(x_k)$. By this we mean that for each x with $P_{c,f}(x) \notin X^*$, there exists $\gamma \in (1, 2)$ such that

$$\min_{x^* \in X^*} \left\| x + \gamma\big(P_{c,f}(x) - x\big) - x^* \right\| < \min_{x^* \in X^*} \left\| P_{c,f}(x) - x^* \right\|. \tag{5.32}$$

This can be seen with a simple geometrical argument (cf. Fig. 5.1.9). Thus *the proximal algorithm can always benefit from overrelaxation, i.e.,* $\gamma_k \in (1, 2)$, if only we knew how to do it effectively. One may consider a trial and error scheme to determine a constant value of $\gamma_k \in (1, 2)$ that accelerates convergence relative to $\gamma_k \equiv 1$; this may work well when c_k is kept constant. More systematic procedures for variable values c_k have been suggested in [Ber75d] and [Ber82a], Section 2.3.1. In the procedure of [Ber82a], the overrelaxation parameter is chosen within $(1, 2)$ as

$$\gamma_k = 2\left(1 - \frac{c_k}{\beta + 2c_k}\right),$$

where β is a positive scalar that is determined experimentally for a given problem.

Generalization of the Proximal Algorithm

The preceding analysis can be generalized further to address the problem of *finding a zero of a multivalued mapping* $M : \Re^n \mapsto 2^{\Re^n}$, which maps vectors $x \in \Re^n$ into subsets $M(x) \subset \Re^n$. By a zero of M we mean a vector

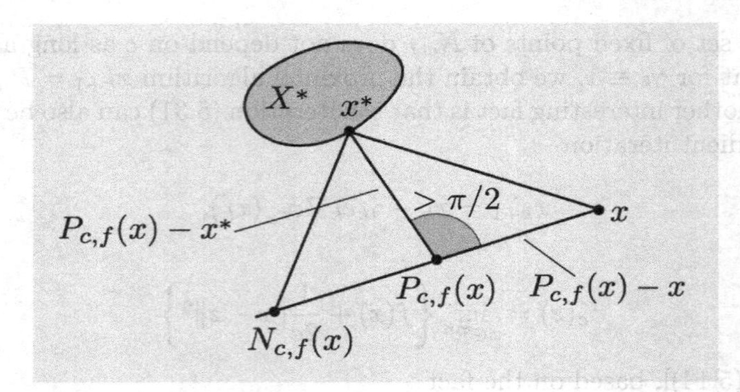

Figure 5.1.9. Geometric proof that overrelaxation in the proximal algorithm can produce iterates that are closer to any point in X^* than $P_{c,f}(x)$, assuming $P_{c,f}(x)$ is not optimal. Let x^* be the point of X^* that is at minimum distance from $P_{c,f}(x)$. The iterate $P_{c,f}(x)$ lies at the midpoint between x and $N_{c,f}(x)$ [cf. Eq. (5.19)]. By Prop. 5.1.1, the vector $\frac{1}{c}\big(x - P_{c,f}(x)\big)$ is a subgradient of f at $P_{c,f}(x)$, so

$$\frac{1}{c}\big(x - P_{c,f}(x)\big)'\big(x^* - P_{c,f}(x)\big) \leq f(x^*) - f\big(P_{c,f}(x)\big) < 0.$$

Thus the angle between $P_{c,f}(x) - x$ and $P_{c,f}(x) - x^*$ is strictly greater than $\pi/2$. From triangle geometry, it follows that there exist points in the interval connecting $P_{c,f}(x)$ and $N_{c,f}(x)$, which are closer to x^* than $P_{c,f}(x)$, so there exists $\gamma \in (1, 2)$ such that Eq. (5.32) holds.

x^* such that $0 \in M(x^*)$. As an example, the problem of minimizing a convex function $f : \Re^n \mapsto (-\infty, \infty]$ is equivalent to finding a zero of the multivalued mapping $M(x) = \partial f(x)$. However, there are other important applications where M is not the subdifferential mapping of a convex function, such as for example the solution of monotone variational inequalities (see the end-of-chapter references).

Looking back into the preceding analysis, we see that it generalizes, essentially verbatim, from the case $M(x) = \partial f(x)$ to the case of a general multivalued mapping $M : \Re^n \mapsto 2^{\Re^n}$, provided M has the following two properties:

(a) Any vector $x \in \Re^n$ can be written in exactly one way as

$$x = \overline{x} + cv \qquad \text{where } \overline{x} \in \Re^n,\ v \in M(\overline{x}), \tag{5.33}$$

[cf. Eq. (5.24)]. This was necessary in order for the mapping $P_{c,f}$ that maps x to $\overline{x}$, and the corresponding mapping

$$N_{c,f}(x) = 2P_{c,f}(x) - x, \qquad x \in \Re^n,$$

[cf. Eq. (5.19)] to be well-defined as a single-valued mapping.

(b) We have

$$(x_1 - x_2)'(v_1 - v_2) \geq 0, \quad \forall\, x_1, x_2 \in \text{dom}(M) \text{ and } v_1 \in M(x_1),\ v_2 \in M(x_2), \tag{5.34}$$

where

$$\text{dom}(M) = \{x \mid M(x) \neq \emptyset\}$$

(assumed nonempty). This property, known as *monotonicity* of M, was used to prove that the mapping $N_{c,f}$ is nonexpansive in Prop. 5.1.8 [cf. Eq. (5.28)].

It can be shown that both of the preceding two properties hold if and only if M is *maximal monotone*, i.e., it is monotone in the sense of Eq. (5.34), and its graph $\{(x, v) \mid v \in M(x)\}$, is not strictly contained in the graph of any other monotone mapping on $\Re^n$ † (the subdifferential mapping can be shown to be maximal monotone; this is shown in several sources, [Roc66], [Roc70], [RoW98], [BaC11]). Maximal monotone mappings, the associated proximal algorithms, and related subjects have been extensively treated in the literature, to which we refer for further discussion; see the end-of-chapter references.

In summary, the proximal algorithm in its full generality applies to the problem of finding a zero of a maximal monotone multivalued mapping $M : \Re^n \mapsto 2^{\Re^n}$ [a vector x^* such that $0 \in M(x^*)$]. It takes the form

$$x_{k+1} = x_k - cv_k,$$

where v_k is the unique point v such that $v \in M(x_{k+1})$; cf. Fig. 5.1.10. If M is a single-valued mapping, we have $x_{k+1} = x_k - cM(x_{k+1})$, or

$$x_{k+1} = (I + cM)^{-1}(x_k),$$

where I is the identity mapping and $(I + cM)^{-1}$ is the inverse of the mapping $I + cM$. Moreover, a more general version of the algorithm is valid, allowing for a stepsize $\gamma_k \in (0, 2)$,

$$x_{k+1} = x_k - \gamma_k c v_k,$$

with the possibility of reduction of the distance to all zeroes of M using an appropriate overrelaxation scheme with $\gamma_k > 1$. The analysis of the present section readily extends to this more general context. There is only one difficult point in this analysis, which we have not addressed and have referred instead to the literature: the equivalence of the properties (a) and (b) above with the maximal monotonicity of the mapping M.

---

† Note that the monotonicity property (5.34) and the existence of the representation (5.33) for some $x \in \Re^n$ imply the uniqueness of this representation [if $x = \overline{x}_1 + cv_1 = \overline{x}_2 + cv_2$, then $0 \leq (\overline{x}_1 - \overline{x}_2)'(v_1 - v_2) = -\frac{1}{c}\|\overline{x}_1 - \overline{x}_2\|^2$, so $\overline{x}_1 = \overline{x}_2$]. Thus maximal monotonicity of M is equivalent to monotonicity and existence of a representation of the form (5.33) for every $x \in \Re^n$, something that can be easily visualized (cf. Fig. 5.1.10) but quite hard to prove (see the original work [Min62], or subsequent sources such as [Bre73], [RoW98], [BaC11]).

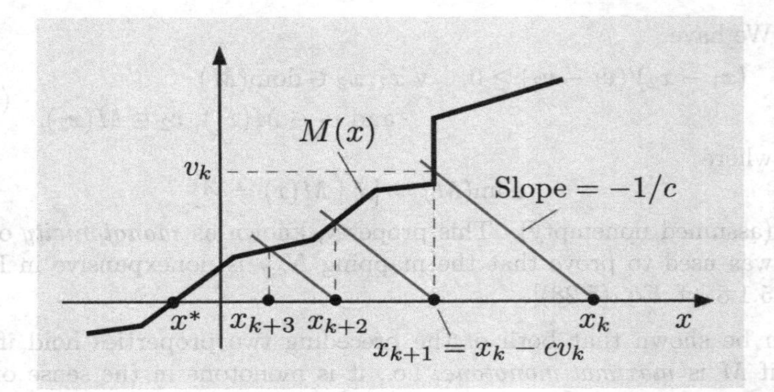

Figure 5.1.10. One-dimensional illustration of the proximal algorithm for finding a zero of a maximal monotone multivalued mapping $M : \Re^n \mapsto 2^{\Re^n}$. The important fact here is that the maximal monotonicity of M implies that every $x_k \in \Re^n$ can be uniquely represented as $x_k = x_{k+1} + cv_k$, where $v_k \in M(x_{k+1})$.

5.2 DUAL PROXIMAL ALGORITHMS

In this section we will develop an equivalent dual implementation of the proximal algorithm, based on Fenchel duality. We will then apply it in a special way to obtain a popular constrained optimization algorithm, the augmented Lagrangian method.

We recall the proximal algorithm of Section 5.1:

$$x_{k+1} \in \arg\min_{x\in\Re^n} \left\{ f(x) + \frac{1}{2c_k}\|x - x_k\|^2 \right\}, \tag{5.35}$$

where $f : \Re^n \mapsto (-\infty, \infty]$ is a closed proper convex function, x_0 is an arbitrary starting point, and $\{c_k\}$ is a positive scalar parameter sequence. We note that the minimization above is in a form suitable for application of the Fenchel duality theory of Section 1.2, with the identifications

$$f_1(x) = f(x), \qquad f_2(x) = \frac{1}{2c_k}\|x - x_k\|^2.$$

We can write the Fenchel dual problem as

$$\begin{aligned} &\text{minimize} \quad f_1^\star(\lambda) + f_2^\star(-\lambda) \\ &\text{subject to} \quad \lambda \in \Re^n, \end{aligned} \tag{5.36}$$

where $f_1^\star$ and $f_2^\star$ are the conjugate functions of f_1 and f_2, respectively. We have

$$f_2^\star(\lambda) = \sup_{x\in\Re^n} \big\{x'\lambda - f_2(x)\big\} = \sup_{x\in\Re^n} \left\{ x'\lambda - \frac{1}{2c_k}\|x - x_k\|^2 \right\} = x_k'\lambda + \frac{c_k}{2}\|\lambda\|^2,$$

where the last equality follows by noting that the supremum over x is attained at $x = x_k + c_k\lambda$. Introducing $f^\star$, the conjugate of f,

$$f_1^\star(\lambda) = f^\star(\lambda) = \sup_{x\in\Re^n}\{x'\lambda - f(x)\},$$

and substituting into Eq. (5.36), we see that the dual problem (5.36) can be written as

$$\begin{aligned}&\text{minimize} \quad f^\star(\lambda) - x_k'\lambda + \frac{c_k}{2}\|\lambda\|^2\\ &\text{subject to} \quad \lambda \in \Re^n.\end{aligned} \tag{5.37}$$

We also note that there is no duality gap, since f_2 and $f_2^\star$ are real-valued, so the relative interior conditions of the Fenchel Duality Theorem [Prop. 1.2.1(a),(b)] are satisfied. In fact there exist unique primal and dual optimal solutions, since both primal and dual problems involve a strictly convex cost function with compact level sets.

Let λ_{k+1} be the unique solution of the minimization (5.37). Then λ_{k+1} together with x_{k+1} satisfy the necessary and sufficient optimality conditions of Prop. 1.2.1(c),

$$\lambda_{k+1} \in \partial f_1(x_{k+1}), \qquad -\lambda_{k+1} \in \partial f_2(x_{k+1}). \tag{5.38}$$

Using the form of f_2, the second relation above yields

$$\lambda_{k+1} = \frac{x_k - x_{k+1}}{c_k}, \tag{5.39}$$

see Fig. 5.2.1. This equation can be used to find the primal proximal iterate x_{k+1} of Eq. (5.35), once λ_{k+1} is known,

$$x_{k+1} = x_k - c_k\lambda_{k+1}. \tag{5.40}$$

We thus obtain a dual implementation of the proximal algorithm. In this algorithm, instead of solving the Fenchel primal problem involved in the proximal iteration (5.35), we first solve the Fenchel dual problem (5.37) to obtain the optimal dual solution λ_{k+1}, and then obtain the optimal primal Fenchel solution x_{k+1} using Eq. (5.40).

Dual Proximal Algorithm:

Find

$$\lambda_{k+1} \in \arg\min_{\lambda\in\Re^n}\left\{f^\star(\lambda) - x_k'\lambda + \frac{c_k}{2}\|\lambda\|^2\right\}, \tag{5.41}$$

and then set

$$x_{k+1} = x_k - c_k\lambda_{k+1}. \tag{5.42}$$

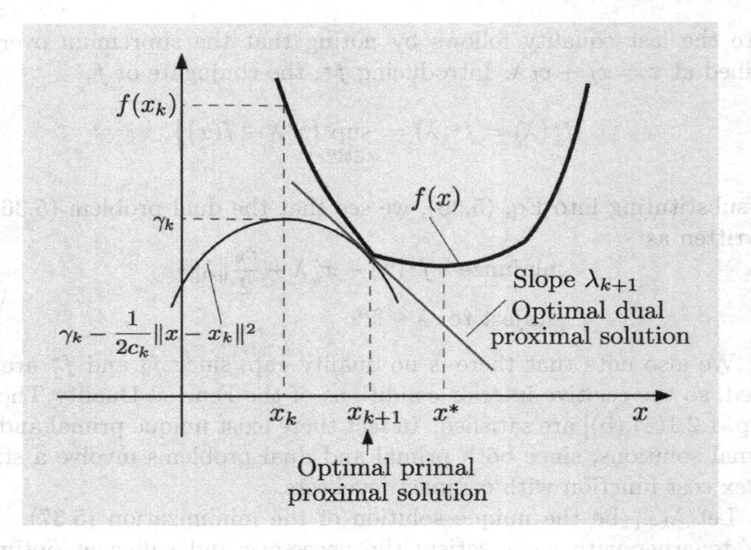

Figure 5.2.1. Illustration of the optimality condition

$$\lambda_{k+1} = \frac{x_k - x_{k+1}}{c_k},$$

cf. Eq. (5.39), and the relation between the primal and dual proximal solutions.

The dual algorithm is illustrated in Fig. 5.2.2. Note that as x_k converges to a minimum x^* of f, λ_k converges to 0. Thus the dual iteration (5.41) does not aim to minimize $f^\star$, but rather to find a subgradient of $f^\star$ at 0, which minimizes f [cf. Prop. 5.4.4(b) in Appendix B]. In particular, we have

$$\lambda_{k+1} \in \partial f(x_{k+1}), \qquad x_{k+1} \in \partial f^\star(\lambda_{k+1}), \qquad \forall\, k \geq 0,$$

[cf. Eq. (5.38) for the left side, and the Conjugate Subgradient Theorem (Prop. 5.4.3 in Appendix B) for the right side], and as λ_k converges to 0 and x_k converges to a minimum x^* of f, we have

$$0 \in \partial f(x^*), \qquad x^* \in \partial f^\star(0).$$

The primal and dual implementations of the proximal algorithm are mathematically equivalent and generate identical sequences $\{x_k\}$, assuming the same starting point x_0 and penalty parameter sequence $\{c_k\}$. Whether one is preferable over the other depends on which of the minimizations (5.35) and (5.41) is easier, i.e., whether f or its conjugate $f^\star$ has more convenient structure. In the next section we will discuss a case where the dual proximal algorithm is more convenient and yields the augmented Lagrangian method.

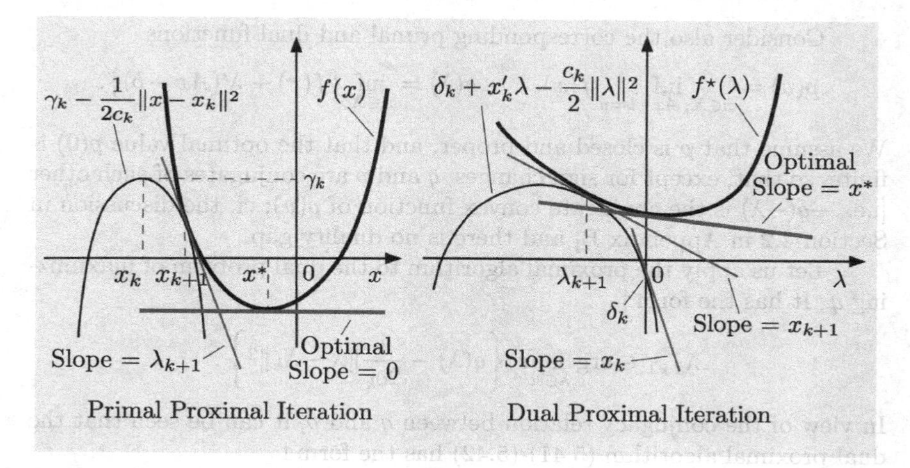

Figure 5.2.2. Illustration of primal and dual proximal algorithms. The primal algorithm aims to find x^*, a minimum of f. The dual algorithm aims to find x^* as a subgradient of $f^\star$ at 0 [cf. Prop. 5.4.4(b) in Appendix B].

5.2.1 Augmented Lagrangian Methods

We will now apply the proximal algorithm to the dual problem of a constrained optimization problem. We will show how the corresponding dual proximal algorithm leads to the class of augmented Lagrangian methods. These methods are popular because they allow the solution of constrained optimization problems, through a sequence of easier unconstrained (or less constrained) optimizations, which can be performed with fast and reliable algorithms, such as Newton, quasi-Newton, and conjugate gradient methods. Augmented Lagrangian methods can also be used for smoothing of nondifferentiable cost functions, as described in Section 2.2.5; see nonlinear programming textbooks, and the monograph [Ber82a], which is a comprehensive reference on augmented Lagrangian, and related smoothing and sequential quadratic programming methods.

Consider the constrained minimization problem

$$\begin{aligned} &\text{minimize} \quad f(x) \\ &\text{subject to} \quad x \in X, \quad Ax = b, \end{aligned} \tag{5.43}$$

where $f : \Re^n \mapsto (-\infty, \infty]$ is a convex function, X is a convex set, A is an $m \times n$ matrix, and $b \in \Re^m$.†

† We focus on linear equality constraints for convenience, but the analysis can be extended to convex inequality constraints as well (see the subsequent discussion). In particular, a linear inequality constraint of the form $a_j'x \leq b_j$ can be converted to an equality constraint $a_j'x + z^j = b_j$ by using an artificial variable z^j, and the constraint $z^j \geq 0$, which can be absorbed into the set X.

Consider also the corresponding primal and dual functions

$$p(u) = \inf_{x \in X,\, Ax-b=u} f(x), \qquad q(\lambda) = \inf_{x \in X} \{f(x) + \lambda'(Ax-b)\}.$$

We assume that p is closed and proper, and that the optimal value $p(0)$ is finite, so that, except for sign changes, q and p are conjugates of each other [i.e., $-q(-\lambda)$ is the conjugate convex function of $p(u)$; cf. the discussion in Section 4.2 in Appendix B] and there is no duality gap.

Let us apply the proximal algorithm to the dual problem of maximizing q. It has the form †

$$\lambda_{k+1} \in \arg\max_{\lambda \in \Re^m} \left\{ q(\lambda) - \frac{1}{2c_k}\|\lambda - \lambda_k\|^2 \right\}.$$

In view of the conjugacy relation between q and p, it can be seen that the dual proximal algorithm (5.41)-(5.42) has the form ‡

$$u_{k+1} \in \arg\min_{u \in \Re^m} \left\{ p(u) + \lambda_k' u + \frac{c_k}{2}\|u\|^2 \right\}, \tag{5.44}$$

which is Eq. (5.41), and

$$\lambda_{k+1} = \lambda_k + c_k u_{k+1}, \tag{5.45}$$

which is Eq. (5.42); see Fig. 5.2.3.

To implement this algorithm, we introduce for any $c > 0$, the *augmented Lagrangian function*

$$L_c(x, \lambda) = f(x) + \lambda'(Ax - b) + \frac{c}{2}\|Ax - b\|^2, \qquad x \in \Re^n,\ \lambda \in \Re^m,$$

and we use the definition of p to write the minimization (5.44) as

$$\begin{aligned}
&\inf_{u \in \Re^m} \left\{ \inf_{x \in X,\, Ax-b=u} \{f(x)\} + \lambda_k' u + \frac{c_k}{2}\|u\|^2 \right\} \\
&\quad = \inf_{u \in \Re^m} \inf_{x \in X,\, Ax-b=u} \left\{ f(x) + \lambda_k'(Ax-b) + \frac{c_k}{2}\|Ax-b\|^2 \right\} \\
&\quad = \inf_{x \in X} \left\{ f(x) + \lambda_k'(Ax-b) + \frac{c_k}{2}\|Ax-b\|^2 \right\} \\
&\quad = \inf_{x \in X} L_{c_k}(x, \lambda_k).
\end{aligned}$$

---

† There is an unfortunate (but hard to avoid) reversal of notation in this section, because the primal proximal algorithm is applied to the dual problem $\max_\lambda q(\lambda)$ (i.e., to minimize the negative dual function $-q$), while the dual proximal algorithm involves the conjugate of $-q$ whose argument is the perturbation vector u. Thus the dual variable λ corresponds to the primal vector x in the preceding section, while the perturbation vector u corresponds to the dual vector λ of the preceding section.

‡ This takes into account the required sign and symbol changes, so that $f \sim -q$, $x \sim -\lambda$, $x_k \sim -\lambda_k$, $f^\star \sim p$, and u is the argument of p.

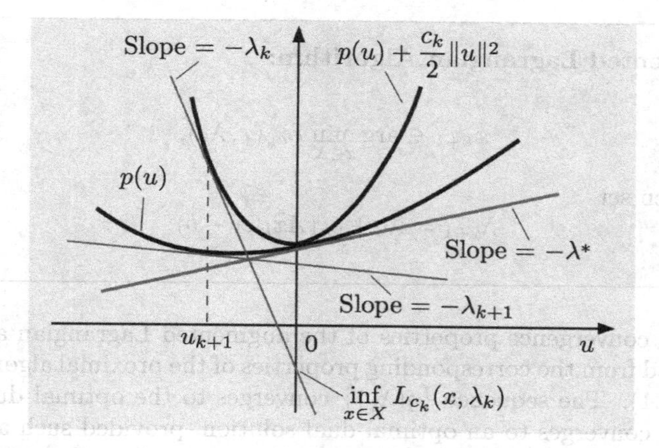

Figure 5.2.3. Geometric interpretation of the dual proximal minimization

$$u_{k+1} \in \arg\min_{u \in \Re^m} \left\{ p(u) + \lambda_k' u + \frac{c_k}{2}\|u\|^2 \right\}, \tag{5.46}$$

and the update

$$\lambda_{k+1} = \lambda_k + c_k u_{k+1}$$

in the augmented Lagrangian algorithm. From the minimization (5.46) we have

$$0 \in \partial p(u_{k+1}) + \lambda_k + c_k u_{k+1},$$

so the vector u_{k+1} is the one for which $-\lambda_k$ is a subgradient of $p(u) + \frac{c_k}{2}\|u\|^2$ at $u = u_{k+1}$, as shown in the figure. By combining the last two relations, we obtain $-\lambda_{k+1} \in \partial p(u_{k+1})$, as shown in the figure. The optimal value in the minimization (5.46) is equal to $\inf_{x \in X} L_{c_k}(x, \lambda_k)$, and can be geometrically interpreted as in the figure.

The minimizing u and x in this equation are related, and we have

$$u_{k+1} = Ax_{k+1} - b,$$

where x_{k+1} is any vector that minimizes $L_{c_k}(x, \lambda_k)$ over X (we assume that such a vector exists – while the existence of the minimizing u_{k+1} is guaranteed, since the minimization (5.44) has a solution, the existence of the minimizing x_{k+1} is not guaranteed, and must be either assumed or verified independently).

Using the preceding expression for u_{k+1}, we see that the dual proximal algorithm (5.44)-(5.45), applied to the maximization of the dual function q, starts with an arbitrary initial vector λ_0, and iterates according to

$$\lambda_{k+1} = \lambda_k + c_k(Ax_{k+1} - b),$$

where x_{k+1} is any vector that minimizes $L_{c_k}(x, \lambda_k)$ over X. This method is known as the *augmented Lagrangian algorithm* or the *method of multipliers*.

Augmented Lagrangian Algorithm:

Find

$$x_{k+1} \in \arg\min_{x \in X} L_{c_k}(x, \lambda_k), \tag{5.47}$$

and then set

$$\lambda_{k+1} = \lambda_k + c_k(Ax_{k+1} - b). \tag{5.48}$$

The convergence properties of the augmented Lagrangian algorithm are derived from the corresponding properties of the proximal algorithm (cf. Section 5.1). The sequence $\{q(\lambda_k)\}$ converges to the optimal dual value, and $\{\lambda_k\}$ converges to an optimal dual solution, provided such a solution exists (cf. Prop. 5.1.3). Furthermore, convergence in a finite number of iterations is obtained when q is polyhedral (cf. Prop. 5.1.5).

Assuming that there exists a dual optimal solution, so that $\{\lambda_k\}$ converges to such a solution, we also claim that every limit point of the generated sequence $\{x_k\}$ is an optimal solution of the primal problem (5.43). To see this, note that from the update formula (5.48) we obtain

$$Ax_{k+1} - b \to 0, \qquad c_k\big(Ax_{k+1} - b\big) \to 0.$$

Furthermore, we have

$$L_{c_k}\big(x_{k+1}, \lambda_k\big) = \min_{x \in X}\left\{ f(x) + \lambda_k'(Ax - b) + \frac{c_k}{2}\|Ax - b\|^2 \right\}.$$

The preceding relations yield

$$\limsup_{k\to\infty} f(x_{k+1}) = \limsup_{k\to\infty} L_{c_k}\big(x_{k+1}, \lambda_k\big) \le f(x), \quad \forall\, x \in X \text{ with } Ax = b,$$

so if $x^* \in X$ is a limit point of $\{x_k\}$, we obtain

$$f(x^*) \le f(x), \qquad \forall\, x \in X \text{ with } Ax = b,$$

as well as $Ax^* = b$ (in view of $Ax_{k+1} - b \to 0$). Therefore any limit point x^* of the generated sequence $\{x_k\}$ is an optimal solution of the primal problem (5.43). We summarize the preceding discussion in the following proposition.

Proposition 5.2.1: (Convergence Properties of Augmented Lagrangian Algorithm) Consider a sequence $\{(x_k, \lambda_k)\}$ generated by the augmented Lagrangian algorithm (5.47), (5.48), applied to problem (5.43), assuming that $\sum_{k=0}^{\infty} c_k = \infty$. Assume further that the primal function p is closed and proper, and that the optimal value $p(0)$ is finite. Then the dual function sequence $\{q(\lambda_k)\}$ converges to the common primal and dual optimal value. Moreover, if the dual problem has at least one optimal solution, the following hold:

(a) The sequence $\{\lambda_k\}$ converges to an optimal dual solution. Furthermore, convergence in a finite number of iterations is obtained if q is polyhedral.

(b) Every limit point of $\{x_k\}$ is an optimal solution of the primal problem (5.43).

Note that there is no guarantee that $\{x_k\}$ has a limit point, and indeed the dual sequence $\{\lambda_k\}$ will converge to a dual optimal solution, if one exists, even if the primal problem (5.43) does not have an optimal solution. As an example, the reader may verify that for the two-dimensional/single constraint problem where $f(x) = e^{x^1}$, $x^1 + x^2 = 0$, $x^1 \in \Re$, $x^2 \geq 0$, the dual optimal solution is $\lambda^* = 0$, but there is no primal optimal solution. For this problem, the augmented Lagrangian algorithm will generate sequences $\{\lambda_k\}$ and $\{x_k\}$ such that $\lambda_k \to 0$ and $x_k^1 \to -\infty$, while $f(x_k) \to f^* = 0$.

Linear and Nonlinear Inequality Constraints

The simplest way to treat inequality constraints in the context of the augmented Lagrangian methodology, is to convert them to equality constraints by using additional nonnegative variables. In particular, consider the version of the problem with linear inequality constraints $Ax \leq b$, which we write as

$$\begin{aligned} &\text{minimize} \quad f(x) \\ &\text{subject to} \quad x \in X, \quad a_1'x \leq b_1, \ldots, a_r'x \leq b_r, \end{aligned} \tag{5.49}$$

where $f : \Re^n \mapsto (-\infty, \infty]$ is a convex function and X is a convex set. We can convert this problem to the equality constrained problem

$$\begin{aligned} &\text{minimize} \quad f(x) \\ &\text{subject to} \quad x \in X, \quad z \geq 0, \quad a_1'x + z^1 = b_1, \ldots, a_r'x + z^r = b_r, \end{aligned} \tag{5.50}$$

where $z = (z^1, \ldots, z^r)$ is a vector of additional artificial variables.

The augmented Lagrangian method for this problem involves minimizations of the form

$$\min_{x \in X,\, z \geq 0} \bar{L}_c(x, z, \mu) = f(x) + \sum_{j=1}^{r} \left\{ \mu^j \left(a_j'x - b_j + z^j \right) + \frac{c}{2} |a_j'x - b_j + z^j|^2 \right\},$$

for a sequence of values of $\mu = (\mu^1, \ldots, \mu^r)$ and $c > 0$. This type of minimization can be done by first minimizing $\bar{L}_c(x, z, \mu)$ over $z \geq 0$, obtaining

$$L_c(x, \mu) = \min_{z \geq 0} \bar{L}_c(x, z, \mu),$$

and then by minimizing $L_c(x,\mu)$ over $x \in X$. A key observation is that *the first minimization with respect to z can be carried out in closed form for each fixed x*, thereby yielding a closed form expression for $L_c(x,\mu)$.

Indeed, we have

$$\min_{z\geq 0} \bar{L}_c(x,z,\mu) = f(x) + \sum_{j=1}^{r} \min_{z^j\geq 0} \left\{ \mu^j\left(a_j'x - b_j + z^j\right) + \frac{c}{2}|a_j'x - b_j + z^j|^2 \right\}. \tag{5.51}$$

The function in braces above is quadratic in z^j. Its constrained minimum is $\hat{z}^j = \max\{0, \tilde{z}^j\}$, where $\tilde{z}^j$ is the unconstrained minimum at which the derivative is zero. The derivative is $\mu^j + c(a_j'x - b_j + \tilde{z}^j)$, so we obtain

$$\hat{z}^j = \max\{0, \tilde{z}^j\} = \max\left\{0, -\left(\frac{\mu^j}{c} + a_j'x - b_j\right)\right\}.$$

Denoting

$$g_j^+(x,\mu^j,c) = \max\left\{a_j'x - b_j, -\frac{\mu^j}{c}\right\}, \tag{5.52}$$

we have $a_j'x - b_j + \hat{z}^j = g_j^+(x,\mu^j,c)$. Substituting in Eq. (5.51), we obtain a closed form expression for $L_c(x,\mu) = \min_{z\geq 0} \bar{L}_c(x,z,\mu)$:

$$L_c(x,\mu) = f(x) + \sum_{j=1}^{r} \left\{ \mu^j g_j^+(x,\mu^j,c) + \frac{c}{2}\left(g_j^+(x,\mu^j,c)\right)^2 \right\}. \tag{5.53}$$

After some calculation, left for the reader, we can also write this expression as

$$L_c(x,\mu) = f(x) + \frac{1}{2c}\sum_{j=1}^{r} \left\{ \left(\max\{0, \mu^j + c(a_j'x - b_j)\}\right)^2 - (\mu^j)^2 \right\}, \tag{5.54}$$

and we can view it as the augmented Lagrangian function for the inequality constrained problem (5.49).

It follows from the preceding transcription that the augmented Lagrangian method for the inequality constrained problem (5.49) consists of a sequence of minimizations of the form

$$\begin{aligned} &\text{minimize} \ \ L_{c_k}(x,\mu_k) \\ &\text{subject to} \ \ x \in X, \end{aligned}$$

followed by the multiplier iterations

$$\mu_{k+1}^j = \mu_k^j + c_k g_j^+(x_k,\mu_k^j,c_k), \qquad j = 1,\ldots,r,$$

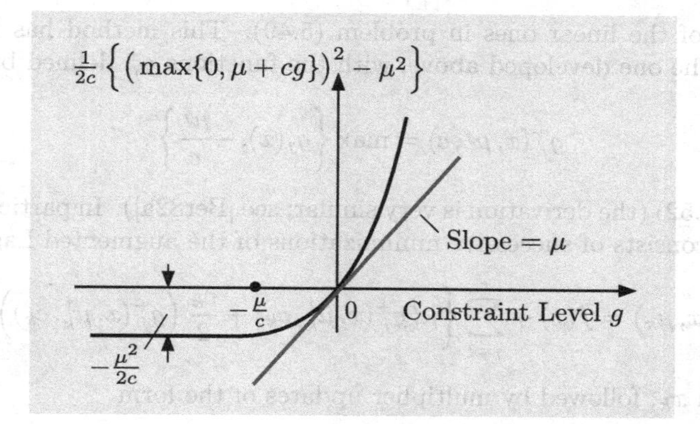

Figure 5.2.4. Form of the quadratic penalty term for a single inequality constraint $g(x) \leq 0$.

which can be equivalently written as

$$\mu_{k+1}^j = \max\left\{0, \mu_k^j + c_k(a_j'x_k - b_j)\right\}, \qquad j = 1, \ldots, r.$$

Note that the penalty term

$$\frac{1}{2c}\left\{\left(\max\{0, \mu^j + c(a_j'x - b_j)\}\right)^2 - (\mu^j)^2\right\}$$

corresponding to the jth inequality constraint in Eq. (5.54) is convex and continuously differentiable in x (see Fig. 5.2.4). However, its Hessian matrix is discontinuous for all x such that $a_j'x - b_j = -\mu^j/c$; this may cause some difficulties in the minimization of $L_c(x, \mu)$, particularly when Newton-like methods are used, and motivates alternative twice differentiable augmented Lagrangian methods for inequality constraints (see Section 6.6).

To summarize, the augmented Lagrangian method for the inequality constrained problem (5.49) consists of a sequence of minimizations

$$\begin{aligned} &\text{minimize} \quad L_{c_k}(x, \mu_k) \\ &\text{subject to} \quad x \in X, \end{aligned}$$

where $L_{c_k}(x, \mu_k)$ is given by Eq. (5.53) or Eq. (5.54), $\{\mu_k\}$ is a sequence updated as above, and $\{c_k\}$ is a positive penalty parameter sequence with $\sum_{k=0}^{\infty} c_k = \infty$. Since this method is equivalent to the equality-constrained method applied to the corresponding equality-constrained problem (5.50), our earlier convergence results (cf. Prop. 5.2.1) apply with the obvious modifications.

We finally note that there is a similar augmented Lagrangian method for problems with nonlinear inequality constraints

$$g_1(x) \leq 0, \ldots, g_r(x) \leq 0,$$

in place of the linear ones in problem (5.49). This method has identical form to the one developed above, with the functions g_j^+ defined by

$$g_j^+(x, \mu^j, c) = \max\left\{g_j(x), -\frac{\mu^j}{c}\right\},$$

cf. Eq. (5.52) (the derivation is very similar; see [Ber82a]). In particular, the method consists of successive minimizations of the augmented Lagrangian

$$L_{c_k}(x, \mu_k) = f(x) + \sum_{j=1}^{r}\left\{\mu_k^j g_j^+(x, \mu_k^j, c_k) + \frac{c_k}{2}\left(g_j^+(x, \mu_k^j, c_k)\right)^2\right\},$$

to obtain x_k, followed by multiplier updates of the form

$$\mu_{k+1}^j = \mu_k^j + c_k g_j^+(x_k, \mu_k^j, c_k), \qquad j = 1, \ldots, r;$$

see the end-of-chapter references. Note that $L_{c_k}(\cdot, \mu_k)$ is continuously differentiable in x if f and g_j are, and is also convex in x if f and g_j are.

Variants of the Augmented Lagrangian Algorithm

The augmented Lagrangian algorithm is an excellent general purpose constrained minimization method, and applies to considerably more general problems than the ones treated here. For example, it can be used for differentiable problems with nonconvex cost functions and constraints. It may also be used in smoothing approaches (cf. Section 2.2.5), for both convex and nonconvex optimization.

These properties and their connections to duality are due to the convexification effect of the quadratic penalty, even in the context of a nonconvex problem. Further discussion is beyond our scope, and we refer to nonlinear programming textbooks and the monograph [Ber82a], which focuses on augmented Lagrangian and other Lagrange multiplier methods.

The algorithm also embodies a rich structure, which lends itself to many variations. In particular, let us consider the "penalized" dual function q_c, given by

$$q_c(\lambda) = \max_{y \in \Re^m}\left\{q(y) - \frac{1}{2c}\|y - \lambda\|^2\right\}. \tag{5.55}$$

Then, according to Prop. 5.1.7, q_c is differentiable, and we have

$$\nabla q_c(\lambda) = \frac{y_c(\lambda) - \lambda}{c}, \tag{5.56}$$

where $y_c(\lambda)$ is the unique vector attaining the maximum in Eq. (5.55). Since $y_{c_k}(\lambda_k) = \lambda_{k+1}$, we have using Eqs. (5.48) and (5.56),

$$\nabla q_{c_k}(\lambda_k) = \frac{\lambda_{k+1} - \lambda_k}{c_k} = Ax_{k+1} - b,$$

so the multiplier iteration (5.48) can be written as a gradient iteration:

$$\lambda_{k+1} = \lambda_k + c_k \nabla q_{c_k}(\lambda_k),$$

[cf. Eq. (5.17)].

This interpretation motivates variations based on faster Newton or quasi-Newton methods for maximizing q_c, whose maxima coincide with the ones of q, for any $c > 0$. (The algorithm described so far is also known as the *first order method of multipliers*, to distinguish it from Newton-like methods, which are also known as *second order methods of multipliers*.) There are many algorithms along this line, some of which involve inexact minimization of the augmented Lagrangian to enhance computational efficiency. We refer to [Ber82a] and other literature cited at the end of the chapter for analysis of such methods.

We finally note that because the proximal algorithm can be generalized to the case where a nonquadratic regularization function is used, the dual proximal algorithm and hence also the augmented Lagrangian method can be accordingly generalized (see Section 6.6 and [Ber82a], Chapter 5).

One difficulty with the method is that even if the cost function is separable, the augmented Lagrangian $L_c(\cdot, \lambda)$ is typically nonseparable because it involves the quadratic term $\|Ax - b\|^2$. With the use of a block coordinate descent method, however, it may be possible to deal to some extent with the loss of separability, as shown in the following example from [BeT89a].

Example 5.2.1: (Additive Cost Problems)

Consider the problem

$$\text{minimize} \sum_{i=1}^{m} f_i(x)$$
$$\text{subject to } x \in \cap_{i=1}^{m} X_i,$$

where $f_i : \Re^n \mapsto \Re$ are convex functions and X_i are closed, convex sets with nonempty intersection. This problem contains as special cases several of the examples given in Section 1.3, such as regularized regression, classification, and maximum likelihood.

We introduce additional artificial variables z^i, $i = 1, \ldots, m$, we consider the equivalent problem

$$\begin{aligned} &\text{minimize} \quad \sum_{i=1}^{m} f_i(z^i) \\ &\text{subject to } x = z^i, \quad z^i \in X_i, \quad i = 1, \ldots, m, \end{aligned} \tag{5.57}$$

and we apply the augmented Lagrangian method to eliminate the constraints $z^i = x$, using corresponding multiplier vectors λ^i. The method takes the form

$$\lambda^i_{k+1} = \lambda^i_k + c_k(x_{k+1} - z^i_{k+1}), \qquad i = 1, \ldots, m, \tag{5.58}$$

where x_{k+1} and z^i_{k+1}, $i=1,\ldots,m$, solve the problem

$$\begin{aligned} &\text{minimize} \quad \sum_{i=1}^{m} \left(f_i(z^i) + \lambda_k^{i\,\prime}(x - z^i) + \frac{c_k}{2}\|x - z^i\|^2 \right) \\ &\text{subject to} \quad x \in \Re^n, \qquad z^i \in X_i, \quad i = 1,\ldots,m. \end{aligned}$$

Note that there is coupling between x and the vectors z^i, so this problem cannot be decomposed into separate minimizations with respect to some of the variables. On the other hand, the problem has a Cartesian product constraint set, and a structure that is suitable for the application of block coordinate descent methods that cyclically minimize the cost function, one component at a time. In particular, we can consider a method that minimizes the augmented Lagrangian with respect to $x \in \Re^n$ with the iteration

$$x := \frac{\sum_{i=1}^{m} z^i}{m} - \frac{\sum_{i=1}^{m} \lambda_k^i}{m c_k}, \tag{5.59}$$

then minimizes the augmented Lagrangian with respect to $z^i \in X_i$, with the iteration

$$z^i \in \arg\min_{z^i \in X_i} \left\{ f_i(z^i) - \lambda_k^{i\,\prime} z^i + \frac{c_k}{2}\|x - z^i\|^2 \right\}, \qquad i = 1,\ldots,m, \tag{5.60}$$

and repeats until convergence to a minimum of the augmented Lagrangian, which is then followed by a multiplier update of the form (5.58).

In the preceding example the minimization of the augmented Lagrangian exploits the problem structure, yet requires an infinite number of cyclic minimization iterations of the form (5.59)-(5.60), before the multiplier update (5.58) can be performed. Actually, exact convergence to a minimum of the augmented Lagrangian is not necessary, only a limited number of minimization cycles in x and z^i may be performed prior to a multiplier update. In particular, there are modified versions of augmented Lagrangian methods, with sound convergence properties, which allow for inexact minimization of the augmented Lagrangian, subject to certain termination criteria (see [Ber82a], and subsequent sources such as [EcB92], [Eck03], [EkS13]). In Section 5.4, we will discuss the alternating direction method of multipliers, a somewhat different type of method, which is based on augmented Lagrangian ideas and performs only one minimization cycle in x and z^i before updating the multiplier vectors.

5.3 PROXIMAL ALGORITHMS WITH LINEARIZATION

In this section we will consider the minimization of a real-valued convex function $f : \Re^n \mapsto \Re$, over a closed convex set X, and we will combine the proximal algorithm and the polyhedral approximation approaches of

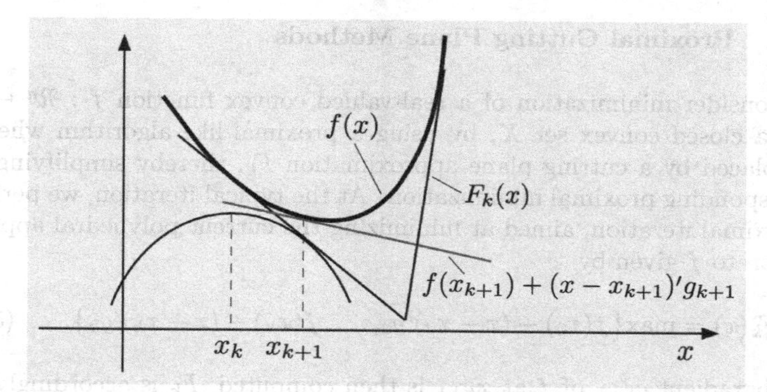

Figure 5.3.1. Illustration of the proximal algorithm with outer linearization. The point x_{k+1} is the one at which the graph of the negative proximal term, raised by some amount, first touches the graph of F_k. A new cutting plane is added, based on a subgradient g_{k+1} of f at x_{k+1}. Note that the proximal term reduces the effect of instability: x_{k+1} tends to be closer to x_k, with the distance $\|x_{k+1} - x_k\|$ depending on the size of the proximal term, i.e., the penalty parameter c_k.

Chapter 4. As discussed in Section 4.1, one of the drawbacks of the cutting plane method is the instability phenomenon, whereby the method can take large steps away from the current point, with significant deterioration of the cost function value. A way to limit the effects of this is to add to the polyhedral function approximation a regularization term $p_k(x)$ that penalizes large deviations from some reference point y_k, similar to the proximal algorithm of Section 5.1. Thus in this method, x_{k+1} is obtained as

$$x_{k+1} \in \arg\min_{x \in X} \{F_k(x) + p_k(x)\}, \tag{5.61}$$

where similar to the cutting plane method,

$$F_k(x) = \max\{f(x_0) + (x - x_0)'g_0, \ldots, f(x_k) + (x - x_k)'g_k\},$$

and similar to the proximal algorithm,

$$p_k(x) = \frac{1}{2c_k}\|x - y_k\|^2,$$

where c_k is a positive scalar parameter; see Fig. 5.3.1 for the case where $y_k = x_k$. We refer to $p_k(x)$ as the *proximal term*, and to its center y_k as the *proximal center* (which may be different than x_k, for reasons to be explained later).

The idea of the method is to provide a measure of stability to the cutting plane method at the expense of solving a more difficult subproblem at each iteration (e.g., a quadratic versus a linear program, in the case where X is polyhedral). We can view iteration (5.61) as a combination of the cutting plane method and the proximal method. We first discuss the case where the proximal center is the current iterate ($y_k = x_k$) in the next section. Then in Section 5.3.2, we discuss alternative choices of y_k, which are aimed at further improvements in stability.

5.3.1 Proximal Cutting Plane Methods

We consider minimization of a real-valued convex function $f : \Re^n \mapsto \Re$, over a closed convex set X, by using a proximal-like algorithm where f is replaced by a cutting plane approximation F_k, thereby simplifying the corresponding proximal minimization. At the typical iteration, we perform a proximal iteration, aimed at minimizing the current polyhedral approximation to f given by

$$F_k(x) = \max\big\{f(x_0) + (x - x_0)'g_0, \ldots, f(x_k) + (x - x_k)'g_k\big\}. \tag{5.62}$$

A subgradient g_{k+1} of f at x_{k+1} is then computed, F_k is accordingly updated, and the process is repeated. We call this the *proximal cutting plane method*. This is the method (5.61) where the proximal center is reset at every iteration to the current point ($y_k = x_k$ for all k); see Fig. 5.3.1.

Proximal Cutting Plane Method:

Find

$$x_{k+1} \in \arg\min_{x \in X} \left\{ F_k(x) + \frac{1}{2c_k}\|x - x_k\|^2 \right\}, \tag{5.63}$$

where F_k is the outer approximation function of Eq. (5.62), and c_k is a positive scalar parameter, and then refine the approximation by introducing a new cutting plane based on $f(x_{k+1})$ and a subgradient

$$g_{k+1} \in \partial f(x_{k+1}).$$

The method terminates if $x_{k+1} = x_k$; in this case, Eqs. (5.62) and (5.63) imply that

$$f(x_k) = F_k(x_k) \le F_k(x) + \frac{1}{2c_k}\|x - x_k\|^2 \le f(x) + \frac{1}{2c_k}\|x - x_k\|^2, \qquad \forall\, x \in X,$$

so x_k is a point where the proximal algorithm terminates, and it must therefore be optimal by Prop. 5.1.3. Note, however, that unless f and X are polyhedral, finite termination is unlikely.

The convergence properties of the method are easy to obtain, based on what we already know. The idea is that F_k asymptotically "converges" to f, at least near the generated iterates, so asymptotically, the algorithm essentially becomes the proximal algorithm, and inherits the corresponding convergence properties. In particular, we can show that if the optimal solution set X^* is nonempty, the sequence $\{x_k\}$ generated by the proximal cutting plane method (5.63) converges to some point in X^*. The proof is based on a combination of the convergence arguments of the cutting plane method (cf. Prop. 4.1.1) and the proximal algorithm (cf. Prop. 5.1.3), and will not be given.

In the case where f and X are polyhedral, convergence to an optimal solution occurs in a finite number of iterations, as shown in the following proposition. This is a consequence of the finite convergence property of both the cutting plane and the proximal methods.

Proposition 5.3.1: (Finite Termination of the Proximal Cutting Plane Method) Consider the proximal cutting plane method for the case where f and X are polyhedral, with

$$f(x) = \max_{i \in I} \{ a_i' x + b_i \},$$

where I is a finite index set, and a_i and b_i are given vectors and scalars, respectively. Assume that the optimal solution set is nonempty and that the subgradient added to the cutting plane approximation at each iteration is one of the vectors a_i, $i \in I$. Then the method terminates finitely with an optimal solution.

Proof: Since there are only finitely many vectors a_i to add, eventually the polyhedral approximation F_k will not change, i.e., $F_k = F_{\bar{k}}$ for all $k > \bar{k}$. Thus, for $k \geq \bar{k}$, the method will become the proximal algorithm for minimizing $F_{\bar{k}}$, so by Prop. 5.1.5, it will terminate with a point $\bar{z}$ that minimizes $F_{\bar{k}}$ subject to $x \in X$. But then, we will have concluded an iteration of the cutting plane method for minimizing f over X, with no new vector added to the approximation F_k. This implies termination of the cutting plane method, necessarily at a minimum of f over X. **Q.E.D.**

The proximal cutting plane method aims at increased stability over the ordinary cutting plane method, but it has some drawbacks:

(a) There is a potentially difficult tradeoff in the choice of the parameter c_k. In particular, stability is achieved only by choosing c_k small, since for large values of c_k the changes $x_{k+1} - x_k$ may be substantial. Indeed for large enough c_k, the method finds the exact minimum of F_k over X in a single minimization (cf. Prop. 5.1.5), so it is identical to the ordinary cutting plane method, and fails to provide any stabilization! On the other hand, small values of c_k lead to slow rate of convergence, even when f is polyhedral, or even linear (cf. Fig. 5.1.2).

(b) The number of subgradients used in the approximation F_k may grow to be very large, in which case the quadratic program solved in Eq. (5.63) may become very time-consuming.

These drawbacks motivate algorithmic variants, called *bundle methods*, which we will discuss next. The main difference is that in order to ensure a certain measure of stability, the proximal center is updated selectively, only after making enough progress in minimizing f.

5.3.2 Bundle Methods

In the basic form of a bundle method, the iterate x_{k+1} is obtained by minimizing over X the sum of F_k, a cutting plane approximation to f, and a quadratic proximal term $p_k(x)$:

$$x_{k+1} \in \arg\min_{x \in X} \{F_k(x) + p_k(x)\}. \tag{5.64}$$

The proximal center of p_k need not be x_k (as in the proximal cutting plane method), but is rather one of the past iterates x_i, $i \leq k$.

In one version of the method, F_k is given by

$$F_k(x) = \max\{f(x_0) + (x - x_0)'g_0, \ldots, f(x_k) + (x - x_k)'g_k\}, \tag{5.65}$$

while $p_k(x)$ is of the form

$$p_k(x) = \frac{1}{2c_k}\|x - y_k\|^2,$$

where $y_k \in \{x_i \mid i \leq k\}$. Following the computation of x_{k+1}, the new proximal center y_{k+1} is set to x_{k+1}, or is left unchanged ($y_{k+1} = y_k$) depending on whether, according to a certain test, "sufficient progress" has been made or not. An example of such a test is

$$f(y_k) - f(x_{k+1}) \geq \beta\delta_k,$$

where β is a fixed scalar with $\beta \in (0, 1)$, and

$$\delta_k = f(y_k) - \big(F_k(x_{k+1}) + p_k(x_{k+1})\big).$$

Thus,

$$y_{k+1} = \begin{cases} x_{k+1} & \text{if } f(y_k) - f(x_{k+1}) \geq \beta\delta_k, \\ y_k & \text{if } f(y_k) - f(x_{k+1}) < \beta\delta_k, \end{cases} \tag{5.66}$$

and initially $y_0 = x_0$. In the parlance of bundle methods, iterations where y_{k+1} is updated to x_{k+1} are called *serious steps*, while iterations where $y_{k+1} = y_k$ are called *null steps*.

The scalar δ_k is illustrated in Fig. 5.3.2. Since $f(y_k) = F_k(y_k)$ [cf. Eq. (5.65)], δ_k *represents the reduction in the proximal objective* $F_k + p_k$ *in moving from* y_k *to* x_{k+1}. If the reduction in the true objective,

$$f(y_k) - f(x_{k+1}),$$

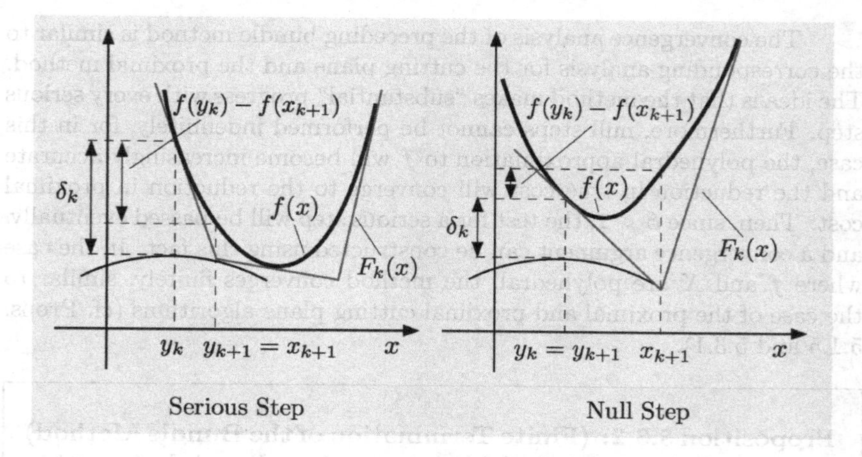

Figure 5.3.2. Illustration of the test (5.66) for a serious or a null step in the bundle method. It is based on

$$\delta_k = f(y_k) - \big(F_k(x_{k+1}) + p_k(x_{k+1})\big) = \big(F_k(y_k) + p_k(y_k)\big) - \big(F_k(x_{k+1}) + p_k(x_{k+1})\big),$$

the reduction in proximal cost, which is always positive, except at termination. A serious step is performed if and only if the reduction in true cost, $f(y_k) - f(x_{k+1})$, exceeds a fraction β of the reduction δ_k in proximal cost.

does not exceed a fraction β of δ_k (or is even negative as in the right-hand side of Fig. 5.3.2), this indicates a large discrepancy between proximal and true objective, and an associated instability. As a result the algorithm foregoes the move from y_k to x_{k+1} with a null step [cf. Eq. (5.66)], but improves the cutting plane approximation by adding the new plane corresponding to x_{k+1}. Otherwise, it performs a serious step, with the guarantee of true cost improvement afforded by the test (5.66).

An important point is that if $x_{k+1} \neq y_k$, then $\delta_k > 0$. Indeed, since

$$F_k(x_{k+1}) + p_k(x_{k+1}) \leq F_k(y_k) + p_k(y_k) = F_k(y_k),$$

and $F_k(y_k) = f(y_k)$, we have

$$0 \leq f(y_k) - \big(F_k(x_{k+1}) + p_k(x_{k+1})\big) = \delta_k,$$

with equality only if $x_{k+1} = y_k$.

The method terminates if $x_{k+1} = y_k$; in this case, Eqs. (5.64) and (5.65) imply that

$$f(y_k) + p_k(y_k) = F_k(y_k) + p_k(y_k) \leq F_k(x) + p_k(x) \leq f(x) + p_k(x), \quad \forall\, x \in X,$$

so y_k is a point where the proximal algorithm terminates, and must therefore be optimal. Of course, finite termination is unlikely, unless f and X are polyhedral.

The convergence analysis of the preceding bundle method is similar to the corresponding analysis for the cutting plane and the proximal method. The idea is that the method makes "substantial" progress with every serious step. Furthermore, null steps cannot be performed indefinitely, for in this case, the polyhedral approximation to f will become increasingly accurate and the reduction in true cost will converge to the reduction in proximal cost. Then, since $\beta < 1$, the test for a serious step will be passed eventually, and a convergence argument can be constructed using this fact. In the case where f and X are polyhedral, the method converges finitely, similar to the case of the proximal and proximal cutting plane algorithms (cf. Props. 5.1.5 and 5.3.1).

Proposition 5.3.2: (Finite Termination of the Bundle Method) Consider the bundle method for the case where X and f are polyhedral, with

$$f(x) = \max_{i \in I} \{a_i'x + b_i\},$$

where I is a finite index set, and a_i and b_i are given vectors and scalars, respectively. Assume that the optimal solution set is nonempty and that the subgradient added to the cutting plane approximation at each iteration is one of the vectors a_i, $i \in I$. Then the method terminates finitely with an optimal solution.

Proof: Since there are only finitely many vectors a_i to add, eventually the polyhedral approximation F_k will not change, i.e., $F_k = F_{\bar{k}}$ for all $k > \bar{k}$. We then have $F_k(x_{k+1}) = f(x_{k+1})$ for all $k > \bar{k}$, since otherwise a new cutting plane would be added to F_k. Thus, for $k > \bar{k}$,

$$\begin{aligned} f(y_k) - f(x_{k+1}) &= f(y_k) - F_k(x_{k+1}) \\ &= f(y_k) - \big(F_k(x_{k+1}) + p_k(x_{k+1})\big) + p_k(x_{k+1}) \\ &= \delta_k + p_k(x_{k+1}) \\ &\geq \beta \delta_k. \end{aligned}$$

Therefore, according to Eq. (5.66), the method will perform serious steps for all $k > \bar{k}$, and become identical to the proximal cutting plane algorithm, which converges finitely by Prop. 5.3.1. **Q.E.D.**

Discarding Old Subgradients

We mentioned earlier that one of the drawbacks of the cutting plane algorithms is that the number of subgradients used in the approximation F_k may grow to be very large. The monitoring of progress through the test (5.66) for serious/null steps can also be used to discard some of the accumulated cutting planes. For example, at the end of a serious step, upon updating the proximal center y_k to $y_{k+1} = x_{k+1}$, we may discard any subset of the cutting planes.

It may of course be useful to retain some of the cutting planes, particularly the ones that are "active" or "nearly active" at y_{k+1}, i.e., those $i \leq k$ for which the linearization error

$$F_k(y_{k+1}) - \big(f(x_i) + (y_{k+1} - x_i)'g_i\big)$$

is 0 or close to 0, respectively. The essential validity of the method is maintained, by virtue of the fact that $\{f(y_k)\}$ is a monotonically decreasing sequence, with "sufficiently large" cost reductions between proximal center updates.

An extreme possibility is to discard all past subgradients following a serious step from y_k to x_{k+1}. Then, after a subgradient g_{k+1} at x_{k+1} is calculated, the next iteration becomes

$$x_{k+2} \in \arg\min_{x \in X} \left\{ f(x_{k+1}) + g'_{k+1}(x - x_{k+1}) + \frac{1}{2c_{k+1}} \|x - x_{k+1}\|^2 \right\}.$$

It can be seen that we have

$$x_{k+2} = P_X(x_{k+1} - c_{k+1} g_{k+1}),$$

where $P_X(\cdot)$ denotes projection on X, so after discarding all past subgradients following a serious step, the next iteration is an ordinary subgradient iteration with stepsize equal to c_{k+1}.

Another possibility is, following a serious step, to replace all the cutting planes with a single cutting plane: the one obtained from the hyperplane that passes through $\big(x_{k+1}, F_k(x_{k+1})\big)$ and separates the epigraphs of the functions $F_k(x)$ and $\gamma_k - \frac{1}{2c_k}\|x - y_k\|^2$, where

$$\gamma_k = F_k(x_{k+1}) + \frac{1}{2c_k} \|x_{k+1} - y_k\|^2,$$

(see Fig. 5.3.3). This is the cutting plane

$$F_k(x_{k+1}) + \hat{g}'_k(x - x_{k+1}), \tag{5.67}$$

where $\hat{g}_k$ is given by

$$\hat{g}_k = \frac{y_k - x_{k+1}}{c_k}. \tag{5.68}$$

The next iteration will then be performed with only two cutting planes: the one just given by Eqs. (5.67)-(5.68) and a new one obtained from x_{k+1},

$$f(x_{k+1}) + g'_{k+1}(x - x_{k+1}),$$

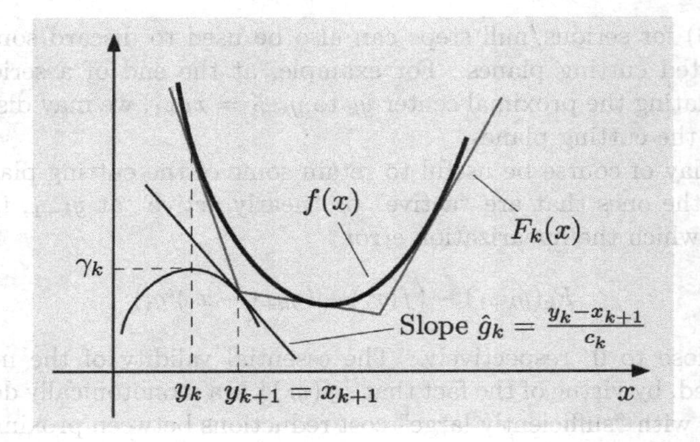

Figure 5.3.3. Illustration of the cutting plane

$$F_k(x_{k+1}) + \hat{g}_k'(x - x_{k+1}),$$

where

$$\hat{g}_k = \frac{y_k - x_{k+1}}{c_k}.$$

The "slope" $\hat{g}_k$ can be shown to be a convex combination of the subgradients that are "active" at x_{k+1}.

where $g_{k+1} \in \partial f(x_{k+1})$.

The vector $\hat{g}_k$ is sometimes called an "aggregate subgradient," because it can be shown to be a convex combination of the past subgradients $g_0, \ldots, g_k$. This is evident from Fig. 5.3.3, and can also be verified by using quadratic programming duality arguments.

There are also many other variants of bundle methods, which aim at increased efficiency and the exploitation of special structure. We refer to the literature for discussion and analysis.

5.3.3 Proximal Inner Linearization Methods

In the preceding section we saw that the proximal algorithm can be combined with outer linearization to yield the proximal cutting plane algorithm and its bundle versions. In this section we use a dual combination, involving the dual proximal algorithm (5.41)-(5.42) and inner linearization (the dual of outer linearization). This yields another method, which is connected to the proximal cutting plane algorithm of Section 5.3.1 by Fenchel duality (see Fig. 5.3.4).

Let us recall the proximal cutting plane method applied to minimizing a real-valued convex function $f : \Re^n \mapsto \Re$, over a closed convex set X. The typical iteration involves a proximal minimization of the current cutting

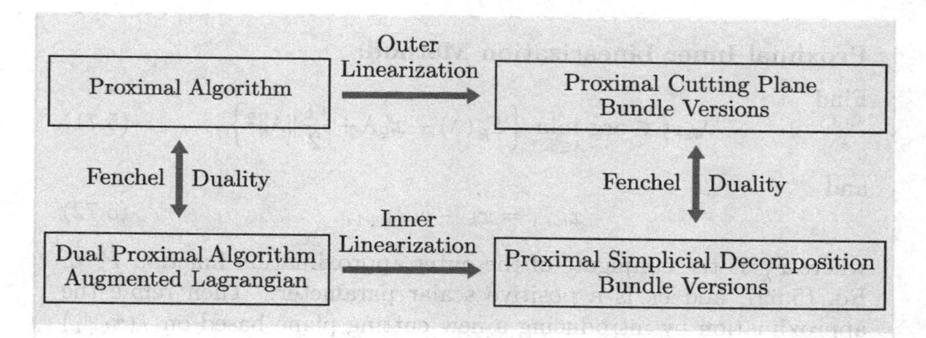

Figure 5.3.4. Relations of the proximal and proximal cutting plane methods, and their duals. The dual algorithms are obtained by application of the Fenchel Duality Theorem (Prop. 1.2.1), taking also into account the conjugacy relation between outer and inner linearization (cf. Section 4.3).

plane approximation to f given by

$$F_k(x) = \max\{f(x_0)+(x-x_0)'g_0, \ldots, f(x_k)+(x-x_k)'g_k\}+\delta_X(x), \quad (5.69)$$

where $g_i \in \partial f(x_i)$ for all i and δ_X is the indicator function of X. Thus,

$$x_{k+1} \in \arg\min_{x\in\Re^n} \left\{ F_k(x) + \frac{1}{2c_k}\|x - x_k\|^2 \right\},$$

where c_k is a positive scalar parameter. A subgradient g_{k+1} of f at x_{k+1} is then computed, F_{k+1} is accordingly updated, and the process is repeated.

There is a dual version of this algorithm, similar to the one of Section 5.2. In particular, we may use Fenchel duality to implement the preceding proximal minimization in terms of conjugate functions [cf. Eq. (5.41)]. Thus, the Fenchel dual of this minimization can be written as [cf. Eq. (5.37)]

$$\begin{aligned} &\text{minimize} \quad F_k^\star(\lambda) - x_k'\lambda + \frac{c_k}{2}\|\lambda\|^2 \\ &\text{subject to} \quad \lambda \in \Re^n, \end{aligned} \quad (5.70)$$

where $F_k^\star$ is the conjugate of F_k. Once λ_{k+1}, the unique minimizer in this dual proximal iteration, is computed, x_k is updated via

$$x_{k+1} = x_k - c_k\lambda_{k+1}$$

[cf. Eq. (5.42)]. Then, a subgradient g_{k+1} of f at x_{k+1} is obtained by "differentiation," and is used to update F_k.

Proximal Inner Linearization Method:

Find

$$\lambda_{k+1} \in \arg \min_{\lambda \in \Re^n} \left\{ F_k^\star(\lambda) - x_k'\lambda + \frac{c_k}{2}\|\lambda\|^2 \right\}, \tag{5.71}$$

and

$$x_{k+1} = x_k - c_k\lambda_{k+1}, \tag{5.72}$$

where $F_k^\star$ is the conjugate of the outer approximation function F_k of Eq. (5.69), and c_k is a positive scalar parameter. Then refine the approximation by introducing a new cutting plane based on $f(x_{k+1})$ and a subgradient

$$g_{k+1} \in \partial f(x_{k+1}),$$

to form F_{k+1} and $F_{k+1}^\star$.

Note that the new subgradient g_{k+1} may also be obtained as a vector attaining the supremum in the conjugacy relation

$$f(x_{k+1}) = \sup_{\lambda \in \Re^n} \{x_{k+1}'\lambda - f^\star(\lambda)\},$$

where $f^\star$ is the conjugate function of f, since we have

$$g_{k+1} \in \partial f(x_{k+1}) \qquad \text{if and only if} \qquad g_{k+1} \in \arg \max_{\lambda \in \Re^n} \{x_{k+1}'\lambda - f^\star(\lambda)\}, \tag{5.73}$$

(cf. the Conjugate Subgradient Theorem of Prop. 5.4.3 in Appendix B). The maximization above may be preferable if the "differentiation" $g_{k+1} \in \partial f(x_{k+1})$ is inconvenient.

Implementation by Simplicial Decomposition

We will now discuss the details of the preceding computations, assuming for simplicity that there are no constraints, i.e., $X = \Re^n$. According to Section 4.3, $F_k^\star$ (the conjugate of the outer linear approximation F_k of f) is the piecewise linear (inner) approximation of $f^\star$ with domain

$$\text{dom}(F_k^\star) = \text{conv}(\{g_0, \ldots, g_k\}),$$

and "break points" at g_i, $i = 0, \ldots, k$. In particular, using the formula of Section 4.3 for the conjugate $F_k^\star$, the dual proximal optimization of Eq. (5.71) takes the form

$$\begin{aligned} &\text{minimize} \quad \sum_{i=0}^{k} \alpha_i f^\star(g_i) - x_k' \sum_{i=0}^{k} \alpha_i g_i + \frac{c_k}{2} \left\| \sum_{i=0}^{k} \alpha_i g_i \right\|^2 \\ &\text{subject to} \quad \sum_{i=0}^{k} \alpha_i = 1, \quad \alpha_i \geq 0, \ i = 0, \ldots, k. \end{aligned} \tag{5.74}$$

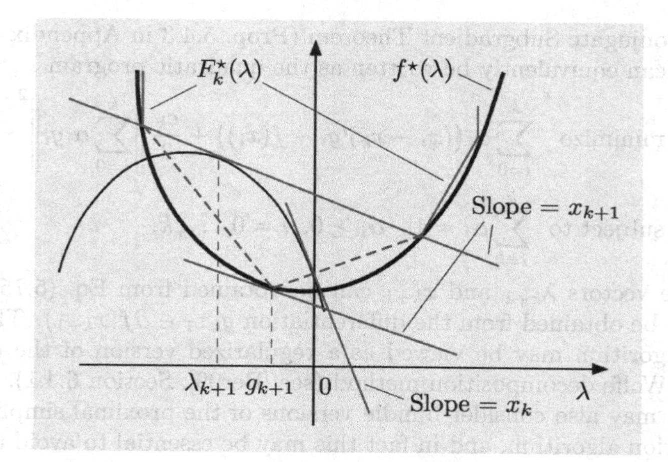

Figure 5.3.5. Illustration of an iteration of the proximal simplicial decomposition algorithm. The proximal minimization determines the "slope" x_{k+1} of $F_k^\star$, which then determines the next subgradient/break point g_{k+1} via the maximization

$$g_{k+1} \in \arg \max_{\lambda \in \Re^n} \big\{ x_{k+1}'\lambda - f^\star(\lambda) \big\},$$

i.e., g_{k+1} is a point at which x_{k+1} is a subgradient of $f^\star$.

If $(\alpha_0^k, \ldots, \alpha_k^k)$ attains the minimum, Eqs. (5.71) and (5.72) yield

$$\lambda_{k+1} = \sum_{i=0}^{k} \alpha_i^k g_i, \qquad x_{k+1} = x_k - c_k \sum_{i=0}^{k} \alpha_i^k g_i. \tag{5.75}$$

The next subgradient $g_{k+1} \in \partial f(x_{k+1})$ may also be obtained from the maximization

$$g_{k+1} \in \arg \max_{\lambda \in \Re^n} \big\{ x_{k+1}'\lambda - f^\star(\lambda) \big\} \tag{5.76}$$

if this is convenient; cf. Eq. (5.73). As Fig. 5.3.5 indicates, g_{k+1} provides a new break point and an improved inner approximation to $f^\star$.

We refer to the algorithm defined by Eqs. (5.74)-(5.76), as the *proximal simplicial decomposition algorithm*. Note that all the computations of the algorithm involve the conjugate $f^\star$ and not f. Thus, if $f^\star$ is more convenient to work with than f, the proximal simplicial decomposition algorithm is preferable to the proximal cutting plane algorithm. Note also that the duality between the two linear approximation versions of the proximal algorithm is a special case of the generalized polyhedral approximation framework of Section 4.4.

The problem (5.74) can also be written without reference to the conjugate $f^\star$. Since g_i is a subgradient of f at x_i, and hence we have

$$f^\star(g_i) = x_i' g_i - f(x_i), \qquad i = 0, \ldots, k,$$

by the Conjugate Subgradient Theorem (Prop. 5.4.3 in Appendix B), this problem can equivalently be written as the quadratic program

$$\begin{aligned} &\text{minimize} \quad \sum_{i=0}^{k} \alpha_i\big((x_i - x_k)'g_i - f(x_i)\big) + \frac{c_k}{2}\left\|\sum_{i=0}^{k} \alpha_i g_i\right\|^2 \\ &\text{subject to} \quad \sum_{i=0}^{k} \alpha_i = 1, \ \ \alpha_i \geq 0, \ i = 0, \ldots, k. \end{aligned}$$

Then the vectors λ_{k+1} and x_{k+1} can be obtained from Eq. (5.75), while g_{k+1} can be obtained from the differentiation $g_{k+1} \in \partial f(x_{k+1})$. This form of the algorithm may be viewed as a regularized version of the classical Dantzig-Wolfe decomposition method (see [Ber99], Section 6.4.1).

We may also consider bundle versions of the proximal simplicial decomposition algorithm, and in fact this may be essential to avoid the kind of difficulties discussed at the end of Section 5.3.1. For this we need a test to distinguish between serious steps, where we update x_k via Eq. (5.75), and null steps, where we leave x_k unchanged, but simply add the pair $\big(g_{k+1}, f^\star(g_{k+1})\big)$ to the current inner approximation of $f^\star$.

Finally, in a different line of extension, it is possible to combine the proximal algorithm and its bundle versions with the generalized polyhedral approximation algorithms of Sections 4.4-4.6 for problems whose cost function has an additive form. Such combinations are clearly valid and useful when both polyhedral approximation and regularization are beneficial, but have not been analyzed or systematically tested so far.

5.4 ALTERNATING DIRECTION METHODS OF MULTIPLIERS

In this section we discuss an algorithm that is related to the augmented Lagrangian method of Section 5.2.1, and is well suited for special structures involving among others separability and large sums of component functions. The algorithm uses alternate minimizations to decouple sets of variables that are coupled within the augmented Lagrangian, and is known as the *alternating direction method of multipliers* or ADMM. The name comes from the similarity with some methods for solving differential equations, known as alternating direction methods (see [FoG83] for an explanation).

The following example, which involves the additive cost problem of Example 5.2.1, illustrates the decoupling process of ADMM.

Example 5.4.1: (Additive Cost Problems – Continued)

Consider the problem

$$\begin{aligned} &\text{minimize} \quad \sum_{i=1}^{m} f_i(x) \\ &\text{subject to} \quad x \in \cap_{i=1}^{m} X_i, \end{aligned} \tag{5.77}$$

where $f_i : \Re^n \mapsto \Re$ are convex functions and X_i are closed convex sets with nonempty intersection. As in Example 5.2.1 we can reformulate this as an equality constrained problem, by introducing additional artificial variables z^i, $i = 1, \ldots, m$, and the equality constraints $x = z^i$:

$$\begin{aligned} &\text{minimize} && \sum_{i=1}^{m} f_i(z^i) \\ &\text{subject to} && x = z^i, \quad z^i \in X_i, \qquad i = 1, \ldots, m, \end{aligned}$$

[cf. Eq. (5.57)].

As motivation for the development of the ADMM for this problem, let us recall the augmented Lagrangian method that uses multipliers λ^i for the constraints $x = z^i$ (cf. Example 5.2.1). At the typical iteration of this method, we find x_{k+1} and z^i_{k+1}, $i = 1, \ldots, m$, that solve the problem

$$\begin{aligned} &\text{minimize} && \sum_{i=1}^{m} \left(f_i(z^i) + \lambda_k^{i\,\prime}(x - z^i) + \frac{c_k}{2}\|x - z^i\|^2 \right) \\ &\text{subject to} && x \in \Re^n, \qquad z^i \in X_i, \quad i = 1, \ldots, m, \end{aligned} \tag{5.78}$$

[cf. Eqs. (5.58), (5.47)], and then update the multipliers according to

$$\lambda^i_{k+1} = \lambda^i_k + c_k(x_{k+1} - z^i_{k+1}), \qquad i = 1, \ldots, m.$$

The minimization in Eq. (5.78) can be done by alternating minimizations of x and z^i (a block coordinate descent method), and the multipliers λ^i may be changed only after (typically) many updates of x and z^i (enough to minimize the augmented Lagrangian within adequate precision).

An interesting variation is to perform only a small number of minimizations with respect to x and z^i before changing the multipliers. In the extreme case, where only one minimization is performed, the method takes the form

$$x_{k+1} = \frac{\sum_{i=1}^{m} z^i_k}{m} - \frac{\sum_{i=1}^{m} \lambda^i_k}{m c_k}, \tag{5.79}$$

$$z^i_{k+1} \in \arg\min_{z^i \in X_i} \left\{ f_i(z^i) - \lambda_k^{i\,\prime} z^i + \frac{c_k}{2}\|x_{k+1} - z^i\|^2 \right\}, \quad i = 1, \ldots, m, \tag{5.80}$$

[cf. Eqs. (5.59), (5.60)], followed by the multiplier update

$$\lambda^i_{k+1} = \lambda^i_k + c_k(x_{k+1} - z^i_{k+1}), \qquad i = 1, \ldots, m, \tag{5.81}$$

[cf. Eq. (5.58)]. Thus the multiplier iteration is performed after just one block coordinate descent iteration on each of the (now decoupled) variables x and $(z^1, \ldots, z^m)$. This is precisely the ADMM specialized to the problem of this example.

The preceding example also illustrates another advantage of ADMM. Frequently the decoupling process results in computations that are well-suited for parallel and distributed processing (see e.g., [BeT89a], [WeO13]).

This will be observed in many of the examples to be presented in this section.

We will now formulate the ADMM and discuss its convergence properties. The starting point is the minimization problem of the Fenchel duality context:

$$\begin{aligned} &\text{minimize} \quad f_1(x) + f_2(Ax) \\ &\text{subject to} \quad x \in \Re^n, \end{aligned} \tag{5.82}$$

where A is an $m \times n$ matrix, $f_1 : \Re^n \mapsto (-\infty, \infty]$ and $f_2 : \Re^m \mapsto (-\infty, \infty]$ are closed proper convex functions, and we assume that there exists a feasible solution. We convert this problem to the equivalent constrained minimization problem

$$\begin{aligned} &\text{minimize} \quad f_1(x) + f_2(z) \\ &\text{subject to} \quad x \in \Re^n,\ z \in \Re^m, \quad Ax = z, \end{aligned} \tag{5.83}$$

and we introduce its augmented Lagrangian function

$$L_c(x, z, \lambda) = f_1(x) + f_2(z) + \lambda'(Ax - z) + \frac{c}{2}\|Ax - z\|^2.$$

The ADMM, given the current iterates $(x_k, z_k, \lambda_k) \in \Re^n \times \Re^m \times \Re^m$, generates a new iterate $(x_{k+1}, z_{k+1}, \lambda_{k+1})$ by first minimizing the augmented Lagrangian with respect to x, then with respect to z, and finally performing a multiplier update:

$$x_{k+1} \in \arg\min_{x \in \Re^n} L_c(x, z_k, \lambda_k), \tag{5.84}$$

$$z_{k+1} \in \arg\min_{z \in \Re^m} L_c(x_{k+1}, z, \lambda_k), \tag{5.85}$$

$$\lambda_{k+1} = \lambda_k + c(Ax_{k+1} - z_{k+1}). \tag{5.86}$$

The penalty parameter c is kept constant in ADMM. Contrary to the case of the augmented Lagrangian method (where c_k is often taken to be increasing with k in order to accelerate convergence), there seems to be no generally good way to adjust c from one iteration to the next. Note that the iteration (5.79)-(5.81), given earlier for the additive cost problem (5.77), is a special case of the preceding iteration, with the identification $z = (z^1, \ldots, z^m)$.

We may also formulate an ADMM that applies to the closely related problem

$$\begin{aligned} &\text{minimize} \quad f_1(x) + f_2(z) \\ &\text{subject to} \quad x \in X, \quad z \in Z, \quad Ax + Bz = d, \end{aligned} \tag{5.87}$$

where $f_1 : \Re^n \mapsto \Re$, $f_2 : \Re^m \mapsto \Re$ are convex functions, X and Z are closed convex sets, and A, B, and d are given matrices and vector, respectively, of appropriate dimensions. Then the corresponding augmented Lagrangian is

$$L_c(x, z, \lambda) = f_1(x) + f_2(z) + \lambda'(Ax + Bz - d) + \frac{c}{2}\|Ax + Bz - d\|^2, \tag{5.88}$$

and the ADMM iteration takes a similar form [cf. Eqs. (5.84)-(5.86)]:

$$x_{k+1} \in \arg\min_{x \in X} L_c(x, z_k, \lambda_k), \tag{5.89}$$

$$z_{k+1} \in \arg\min_{z \in Z} L_c(x_{k+1}, z, \lambda_k), \tag{5.90}$$

$$\lambda_{k+1} = \lambda_k + c(Ax_{k+1} + Bz_{k+1} - d). \tag{5.91}$$

For some problems, this form may be more convenient than the ADMM of Eqs. (5.84)-(5.86), although the two forms are essentially equivalent.

The important advantage that the ADMM may offer over the augmented Lagrangian method is that it involves a separate minimization with respect to x and with respect to z. Thus the complications resulting from the coupling of x and z in the penalty term $\|Ax - z\|^2$ or the penalty term $\|Ax + Bz - d\|^2$ are eliminated. Here is another illustration of this advantage.

Example 5.4.2 (Finding a Point in a Set Intersection)

We are given m closed convex sets $X_1, \ldots, X_m$ in $\Re^n$, and we want to find a point in their intersection. We write this problem in the form (5.87), with x defined as $x = (x^1, \ldots, x^m)$,

$$f_1(x) \equiv 0, \qquad f_2(z) \equiv 0, \qquad x \in X_1 \times \cdots \times X_m, \qquad Z = \Re^n,$$

and with the constraint $Ax + Bz = d$ representing the system of equations

$$x^i = z, \qquad i = 1, \ldots, m.$$

The augmented Lagrangian is

$$L_c(x^1, \ldots, x^m, z, \lambda^1, \ldots, \lambda^m) = \sum_{i=1}^{m} \lambda^{i\prime}(x^i - z) + \frac{c}{2} \sum_{i=1}^{m} \|x^i - z\|^2.$$

The parameter c does not influence the algorithm, because it simply introduces scaling of λ^i by $1/c$, so we may assume with no loss of generality that $c = 1$. Then, by completing the square, we may write the augmented Lagrangian as

$$L_1(x^1, \ldots, x^m, z, \lambda^1, \ldots, \lambda^m) = \frac{1}{2} \sum_{i=1}^{m} \|x^i - z + \lambda^i\|^2 - \frac{1}{2} \sum_{i=1}^{m} \|\lambda^i\|^2.$$

Using Eqs. (5.89)-(5.91), we see that the corresponding ADMM iterates for x^i according to

$$x^i_{k+1} = P_{X_i}(z_k - \lambda^i_k), \qquad i = 1, \ldots, m,$$

then iterates for z according to

$$z_{k+1} = \frac{x_{k+1}^1 + \lambda_k^1 + \cdots + x_{k+1}^m + \lambda_k^m}{m},$$

and finally iterates for the multipliers according to

$$\lambda_{k+1}^i = \lambda_k^i + x_{k+1}^i - z_{k+1}, \qquad i = 1, \ldots, m.$$

Aside from the decoupling of the iterations of the variables x^i and z, notice that the projections on X_i can be carried out in parallel.

In the special case where $m = 2$, we can write the constraint more simply as $x^1 = x^2$, in which case the augmented Lagrangian takes the form

$$L_c(x^1, x^2, \lambda) = \lambda'(x^1 - x^2) + \frac{c}{2}\|x^1 - x^2\|^2.$$

Assuming as above that $c = 1$, the corresponding ADMM is

$$x_{k+1}^1 = P_{X_1}(x_k^2 - \lambda_k),$$

$$x_{k+1}^2 = P_{X_2}(x_{k+1}^1 + \lambda_k),$$

$$\lambda_{k+1} = \lambda_k + x_{k+1}^1 - x_{k+1}^2.$$

On the other hand there is a price for the flexibility that the ADMM provides. A major drawback is a much slower practical convergence rate relative to the augmented Lagrangian method of the preceding subsection. Both methods can be shown to have a linear convergence rate for the multiplier updates under favorable circumstances (see e.g., [Ber82a] for augmented Lagrangian, and [HoL13], [DaY14a], [DaY14b], [GiB14] for ADMM). However, it seems difficult to compare them on the basis of theoretical results alone, because the geometric progression rate at which they converge is different and also because the amount of work between multiplier updates must be properly taken into account. A corollary of this is that just because the ADMM updates the multipliers more often than the augmented Lagrangian method, it does not necessarily require less computation time to solve a problem. A further consideration in comparing the two types of methods is that while ADMM effectively decouples the minimizations with respect to x and z, augmented Lagrangian methods allow for some implementation flexibility that may be exploited by taking advantage of the structure of the given problem:

(a) The minimization of the augmented Lagrangian can be done with a broad variety of methods (not just block coordinate descent). Some of these methods may be well suited for the problem's structure.

(b) The minimization of the augmented Lagrangian need not be done exactly, and its accuracy can be readily controlled through theoretically sound and easily implementable termination criteria.

(c) The adjustment of the penalty parameter c can be used with advantage in the augmented Lagrangian method, but there is apparently no general way to do this in ADMM. In particular, by increasing c to ∞, superlinear or finite convergence can often be achieved in the augmented Lagrangian method [cf. Props. 5.1.4(b) and 5.1.5].

Thus, on balance, it appears that the relative performance merits of ADMM and augmented Lagrangian methods are problem-dependent in practice.

Convergence Analysis

We now turn to the convergence analysis of ADMM of Eqs. (5.84)-(5.86). The following proposition gives the main convergence result. The proof of the proposition is long and not very insightful. It may be found in Section 3.4 (Prop. 4.2) of [BeT89a] (which can be accessed on-line). A variant of this proof for the ADMM of Eqs. (5.89)-(5.91) is given in [BPC11], and essentially the same convergence result is shown.

Proposition 5.4.1: (ADMM Convergence) Consider problem (5.82) and assume that there exists a primal and dual optimal solution pair, and that either $\text{dom}(f_1)$ is compact or else $A'A$ is invertible. Then:

(a) The sequence $\{x_k, z_k, \lambda_k\}$ generated by the ADMM (5.84)-(5.86) is bounded, and every limit point of $\{x_k\}$ is an optimal solution of problem (5.83). Furthermore $\{\lambda_k\}$ converges to an optimal dual solution.

(b) The residual sequence $\{Ax_k - z_k\}$ converges to 0, and if $A'A$ is invertible, then $\{x_k\}$ converges to an optimal primal solution.

There is an alternative line of analysis, given in [EcB92], which connects the ADMM with the generalized proximal point algorithm of Section 5.1.4. This line of analysis is more insightful, and is based on the fixed point view of the proximal algorithm discussed in Section 5.1.4. In particular, it treats the ADMM as an algorithm for finding a fixed point of the composition of reflection operators corresponding to the conjugate functions $f_1^\star$ and $f_2^\star$. The Krasnosel'skii-Mann theorem (Prop. 5.1.9), which establishes the convergence of interpolated nonexpansive iterations, applies to this fixed point algorithm (each of the reflection operators is nonexpansive by Prop. 5.1.8).

Among others, this line of analysis shows that despite similarities, the ADMM is not really an approximate version of an augmented Lagrangian method that uses cyclic minimization with respect to the vectors x and z. Instead both methods may be viewed as exact versions of the proximal algorithm, involving the same fixed point convergence mechanism, but different mappings (and hence also different convergence rate). The full proof is somewhat lengthy, but we will provide an outline and some of the key points in Exercise 5.5.

5.4.1 Applications in Machine Learning

We noted earlier the application of ADMM to the additive cost problem of Example 5.4.1. This problem contains as special cases important machine learning contexts, such as the ℓ_1-regularization and maximum likelihood examples of Section 1.3. Here are some more examples with similar flavor.

Example 5.4.3 (Basis Pursuit)

Consider the problem

$$\begin{aligned} &\text{minimize} \quad \|x\|_1 \\ &\text{subject to} \quad Cx = b, \end{aligned}$$

where $\|\cdot\|_1$ is the ℓ_1 norm in $\Re^n$, C is a given $m \times n$ matrix and b is a vector in $\Re^m$. This is the basis pursuit problem of Example 1.4.2. We reformulate it as

$$\begin{aligned} &\text{minimize} \quad f_1(x) + f_2(z) \\ &\text{subject to} \quad x = z, \end{aligned}$$

where f_1 is the indicator function of the set $\{x \mid Cx = b\}$ and $f_2(z) = \|z\|_1$. The augmented Lagrangian is

$$L_c(x, z, \lambda) = \begin{cases} \|z\|_1 + \lambda'(x - z) + \frac{c}{2}\|x - z\|^2 & \text{if } Cx = b, \\ \infty & \text{if } Cx \neq b. \end{cases}$$

The ADMM iteration (5.84)-(5.86) takes the form

$$x_{k+1} \in \arg\min_{Cx=b} \left\{ \lambda_k' x + \frac{c}{2}\|x - z_k\|^2 \right\},$$

$$z_{k+1} \in \arg\min_{z\in\Re^n} \left\{ \|z\|_1 - \lambda_k' z + \frac{c}{2}\|x_{k+1} - z\|^2 \right\},$$

$$\lambda_{k+1} = \lambda_k + c(x_{k+1} - z_{k+1}).$$

The iteration for z can also be written as

$$z_{k+1} \in \arg\min_{z\in\Re^n} \left\{ \|z\|_1 + \frac{c}{2} \left\| x_{k+1} - z + \frac{\lambda_k}{c} \right\|^2 \right\}. \tag{5.92}$$

The type of minimization over z in Eq. (5.92) arises often in ℓ_1-regularization problems. It is straightforward to verify that the solution is given by the so-called *shrinkage operation*, which for any $\alpha > 0$ and $w = (w^1, \ldots, w^m) \in \Re^m$, is defined as

$$S(\alpha, w) \in \arg \min_{z \in \Re^m} \left\{ \|z\|_1 + \frac{1}{2\alpha} \|z - w\|^2 \right\}, \tag{5.93}$$

and has components given by

$$S^i(\alpha, w) = \begin{cases} w^i - \alpha & \text{if } w^i > \alpha, \\ 0 & \text{if } |w^i| \leq \alpha, \\ w^i + \alpha & \text{if } w^i < -\alpha, \end{cases} \qquad i = 1, \ldots, m. \tag{5.94}$$

Thus the minimization over z in Eq. (5.92) is expressed in terms of the shrinkage operation as

$$z_{k+1} = S\left(\frac{1}{c},\, x_{k+1} + \frac{\lambda_k}{c}\right).$$

Example 5.4.4 (ℓ_1-Regularization)

Consider the problem

$$\begin{aligned} &\text{minimize} \quad f(x) + \gamma \|x\|_1 \\ &\text{subject to} \quad x \in \Re^n, \end{aligned}$$

where $f : \Re^n \mapsto (-\infty, \infty]$ is a closed proper convex function and γ is a positive scalar. This includes as special cases the ℓ_1-regularization problem of Example 1.3.2, including the lasso formulation where f is a quadratic function. We reformulate the problem as

$$\begin{aligned} &\text{minimize} \quad f_1(x) + f_2(z) \\ &\text{subject to} \quad x = z, \end{aligned}$$

where $f_1(x) = f(x)$ and $f_2(z) = \gamma \|z\|_1$. The augmented Lagrangian is

$$L_c(x, z, \lambda) = f(x) + \gamma \|z\|_1 + \lambda'(x - z) + \frac{c}{2} \|x - z\|^2.$$

The ADMM iteration (5.84)-(5.86) takes the form

$$x_{k+1} \in \arg \min_{x \in \Re^n} \left\{ f(x) + \lambda_k' x + \frac{c}{2} \|x - z_k\|^2 \right\},$$

$$z_{k+1} \in \arg \min_{z \in \Re^n} \left\{ \gamma \|z\|_1 - \lambda_k' z + \frac{c}{2} \|x_{k+1} - z\|^2 \right\},$$

$$\lambda_{k+1} = \lambda_k + c(x_{k+1} - z_{k+1}).$$

The iteration for z can also be written in closed form, in terms of the shrinkage operation (5.93)-(5.94):

$$z_{k+1} = S\left(\frac{\gamma}{c},\, x_{k+1} + \frac{\lambda_k}{c}\right).$$

Example 5.4.5 (Least Absolute Deviations Problem)

Consider the problem

$$\begin{aligned} &\text{minimize} \quad \|Cx - b\|_1 \\ &\text{subject to} \quad x \in \Re^n, \end{aligned}$$

where C is an $m \times n$ matrix of rank n, and $b \in \Re^m$ is a given vector. This is the least absolute deviations problem of Example 1.3.1. We reformulate it as

$$\begin{aligned} &\text{minimize} \quad f_1(x) + f_2(z) \\ &\text{subject to} \quad Cx - b = z, \end{aligned}$$

where

$$f_1(x) \equiv 0, \qquad f_2(z) = \|z\|_1.$$

Here the augmented Lagrangian function is modified to include the constant vector b [cf. Eq. (5.88)]. It is given by

$$L_c(x, z, \lambda) = \|z\|_1 + \lambda'(Cx - z - b) + \frac{c}{2}\|Cx - z - b\|^2.$$

The ADMM iteration (5.89)-(5.91) takes the form

$$x_{k+1} = (C'C)^{-1}C'\left(z_k + b - \frac{\lambda_k}{c}\right),$$

$$z_{k+1} \in \arg\min_{z \in \Re^m} \left\{ \|z\|_1 - \lambda_k' z + \frac{c}{2}\|Cx_{k+1} - z - b\|^2 \right\},$$

$$\lambda_{k+1} = \lambda_k + c(Cx_{k+1} - z_{k+1} - b).$$

Setting $\bar{\lambda}_k = \lambda_k / c$, the iteration can be written in the notationally simpler form

$$x_{k+1} = (C'C)^{-1}C'(z_k + b - \bar{\lambda}_k),$$

$$z_{k+1} \in \arg\min_{z \in \Re^m} \left\{ \|z\|_1 + \frac{c}{2}\|Cx_{k+1} - z - b + \bar{\lambda}_k\|^2 \right\}, \tag{5.95}$$

$$\bar{\lambda}_{k+1} = \bar{\lambda}_k + Cx_{k+1} - z_{k+1} - b.$$

The minimization over z in Eq. (5.95) is expressed in terms of the shrinkage operation as

$$z_{k+1} = S\left(\frac{1}{c}, Cx_{k+1} - b + \bar{\lambda}_k\right).$$

5.4.2 ADMM Applied to Separable Problems

In this section we consider a separable problem of the form

$$\begin{aligned} &\text{minimize} \quad \sum_{i=1}^{m} f_i(x^i) \\ &\text{subject to} \quad \sum_{i=1}^{m} A_i x^i = b, \qquad x^i \in X_i, \quad i = 1, \ldots, m, \end{aligned} \tag{5.96}$$

where $f_i : \Re^{n_i} \mapsto \Re$ are convex functions, X_i are closed convex sets, and A_i and b are given. We have often noted that this problem has a favorable structure that is well-suited for the application of decomposition approaches. Since the primary attractive feature of ADMM is that it decouples the augmented Lagrangian optimization calculations, it is natural to consider its application to this problem.

An idea that readily comes to mind is to form the augmented Lagrangian

$$L_c(x^1, \ldots, x^m, \lambda) = \sum_{i=1}^{m} f_i(x^i) + \lambda' \left(\sum_{i=1}^{m} A_i x^i - b \right) + \frac{c}{2} \left\| \sum_{i=1}^{m} A_i x^i - b \right\|^2,$$

and use an ADMM-like iteration, whereby we minimize L_c sequentially with respect to $x^1, \ldots, x^m$, i.e.,

$$x^i_{k+1} \in \arg \min_{x^i \in X_i} L_c(x^1_{k+1}, \ldots, x^{i-1}_{k+1}, x^i, x^{i+1}_k, \ldots, x^m_k, \lambda_k), \quad i = 1, \ldots, m, \tag{5.97}$$

and follow these minimizations with the multiplier iteration

$$\lambda_{k+1} = \lambda_k + c \left(\sum_{i=1}^{m} A_i x^i_{k+1} - b \right). \tag{5.98}$$

Methods of this type have been proposed in various forms long time ago, starting with the important paper [StW75], which stimulated considerable further research. The context was unrelated to ADMM (which was unknown at that time), but the motivation was similar to the one of the ADMM: working around the coupling of variables induced by the penalty term in the augmented Lagrangian method.

When there is only one component, $m = 1$, we obtain the augmented Lagrangian method. When there are only two components, $m = 2$, the above method is equivalent to the ADMM of Eqs. (5.89)-(5.91), so it has the corresponding convergence properties. On the other hand, when $m > 2$, the method is not a special case of the ADMM that we have discussed and is not covered by similar convergence guarantees. In fact a convergence counterexample has been given for $m = 3$ in [CHY13]. The same reference shows that the iteration (5.97)-(5.98) is convergent under additional, but substantially stronger assumptions. Related convergence results are proved in [HoL13] under alternative but also strong assumptions; see also [WHM13].

In what follows we will develop an ADMM (first given in [BeT89a], Section 3.4 and Example 4.4), which is similar to the iteration (5.97)-(5.98), and is covered by the convergence guarantees of Prop. 5.4.1 without any assumptions other than the ones given in the proposition. We will derive this algorithm, by formulating the separable problem as a special case of the Fenchel framework problem (5.82), and by applying the convergent ADMM (5.84)-(5.86).

We reformulate problem (5.96) by introducing additional variables $z^1, \ldots, z^m$ as follows:

$$\begin{aligned} &\text{minimize} \quad \sum_{i=1}^m f_i(x^i) \\ &\text{subject to} \quad A_i x^i = z^i, \quad x^i \in X_i, \qquad i = 1, \ldots, m, \\ &\qquad\qquad\quad \sum_{i=1}^m z^i = b. \end{aligned} \tag{5.99}$$

We denote $x = (x^1, \ldots, x^m)$, $z = (z^1, \ldots, z^m)$, we view $X = X_1 \times \ldots \times X_m$ as a constraint set for x, we view

$$Z = \left\{ z \;\middle|\; \sum_{i=1}^m z^i = b \right\}$$

as a constraint set for z, and we introduce a multiplier vector p^i for each of the equality constraints $A_i x^i = z^i$. The augmented Lagrangian has the form

$$L_c(x, z, p) = \sum_{i=1}^m \left(f_i(x^i) + (A_i x^i - z^i)' p^i + \frac{c}{2} \|A_i x^i - z^i\|^2 \right), \; x \in X, \, z \in Z,$$

and the ADMM (5.84)-(5.86) is given by

$$x_{k+1}^i \in \arg \min_{x^i \in X_i} \left\{ f_i(x^i) + (A_i x^i - z_k^i)' p_k^i + \frac{c}{2} \|A_i x^i - z_k^i\|^2 \right\}, \; i = 1, \ldots, m, \tag{5.100}$$

$$z_{k+1} \in \arg \min_{\sum_{i=1}^m z^i = b} \left\{ \sum_{i=1}^m (A_i x_{k+1}^i - z^i)' p_k^i + \frac{c}{2} \|A_i x_{k+1}^i - z^i\|^2 \right\}, \tag{5.101}$$

$$p_{k+1}^i = p_k^i + c(A_i x_{k+1}^i - z_{k+1}^i), \qquad i = 1, \ldots, m. \tag{5.102}$$

We will now show how to simplify this algorithm. We will first obtain the minimization (5.101) for z in closed form by introducing a multiplier vector λ_{k+1} for the constraint $\sum_{i=1}^m z^i = b$, and then show that the multipliers p^i_{k+1} obtained from the update (5.102) are all equal to λ_{k+1}. To this end we note that the Lagrangian function corresponding to the minimization (5.101), is given by

$$\sum_{i=1}^{m} \left((A_i x^i_{k+1} - z^i)' p^i_k + \frac{c}{2} \|A_i x^i_{k+1} - z^i\|^2 + \lambda'_{k+1} z^i \right) - \lambda'_{k+1} b.$$

By setting to zero its gradient with respect to z^i, we see that the minimizing vectors z^i_{k+1} are given in terms of λ_{k+1} by

$$z^i_{k+1} = A_i x^i_{k+1} + \frac{p^i_k - \lambda_{k+1}}{c}. \tag{5.103}$$

A key observation is that we can write this equation as

$$\lambda_{k+1} = p^i_k + c(A_i x^i_{k+1} - z^i_{k+1}), \qquad i = 1, \ldots, m,$$

so from Eq. (5.102), we have

$$p^i_{k+1} = \lambda_{k+1}, \qquad i = 1, \ldots, m.$$

Thus during the algorithm *all the multipliers p^i are updated to a common value: the multiplier λ_{k+1} of the constraint $\sum_{i=1}^m z^i = b$ of problem (5.101).*

We now use this fact to simplify the ADMM (5.100)-(5.102). Given z_k and λ_k (which is equal to p^i_k for all i), we first obtain x_{k+1} from the augmented Lagrangian minimization (5.100) as

$$x^i_{k+1} \in \arg \min_{x^i \in X_i} \left\{ f_i(x^i) + (A_i x^i - z^i_k)' \lambda_k + \frac{c}{2} \|A_i x^i - z^i_k\|^2 \right\}. \tag{5.104}$$

To obtain z_{k+1} and λ_{k+1} from the augmented Lagrangian minimization (5.101), we express z^i_{k+1} in terms of the unknown λ_{k+1} as

$$z^i_{k+1} = A_i x^i_{k+1} + \frac{\lambda_k - \lambda_{k+1}}{c}, \tag{5.105}$$

[cf. Eq. (5.103)], and then obtain λ_{k+1} by requiring that the constraint of the minimization (5.101), $\sum_{i=1}^m z^i_{k+1} = b$, is satisfied. Thus, by adding Eq. (5.105) over $i = 1, \ldots, m$, we have

$$\lambda_{k+1} = \lambda_k + \frac{c}{m} \left(\sum_{i=1}^{m} A_i x^i_{k+1} - b \right). \tag{5.106}$$

We then obtain z_{k+1} using Eq. (5.105).

In summary, the iteration of the algorithm consists of the three equations (5.104), (5.106), and (5.105), applied in that order. The vectors λ_0 and z_0, which are needed at the first iteration, can be chosen arbitrarily. This is a correct ADMM, mathematically equivalent to the algorithm (5.100)-(5.102), which has guaranteed convergence as per Prop. 5.4.1. It is as simple as the iteration (5.97)-(5.98), which, however, is not theoretically sound for $m > 2$ as we noted earlier.

Example 5.4.6 (Constrained ADMM)

Consider the problem

$$\begin{aligned} &\text{minimize} \quad f_1(x) + f_2(Ax) \\ &\text{subject to} \quad Ex = d, \quad x \in X, \end{aligned}$$

which differs from the standard format (5.82) in that it includes the convex set constraint $x \in X$, and the linear equality constraint $Ex = d$, where E and d are given matrix and vector, respectively. We convert this problem into the separable form (5.96) as follows:

$$\begin{aligned} &\text{minimize} \quad f_1(x^1) + f_2(x^2) \\ &\text{subject to} \quad \begin{pmatrix} A \\ E \end{pmatrix} x^1 + \begin{pmatrix} -I \\ 0 \end{pmatrix} x^2 = \begin{pmatrix} 0 \\ d \end{pmatrix}, \quad x^1 \in X, \; x^2 \in \Re^m. \end{aligned}$$

Assigning multipliers λ and μ to the two equality constraints, and applying the algorithm of Eqs. (5.104)-(5.106), with the notation

$$z_k^1 = \begin{pmatrix} z_k^{11} \\ z_k^{12} \end{pmatrix}, \qquad z_k^2 = \begin{pmatrix} z_k^{21} \\ z_k^{22} \end{pmatrix},$$

we obtain the following iteration:

$$\begin{aligned} x_{k+1}^1 \in \arg \min_{x^1 \in X} \Big\{ & f_1(x^1) + (Ax^1 - z_k^{11})'\lambda_k + (Ex^1 - z_k^{12})'\mu_k \\ & + \frac{c}{2}\big(\|Ax^1 - z_k^{11}\|^2 + \|Ex^1 - z_k^{12}\|^2\big) \Big\}, \end{aligned}$$

$$\begin{aligned} x_{k+1}^2 \in \arg \min_{x^2 \in \Re^n} \Big\{ & f_2(x^2) - (x^2 + z_k^{21})'\lambda_k - (z_k^{22})'\mu_k \\ & + \frac{c}{2}\big(\|x^2 + z_k^{21}\|^2 + \|z_k^{22}\|^2\big) \Big\}, \end{aligned}$$

$$\lambda_{k+1} = \lambda_k + \frac{c}{2}(Ax_{k+1}^1 - x_{k+1}^2),$$

$$\mu_{k+1} = \mu_k + \frac{c}{2}(Ex_{k+1}^1 - d),$$

$$z_{k+1}^1 = \begin{pmatrix} A \\ E \end{pmatrix} x_{k+1}^1 + \frac{1}{c}\begin{pmatrix} \lambda_k - \lambda_{k+1} \\ \mu_k - \mu_{k+1} \end{pmatrix},$$

$$z_{k+1}^2 = \begin{pmatrix} -I \\ 0 \end{pmatrix} x_{k+1}^2 + \frac{1}{c}\begin{pmatrix} \lambda_k - \lambda_{k+1} \\ \mu_k - \mu_{k+1} \end{pmatrix}.$$

Note that this algorithm maintains the main attractive characteristic of the ADMM: the components f_1 and f_2 of the cost function are decoupled in the augmented Lagrangian minimizations.

We mention a more refined form of the multiplier iteration (5.106), also from [BeT89a], Example 4.4, whereby the coordinates λ_k^j of the multiplier vector λ_k are updated according to

$$\lambda_{k+1}^j = \lambda_k^j + \frac{c}{m_j}\left(\sum_{i=1}^{m} A_i x_{k+1}^i - b\right), \qquad j = 1, \ldots, r, \tag{5.107}$$

where r is the row dimension of the matrices A_i, and m_j is the number of submatrices A_i that have nonzero jth row. Using the j-dependent stepsize c/m_j of Eq. (5.107) in place of the stepsize c/m of Eq. (5.106) may be viewed as a form of diagonal scaling. The derivation of the algorithm of Eqs. (5.104), (5.105), and (5.107) is nearly identical to the one given for the algorithm (5.104)-(5.106). The idea is that the components of the vector z_k^i represent estimates of the corresponding components of $A_i x^i$ at the optimum. However, if some of these components are known to be 0 because some of the rows of A_i are 0, then the corresponding values of z_k^i might as well be set to 0 rather than be estimated. If we repeat the preceding derivation of the algorithm (5.104)-(5.106), but without introducing the components of z^i that are known to be 0, we obtain by a straightforward calculation the multiplier iteration (5.107).

Finally let us note that the ADMM can be applied to the dual of the separable problem (5.96), and yield a similar decomposition algorithm. The idea is that the dual problem has the form discussed in Example 5.4.1, for which the ADMM can be conveniently applied. This approach is applicable even in the case where the primal problem has some nonlinear inequality constraints; see [Fuk92], which also discusses a connection with the method of partial inverses of [Spi83], [Spi85].

5.5 NOTES, SOURCES, AND EXERCISES

Section 5.1: The proximal algorithm was introduced by Martinet [Mar70], [Mar72] in a form that applies to convex optimization and monotone variational inequality problems. The generalized version that aims to find a zero of a maximal monotone operator (cf. Section 5.1.4) received wide attention following the work of Rockafellar [Roc76a], [Roc76b], which was based on the fundamental work of Minty [Min62], [Min64].

Together with its special cases and variations, the proximal algorithm has been analyzed by several authors, including connections with the augmented Lagrangian method, convergence and rate of convergence issues, and various extensions dealing with alternative stepsize rules, approximate implementations, special cases, and generalizations to nonconvex problems (see e.g., Brezis and Lions [BrL78], Spingarn [Spi83], [Spi785], Luque [Luq84], Golshtein [Gol85], Lawrence and Spingarn [LaS87], Lemaire [Lem89], Rockafellar and Wets [RoW91], Tseng [Tse91b], Eckstein and

Bertsekas [EcB92], Guler [Gul92]). For textbook discussions, see Rockafellar and Wets [RoW98], Facchinei and Pang [FaP03], and Bauschke and Combettes [BaC11], which include many references.

The rate of convergence analysis given here (Prop. 5.1.4) is due to Kort and Bertsekas [KoB76], and has been extensively discussed in the monograph [Ber82a] (Section 5.4) in a more general form where the regularization term may be nonquadratic (in this case the order of growth of the regularization term also affects the convergence rate; see Section 6.6). In particular, if a regularization term $\|x - x_k\|^\rho$ with $1 < \rho < 2$ is used in place of the quadratic $\|x - x_k\|^2$, the order of convergence is $1/(\rho-1)(\gamma-1)$, where γ is the order of growth of f (cf. Prop. 5.1.4). Thus superlinear convergence is achieved when $\gamma = 2$. Limited computational experimentation, some of which is described in [KoB76], suggests that values $\rho < 2$ may be beneficial in some problem contexts. For an extension of the linear convergence result of Prop. 5.1.4(c), which applies to finding a zero of a maximal monotone operator, see Luque [Luq84].

The finite termination of the proximal algorithm when applied to polyhedral functions (Prop. 5.1.5) was shown independently, using different methods and assumptions, by Poljak and Tretjakov [PoT74], and by Bertsekas [Ber75a], which we follow in our analysis here.

Section 5.2: The augmented Lagrangian algorithm was independently proposed in the papers by Hestenes [Hes69], Powell [Pow69], and Haarhoff and Buys [HaB70] in a nonlinear programming context where convexity played no apparent role. These papers contained little analysis, and did not suggest any relation to duality and the proximal algorithm. This relation was clarified and analyzed for convex programming problems by Rockafellar [Roc73], [Roc76b]. The convergence and rate of convergence of the algorithm for nonconvex differentiable problems were investigated by the author in [Ber75c], [Ber76b], [Ber76c], along with variants including inexact augmented Lagrangian minimization and second order methods of multipliers. A related and contemporary development was the introduction of nonquadratic augmented Lagrangians in the papers by Kort and Bertsekas [KoB72], [KoB76], including the exponential method, which will be discussed in Section 6.6. There has been much subsequent work; representative references, some relating to the proximal algorithm, include Robinson [Rob99], Pennanen [Pen02], Eckstein [Eck03], Iusem, Pennanen, and Svaiter [IPS03], and Eckstein and Silva [EcS13].

For surveys of the augmented Lagrangian literature, see Bertsekas [Ber76b], Rockafellar [Roc76c], and Iusem [Ius99]. Discussions of augmented Lagrangian methods of varying levels of detail may also be found in most nonlinear programming textbooks. An extensive treatment of augmented Lagrangian methods, for both convex and nonconvex constrained problems, is given in the author's research monograph [Ber82a], including an analysis of smoothing and sequential quadratic programming algorithms, and detailed references on the early history of the subject. The distributed algorithms monograph by Bertsekas and Tsitsiklis [BeT89a] discusses applications of augmented Lagrangians to classes of large-scale problems with special structure, including separable problems and problems with additive cost functions. There has been considerable recent interest in using augmented Lagrangian, proximal, and smoothing methods for machine learning and signal processing applications; see e.g., Osher et al. [OBG05], Yin et al. [YOG08], and Goldstein and Osher [GoO09].

Section 5.3: The proximal cutting plane and simplicial decomposition algorithms of Sections 5.3.1 and 5.3.3, may be viewed as regularized versions of the classical Dantzig-Wolfe decomposition algorithm (see e.g., [Las70], [BeT97], [Ber99]). The latter algorithm is obtained in the limit, as the regularization term diminishes to 0 ($c_k \to \infty$).

For presentations of bundle methods, see the books by Hiriart-Urrutu and Lemarechal [HiL93], and Bonnans et al. [BGL06], which give many references. For related methods, see Ruszczynski [Rus86], Lemaréchal and Sagastizábal [LeS93], Mifflin [Mif96], Burke and Qian [BuQ98], Mifflin, Sun, and Qi [MSQ98], Frangioni [Fra02], and Teo et al. [TVS10].

The term "bundle" has been used with a few different meanings in the convex algorithmic optimization literature, with some confusion resulting. To our knowledge, it was first introduced in the 1975 paper by Wolfe [Wol75] to describe a collection of subgradients used for calculating a descent direction in the context of a specific algorithm of the descent type – a context with no connection to cutting planes or proximal minimization. It subsequently appeared in related descent contexts through the 70s and early 80s. The meaning of the term "bundle method" shifted gradually in the 80s, and it is now commonly associated with the stabilized proximal cutting plane methods that we have described in Section 5.3.2.

Section 5.4: The ADMM was first proposed by Glowinskii and Morocco [GIM75], and Gabay and Mercier [GaM76], and was further developed by Gabay [Gab79], [Gab83]. It was generalized by Lions and Mercier [LiM79], where the connection with alternating direction methods for solving differential equations was pointed out. The method and its applications in large boundary-value problems were discussed by Fortin and Glowinskii [FoG83].

The recent literature on the ADMM is voluminous and cannot be surveyed here (a Google Scholar search produced thousands of papers appearing in the two years preceding the publication of this book). The surge of interest is largely due to the flexibility that the ADMM provides in exploiting special problem structures, such as for example the ones from machine learning that we have discussed in Section 5.4.1.

In our discussion we have followed the analysis of the book by Bertsekas and Tsitsiklis [BeT89a] (which among others gave the ADMM for separable problems of Section 5.4.2), and in part the paper by Eckstein and Bertsekas [EcB92] (which established the connection of the ADMM with the general form of the proximal algorithm of Section 5.1.4, and gave extensions involving, among others, extrapolation and inexact minimization). In particular, the paper [EcB92] showed that the general form of the proximal algorithm contains as a special case the Douglas-Ratchford splitting algorithm for finding a zero of the sum of two maximal monotone operators, proposed by Lions and Mercier [LiM79]. The latter algorithm contains in turn as a special case the ADMM, as shown by Gabay [Gab83].

EXERCISES

5.1 (Proximal Algorithm via Trust Regions)

Consider using in place of the proximal algorithm the following iteration:

$$x_{k+1} \in \arg \min_{\|x-x_k\| \leq \gamma_k} f(x),$$

where $\{\gamma_k\}$ is a sequence of positive scalars. Use dual variables to relate this algorithm with the proximal algorithm. In particular, provide conditions under which there is a proximal algorithm, with an appropriate sequence of penalty parameters $\{c_k\}$, which generates the same iterate sequence $\{x_k\}$ starting from the same point x_0.

5.2 (Contraction Properties of the Proximal Operator [Roc76a])

Consider a multivalued mapping $M : \Re^n \mapsto 2^{\Re^n}$, which maps vectors $x \in \Re^n$ into subsets $M(x) \subset \Re^n$. Assume that M has the following two properties:

(1) Any vector $z \in \Re^n$ can be written in exactly one way as

$$z = \overline{z} + cv \quad \text{where } \overline{z} \in \Re^n,\ v \in M(\overline{z}), \tag{5.108}$$

[cf. Eq. (5.33)]. (As noted in Section 5.1.4, this is true if M has a maximal monotonicity property.)

(2) For some $\sigma > 0$, we have

$$(x_1 - x_2)'(v_1 - v_2) \geq \sigma \|x_1 - x_2\|^2, \quad \forall\, x_1, x_2 \in \text{dom}(M)$$
$$\text{and } v_1 \in M(x_1),\ v_2 \in M(x_2),$$

where $\text{dom}(M) = \{x \mid M(x) \neq \emptyset\}$ (assumed nonempty). (This is called the *strong monotonicity* condition.)

Show that the proximal operator $P_{c,f}$, which maps z to the unique vector $\overline{z}$ of Eq. (5.108) is a contraction mapping with respect to the Euclidean norm, and in fact

$$\big\|P_{c,f}(z_1) - P_{c,f}(z_2)\big\| \leq \frac{1}{1+c\sigma}\|z_1 - z_2\|, \qquad \forall\, z_1, z_2 \in \Re^n.$$

Note: While a fixed point iteration involving a contraction has a linear convergence rate, the reverse is not true. In particular, Prop. 5.1.4(c) gives a condition under which the proximal algorithm has a linear convergence rate. However, this condition does not guarantee that the proximal operator $P_{c,f}$ is a contraction with respect to any particular norm. For example, all the minimizing points of f are fixed points of $P_{c,f}$, but the condition of Prop. 5.1.4(c) does not preclude the possibility of f having multiple minimizing points. See also [Luq84] for an extension of Prop. 5.1.4(c), which applies to the case of a maximal monotone operator, and other related convergence rate results. *Hint*: Consider the multi-valued mapping

$$\overline{M}(z) = M(z) - \sigma z,$$

and for any $\overline{c} > 0$, let $\overline{P}_{\overline{c},f}(z)$ be the unique vector $\overline{z}$ such that $z = \overline{z} + \overline{c}(v - \sigma z)$ [cf. Eq. (5.108)]. Note that $\overline{M}$ is monotone, and hence by the theory of Section 5.1.4, $\overline{P}_{\overline{c},f}$ is nonexpansive. Verify that

$$P_{c,f}(z) = \overline{P}_{\overline{c},f}\big((1 + c\sigma)^{-1}z\big), \qquad z \in \Re^n,$$

where $\overline{c} = c(1 + c\sigma)^{-1}$, and use the nonexpansiveness of $\overline{P}_{\overline{c},f}$.

5.3 (Partial Proximal Algorithm [Ha90], [BeT94a], [IbF96])

The purpose of this exercise is to develop the elements of an algorithm that is similar to the proximal, but uses partial regularization, that is, a quadratic regularization term that involves only a subset of the coordinates of x. For $c > 0$, let ϕ_c be the real-valued convex function on $\Re^n$ defined by

$$\phi_c(z) = \min_{x \in X} \left\{ f(x) + \frac{1}{2c}\|x - z\|^2 \right\},$$

where f is a convex function over the closed convex set X. Let $x^1, \ldots, x^n$ denote the scalar components of the vector x and let I be a subset of the index set $\{1, \ldots, n\}$. For any $z = (z^1, \ldots, z^n) \in \Re^n$, consider a vector $\overline{z}$ satisfying

$$\overline{z} \in \arg\min_{x \in X} \left\{ f(x) + \frac{1}{2c} \sum_{i \in I} |x^i - z^i|^2 \right\}. \qquad (5.109)$$

(a) Show that for a given z, the iterate $\overline{z}$ can be obtained by the two-step process

$$\tilde{z} \in \arg \min_{\{x \mid x^i = z^i,\, i \in I\}} \phi_c(x),$$

$$\overline{z} \in \arg\min_{x \in X} \left\{ f(x) + \frac{1}{2c}\|x - \tilde{z}\|^2 \right\}.$$

(b) Interpret $\overline{z}$ as the result of a block coordinate descent iteration corresponding to the components z^i, $i \notin I$, followed by a proximal iteration, and show that

$$\phi_c(\overline{z}) \le f(\overline{z}) \le \phi_c(z) \le f(z).$$

Note: Partial regularization, as in iteration (5.109), may yield better approximation to the original problem, and accelerated convergence if f is "well-behaved" with respect to some of the components of x (the components x^i with $i \notin I$).

5.4 (Convergence of Proximal Cutting Plane Method)

Show that if the optimal solution set X^* is nonempty, the sequence $\{x_k\}$ generated by the proximal cutting plane method (5.69) converges to some point in X^*.

5.5 (Fixed Point Interpretation of ADMM)

Consider the ADMM framework of Section 5.4. Let $d_1 : \Re^m \mapsto (-\infty, \infty]$ and $d_2 : \Re^m \mapsto (-\infty, \infty]$ be the functions

$$d_1(\lambda) = \sup_{x\in\Re^n} \{\lambda' Ax - f_1(x)\}, \qquad d_2(\lambda) = \sup_{z\in\Re^m} \{(-\lambda)'z - f_2(z)\},$$

and note that the dual to the Fenchel problem (5.82) is to minimize $d_1 + d_2$ [cf. Eq. (5.36) or Prop. 1.2.1]. Let $N_1 : \Re^m \mapsto \Re^m$ and $N_2 : \Re^m \mapsto \Re^m$ be the reflection operators corresponding to d_1 and d_2, respectively [cf. Eq. (5.19)].

(a) Show that the set of fixed points of the composition $N_1 \cdot N_2$ is the set

$$\{\lambda - cv \mid v \in \partial d_1(\lambda), -v \in \partial d_2(\lambda)\}; \tag{5.110}$$

see Fig. 5.5.1.

(b) Show that the interpolated fixed point iteration

$$y_{k+1} = (1 - \alpha_k)y_k + \alpha_k(N_1 \cdot N_2)(y_k), \tag{5.111}$$

where $\alpha_k \in [0, 1]$ for all k and $\sum_{k=0}^{\infty} \alpha_k(1 - \alpha_k) = \infty$, converges to a fixed point of $N_1 \cdot N_2$. Moreover, when $\alpha_k \equiv 1/2$, this iteration is equivalent to the ADMM; see Fig. 5.5.2.

Hints and Notes: We have $N_1(z) = \overline{z} - cv$, where $\overline{z} \in \Re^m$ and $v \in \partial d_1(\overline{z})$ are obtained from the unique decomposition $z = \overline{z} + cv$, and $N_2(z) = \overline{z} - cv$, where $\overline{z} \in \Re^m$ and $v \in \partial d_2(\overline{z})$ are obtained from the unique decomposition $z = \overline{z} + cv$ [cf. Eq. (5.25)]. Part (a) shows that finding a fixed point of $N_1 \cdot N_2$ is equivalent to finding two vectors λ^* and v^* that satisfy the optimality conditions for the dual problem of minimizing $d_1 + d_2$, and then computing $\lambda^* - cv^*$ (assuming that the conditions for strong duality are satisfied). In terms of the primal problem, we will then have

$$A'\lambda^* \in \partial f_1(x^*) \quad \text{and} \quad -\lambda^* \in \partial f_2(Ax^*),$$

as well as the equivalent condition

$$x^* \in \partial f_1^\star(A'\lambda^*) \quad \text{and} \quad Ax^* \in \partial f_2^\star(-\lambda^*),$$

where x^* is any optimal primal solution [cf. Prop. 1.2.1(c)]. Moreover we will have $v^* = -Ax^*$.

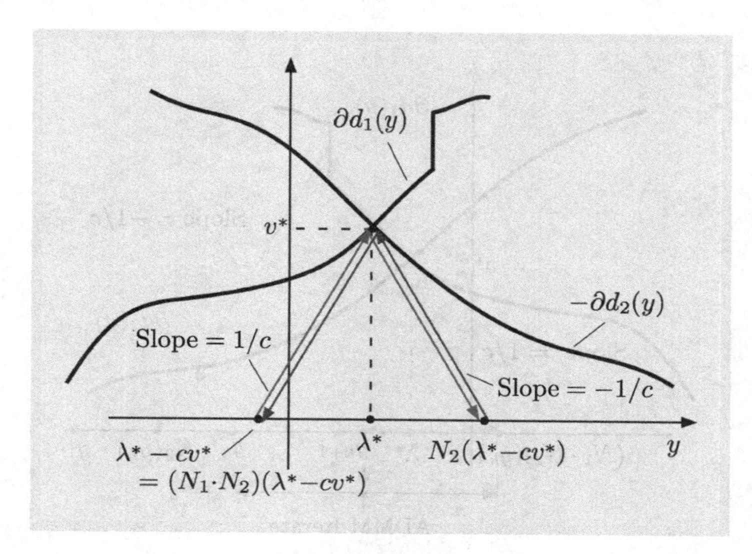

Figure 5.5.1. Illustration of the mapping $N_1 \cdot N_2$ and its fixed points (cf. Exercise 5.5). The vector λ^* shown is an optimal solution of the dual problem of minimizing $d_1 + d_2$, and according to the optimality conditions we have $v^* \in \partial d_1(\lambda^*)$ and $v^* \in -\partial d_2(\lambda^*)$ for some v^*. It can be seen then that $\lambda^* - cv^*$ is a fixed point of $N_1 \cdot N_2$ and conversely (in the figure, by applying N_2 to $\lambda^* - cv^*$ using the graphical process of Fig. 5.1.8, and by applying N_1 to the result, we end back at $\lambda^* - cv^*$).

For a proof of part (a), note that for any z we have

$$N_2(z) = \overline{z}_2 - cv_2 \text{ with } z = \overline{z}_2 + cv_2 \text{ and } v_2 \in \partial d_2(\overline{z}_2), \tag{5.112}$$

which also implies that

$$N_1\big(N_2(z)\big) = \overline{z}_1 - cv_1 \text{ with } \overline{z}_2 - cv_2 = \overline{z}_1 + cv_1 \text{ and } v_1 \in \partial d_1(\overline{z}_1). \tag{5.113}$$

Thus from Eqs. (5.112) and (5.113), z is a fixed point of $N_1 \cdot N_2$ if and only if

$$z = \overline{z}_2 + cv_2 = \overline{z}_1 - cv_1,$$

while from Eq. (5.113), we have

$$\overline{z}_2 - cv_2 = \overline{z}_1 + cv_1.$$

The last two relations yield $\overline{z}_1 = \overline{z}_2$ and $v_1 = -v_2$. Thus, denoting $\lambda = \overline{z}_1 = \overline{z}_2$ and $v = v_1 = -v_2$, we have that z is a fixed point of $N_1 \cdot N_2$ if and only if it has the form $\lambda + cv$, with $v \in \partial d_1(\lambda)$ and $-v \in \partial d_2(\lambda)$, verifying Eq. (5.110) for the set of fixed points.

For part (b), note that since both N_1 and N_2 are nonexpansive (cf. Prop. 5.1.8), the composition $N_1 \cdot N_2$ is nonexpansive. Therefore, based on the Krasnosel'skii-Mann theorem (Prop. 5.1.9), the interpolated fixed point iteration (5.111)

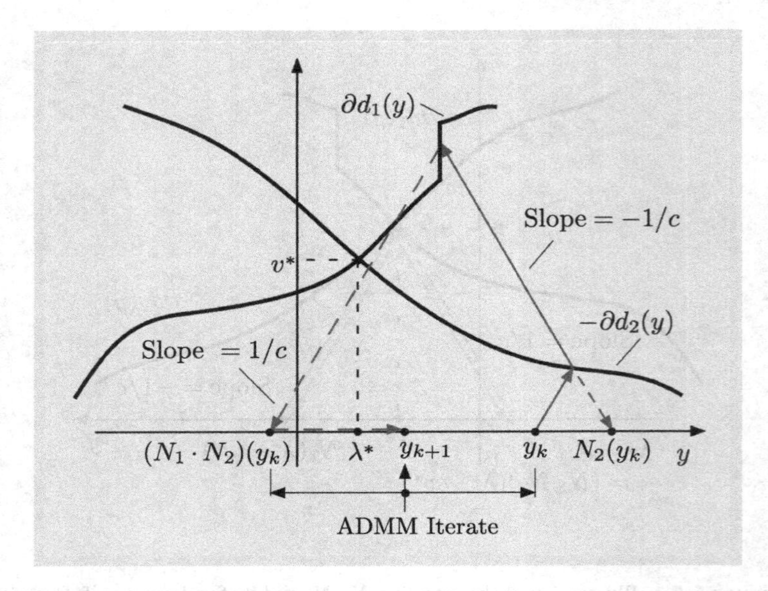

Figure 5.5.2. Illustration of the interpolated fixed point iteration (5.111). Starting from y_k, we obtain $(N_1 \cdot N_2)(y_k)$ using the process illustrated in the figure: first compute $N_2(y_k)$ as shown (cf. Fig. 5.1.8), then apply N_1 to compute $(N_1 \cdot N_2)(y_k)$, and finally interpolate between y_k and $(N_1 \cdot N_2)(y_k)$ using a parameter $\alpha_k \in (0,1)$. When the interpolation parameter is $\alpha_k \equiv 1/2$, we obtain the ADMM iterate, which is the midpoint between y_k and $(N_1 \cdot N_2)(y_k)$, denoted by y_{k+1} in the figure. The iteration converges to a fixed point y^* of $N_1 \cdot N_2$, which when written in the form $y^* = \lambda^* - cv^*$, yields a dual optimal solution λ^*.

converges to a fixed point of $N_1 \cdot N_2$, starting from any $y_0 \in \Re^n$ and assuming that $N_1 \cdot N_2$ has at least one point (see Fig. 5.5.2).

The verification that for $\alpha_k \equiv 1/2$ we obtain the ADMM, is a little complicated, but the idea is clear. As shown in Section 5.2.1, generally, a proximal iteration can be dually implemented as an augmented Lagrangian minimization involving the conjugate function. Thus a proximal iteration for d_2 (or d_1) is the dual of an augmented Lagrangian iteration involving f_2 (or f_1, respectively). Therefore, a fixed point iteration for the mapping $N_1 \cdot N_2$, which is the composition of these proximal iterations, can be implemented with an augmented Lagrangian iteration for f_1, followed by an augmented Lagrangian iteration for f_2, consistently with the character of the ADMM. See [EcB92] or [Eck12] for a detailed derivation.

6

Additional Algorithmic Topics

Contents

In this chapter, we consider a variety of algorithmic topics that supplement our discussion of the descent and approximation methods of the preceding chapters. In Section 6.1, we expand our discussion of descent algorithms for differentiable optimization of Sections 2.1.1 and 2.1.2, and the gradient projection method in particular. We then focus on clarifying some computational complexity issues relating to convex optimization problems, and in Section 6.2 we derive algorithms based on extrapolation ideas, which are optimal in terms of iteration complexity. These algorithms are developed for differentiable problems, and can be extended to the nondifferentiable case by means of a smoothing scheme.

Following the analysis of optimal algorithms in Section 6.2, we discuss in Sections 6.3-6.5 additional algorithms that are largely based on the descent approach. A common theme in these sections is to split the optimization process into a sequence of simpler optimizations, with the motivation to exploit the special structure of the problem. The simpler optimizations handle incrementally the components of the cost function, or the components of the constraint set, or the components of the optimization vector x. In particular, in Section 6.3, we discuss proximal gradient methods, for minimizing the sum of a differentiable function (handled by a gradient method) and a nondifferentiable function (handled by a proximal method). In Section 6.4, we discuss combinations of the incremental subgradient algorithms of Section 3.3.1 with proximal methods that enhance the flexibility and reliability of the incremental approach. In Section 6.5, we discuss the classical block coordinate descent approach, its many variations, and related issues of distributed asynchronous computation.

In Sections 6.6-6.8, we selectively describe a variety of algorithms that supplement the ones of the preceding sections. In particular, in Section 6.6, we generalize the proximal algorithm to use a nonquadratic regularization term, such as an entropy function, and we obtain corresponding augmented Lagrangian algorithms. In Section 6.7, we focus on the ϵ-descent method, a descent algorithm based on ϵ-subgradients, which among others can be used to establish strong duality properties for the extended monotropic programming problem that we addressed algorithmically in Section 4.4, using polyhedral approximation schemes. Finally, in Section 6.8, we discuss interior point methods and their application in linear and conic programming.

Our coverage of the various methods in this chapter is not always as comprehensive as in earlier chapters, and occasionally it takes the character of a survey. Part of the rationale for this is to limit the size of the book. Another reason is that some of the methods are under active development, and views about their character and merits have not settled.

6.1 GRADIENT PROJECTION METHODS

In this section we focus on a fundamental method that we briefly discussed at several points earlier. In particular, we consider the gradient projection

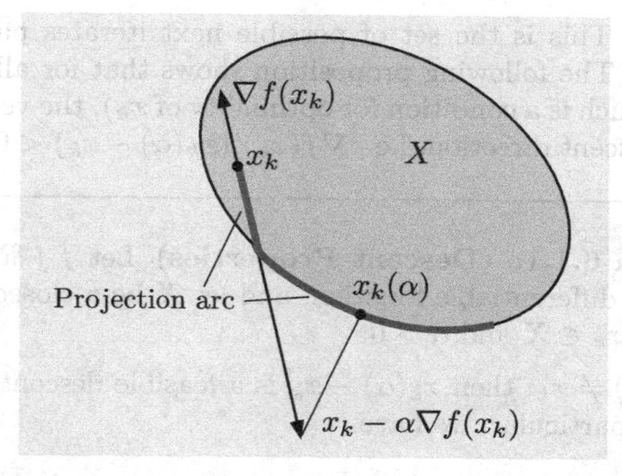

Figure 6.1.1. Illustration of the projection arc

$$x_k(\alpha) = P_X\big(x_k - \alpha\nabla f(x_k)\big), \qquad \alpha > 0.$$

It starts at x_k and defines a curve continuously parametrized by $\alpha \in (0, \infty)$.

method,

$$x_{k+1} = P_X\big(x_k - \alpha_k\nabla f(x_k)\big), \tag{6.1}$$

where $f : \Re^n \mapsto \Re$ is a continuously differentiable function, X is a closed convex set, and $\alpha_k > 0$ is a stepsize. We have outlined in Section 2.1.2 some of its characteristics, and its connection to feasible direction methods. In this section we take a closer look at its convergence and rate of convergence properties, and its implementation.

Descent Properties of the Gradient Projection Method

The gradient projection method can be viewed as the specialization of the subgradient method of Section 3.2, for the case where f is differentiable, so it is covered by the convergence analysis given there. However, when f is differentiable, stronger convergence and rate of convergence results can be shown than for the nondifferentiable case. Furthermore, there is more flexibility in applying the method because there is a greater variety of stepsize rules that can be used. The fundamental reason is that we can operate the gradient projection method based on iterative cost function descent, i.e., it is possible to select α_k so that

$$f(x_{k+1}) < f(x_k),$$

if x_k is not optimal. By contrast, this is not possible when f is nondifferentiable, and an arbitrary subgradient is used in place of the gradient, as we have seen in Section 3.2.

For a given $x_k \in X$, let us consider the *projection arc*, defined by

$$x_k(\alpha) = P_X\big(x_k - \alpha\nabla f(x_k)\big), \qquad \alpha > 0; \tag{6.2}$$

see Fig. 6.1.1. This is the set of possible next iterates parametrized by the stepsize α. The following proposition shows that for all $\alpha > 0$, unless $x_k(\alpha) = x_k$ (which is a condition for optimality of x_k), the vector $x_k(\alpha) - x_k$ is a feasible descent direction, i.e., $\nabla f(x_k)'\big(x_k(\alpha) - x_k\big) < 0$.

Proposition 6.1.1: (Descent Properties) Let $f : \Re^n \mapsto \Re$ be a continuously differentiable function, and let X be a closed convex set. Then for all $x_k \in X$ and $\alpha > 0$:

(a) If $x_k(\alpha) \neq x_k$, then $x_k(\alpha) - x_k$ is a feasible descent direction at x_k. In particular, we have

$$\nabla f(x_k)'\big(x_k(\alpha) - x_k\big) \leq -\frac{1}{\alpha}\,\big\|x_k(\alpha) - x_k\big\|^2, \qquad \forall\, \alpha > 0. \tag{6.3}$$

(b) If $x_k(\alpha) = x_k$ for some $\alpha > 0$, then x_k satisfies the necessary condition for minimizing f over X,

$$\nabla f(x_k)'(x - x_k) \geq 0, \qquad \forall\, x \in X. \tag{6.4}$$

Proof: (a) From the Projection Theorem (Prop. 1.1.9 in Appendix B), we have

$$\big(x_k - \alpha\nabla f(x_k) - x_k(\alpha)\big)'\big(x - x_k(\alpha)\big) \leq 0, \qquad \forall\, x \in X, \tag{6.5}$$

so by setting $x = x_k$, we obtain Eq. (6.3).

(b) If $x_k(\alpha) = x_k$ for some $\alpha > 0$, Eq. (6.5) becomes Eq. (6.4). **Q.E.D.**

Convergence for a Constant Stepsize

The simplest way to guarantee cost function descent in the gradient projection method is to keep the stepsize fixed at a constant but sufficiently small value $\alpha > 0$. In this case, however, it is necessary to assume that f has Lipschitz continuous gradient, i.e., for some constant L, we have†

$$\big\|\nabla f(x) - \nabla f(y)\big\| \leq L\,\|x - y\|, \qquad \forall\, x, y \in X. \tag{6.6}$$

† Without this condition the method may not converge, for any constant stepsize choice α, as can be seen with the scalar example where $f(x) = |x|^{3/2}$ (the method oscillates around the minimum $x^* = 0$, because the gradient grows too fast around x^*). A different stepsize rule that ensures cost function descent is needed for convergence.

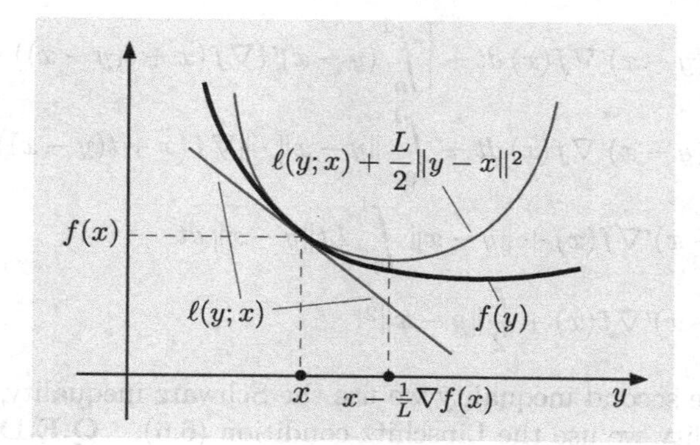

Figure 6.1.2. Visualization of the descent lemma (cf. Prop. 6.1.2). The Lipschitz constant L serves as an upper bound to the "curvature" of f along directions, so the quadratic function $\ell(y;x)+\frac{L}{2}\|y-x\|^2$ is an upper bound to $f(y)$. The steepest descent iterate $x - \frac{1}{L}\nabla f(x)$, with stepsize $\alpha = 1/L$, minimizes this upper bound.

This condition can be used to provide a quadratic function/upper bound to f in terms of the linear approximation of f based on the gradient at any $x \in \Re^n$, defined as

$$\ell(y;x) = f(x) + \nabla f(x)'(y-x), \qquad x, y \in \Re^n. \tag{6.7}$$

This is shown in the following proposition and is illustrated in Fig. 6.1.2.

Proposition 6.1.2: (Descent Lemma) Let $f : \Re^n \mapsto \Re$ be a continuously differentiable function, with gradient satisfying the Lipschitz condition (6.6), and let X be a closed convex set. Then for all $x, y \in X$, we have

$$f(y) \le \ell(y;x) + \frac{L}{2}\|y-x\|^2. \tag{6.8}$$

Proof: Let t be a scalar parameter and let $g(t) = f\big(x + t(y-x)\big)$. The chain rule yields

$$(dg/dt)(t) = (y-x)'\nabla f\big(x + t(y-x)\big).$$

Thus, we have

$$\begin{aligned} f(y) - f(x) &= g(1) - g(0) \\ &= \int_0^1 \frac{dg}{dt}(t)\,dt \\ &= \int_0^1 (y-x)'\nabla f\big(x+t(y-x)\big)\,dt \end{aligned}$$

$$
\begin{aligned}
&\le \int_0^1 (y-x)'\nabla f(x)\,dt + \left| \int_0^1 (y-x)'\big(\nabla f(x+t(y-x)) - \nabla f(x)\big)\,dt \right| \\
&\le \int_0^1 (y-x)'\nabla f(x)\,dt + \int_0^1 \|y-x\| \cdot \|\nabla f(x+t(y-x)) - \nabla f(x)\|dt \\
&\le (y-x)'\nabla f(x) + \|y-x\| \int_0^1 Lt\|y-x\|\,dt \\
&= (y-x)'\nabla f(x) + \frac{L}{2}\|y-x\|^2,
\end{aligned}
$$

where for the second inequality we use the Schwarz inequality, and for the third inequality we use the Lipschitz condition (6.6). **Q.E.D.**

We can use the preceding descent lemma to assert that a constant stepsize in the range $(0, 2/L)$ leads to cost function descent, and guarantees convergence. This is shown in the following classical convergence result, which does not require convexity of f.

Proposition 6.1.3: Let $f : \Re^n \mapsto \Re$ be a continuously differentiable function, and let X be a closed convex set. Assume that ∇f satisfies the Lipschitz condition (6.6), and consider the gradient projection iteration

$$x_{k+1} = P_X\big(x_k - \alpha\nabla f(x_k)\big),$$

with a constant stepsize α in the range $(0, 2/L)$. Then every limit point $\overline{x}$ of the generated sequence $\{x_k\}$ satisfies the optimality condition

$$\nabla f(\overline{x})'(x-\overline{x}) \ge 0, \qquad \forall\, x \in X.$$

Proof: From Eq. (6.8) with $y = x_{k+1}$, we have

$$
\begin{aligned}
f(x_{k+1}) &\le \ell(x_{k+1}; x_k) + \frac{L}{2}\|x_{k+1} - x_k\|^2 \\
&= f(x_k) + \nabla f(x_k)'(x_{k+1} - x_k) + \frac{L}{2}\|x_{k+1} - x_k\|^2,
\end{aligned}
$$

while from Prop. 6.1.1 [cf. Eq. (6.3)], we have

$$\nabla f(x_k)'(x_{k+1} - x_k) \le -\frac{1}{\alpha}\big\|x_{k+1} - x_k\big\|^2.$$

By combining the preceding two relations,

$$f(x_{k+1}) \le f(x_k) - \left(\frac{1}{\alpha} - \frac{L}{2}\right)\|x_{k+1} - x_k\|^2, \tag{6.9}$$

so since $\alpha \in \big(0, 2/L\big)$, the gradient projection method (6.1) reduces the cost function value at each iteration. It follows that if $\overline{x}$ is the limit of a subsequence $\{x_k\}_\mathcal{K}$, we have $f(x_k) \downarrow f(\overline{x})$, and by Eq. (6.9),

$$\|x_{k+1} - x_k\| \to 0.$$

Hence

$$P_X\big(\overline{x} - \alpha \nabla f(\overline{x})\big) - \overline{x} = \lim_{k\to\infty,\, k\in\mathcal{K}} (x_{k+1} - x_k) = 0,$$

which, by Prop. 6.1.1(b), implies that $\overline{x}$ satisfies the optimality condition. **Q.E.D.**

Connection with the Proximal Algorithm

We will now consider the convergence rate of the gradient projection method, and as a first step in this direction, we develop a connection to the proximal algorithm of Section 5.1. The following lemma shows that the gradient projection iteration (6.1) may also be viewed as an iteration of the proximal algorithm, applied to the linear approximation function $\ell(\cdot; x_k)$ (plus the indicator function of X), with penalty parameter equal to the stepsize α_k (see Fig. 6.1.3).

Proposition 6.1.4: Let $f : \Re^n \mapsto \Re$ be a continuously differentiable function, and let X be a closed convex set. Then for all $x \in X$ and $\alpha > 0$,

$$P_X\big(x - \alpha \nabla f(x)\big)$$

is the unique vector that attains the minimum in

$$\min_{y\in X} \left\{ \ell(y; x) + \frac{1}{2\alpha}\|y - x\|^2 \right\}.$$

Proof: Using the definition of ℓ [cf. Eq. (6.7)], we have for all $x, y \in \Re^n$ and $\alpha > 0$,

$$\begin{aligned}
\ell(y; x) + \frac{1}{2\alpha}\|y - x\|^2 &= f(x) + \nabla f(x)'(y - x) + \frac{1}{2\alpha}\|y - x\|^2 \\
&= f(x) + \frac{1}{2\alpha}\big\|y - \big(x - \alpha\nabla f(x)\big)\big\|^2 - \frac{\alpha}{2}\big\|\nabla f(x)\big\|^2.
\end{aligned}$$

The gradient projection iterate $P_X\big(x - \alpha\nabla f(x)\big)$ minimizes the right-hand side over $y \in X$, so it minimizes the left hand side. **Q.E.D.**

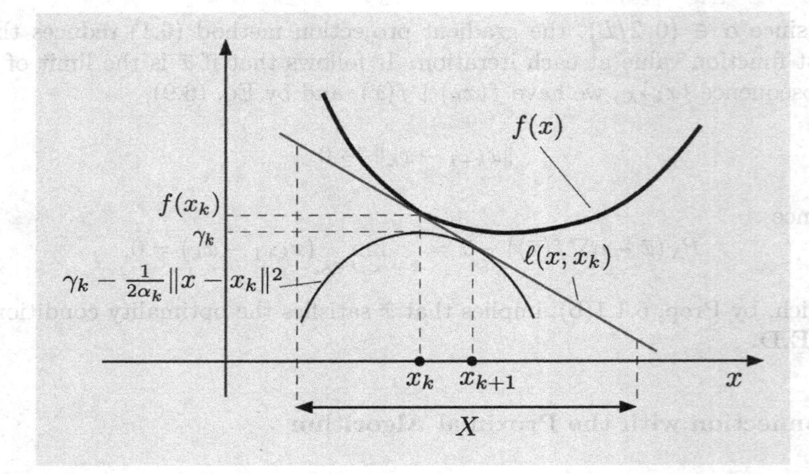

Figure 6.1.3. Illustration of the relation of the gradient projection method and the proximal algorithm, as per Prop. 6.1.4. The gradient projection iterate x_{k+1} is the same as the proximal iterate with f replaced by $\ell(\cdot; x_k)$.

A useful consequence of the connection with the proximal algorithm is the following adaptation of the three-term inequality of Prop. 5.1.2.

Proposition 6.1.5: Let $f : \Re^n \mapsto \Re$ be a continuously differentiable function, and let X be a closed convex set. For a given iterate x_k of the gradient projection method (6.1), consider the arc of points

$$x_k(\alpha) = P_X\big(x_k - \alpha \nabla f(x_k)\big), \qquad \alpha > 0.$$

Then for all $y \in \Re^n$ and $\alpha > 0$, we have

$$\big\|x_k(\alpha) - y\big\|^2 \leq \|x_k - y\|^2 - 2\alpha\big(\ell(x_k(\alpha); x_k) - \ell(y; x_k)\big) - \big\|x_k - x_k(\alpha)\big\|^2. \tag{6.10}$$

Proof: Based on the proximal interpretation of Prop. 6.1.4, we can apply Prop. 5.1.2, with f replaced by $\ell(\cdot; x_k)$ plus the indicator function of X, and c_k replaced by α. The three-term inequality of this proposition yields Eq. (6.10). **Q.E.D.**

Eventually Constant Stepsize Rule

While the constant stepsize rule is simple, it requires the knowledge of the Lipschitz constant L. There is an alternative stepsize rule that aims to deal with the practically common situation where L, and hence also the range $(0, 2/L)$ of stepsizes that guarantee cost reduction, are unknown. Here we start with a stepsize $\alpha > 0$ that is a guess for the midpoint $1/L$ of the range $(0, 2/L)$. Then we keep using α, and generate iterates according to

$$x_{k+1} = P_X\big(x_k - \alpha \nabla f(x_k)\big),$$

as long as the condition

$$f(x_{k+1}) \leq \ell(x_{k+1}; x_k) + \frac{1}{2\alpha}\|x_{k+1} - x_k\|^2 \tag{6.11}$$

is satisfied. As soon as this condition is violated at some iteration, we reduce α by a certain factor, and repeat the iteration as many times as is necessary for Eq. (6.11) to hold.

From the descent lemma (Prop. 6.1.2), the condition (6.11) will be satisfied if $\alpha \leq 1/L$. Thus, after a finite number of reductions, the test (6.11) will be passed at every subsequent iteration and α will stay constant at some value $\overline{\alpha} > 0$. We refer to this as the *eventually constant* stepsize rule. The next proposition shows that cost function descent is guaranteed with this rule.

Proposition 6.1.6: Let $f : \Re^n \mapsto \Re$ be a continuously differentiable function, and let X be a closed convex set. Assume that ∇f satisfies the Lipschitz condition (6.6). For a given iterate x_k of the gradient projection method (6.1), consider the arc of points

$$x_k(\alpha) = P_X\big(x_k - \alpha \nabla f(x_k)\big), \qquad \alpha > 0.$$

Then the inequality

$$f\big(x_k(\alpha)\big) \leq \ell\big(x_k(\alpha); x_k\big) + \frac{1}{2\alpha}\big\|x_k(\alpha) - x_k\big\|^2 \tag{6.12}$$

implies the cost reduction property

$$f\big(x_k(\alpha)\big) \leq f(x_k) - \frac{1}{2\alpha}\big\|x_k(\alpha) - x_k\big\|^2. \tag{6.13}$$

Moreover the inequality (6.12) is satisfied for all $\alpha \in (0, 1/L]$.

Proof: By setting $y = x_k$ in Eq. (6.10), using the fact $\ell(x_k; x_k) = f(x_k)$, and rearranging, we have

$$\ell\big(x_k(\alpha); x_k\big) \leq f(x_k) - \frac{1}{\alpha}\big\|x_k(\alpha) - x_k\big\|^2.$$

If the inequality (6.12) holds, it can be added to the preceding relation to yield Eq. (6.13). Also, the descent lemma (Prop. 6.1.2) implies that the inequality (6.12) is satisfied for all $\alpha \in (0, 1/L]$. **Q.E.D.**

Convergence and Convergence Rate for a Convex Cost Function

We will now show convergence and derive the convergence rate of the gradient projection method for a convex cost function, under the gradient Lipschitz condition (6.6). It turns out that an additional benefit of convexity is that we can show stronger convergence results than for the nonconvex case (cf. Prop. 6.1.3). In particular, we will show convergence to an optimal solution of the entire sequence of iterates $\{x_k\}$ as long as the optimal solution set is nonempty.† By contrast, in the absence of convexity of f, Prop. 6.1.3 asserts that all limit points of $\{x_k\}$ satisfy the optimality condition, but there is no assertion of uniqueness of limit point.

We will assume that the stepsize satisfies some conditions, which hold in particular if it is either a constant in the range $(0, 1/L]$ or it is chosen according to the eventually constant stepsize rule described earlier.

Proposition 6.1.7: Let $f : \Re^n \mapsto \Re$ be a convex differentiable function, and let X be a closed convex set. Assume that ∇f satisfies the Lipschitz condition (6.6), and that the set of minima X^* of f over X is nonempty. Let $\{x_k\}$ be a sequence generated by the gradient projection method (6.1) using the eventually constant stepsize rule, or more generally, any stepsize rule such that $\alpha_k \downarrow \overline{\alpha}$ for some $\overline{\alpha} > 0$, and for all k, we have

$$f(x_{k+1}) \leq \ell(x_{k+1}; x_k) + \frac{1}{2\alpha_k}\|x_{k+1} - x_k\|^2. \tag{6.14}$$

Then $\{x_k\}$ converges to some point in X^* and

$$f(x_{k+1}) - f^* \leq \frac{\min_{x^* \in X^*} \|x_0 - x^*\|^2}{2(k+1)\overline{\alpha}}, \qquad k = 0, 1, \ldots. \tag{6.15}$$

Proof: By applying Prop. 6.1.5, with $\alpha = \alpha_k$ and $y = x^*$, where $x^* \in X^*$,

---

† We have seen a manifestation of this result in the analysis of Section 3.2, where we showed convergence of the subgradient method with diminishing stepsize to an optimal solution. In that section we used ideas of supermartingale and Fejér convergence. Similar ideas apply for gradient projection methods as well.

we have

$$\ell(x_{k+1};x_k)+\frac{1}{2\alpha_k}\|x_{k+1}-x_k\|^2$$
$$\leq \ell(x^*;x_k)+\frac{1}{2\alpha_k}\|x^*-x_k\|^2-\frac{1}{2\alpha_k}\|x^*-x_{k+1}\|^2.$$

By adding Eq. (6.14), we obtain for all $x^* \in X^*$,

$$\begin{aligned} f(x_{k+1}) &\leq \ell(x^*;x_k)+\frac{1}{2\alpha_k}\|x^*-x_k\|^2-\frac{1}{2\alpha_k}\|x^*-x_{k+1}\|^2 \\ &\leq f(x^*)+\frac{1}{2\alpha_k}\|x^*-x_k\|^2-\frac{1}{2\alpha_k}\|x^*-x_{k+1}\|^2, \end{aligned} \tag{6.16}$$

where for the last inequality we use the convexity of f and the gradient inequality:

$$f(x^*)-\ell(x^*;x_k)=f(x^*)-f(x_k)-\nabla f(x_k)'(x^*-x_k)\geq 0.$$

Thus, denoting $e_k=f(x_k)-f^*$, we have from Eq. (6.16) for all k and $x^* \in X^*$,

$$\|x^*-x_{k+1}\|^2 \leq \|x^*-x_k\|^2-2\alpha_k e_{k+1}. \tag{6.17}$$

Using repeatedly this relation with k replaced by $k-1,\ldots,0$, and adding, we obtain for all k and $x^* \in X^*$,

$$0\leq \|x^*-x_{k+1}\|^2 \leq \|x^*-x_0\|^2-2(\alpha_0 e_1+\cdots+\alpha_k e_{k+1}),$$

so since $\alpha_0 \geq \alpha_1 \geq \cdots \geq \alpha_k \geq \overline{\alpha}$ and $e_1 \geq e_2 \geq \cdots \geq e_{k+1}$ (cf. Prop. 6.1.6), we have

$$2\overline{\alpha}(k+1)e_{k+1}\leq \|x^*-x_0\|^2.$$

Taking the minimum over $x^* \in X^*$ in the preceding relation, we obtain Eq. (6.15). Finally, by Eq. (6.17), $\{x_k\}$ is bounded, and each of its limit points must belong to X^* since $e_k \to 0$. Also by Eq. (6.17), the distance of x_k to each $x^* \in X^*$ is monotonically nonincreasing, so $\{x_k\}$ cannot have multiple limit points, and it must converge to some point in X^*. **Q.E.D.**

The monotone decrease property of $\{f(x_k)\}$, shown in Prop. 6.1.6 [cf. Eq. (6.13)], has an interesting consequence. It can be used to show the convergence result of Prop. 6.1.7 assuming that the Lipschitz condition (6.6) holds just within the level set

$$X_0=\{x\in X \mid f(x)\leq f(x_0)\}$$

rather than within X. The reason is that under the assumptions of the proposition the iterates x_k are guaranteed to stay within the initial level set X_0, and the preceding analysis still goes through. This allows the application of the proposition to cost functions such as $|x|^3$, for which the Lipschitz condition (6.6) does not hold when X is unbounded.

Convergence Rate Under Strong Convexity

Proposition 6.1.7 shows that the cost function error of the gradient projection method converges to 0 as $O(1/k)$. However, this is true without assuming any condition other than Lipschitz continuity of ∇f. When f satisfies a growth condition in the neighborhood of the optimal set, a faster convergence rate can be proved as noted in Section 2.1.1. In the case where the growth of f is at least quadratic, a linear convergence rate can be shown, and in fact it turns out that *if f is strongly convex, the gradient projection mapping*

$$G_\alpha(x) = P_X\big(x - \alpha\nabla f(x)\big) \tag{6.18}$$

is a contraction when $0 < \alpha < 2/L$. Let us recall here that the differentiable convex function f is strongly convex over $\Re^n$ with a coefficient $\sigma > 0$ if

$$\big(\nabla f(x) - \nabla f(y)\big)'(x - y) \geq \sigma\|x - y\|^2, \qquad \forall\, x, y \in \Re^n; \tag{6.19}$$

cf. Section 1.1 of Appendix B. Note that by using the Schwarz inequality to bound the inner product on the left above, this condition implies that

$$\big\|\nabla f(x) - \nabla f(y)\big\| \geq \sigma\|x - y\|, \qquad \forall\, x, y \in \Re^n, \tag{6.20}$$

so that if in addition ∇f satisfies a Lipschitz condition with Lipschitz constant L, we must have $L \geq \sigma$.†

The following proposition shows the contraction property of the gradient projection mapping.

Proposition 6.1.8: (Contraction Property under Strong Convexity) Let $f : \Re^n \mapsto \Re$ be a convex differentiable function, and let X be a closed convex set. Assume that ∇f satisfies the Lipschitz condition

$$\big\|\nabla f(x) - \nabla f(y)\big\| \leq L\,\|x - y\|, \qquad \forall\, x, y \in \Re^n, \tag{6.21}$$

for some $L > 0$, and that it is strongly convex over $\Re^n$ in the sense that for some $\sigma \in (0, L]$ it satisfies Eq. (6.19). Then the gradient projection mapping G_α of Eq. (6.18) satisfies

$$\big\|G_\alpha(x) - G_\alpha(y)\big\| \leq \max\big\{|1 - \alpha L|, |1 - \alpha\sigma|\big\}\,\|x - y\|, \qquad \forall\, x, y \in \Re^n,$$

and is a contraction for all $\alpha \in (0, 2/L)$.

---

† A related but different property is that strong convexity of f is equivalent to Lipschitz continuity of the gradient of the conjugate $f^\star$ when f and $f^\star$ are real-valued (see [RoW98], Prop. 12.60, for a more general result).

We first show the following preliminary result, which provides several useful properties of differentiable convex functions relating to Lipschitz continuity of the gradient and strong convexity.

Proposition 6.1.9: Let $f : \Re^n \mapsto \Re$ be a convex differentiable function, and assume that ∇f satisfies the Lipschitz condition (6.21).

(a) We have for all $x, y \in \Re^n$

(i) $f(x) + \nabla f(x)'(y - x) + \frac{1}{2L}\|\nabla f(x) - \nabla f(y)\|^2 \leq f(y)$.

(ii) $\big(\nabla f(x) - \nabla f(y)\big)'(x - y) \geq \frac{1}{L}\|\nabla f(x) - \nabla f(y)\|^2$.

(iii) $\big(\nabla f(x) - \nabla f(y)\big)'(x - y) \leq L\|x - y\|^2$.

(b) If f is strongly convex in the sense that for some $\sigma \in (0, L]$ it satisfies Eq. (6.19), we have for all $x, y \in \Re^n$

$$\big(\nabla f(x) - \nabla f(y)\big)'(x - y) \geq \frac{\sigma L}{\sigma + L}\|x - y\|^2 + \frac{1}{\sigma + L}\|\nabla f(x) - \nabla f(y)\|^2.$$

Proof: (a) To show (i), fix $x \in \Re^n$ and let ϕ be the function

$$\phi(y) = f(y) - \nabla f(x)'y, \qquad y \in \Re^n. \tag{6.22}$$

We have

$$\nabla\phi(y) = \nabla f(y) - \nabla f(x), \tag{6.23}$$

so ϕ is minimized over y at $y = x$, and we have

$$\phi(x) \leq \phi\left(y - \frac{1}{L}\nabla\phi(y)\right), \qquad \forall\, y \in \Re^n. \tag{6.24}$$

Moreover from Eq. (6.23), $\nabla\phi$ is Lipschitz continuous with constant L, so by applying the descent lemma (Prop. 6.1.2) with ϕ in place of f, $y - \frac{1}{L}\nabla\phi(y)$ in place of y, and y in place of x, we have

$$\begin{aligned}\phi\left(y - \frac{1}{L}\nabla\phi(y)\right) &\leq \phi(y) + \nabla\phi(y)'\left(-\frac{1}{L}\nabla\phi(y)\right) + \frac{L}{2}\left\|\frac{1}{L}\nabla\phi(y)\right\|^2 \\ &= \phi(y) - \frac{1}{2L}\|\nabla\phi(y)\|^2.\end{aligned}$$

By combining this inequality with Eq. (6.24), we obtain

$$\phi(x) + \frac{1}{2L}\|\nabla\phi(y)\|^2 \leq \phi(y),$$

and by using the expressions (6.22) and (6.23), we obtain the desired result.

To show (ii), we use (i) twice, with the roles of x and y interchanged, and add to obtain the desired relation. We similarly use the descent lemma (Prop. 6.1.2) twice, with the roles of x and y interchanged, and add to obtain (iii).

(b) If $\sigma = L$, the result follows by combining (ii) of part (a) and Eq. (6.20), which is a consequence of the strong convexity assumption. For $\sigma < L$ consider the function

$$\phi(x) = f(x) - \frac{\sigma}{2}\|x\|^2.$$

We will show that $\nabla\phi$, which is given by

$$\nabla\phi(x) = \nabla f(x) - \sigma x, \tag{6.25}$$

is Lipschitz continuous with constant $L - \sigma$. To this end, based on the equivalence of statements (i) and (v) of Exercise 6.1, it is sufficient to show that

$$\big(\nabla\phi(x) - \nabla\phi(y)\big)'(x-y) \le (L-\sigma)\|x-y\|^2, \qquad \forall\, x, y \in \Re^n,$$

or, using the expression (6.25) for $\nabla\phi$,

$$\big(\nabla f(x) - \nabla f(y) - \sigma(x-y)\big)'(x-y) \le (L-\sigma)\|x-y\|^2, \qquad \forall\, x, y \in \Re^n.$$

This relation is equivalently written as

$$\big(\nabla f(x) - \nabla f(y)\big)'(x-y) \le L\|x-y\|^2, \qquad \forall\, x, y \in \Re^n,$$

and is true by (iii) of part (a).

Having shown that $\nabla\phi$ is Lipschitz continuous with constant $L - \sigma$, we apply (ii) of part (a) to the function ϕ and obtain

$$\big(\nabla\phi(x) - \nabla\phi(y)\big)'(x-y) \ge \frac{1}{L-\sigma}\big\|\nabla\phi(x) - \nabla\phi(y)\big\|^2.$$

Using the expression (6.25) for $\nabla\phi$ in this relation, we have

$$\big(\nabla f(x) - \nabla f(y) - \sigma(x-y)\big)'(x-y) \ge \frac{1}{L-\sigma}\big\|\nabla f(x) - \nabla f(y) - \sigma(x-y)\big\|^2,$$

which after expanding the quadratic and collecting terms, can be verified to be equivalent to the desired relation. **Q.E.D.**

We note that a stronger version of Prop. 6.1.9(a) holds, namely that each of the properties (i)-(iii) is equivalent to ∇f satisfying the Lipschitz condition (6.21); see Exercise 6.1. We now complete the proof of the contraction property of the gradient projection mapping.

Proof of Prop. 6.1.8: For all $x, y \in \Re^n$, we have by using the nonexpansive property of the projection (cf. Prop. 3.2.1)

$$\begin{aligned}\big\|G_\alpha(x) - G_\alpha(y)\big\|^2 &= \big\|P_X\big(x - \alpha\nabla f(x)\big) - P_X\big(y - \alpha\nabla f(y)\big)\big\|^2 \\ &\le \big\|\big(x - \alpha\nabla f(x)\big) - \big(y - \alpha\nabla f(y)\big)\big\|^2.\end{aligned}$$

Expanding the quadratic on the right-hand side, and using Prop. 6.1.9(b), the Lipschitz condition (6.6), and the strong convexity condition (6.20), we obtain

$$\begin{aligned}\big\|G_\alpha(x) - G_\alpha(y)\big\|^2 &\le \|x - y\|^2 - 2\alpha\big(\nabla f(x) - \nabla f(y)\big)'(x - y) \\ &\quad + \alpha^2\big\|\nabla f(x) - \nabla f(y)\big\|^2 \\ &\le \|x - y\|^2 - \frac{2\alpha\sigma L}{\sigma + L}\|x - y\|^2 - \frac{2\alpha}{\sigma + L}\big\|\nabla f(x) - \nabla f(y)\big\|^2 \\ &\quad + \alpha^2\big\|\nabla f(x) - \nabla f(y)\big\|^2 \\ &= \left(1 - \frac{2\alpha\sigma L}{\sigma + L}\right)\|x - y\|^2 \\ &\quad + \alpha\left(\alpha - \frac{2}{\sigma + L}\right)\big\|\nabla f(x) - \nabla f(y)\big\|^2 \\ &\le \left(1 - \frac{2\alpha\sigma L}{\sigma + L}\right)\|x - y\|^2 \\ &\quad + \alpha\max\left\{L^2\left(\alpha - \frac{2}{\sigma + L}\right), \sigma^2\left(\alpha - \frac{2}{\sigma + L}\right)\right\}\|x - y\|^2 \\ &= \max\big\{(1 - \alpha L)^2, (1 - \alpha\sigma)^2\big\}\|x - y\|^2,\end{aligned} \tag{6.26}$$

from which the desired inequality follows. **Q.E.D.**

Note from the last equality of Eq. (6.26) that the smallest modulus of contraction is obtained when

$$\alpha = \frac{2}{\sigma + L}.$$

When this optimal value of stepsize α is used, it can be seen by substitution in Eq. (6.26) that for all $x, y \in \Re^n$,

$$\big\|G_\alpha(x) - G_\alpha(y)\big\| \le \sqrt{1 - \frac{4\sigma L}{(\sigma + L)^2}}\,\|x - y\| = \left(\frac{\frac{L}{\sigma} - 1}{\frac{L}{\sigma} + 1}\right)\|x - y\|.$$

We can observe the similarity of this convergence rate estimate with the one of Section 2.1.1 and Exercise 2.1 for quadratic functions: the ratio L/σ plays the role of the *condition number* of the problem. Indeed for a positive definite quadratic function f we can use as L and σ the maximum and minimum eigenvalues of the Hessian of f, respectively.

Alternative Stepsize Rules

In addition to the constant and eventually constant stepsize rules, there are several other stepsize rules for gradient projection that are often used in practice, and do not require that ∇f satisfies the Lipschitz condition (6.6). We will describe some of them and summarize their properties.

One possibility is *a diminishing stepsize* α_k, satisfying the conditions

$$\lim_{k\to\infty} \alpha_k = 0, \qquad \sum_{k=0}^{\infty} \alpha_k = \infty, \qquad \sum_{k=0}^{\infty} \alpha_k^2 < \infty.$$

With this rule, the convergence behavior of the method is very similar to the one of the corresponding subgradient method. In particular, by Prop. 3.2.6 and the discussion following that proposition (cf. Exercise 3.6), if there is a scalar c such that

$$c^2 \left(1 + \min_{x^* \in X^*} \|x_k - x^*\|^2\right) \geq \sup\{\|\nabla f(x_k)\|^2 \mid k = 0, 1, \ldots\},$$

the gradient projection method converges to some $x^* \in X^*$ (assuming X^* is nonempty). The gradient Lipschitz condition (6.6) is not required for this property; as an example, the method converges to $x^* = 0$ for the scalar function $f(x) = |x|^{3/2}$ with a diminishing stepsize, but not with a constant stepsize. Note that if f is not convex, the standard result for convergence with a diminishing stepsize is that every limit point $\overline{x}$ of $\{x_k\}$ satisfies $\nabla f(\overline{x}) = 0$, and there is no assertion of existence or uniqueness of the limit point. A drawback of the diminishing stepsize rule is that it leads to a sublinear convergence rate even under the most favorable conditions, e.g., when f is a positive definite quadratic function, in which case the constant stepsize rule can be shown to attain a linear convergence rate (see Exercise 2.1 in Chapter 2).

Another possibility is stepsize reduction and line search, based on cost function descent, i.e.,

$$f(x_{k+1}) < f(x_k).$$

Indeed, if $x_k(\alpha) \neq x_k$ for some $\alpha > 0$, it can be shown (see [Ber99], Section 2.3.2) that there exists $\overline{\alpha}_k > 0$ such that

$$f\big(x_k(\alpha)\big) < f(x_k), \qquad \forall\, \alpha \in (0, \overline{\alpha}_k]. \tag{6.27}$$

Thus there is an interval of stepsizes $\alpha \in (0, \overline{\alpha}_k]$ that lead to reduction of the cost function value. Stepsize reduction and line search rules are motivated by some of the drawbacks of the constant and eventually constant stepsize rules: along some directions the growth rate of ∇f may be fast requiring a small stepsize for guaranteed cost function descent, while in other directions the growth rate of ∇f may be slow, requiring a large stepsize for substantial progress. A form of line search may deal adequately with this difficulty.

There are many variants of line search rules. Some rules use an exact line search, aiming to find the stepsize α_k that yields the maximum possible cost improvement; these rules are practical mostly for the unconstrained case, where they can be implemented via one of the several possible interpolation and other one-dimensional algorithms (see nonlinear programming texts such as [Ber99], [Lue84], [NoW06]). For constrained problems, stepsize reduction rules are primarily used: an initial stepsize is chosen through some heuristic procedure (possibly a fixed constant, obtained through some experimentation, or a crude line search based on some polynomial interpolation scheme). This stepsize is then successively reduced by a certain factor until a cost reduction test is passed.

One of the most popular stepsize reduction rules searches for a stepsize along the set of points

$$x_k(\alpha) = P_X\big(x_k - \alpha\nabla f(x_k)\big), \qquad \alpha > 0,$$

cf. Eq. (6.2). This is the *Armijo rule along the projection arc*, proposed in [Ber76a], which is a generalization of the Armijo rule for unconstrained problems, given in Section 2.1. It has the form

$$\alpha_k = \beta^{m_k} s_k,$$

where m_k is the first integer m such that

$$f(x_k) - f\big(x_k(\beta^m s_k)\big) \geq \sigma\nabla f(x_k)'\big(x_k - x_k(\beta^m s_k)\big), \tag{6.28}$$

with $\beta \in (0,1)$ and $\sigma \in (0,1)$ being some constants, and $s_k > 0$ being an initial stepsize. Thus, the stepsize α_k is obtained by reducing s_k as many times as necessary for the inequality (6.28) to be satisfied; see Fig. 6.1.4. This stepsize rule has strong convergence properties. In particular, it can be shown that for a convex f with nonempty set X^* of minima over X, and with initial stepsize s_k that is bounded away from 0, it leads to convergence to some $x^* \in X^*$, without requiring the gradient Lipschitz condition (6.6). The proof of this is nontrivial, and was given in [GaB84]; see [Ber99], Section 2.3.2 for a textbook account [the original paper [Ber76a] gave an easier convergence proof for the special case where X is the nonnegative orthant, and also for the case where the Lipschitz condition (6.6) holds and X is any closed convex set]. Related asymptotic convergence rate results that involve the rate of growth of f, suitably modified for the presence of the constraint set, are given in [Dun81], [Dun87].

The preceding Armijo rule requires that with each reduction of the trial stepsize, a projection operation on X is performed. While this may not involve much overhead in cases where X is simple, such as for example a box

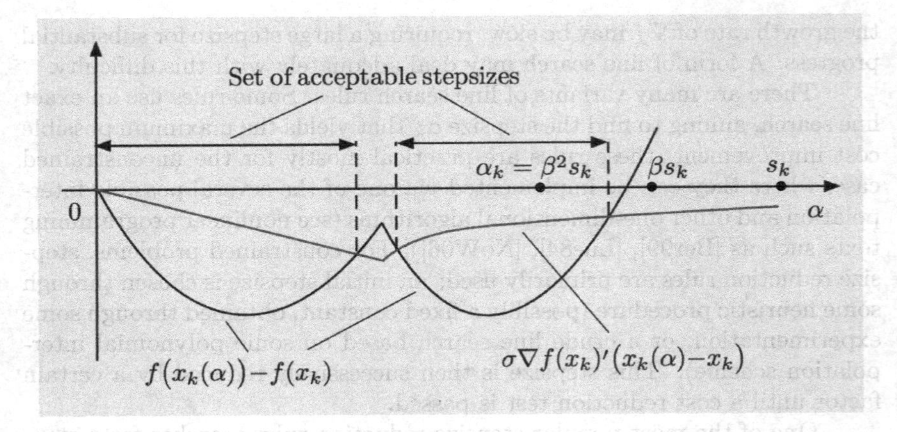

Figure 6.1.4. Illustration of the successive points tested by the Armijo rule along the projection arc. In this figure, α_k is obtained as $\beta^2 s_k$ after two unsuccessful trials.

constraint consisting of lower and/or upper bounds on the variables, there are other cases where X is more complicated. In such cases an alternative Armijo-like rule may be used. Here we first use gradient projection to determine a feasible descent direction d_k according to

$$d_k = P_X\big(x_k - s\nabla f(x_k)\big) - x_k,$$

where s is a fixed positive scalar [cf. Prop. 6.1.1(a)], and then we set

$$x_{k+1} = x_k + \beta^{m_k} d_k,$$

where $\beta \in (0,1)$ is a fixed scalar, and m_k is the first nonnegative integer m for which

$$f(x_k) - f(x_k + \beta^m d_k) \geq -\beta^m \nabla f(x_k)' d_k.$$

In other words, we search along the line $\{x_k + \gamma d_k \mid \gamma > 0\}$ by successively trying the stepsizes $\gamma = 1, \beta, \beta^2, \ldots$, until the above inequality is satisfied for $m = m_k$. While this rule may be simpler to implement,† the earlier rule of Eq. (6.28), which operates on the projection arc, has the advantage that it tends to keep the iterates on the boundary of the constraint set, and thus tends to identify constraints that are active earlier. When $X = \Re^n$ the two Armijo rules described above coincide, and are identical to the Armijo rule given for unconstrained minimization in Section 2.1.1.

† An example where search along a line is considerably simpler than search along the projection arc is when the cost function is of the form $f(x) = h(Ax)$, where A is a matrix such that the calculation of the vector $y = Ax$ for a given x is far more expensive than the calculation of $h(y)$.

For a convex cost function, both Armijo rules can be shown to guarantee convergence to a unique limit point/optimal solution, without requiring the gradient Lipschitz condition (6.6); see [Ius03] and compare also with the comments in Exercise 2.5. When f is not convex but differentiable, the standard convergence results with these rules state that every limit point $\overline{x}$ of $\{x_k\}$ satisfies the optimality condition

$$\nabla f(\overline{x})'(x - \overline{x}) \geq 0, \qquad \forall\, x \in X,$$

but there is no assertion of existence or uniqueness of the limit point; cf. Prop. 6.1.3. We refer to the nonlinear programming literature for further discussion.

Complexity Issues and Gradient Projection

Let us now consider in some generality computational complexity issues relating to optimization problems of the form

$$\begin{aligned} &\text{minimize} \quad f(x) \\ &\text{subject to} \quad x \in X, \end{aligned}$$

where $f : \Re^n \mapsto \Re$ is convex and X is a closed convex set. We denote by f^* the optimal value, and we assume throughout that there exists an optimal solution. We will aim to delineate algorithms that have good performance guarantees, in the sense that they require a relatively low number of iterations (in the worst case) to achieve a given optimal solution tolerance.

Given some $\epsilon > 0$, suppose we want to estimate the number of iterations required by a particular algorithm to obtain a solution with cost that is within ϵ of the optimal. If we can show that any sequence $\{x_k\}$ generated by a method has the property that for any $\epsilon > 0$, we have

$$\min_{k \leq c/\epsilon^p} f(x_k) \leq f^* + \epsilon,$$

where c and p are positive constants, we say that the method has *iteration complexity* $O\left(\frac{1}{\epsilon^p}\right)$ (the constant c may depend on the problem data and the starting point x_0). Alternatively, if we can show that

$$\min_{\ell \leq k} f(x_\ell) \leq f^* + \frac{c}{k^q},$$

where c and q are positive constants, we say that the method involves *cost function error of order* $O\left(\frac{1}{k^q}\right)$.

It is generally thought that if the constant c does not depend on the dimension n of the problem, then the algorithm holds some advantage for problems where n is large. This view favors simple gradient/subgradient-like methods over sophisticated conjugate direction or Newton-like methods, whose overhead per iteration increases at an order up to $O(n^2)$ or

$O(n^3)$. In this chapter, we will focus on algorithms with iteration complexity that is independent of n, and all our subsequent references to complexity estimates implicitly assume this.†

As an example, we mention the subgradient method for which an $O(1/\epsilon^2)$ complexity, or equivalently error $O\left(1/\sqrt{k}\right)$, can be shown (cf., the discussion following Prop. 3.2.3). On the other hand, Prop. 6.1.7 shows that in order for the algorithm to attain a vector x_k with

$$f(x_k) \leq f^* + \epsilon,$$

it requires a number of iterations $k \geq O(1/\epsilon)$, for an error $O(1/k)$. Thus the gradient projection method when applied to convex cost functions with Lipschitz continuous gradient, has iteration complexity $O(1/\epsilon)$.‡

---

† Some caution is necessary when considering the relative advantages of the various gradient and subgradient methods of this and the next section, and comparing them with other methods, such as conjugate direction, Newton-like or incremental, based on the complexity and error estimates that we provide. One reason is that our complexity estimates involve unknown constants, whose size may affect the theoretical comparisons between various methods. Moreover, experience with linear programming methods, such as simplex and ellipsoid, has shown that good (or bad) theoretical complexity may not translate into good (or bad, respectively) practical performance, at least in the context of continuous optimization using worst case measures of complexity (rather than some form of average complexity).

A further weakness of our analysis is that it does not take into account the special structure that is typically present in large-scale problems. An example of such special structure is cost functions of the form $f(x) = h(Ax)$ where A is a matrix such that the calculation of the vector $y = Ax$ for a given x is far more expensive than the calculation of $h(y)$. For such problems, calculation of a stepsize by line minimization, as in conjugate direction methods, as well as solving the low-dimensional problems arising in simplicial decomposition methods, is relatively inexpensive.

Another interesting example of special structure, which is not taken into account by the rate of convergence estimates of this section, is an additive cost function with a large number of components. We have argued in Section 2.1.5 that incremental methods are well suited for such cost functions, while the nonincremental methods of this and the next two sections may not be. Indeed, for a very large number of components in the cost function, it is not uncommon for an incremental method to reach practical convergence after one or very few passes through the components. The notion of iteration complexity loses its meaning when so few iterations are involved. For a discussion of the subtleties underlying the complexity analysis of incremental and nonincremental methods, see [AgB14].

‡ As we have noted in Sections 2.1.1 and 2.1.2, the asymptotic convergence rate of steepest descent and gradient projection also depends on the rate of growth of the cost function in the neighborhood of a minimum. The $O(1/\epsilon)$ iteration

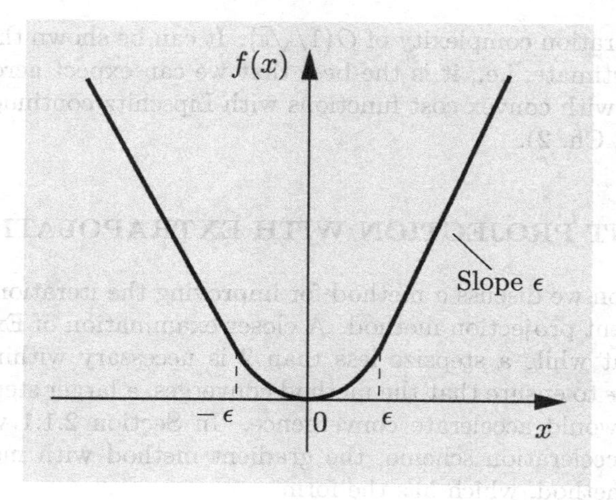

Figure 6.1.5. The differentiable scalar cost function f of Example 6.1.1. It is quadratic for $|x| \leq \epsilon$ and linear for $|x| > \epsilon$. The gradient Lipschitz constant is $L = 1$.

Example 6.1.1

Consider the unconstrained minimization of the scalar function f given by

$$f(x) = \begin{cases} \frac{1}{2}|x|^2 & \text{if } |x| \leq \epsilon, \\ \epsilon|x| - \frac{\epsilon^2}{2} & \text{if } |x| > \epsilon, \end{cases}$$

with $\epsilon > 0$ (cf. Fig. 6.1.5). Here the constant in the Lipschitz condition (6.6) is $L = 1$, and for any $x_k > \epsilon$, we have $\nabla f(x_k) = \epsilon$. Thus the gradient iteration with stepsize $\alpha = 1/L = 1$ takes the form

$$x_{k+1} = x_k - \frac{1}{L}\nabla f(x_k) = x_k - \epsilon.$$

It follows that the number of iterations to get within an ϵ-neighborhood of $x^* = 0$ is $|x_0|/\epsilon$. The number of iterations to get to within ϵ of the optimal cost $f^* = 0$, is also proportional to $1/\epsilon$.

In the next section, we will discuss a variant of the gradient projection method that employs an intricate extrapolation device, and has the

---

complexity estimate assumes the worst case where there is no positive order of growth. For the case where there is a unique minimum x^*, this means that there are no scalars $\beta > 0$, $\delta > 0$, and $\gamma > 1$ such that

$$\beta\|x - x^*\|^\gamma \leq f(x) - f(x^*), \qquad \forall\, x \text{ with } \|x - x^*\| \leq \delta.$$

This is not likely to be true in the context of a practical problem.

improved iteration complexity of $O(1/\sqrt{\epsilon})$. It can be shown that $O(1/\sqrt{\epsilon})$ is a sharp estimate, i.e., it is the best that we can expect across the class of problems with convex cost functions with Lipschitz continuous gradient (see [Nes04], Ch. 2).

6.2 GRADIENT PROJECTION WITH EXTRAPOLATION

In this section we discuss a method for improving the iteration complexity of the gradient projection method. A closer examination of Example 6.1.1 suggests that while a stepsize less than 2 is necessary within the region where $|x| \leq \epsilon$ to ensure that the method converges, a larger stepsize outside this region would accelerate convergence. In Section 2.1.1 we discussed briefly an acceleration scheme, the gradient method with momentum or heavy-ball method, which has the form

$$x_{k+1} = x_k - \alpha \nabla f(x_k) + \beta(x_k - x_{k-1}),$$

and adds the extrapolation term $\beta(x_k - x_{k-1})$ to the gradient increment, where $x_{-1} = x_0$ and β is a scalar with $0 < \beta < 1$.

A variant of this scheme with similar properties separates the extrapolation and the gradient steps as follows:

$$\begin{aligned} y_k &= x_k + \beta(x_k - x_{k-1}), && \text{(extrapolation step)}, \\ x_{k+1} &= y_k - \alpha \nabla f(y_k), && \text{(gradient step)}. \end{aligned} \tag{6.29}$$

When applied to the function of the preceding example, the method converges to the optimum, and reaches a neighborhood of the optimum more quickly: it can be verified that for a starting point $x_0 >> 1$ and $x_k > \epsilon$, it has the form $x_{k+1} = x_k - \epsilon_k$, with $\epsilon \leq \epsilon_k < \epsilon/(1-\beta)$. In fact it is generally true that with extrapolation, the practical performance of the gradient projection method is typically improved. However, for this example the method still has an $O(1/\epsilon)$ iteration complexity, since for $x_0 >> 1$, the number of iterations needed to obtain $x_k < \epsilon$ is $O((1-\beta)/\epsilon)$. This can be seen by verifying that for $x_k >> 1$ we have

$$x_{k+1} - x_k = \beta(x_k - x_{k-1}) - \epsilon,$$

so approximately $|x_{k+1} - x_k| \approx \epsilon/(1-\beta)$.

It turns out that for convex cost functions that have a Lipschitz continuous gradient a better iteration complexity is possible with more vigorous extrapolation. We will show next that what is needed is to replace the constant extrapolation factor β with a variable factor β_k that converges to 1 at a properly selected rate. Unfortunately, it is very difficult to obtain strong intuition about the mechanism by which this remarkable phenomenon occurs, at least based on the analytical framework of this section (see [Nes04] for a line of development that provides intuition from a different point of view).

6.2.1 An Algorithm with Optimal Iteration Complexity

We will consider a constrained version of the gradient/extrapolation method (6.29) with a variable value of β for the problem

$$\begin{aligned} &\text{minimize} \quad f(x) \\ &\text{subject to} \quad x \in X, \end{aligned} \tag{6.30}$$

where $f : \Re^n \mapsto \Re$ is convex and differentiable, and X is a closed convex set. We will assume that f has Lipschitz continuous gradient [cf. Eq. (6.6)], and that X^*, the set of minima of f over X, is nonempty.

The method has the form

$$\begin{aligned} y_k &= x_k + \beta_k(x_k - x_{k-1}), \qquad \text{(extrapolation step)}, \\ x_{k+1} &= P_X\big(y_k - \alpha \nabla f(y_k)\big), \qquad \text{(gradient projection step)}, \end{aligned} \tag{6.31}$$

where $P_X(\cdot)$ denotes projection on X, $x_{-1} = x_0$, and $\beta_k \in (0,1)$; see Fig. 6.2.1. The method has a similar flavor with the heavy ball and PARTAN methods discussed in Section 2.1.1, but with some important differences: it applies to constrained problems, and it also reverses the order of extrapolation and gradient projection within an iteration.

The following proposition shows that with proper choice of β_k, the method has iteration complexity $O(1/\sqrt{\epsilon})$ or equivalently error $O(1/k^2)$. In particular, we use

$$\beta_k = \frac{\theta_k(1 - \theta_{k-1})}{\theta_{k-1}}, \qquad k = 0, 1, \ldots, \tag{6.32}$$

where the sequence $\{\theta_k\}$ satisfies $\theta_0 = \theta_1 \in (0,1]$, and

$$\frac{1 - \theta_{k+1}}{\theta_{k+1}^2} \leq \frac{1}{\theta_k^2}, \qquad k = 0, 1, \ldots. \tag{6.33}$$

As an example, it can be verified that one possible choice is

$$\beta_k = \begin{cases} 0 & \text{if } k = 0, \\ \frac{k-1}{k+2} & \text{if } k = 1, 2, \ldots, \end{cases} \qquad \theta_k = \begin{cases} 1 & \text{if } k = -1, \\ \frac{2}{k+2} & \text{if } k = 0, 1, \ldots. \end{cases}$$

We will assume a stepsize $\alpha = 1/L$, but the result can be extended to stepsize reduction rules along the lines of Prop. 6.1.7. One possibility is the eventually constant stepsize rule of the preceding section, whereby we start with some stepsize $\alpha > 0$ and we keep using α, and generate iterates according to

$$x_{k+1} = P_X\big(x_k - \alpha \nabla f(x_k)\big),$$

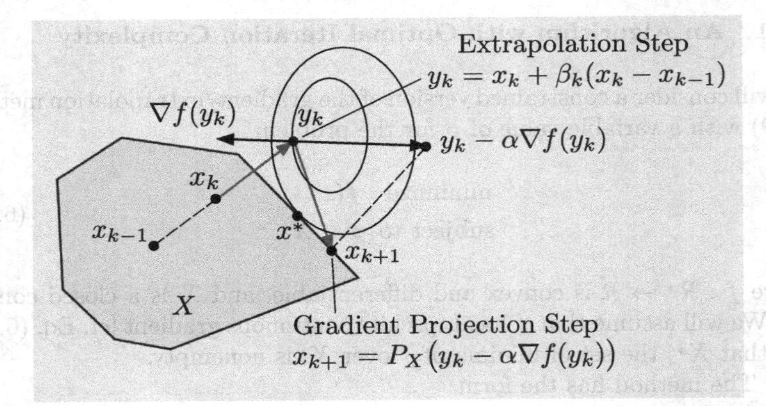

Figure 6.2.1. Illustration of the two-step method (6.31) with extrapolation.

as long as the condition

$$f(x_{k+1}) \leq \ell(x_{k+1}; y_k) + \frac{1}{2\alpha}\|x_{k+1} - y_k\|^2, \tag{6.34}$$

is satisfied. As soon as this condition is violated at some iteration, we reduce α by a certain factor, and repeat the iteration as many times as is necessary for Eq. (6.34) to hold. Similar to the gradient projection case, the condition (6.34) will be satisfied if $\alpha \leq 1/L$, and after a finite number of reductions, the test (6.34) will be passed at every subsequent iteration and α will stay constant at some value $\overline{\alpha} > 0$. With this stepsize rule we can handle the case where the constant L is not known, and the following proof can then be modified to show that the variant has error $O(1/k^2)$.

Proposition 6.2.1: Let $f : \Re^n \mapsto \Re$ be a convex differentiable function, and let X be a closed convex set. Assume that ∇f satisfies the Lipschitz condition (6.6), and that the set of minima X^* of f over X is nonempty. Let $\{x_k\}$ be a sequence generated by the algorithm (6.31), where $\alpha = 1/L$ and β_k satisfies Eqs. (6.32)-(6.33). Then $\lim_{k\to\infty} d(x_k) = 0$, and

$$f(x_k) - f^* \leq \frac{2L}{(k+1)^2}\big(d(x_0)\big)^2, \qquad k = 1, 2, \ldots,$$

where we denote

$$d(x) = \min_{x^*\in X^*} \|x - x^*\|, \qquad x \in \Re^n.$$

Proof: We introduce the sequence

$$z_k = x_{k-1} + \theta_{k-1}^{-1}(x_k - x_{k-1}), \qquad k = 0, 1, \ldots, \tag{6.35}$$

where $x_{-1} = x_0$, so that $z_0 = x_0$. We note that by using Eqs. (6.31), (6.32), z_k can also be rewritten as

$$z_k = x_k + \theta_k^{-1}(y_k - x_k), \qquad k = 1, 2, \ldots. \tag{6.36}$$

Fix $k \geq 0$ and $x^* \in X^*$, and let

$$y^* = (1 - \theta_k)x_k + \theta_k x^*.$$

Using Eq. (6.8), we have

$$f(x_{k+1}) \leq \ell(x_{k+1}; y_k) + \frac{L}{2}\|x_{k+1} - y_k\|^2, \tag{6.37}$$

where we use the notation

$$\ell(u; w) = f(w) + \nabla f(w)'(u - w), \qquad \forall\, u, w \in \Re^n.$$

Since x_{k+1} is the projection of $y_k - (1/L)\nabla f(y_k)$ on X, it minimizes

$$\ell(y; y_k) + \frac{L}{2}\|y - y_k\|^2$$

over $y \in X$, so using Prop. 6.1.5, we have

$$\ell(x_{k+1}; y_k) + \frac{L}{2}\|x_{k+1} - y_k\|^2 \leq \ell(y^*; y_k) + \frac{L}{2}\|y^* - y_k\|^2 - \frac{L}{2}\|y^* - x_{k+1}\|^2.$$

Combining this relation with Eq. (6.37), we obtain

$$\begin{aligned}
f(x_{k+1}) &\leq \ell(y^*; y_k) + \frac{L}{2}\|y^* - y_k\|^2 - \frac{L}{2}\|y^* - x_{k+1}\|^2 \\
&= \ell\big((1 - \theta_k)x_k + \theta_k x^*; y_k\big) + \frac{L}{2}\|(1 - \theta_k)x_k + \theta_k x^* - y_k\|^2 \\
&\quad - \frac{L}{2}\|(1 - \theta_k)x_k + \theta_k x^* - x_{k+1}\|^2 \\
&= \ell\big((1 - \theta_k)x_k + \theta_k x^*; y_k\big) + \frac{\theta_k^2 L}{2}\|x^* + \theta_k^{-1}(x_k - y_k) - x_k\|^2 \\
&\quad - \frac{\theta_k^2 L}{2}\|x^* + \theta_k^{-1}(x_k - x_{k+1}) - x_k\|^2 \\
&= \ell\big((1 - \theta_k)x_k + \theta_k x^*; y_k\big) + \frac{\theta_k^2 L}{2}\|x^* - z_k\|^2 \\
&\quad - \frac{\theta_k^2 L}{2}\|x^* - z_{k+1}\|^2 \\
&\leq (1 - \theta_k)\ell(x_k; y_k) + \theta_k \ell(x^*; y_k) + \frac{\theta_k^2 L}{2}\|x^* - z_k\|^2 \\
&\quad - \frac{\theta_k^2 L}{2}\|x^* - z_{k+1}\|^2,
\end{aligned}$$

where the last equality follows from Eqs. (6.35) and (6.36), and the last inequality follows from the convexity of $\ell(\cdot; y_k)$. Using the inequality

$$\ell(x_k; y_k) \le f(x_k),$$

we have

$$f(x_{k+1}) \le (1-\theta_k)f(x_k) + \theta_k \ell(x^*; y_k) + \frac{\theta_k^2 L}{2}\|x^* - z_k\|^2 - \frac{\theta_k^2 L}{2}\|x^* - z_{k+1}\|^2.$$

Finally, by rearranging terms, we obtain

$$\begin{aligned} &\frac{1}{\theta_k^2}\big(f(x_{k+1}) - f^*\big) + \frac{L}{2}\|x^* - z_{k+1}\|^2 \\ &\qquad \le \frac{1-\theta_k}{\theta_k^2}\big(f(x_k) - f^*\big) + \frac{L}{2}\|x^* - z_k\|^2 - \frac{f^* - \ell(x^*; y_k)}{\theta_k}. \end{aligned}$$

By adding this inequality for $k = 0, 1, \ldots,$ while using the inequality

$$\frac{1-\theta_{k+1}}{\theta_{k+1}^2} \le \frac{1}{\theta_k^2},$$

we obtain

$$\frac{1}{\theta_k^2}\big(f(x_{k+1}) - f^*\big) + \sum_{i=0}^{k} \frac{f^* - \ell(x^*; y_i)}{\theta_i} \le \frac{L}{2}\|x^* - z_0\|^2.$$

Using the facts $x_0 = z_0$, $f^* - \ell(x^*; y_i) \ge 0$, and $\theta_k \le 2/(k+2)$, and taking the minimum over all $x^* \in X^*$, we obtain

$$f(x_{k+1}) - f^* \le \frac{2L}{(k+2)^2}\big(d(x_0)\big)^2,$$

from which the desired result follows. **Q.E.D.**

6.2.2 Nondifferentiable Cost – Smoothing

The preceding analysis applies to differentiable cost functions. However, it can be extended to cases where f is real-valued and convex but nondifferentiable by using a smoothing technique to convert the nondifferentiable problem to a differentiable one.† In this way an iteration complexity of

---

† As noted in Section 2.2.5, smoothing is a general and often very effective technique to deal with nondifferentiabilities. It can be based on differentiable penalty methods and augmented Lagrangian methods (see the papers [Ber75b], [Ber77], and the textbook account of [Ber82a], Chapters 3 and 5). In this section, however, smoothing is used (in combination with the gradient projection method with extrapolation of Section 6.2.1) as an aid for the complexity analysis, and it is not necessarily recommended as an effective practical algorithm.

$O(1/\epsilon)$ can be attained, which is much faster than the $O(1/\epsilon^2)$ complexity of the subgradient method. The idea is to replace a nondifferentiable convex cost function by a smooth ϵ-approximation whose gradient is Lipschitz continuous with constant $L = O(1/\epsilon)$. By applying the optimal method given earlier, we obtain an ϵ-optimal solution with iteration complexity $O(1/\epsilon)$ or equivalently error $O(1/k)$.

We will consider the smoothing technique for the special class of convex functions $f_0 : \Re^n \mapsto \Re$ of the form

$$f_0(x) = \max_{u \in U} \{u'Ax - \phi(u)\}, \tag{6.38}$$

where U is a convex and compact subset of $\Re^m$, $\phi : U \mapsto \Re$ is convex and continuous over U, and A is an $m \times n$ matrix. Note that f_0 is just the composition of the matrix A and the conjugate function of

$$\tilde{\phi}(u) = \begin{cases} \phi(u) & \text{if } u \in U, \\ \infty & \text{if } u \notin U, \end{cases}$$

so the class of convex functions f_0 of the form (6.38) is quite broad. We introduce a function $p : \Re^m \mapsto \Re$ that is strictly convex and differentiable. Let u_0 be the unique minimum of p over U, i.e.,

$$u_0 \in \arg\min_{u \in U} p(u).$$

We assume that $p(u_0) = 0$ and that p is strongly convex over U with modulus of strong convexity σ, i.e., that

$$p(u) \geq \frac{\sigma}{2} \|u - u_0\|^2.$$

An example is the quadratic function $p(u) = \frac{\sigma}{2}\|u - u_0\|^2$, but there are also other functions of interest (see the paper by [Nes05] for some other examples, which also allow p to be nondifferentiable and to be defined only on U).

For a parameter $\epsilon > 0$, consider the function

$$f_\epsilon(x) = \max_{u \in U} \{u'Ax - \phi(u) - \epsilon p(u)\}, \qquad x \in \Re^n, \tag{6.39}$$

and note that f_ϵ is a uniform approximation of f_0 in the sense that

$$f_\epsilon(x) \leq f_0(x) \leq f_\epsilon(x) + p^*\epsilon, \qquad \forall\, x \in \Re^n, \tag{6.40}$$

where

$$p^* = \max_{u \in U} p(u).$$

The following proposition shows that f_ϵ is also smooth and its gradient is Lipschitz continuous with Lipschitz constant that is proportional to $1/\epsilon$.

Proposition 6.2.2: For all $\epsilon > 0$, the function f_ϵ of Eq. (6.39) is convex and differentiable over $\Re^n$, with gradient given by

$$\nabla f_\epsilon(x) = A' u_\epsilon(x),$$

where $u_\epsilon(x)$ is the unique vector attaining the maximum in Eq. (6.39). Furthermore, we have

$$\|\nabla f_\epsilon(x) - \nabla f_\epsilon(y)\| \le \frac{\|A\|^2}{\epsilon\sigma}\|x - y\|, \quad \forall\ x, y \in \Re^n.$$

Proof: We first note that the maximum in Eq. (6.39) is uniquely attained in view of the strong convexity of p (which implies that p is strictly convex). Furthermore, f_ϵ is equal to $f^\star(A'x)$, where $f^\star$ is the conjugate of the function

$$\phi(u) + \epsilon p(u) + \delta_U(u),$$

with δ_U being the indicator function of U. It follows that f_ϵ is convex, and it is also differentiable with gradient

$$\nabla f_\epsilon(x) = A' u_\epsilon(x)$$

by the Conjugate Subgradient Theorem (Prop. 5.4.3 in Appendix B).

Consider any vectors $x, y \in \Re^n$, and let g_x and g_y be subgradients of ϕ at $u_\epsilon(x)$ and $u_\epsilon(y)$, respectively. From the subgradient inequality, we have

$$\phi\big(u_\epsilon(y)\big) - \phi\big(u_\epsilon(x)\big) \ge g_x'\big(u_\epsilon(y) - u_\epsilon(x)\big),$$
$$\phi\big(u_\epsilon(x)\big) - \phi\big(u_\epsilon(y)\big) \ge g_y'\big(u_\epsilon(x) - u_\epsilon(y)\big),$$

so by adding these two inequalities, we obtain

$$(g_x - g_y)'\big(u_\epsilon(x) - u_\epsilon(y)\big) \ge 0. \tag{6.41}$$

By using the optimality condition for the maximization (6.39), we have

$$\Big(Ax - g_x - \epsilon\nabla p\big(u_\epsilon(x)\big)\Big)'\big(u_\epsilon(y) - u_\epsilon(x)\big) \le 0,$$

$$\Big(Ay - g_y - \epsilon\nabla p\big(u_\epsilon(y)\big)\Big)'\big(u_\epsilon(x) - u_\epsilon(y)\big) \le 0.$$

Adding the two inequalities, and using the convexity of ϕ and the strong convexity of p, we obtain

$$
\begin{aligned}
(x-y)'A'\big(u_\epsilon(x)-u_\epsilon(y)\big) &\geq \Big(g_x - g_y + \epsilon\Big(\nabla p\big(u_\epsilon(x)\big) - \nabla p\big(u_\epsilon(y)\big)\Big)'\\
&\qquad \big(u_\epsilon(x)-u_\epsilon(y)\big)\\
&\geq \epsilon\Big(\nabla p\big(u_\epsilon(x)\big) - \nabla p\big(u_\epsilon(y)\big)\Big)'\big(u_\epsilon(x)-u_\epsilon(y)\big)\\
&\geq \epsilon\sigma\big\|u_\epsilon(x)-u_\epsilon(y)\big\|^2,
\end{aligned}
$$

where for the second inequality we used Eq. (6.41), and for the third inequality we used a standard property of strongly convex functions. Thus,

$$
\begin{aligned}
\big\|\nabla f_\epsilon(x) - \nabla f_\epsilon(y)\big\|^2 &= \big\|A'\big(u_\epsilon(x)-u_\epsilon(y)\big)\big\|^2\\
&\leq \|A'\|^2\big\|u_\epsilon(x)-u_\epsilon(y)\big\|^2\\
&\leq \frac{\|A'\|^2}{\epsilon\sigma}(x-y)'A'\big(u_\epsilon(x)-u_\epsilon(y)\big)\\
&\leq \frac{\|A'\|^2}{\epsilon\sigma}\|x-y\|\,\big\|A'\big(u_\epsilon(x)-u_\epsilon(y)\big)\big\|\\
&= \frac{\|A\|^2}{\epsilon\sigma}\|x-y\|\,\big\|\nabla f_\epsilon(x)-\nabla f_\epsilon(y)\big\|,
\end{aligned}
$$

from which the result follows. **Q.E.D.**

We now consider the minimization over a closed convex set X of the function

$$f(x) = F(x) + f_0(x),$$

where f_0 is given by Eq. (6.38), and $F : \Re^n \mapsto \Re$ is convex and differentiable, with gradient satisfying the Lipschitz condition

$$\big\|\nabla F(x) - \nabla F(y)\big\| \leq L\,\|x-y\|, \qquad \forall\, x,y \in X. \tag{6.42}$$

We replace f with the smooth approximation

$$\tilde{f}(x) = F(x) + f_\epsilon(x),$$

and note that $\tilde{f}$ uniformly differs from f by at most $p^*\epsilon$ [cf. Eq. (6.40)], and has Lipschitz continuous gradient with Lipschitz constant $L + L_\epsilon = O(1/\epsilon)$. Thus, by applying the algorithm (6.31) and by using Prop. 6.2.1, we see that we can obtain a solution $\tilde{x} \in X$ such that $f(\tilde{x}) \leq f^* + p^*\epsilon$ with

$$O\Big(\sqrt{(L + \|A\|^2/\epsilon\sigma)/\epsilon}\Big) = O(1/\epsilon)$$

iterations.

6.3 PROXIMAL GRADIENT METHODS

In this section we consider the problem

$$\begin{array}{ll} \text{minimize} & f(x) + h(x) \\ \text{subject to } x \in \Re^n, \end{array} \tag{6.43}$$

where $f : \Re^n \mapsto \Re$ is a differentiable convex function, and $h : \Re^n \mapsto (-\infty, \infty]$ is a closed proper convex function. We will for the most part assume that f has Lipschitz continuous gradient, i.e., for some $L > 0$,

$$\big\|\nabla f(x) - \nabla f(y)\big\| \leq L\,\|x - y\|, \qquad \forall\ x, y \in \Re^n. \tag{6.44}$$

We will discuss an algorithm called *proximal gradient*, which combines ideas from the gradient projection method and the proximal algorithm. It replaces f with its linear approximation in the proximal minimization, i.e.,

$$x_{k+1} \in \arg\min_{x\in\Re^n} \left\{ \ell(x; x_k) + h(x) + \frac{1}{2\alpha_k}\|x - x_k\|^2 \right\}, \tag{6.45}$$

where $\alpha_k > 0$ is a scalar parameter, and as in Sections 6.1 and 6.2, we denote by

$$\ell(y; x) = f(x) + \nabla f(x)'(y - x), \qquad x,\, y \in \Re^n,$$

the linear approximation of f at x. Thus if f is a linear function, we obtain the proximal algorithm for minimizing $f + h$. If h is the indicator function of a closed convex set, then by Prop. 6.1.4 we obtain the gradient projection method. Because of the quadratic term in the proximal minimization (6.45), the minimum is attained uniquely and the algorithm is well defined.

The method exploits the structure of problems where $f + h$ is not suitable for proximal minimization, while h alone (plus a linear function) is. The resulting benefit is that we can treat as much as possible of the cost function with proximal minimization, which is more general (admits nondifferentiable and/or extended real-valued cost) as well as more "stable" than the gradient method (it involves essentially no restriction on the stepsize α). A typical example is the case where h is a regularization function such as the ℓ_1 norm, for which the proximal iteration can be done in essentially closed form by means of the shrinkage operation (cf. the discussion in Section 5.4.1).

One way to view the iteration (6.45) is to write it as a two-step process:

$$z_k = x_k - \alpha_k \nabla f(x_k), \qquad x_{k+1} \in \arg\min_{x\in\Re^n} \left\{ h(x) + \frac{1}{2\alpha_k}\|x - z_k\|^2 \right\}; \tag{6.46}$$

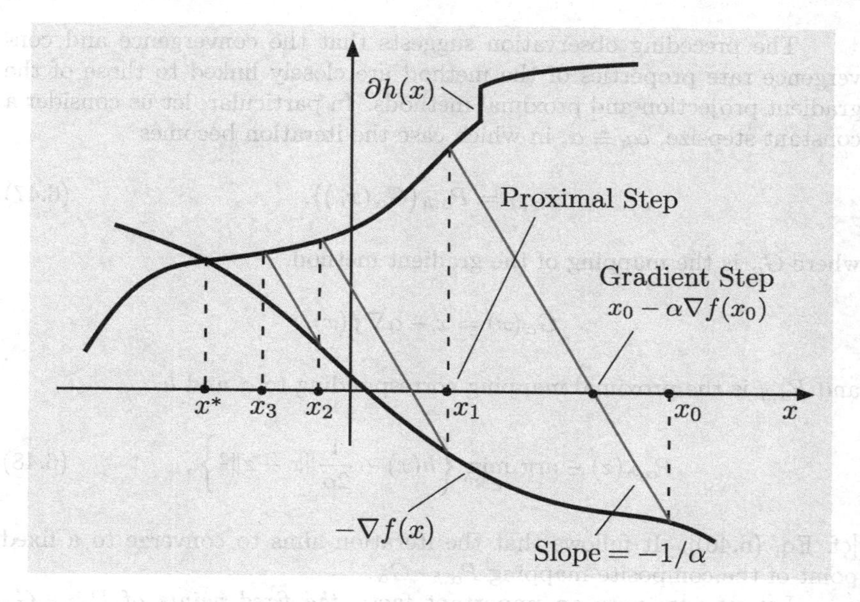

Figure 6.3.1. Illustration of the implementation of the proximal gradient method with the two-step process (6.46). Starting with x_0, we compute

$$x_0 - \alpha \nabla f(x_0)$$

with the gradient step as shown. Starting from that vector, we compute x_1 with a proximal step as shown (cf. the geometric interpretation of the proximal iteration of Fig. 5.1.8 in Section 5.1.4). Assuming that ∇f is Lipschitz continuous (so ∇f has bounded slope along directions) and α is sufficiently small, the method makes progress towards the optimal solution x^*. When $f(x) \equiv 0$ we obtain the proximal algorithm, and when h is the indicator function of a closed convex set, we obtain the gradient projection method.

this can be verified by expanding the quadratic

$$\|x - z_k\|^2 = \big\|x - x_k + \alpha_k \nabla f(x_k)\big\|^2.$$

Thus *the method alternates gradient steps on f with proximal steps on h.* Figure 6.3.1 illustrates the two-step process.†

---

† Iterative methods based on algorithmic mappings that are formed by composition of two or more simpler mappings, each possibly involving only part of the problem data, are often called *splitting algorithms*. They find wide application in optimization and the solution of linear and nonlinear equations, as they are capable of exploiting the special structures of many types of practical problems. Thus the proximal gradient algorithm, as well as other methods that we have discussed such as the ADMM, may be viewed as examples of splitting algorithms.

The preceding observation suggests that the convergence and convergence rate properties of the method are closely linked to those of the gradient projection and proximal methods. In particular, let us consider a constant stepsize, $\alpha_k \equiv \alpha$, in which case the iteration becomes

$$x_{k+1} = P_{\alpha,h}\big(G_\alpha(x_k)\big), \tag{6.47}$$

where G_α is the mapping of the gradient method,

$$G_\alpha(x) = x - \alpha\nabla f(x),$$

and $P_{\alpha,h}$ is the proximal mapping corresponding to α and h,

$$P_{\alpha,h}(z) \in \arg\min_{x\in\Re^n}\left\{h(x) + \frac{1}{2\alpha}\|x-z\|^2\right\}, \tag{6.48}$$

[cf. Eq. (6.46)]. It follows that the iteration aims to converge to a fixed point of the composite mapping $P_{\alpha,h}\cdot G_\alpha$.

Let us now note an important fact: *the fixed points of $P_{\alpha,h}\cdot G_\alpha$ coincide with the minima of $f+h$.* This is guaranteed by the fact that *the same parameter α is used in the gradient and the proximal mappings in the composite iteration (6.47).* To see this, note that x^* is a fixed point of $P_{\alpha,h}\cdot G_\alpha$ if and only if

$$\begin{aligned} x^* &\in \arg\min_{x\in\Re^n}\left\{h(x) + \frac{1}{2\alpha}\big\|x - G_\alpha(x^*)\big\|^2\right\} \\ &= \arg\min_{x\in\Re^n}\left\{h(x) + \frac{1}{2\alpha}\big\|x - \big(x^* - \alpha\nabla f(x^*)\big)\big\|^2\right\}, \end{aligned}$$

which is true if and only if the subdifferential at x^* of the function minimized above contains 0:

$$0 \in \partial h(x^*) + \frac{1}{2\alpha}\nabla\left(\big\|x - \big(x^* - \alpha\nabla f(x^*)\big)\big\|^2\right)\bigg|_{x=x^*} = \partial h(x^*) + \nabla f(x^*).$$

This is the necessary and sufficient condition for x^* to minimize $f+h$.†

Turning to the convergence properties of the algorithm, we know from Prop. 5.1.8 that $P_{\alpha,h}$ is a nonexpansive mapping, so if the mapping G_α corresponds to a convergent algorithm, it is plausible that the proximal

† Among others, this argument shows that it is not correct to use different parameters α in the gradient and the proximal portions of the proximal gradient method. This also highlights the restrictions that must be observed when replacing G_α and $P_{\alpha,h}$ with other mappings (involving for example diagonal or Newton scaling, or extrapolation), with the aim to accelerate convergence.

gradient algorithm is convergent, as it involves the composition of a nonexpansive mapping with a convergent algorithm. This is true in particular, if G_α is a Euclidean contraction (as it is under a strong convexity assumption on f, cf. Prop. 6.1.8), in which case it follows that $P_{\alpha,h} \cdot G_\alpha$ is also a Euclidean contraction. Then the algorithm (6.47) converges to the unique minimum of $f+h$ at a linear rate (determined by the product of the modulii of contraction of G_α and $P_{\alpha,h}$). Still, however, just like the unscaled gradient and gradient projection methods, the proximal gradient method can be very slow for some problems, even in the presence of strong convexity. Indeed this can be easily shown with simple examples.

Convergence Analysis

The analysis of the proximal gradient method combines elements of the analyses of the proximal and the gradient projection algorithms. In particular, using Prop. 5.1.2, we have the following three-term inequality.

Proposition 6.3.1: Let $f : \Re^n \mapsto \Re$ be a differentiable convex function, and let $h : \Re^n \mapsto (-\infty, \infty]$ be a closed proper convex function. For a given iterate x_k of the proximal gradient method (6.45), consider the arc of points

$$x_k(\alpha) \in \arg\min_{x \in \Re^n} \left\{ \ell(x; x_k) + h(x) + \frac{1}{2\alpha}\|x - x_k\|^2 \right\}, \quad \alpha > 0.$$

Then for all $y \in \Re^n$ and $\alpha > 0$, we have

$$\begin{aligned} \big\|x_k(\alpha) - y\big\|^2 \le \|x_k - y\|^2 & \\ - 2\alpha\Big(\ell\big(x_k(\alpha); x_k\big) + h\big(x_k(\alpha)\big) - \ell(y; x_k) - h(y)\Big)& \\ - \big\|x_k - x_k(\alpha)\big\|^2. & \end{aligned} \tag{6.49}$$

The next proposition shows that cost function descent can be guaranteed by a certain inequality test, which is automatically satisfied for all stepsizes in the range $(0, 1/L]$.

Proposition 6.3.2: Let $f : \Re^n \mapsto \Re$ be a differentiable convex function, let $h : \Re^n \mapsto (-\infty, \infty]$ be a closed proper convex function, and assume that ∇f satisfies the Lipschitz condition (6.44). For a given iterate x_k of the proximal gradient method (6.45), consider the arc of points

$$x_k(\alpha) \in \arg\min_{x\in\Re^n}\left\{\ell(x;x_k)+h(x)+\frac{1}{2\alpha}\|x-x_k\|^2\right\}, \qquad \alpha>0.$$

Then for all $\alpha > 0$, the inequality

$$f\big(x_k(\alpha)\big) \le \ell\big(x_k(\alpha);x_k\big)+\frac{1}{2\alpha}\big\|x_k(\alpha)-x_k\big\|^2 \tag{6.50}$$

implies the cost reduction property

$$f\big(x_k(\alpha)\big)+h\big(x_k(\alpha)\big) \le f(x_k)+h(x_k)-\frac{1}{2\alpha}\big\|x_k(\alpha)-x_k\big\|^2. \tag{6.51}$$

Moreover the inequality (6.50) is satisfied for all $\alpha \in (0, 1/L]$.

Proof: By setting $y = x_k$ in Eq. (6.49), using the fact $\ell(x_k;x_k) = f(x_k)$, and rearranging, we have

$$\ell\big(x_k(\alpha);x_k\big)+h\big(x_k(\alpha)\big) \le f(x_k)+h(x_k)-\frac{1}{\alpha}\big\|x_k(\alpha)-x_k\big\|^2.$$

If the inequality (6.50) holds, it can be added to the preceding relation to yield Eq. (6.51). Also, the descent lemma (Prop. 6.1.2) implies that the inequality (6.50) is satisfied for all $\alpha \in (0, 1/L]$. **Q.E.D.**

The proofs of the preceding two propositions go through even if f is nonconvex (but still continuously differentiable). This suggests that the proximal gradient method has good convergence properties even in the absence of convexity of f. We will now show a convergence and rate of convergence result that parallels Prop. 6.1.7 for the gradient projection method, and admits a similar proof (some other convergence rate results are given in the exercises). The following result applies to a constant stepsize rule with $\alpha_k \equiv \alpha \le 1/L$ and an *eventually constant stepsize rule*, similar to the one of Section 6.1. With this rule we start with a stepsize $\alpha_0 > 0$ that is a guess for $1/L$, and keep this stepsize unchanged while generating iterates according to

$$x_{k+1} \in \arg\min_{x\in\Re^n}\left\{\ell(x;x_k)+h(x)+\frac{1}{2\alpha_k}\|x-x_k\|^2\right\},$$

as long as the condition

$$f(x_{k+1}) \le \ell(x_{k+1};x_k)+\frac{1}{2\alpha_k}\|x_{k+1}-x_k\|^2 \tag{6.52}$$

is satisfied. As soon as this condition is violated at some iteration k, we reduce α_k by a certain factor, and repeat the iteration as many times as is necessary for Eq. (6.52) to hold. According to Prop. 6.3.2, this rule guarantees cost function descent, and guarantees that α_k will stay constant after finitely many iterations.

Proposition 6.3.3: Let $f : \Re^n \mapsto \Re$ be a differentiable convex function, and let $h : \Re^n \mapsto (-\infty, \infty]$ be a closed proper convex function. Assume that ∇f satisfies the Lipschitz condition (6.44), and that the set of minima X^* of $f + h$ is nonempty. Let $\{x_k\}$ be a sequence generated by the proximal gradient method (6.45) using the eventually constant stepsize rule or any stepsize rule such that $\alpha_k \downarrow \overline{\alpha}$, for some $\overline{\alpha} > 0$, and for all k,

$$f(x_{k+1}) \leq \ell(x_{k+1}; x_k) + \frac{1}{2\alpha_k}\|x_{k+1} - x_k\|^2. \tag{6.53}$$

Then $\{x_k\}$ converges to some point of X^*, and for all $k = 0, 1, \ldots$, we have

$$f(x_{k+1}) + h(x_{k+1}) - \min_{x \in \Re^n} \{f(x) + h(x)\} \leq \frac{\min_{x^* \in X^*} \|x_0 - x^*\|^2}{2(k+1)\overline{\alpha}}. \tag{6.54}$$

Proof: By using Prop. 6.3.1, we have for all $x \in \Re^n$

$$\begin{aligned}\ell(x_{k+1}; x_k) &+ h(x_{k+1}) + \frac{1}{2\alpha_k}\|x_{k+1} - x_k\|^2 \\ &\leq \ell(x; x_k) + h(x) + \frac{1}{2\alpha_k}\|x - x_k\|^2 - \frac{1}{2\alpha_k}\|x - x_{k+1}\|^2.\end{aligned}$$

By setting $x = x^*$, where $x^* \in X^*$, and adding Eq. (6.53), we obtain

$$\begin{aligned}f(x_{k+1}) + h(x_{k+1}) &\leq \ell(x^*; x_k) + h(x^*) \\ &\quad + \frac{1}{2\alpha_k}\|x^* - x_k\|^2 - \frac{1}{2\alpha_k}\|x^* - x_{k+1}\|^2 \\ &\leq f(x^*) + h(x^*) \\ &\quad + \frac{1}{2\alpha_k}\|x^* - x_k\|^2 - \frac{1}{2\alpha_k}\|x^* - x_{k+1}\|^2,\end{aligned} \tag{6.55}$$

where for the last inequality we use the fact

$$f(x^*) - \ell(x^*; x_k) = f(x^*) - f(x_k) - \nabla f(x_k)'(x^* - x_k) \geq 0.$$

Thus, denoting

$$e_k = f(x_k) + h(x_k) - \min_{x \in \Re^n} \{f(x) + h(x)\},$$

we have from Eq. (6.55), for all k and $x^* \in X^*$,

$$\|x^* - x_{k+1}\|^2 \leq \|x^* - x_k\|^2 - 2\alpha_k e_{k+1}. \tag{6.56}$$

Using repeatedly this relation with k replaced by $k-1,\ldots,0$, we obtain

$$0 \leq \|x^* - x_{k+1}\|^2 \leq \|x^* - x_0\|^2 - 2(\alpha_0 e_1 + \cdots + \alpha_k e_{k+1}),$$

so since $\alpha_0 \geq \alpha_1 \geq \cdots \geq \alpha_k \geq \overline{\alpha}$ and $e_1 \geq e_2 \geq \cdots \geq e_{k+1}$ (cf. Prop. 6.3.2), we have

$$2\overline{\alpha}(k+1)e_{k+1} \leq \|x^* - x_0\|^2.$$

Taking the minimum over $x^* \in X^*$, we obtain Eq. (6.54). Finally, by Eq. (6.56), $\{x_k\}$ is bounded, and each of its limit points must belong to X^* since $e_k \to 0$. Moreover by Eq. (6.56), the distance of x_k to each $x^* \in X^*$ is monotonically nonincreasing, so $\{x_k\}$ cannot have multiple limit points, and it must converge to some point in X^*. **Q.E.D.**

Dual Proximal Gradient Algorithm

The proximal gradient algorithm may also be applied to the Fenchel dual problem

$$\begin{aligned} &\text{minimize} \quad f_1^\star(-A'\lambda) + f_2^\star(\lambda) \\ &\text{subject to } \lambda \in \Re^m, \end{aligned} \tag{6.57}$$

where f_1 and f_2 are closed proper convex functions, $f_1^\star$ and $f_2^\star$ are their conjugates

$$f_1^\star(-A'\lambda) = \sup_{x\in\Re^n} \big\{(-\lambda)'Ax - f_1(x)\big\}, \tag{6.58}$$

$$f_2^\star(\lambda) = \sup_{x\in\Re^n} \big\{\lambda'x - f_2(x)\big\}, \tag{6.59}$$

and A is an $m\times n$ matrix. Note that we have reversed the sign of λ relative to the formulation of Sections 1.2 and 5.4 [the problem has not changed, it is still dual to minimizing $f_1(x)+f_2(Ax)$, but with this reversal of sign, we will obtain more convenient formulas]. The proximal gradient method for the dual problem consists of first applying a gradient step using the function $f_1^\star(-A'\lambda)$ and then applying a proximal step using the function $f_2^\star(\lambda)$ [cf. Eq. (6.46)]. We refer to this as the *dual proximal gradient algorithm*, and we will show that it admits a primal implementation that resembles the ADMM of Section 5.4.

To apply the algorithm it is of course necessary to assume that the function $f_1^\star(-A'\lambda)$ is differentiable. Using Prop. 5.4.4(a) of Appendix B, it can be seen that this is equivalent to requiring that the supremum in Eq. (6.58) is uniquely attained for all $\lambda \in \Re^m$. Moreover, using the chain rule, the gradient of $f_1^\star(-A'\lambda)$, evaluated at any $\lambda \in \Re^m$, is

$$-A\left(\arg\min_{x\in\Re^n} \big\{f_1(x) + \lambda'Ax\big\}\right).$$

Thus, the gradient step of the proximal gradient algorithm is given by

$$\overline{\lambda}_k = \lambda_k + \alpha_k A x_{k+1}, \tag{6.60}$$

where α_k is the stepsize, and

$$x_{k+1} = \arg\min_{x \in \Re^n} \big\{ f_1(x) + \lambda_k' A x \big\}. \tag{6.61}$$

The proximal step of the algorithm has the form

$$\lambda_{k+1} \in \arg\min_{\lambda \in \Re^m} \left\{ f_2^\star(\lambda) + \frac{1}{2\alpha_k} \|\lambda - \overline{\lambda}_k\|^2 \right\}.$$

According to the theory of the dual proximal algorithm (cf. Section 5.2), the proximal step can be dually implemented using an augmented Lagrangian-type minimization: first find

$$z_{k+1} \in \arg\min_{z \in \Re^m} \left\{ f_2(z) - \overline{\lambda}_k' z + \frac{\alpha_k}{2} \|z\|^2 \right\}, \tag{6.62}$$

and then obtain λ_{k+1} using the iteration

$$\lambda_{k+1} = \overline{\lambda}_k - \alpha_k z_{k+1}. \tag{6.63}$$

The dual proximal gradient algorithm (6.60)-(6.63) is a valid implementation of the proximal gradient algorithm, applied to the Fenchel dual problem (6.57). Its convergence is guaranteed by Prop. 6.3.3, provided the gradient of $f_1^\star(-A'\lambda)$ is Lipschitz continuous, and α_k is a sufficiently small constant or is chosen by the eventually constant stepsize rule.

It is interesting to note that this algorithm bears similarity to the ADMM for minimizing $f_1(x) + f_2(Ax)$ (which applies more generally, as it does not require that $f_1^\star$ is differentiable). Indeed we may rewrite the algorithm (6.60)-(6.63) by combining Eqs. (6.60) and (6.63), so that

$$\lambda_{k+1} = \lambda_k + \alpha_k (A x_{k+1} - z_{k+1}),$$

where x_{k+1} minimizes the Lagrangian,

$$x_{k+1} = \arg\min_{x \in \Re^n} \big\{ f_1(x) + \lambda_k' (Ax - z_k) \big\}, \tag{6.64}$$

while by using Eqs. (6.60) and (6.62), we can verify that z_{k+1} minimizes the augmented Lagrangian

$$z_{k+1} \in \arg\min_{z \in \Re^m} \left\{ f_2(z) + \lambda_k' (A x_{k+1} - z) + \frac{\alpha_k}{2} \|A x_{k+1} - z\|^2 \right\}.$$

Other than minimizing with respect to x the Lagrangian in Eq. (6.64), instead of the augmented Lagrangian, the only other difference of this dual proximal gradient algorithm from the ADMM is that there is a restriction on the magnitude of the stepsize [it is limited by the size of the Lipschitz constant of the gradient of the function $f_1^\star(-A'\lambda)$, as per Prop. 6.3.2]. Note that in the ADMM the penalty parameter can be chosen freely, but (contrary to the augmented Lagrangian method) it may not be clear how to choose it in order to accelerate convergence. Thus all three proximal-type methods, proximal gradient, ADMM, and augmented Lagrangian, have similarities, and relative strengths and weaknesses. The choice between them hinges largely on the given problem's structure.

Proximal Newton Methods

The proximal gradient method admits a straightforward generalization to incorporate Newton and quasi-Newton scaling, similar to the steepest descent and gradient projection methods. The scaled version takes the form

$$x_{k+1} \in \arg\min_{x\in\Re^n} \left\{\ell(x;x_k) + h(x) + \tfrac{1}{2}(x-x_k)'H_k(x-x_k)\right\}, \qquad (6.65)$$

where H_k is a positive definite symmetric matrix.

This iteration can be interpreted by a two-step process, as shown in Fig. 6.3.2 for the special case where $H_k = \nabla^2 f(x_k)$ (the Hessian is assumed to exist and be positive definite). In this case, we obtain a Newton-like iteration, sometimes referred to as the *proximal Newton method*. In the case where H_k is a multiple of the identity matrix: $H_k = \frac{1}{\alpha_k} I$, we obtain the earlier proximal gradient method (6.45).

Note that when $H_k = \nabla^2 f(x_k)$ and h is the indicator function of a closed convex set X, the algorithm (6.65) reduces to a constrained version of Newton's method, and when in addition $X = \Re^n$ it coincides with the classical form of Newton's method. It is also possible to use scaling that is simpler (such as diagonal) or that does not require the computation of second derivatives, such as limited memory quasi-Newton schemes. Generally, the machinery of unconstrained and constrained Newton-like schemes, developed for the case where $h(x) \equiv 0$ or h is the indicator function of a given set can be fruitfully brought to bear.

A variety of stepsize rules may also be incorporated into the scaling matrix H_k, to ensure convergence based for example on cost function descent, and under certain conditions, superlinear convergence. In particular, when f is a positive definite quadratic and H_k is chosen to be the Hessian of f (with unit stepsize), the method finds the optimal solution in a single iteration. We refer to the literature for analysis and further discussion.

Proximal Gradient Methods with Extrapolation

There is a proximal gradient method with extrapolation, along the lines of the corresponding optimal complexity gradient projection method of Section 6.2 [cf. Eq. (6.31)]. The method takes the form

$$y_k = x_k + \beta_k(x_k - x_{k-1}), \qquad \text{(extrapolation step)},$$

$$z_k = y_k - \alpha_k \nabla f(y_k), \qquad \text{(gradient step)},$$

$$x_{k+1} \in \arg\min_{x\in\Re^n} \left\{ h(x) + \frac{1}{2\alpha_k}\|x - z_k\|^2 \right\}, \qquad \text{(proximal step)},$$

where $x_{-1} = x_0$, and $\beta_k \in (0,1)$. The extrapolation parameter β_k is selected as in Section 6.2.

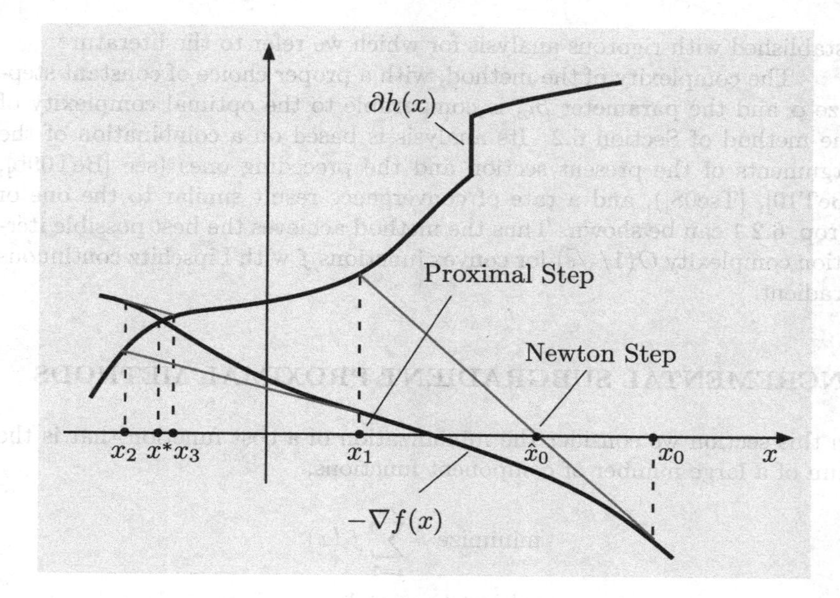

Figure 6.3.2. Geometric interpretation of the proximal Newton iteration. Given x_k, the next iterate x_{k+1} is obtained by minimizing $\hat{f}(x, x_k) + h(x)$, where

$$\hat{f}(x, x_k) = \ell(x; x_k) + \tfrac{1}{2}(x - x_k)'\nabla^2 f(x_k)(x - x_k)$$

is the quadratic approximation of f at x_k. Thus x_{k+1} satisfies

$$0 \in \partial h(x_{k+1}) + \nabla f(x_k) + \nabla^2 f(x_k)(x_{k+1} - x_k). \qquad (6.66)$$

It can be shown that x_{k+1} can be generated by a two-step process: first perform a Newton step, obtaining $\hat{x}_k = x_k - \big(\nabla^2 f(x_k)\big)^{-1}\nabla f(x_k)$, and then a proximal step

$$x_{k+1} \in \arg\min_{x \in \Re^n} \big\{ h(x) + \tfrac{1}{2}(x - \hat{x}_k)'\nabla^2 f(x_k)(x - \hat{x}_k) \big\}.$$

To see this, write the optimality condition for the above minimization and show that it coincides with Eq. (6.66). Note that when $h(x) \equiv 0$ we obtain the pure form of Newton's method. If $\nabla^2 f(x_k)$ is replaced by a positive definite symmetric matrix H_k, we obtain a proximal quasi-Newton method.

The method can be viewed as a splitting algorithm, and can be justified with similar arguments and analysis as the method without extrapolation. In particular, it can be seen that x^* minimizes $f + h$ if and only if (x^*, x^*) is a fixed point of the preceding iteration [viewed as an operation that maps (x_k, x_{k-1}) to (x_{k+1}, x_k)]. Since the algorithm consists of the composition of the nonexpansive proximal mapping with the convergent gradient method with extrapolation, it makes sense that the algorithm is convergent similar to the case of no extrapolation. Indeed, this can be established with rigorous analysis for which we refer to the literature.

The complexity of the method, with a proper choice of constant stepsize α and the parameter β_k, is comparable to the optimal complexity of the method of Section 6.2. Its analysis is based on a combination of the arguments of the present section and the preceding ones (see [BeT09b], [BeT10], [Tse08]), and a rate of convergence result similar to the one of Prop. 6.2.1 can be shown. Thus the method achieves the best possible iteration complexity $O(1/\sqrt{\epsilon})$ for convex functions f with Lipschitz continuous gradient.

6.4 INCREMENTAL SUBGRADIENT PROXIMAL METHODS

In this section we consider the minimization of a cost function that is the sum of a large number of component functions,

$$\text{minimize} \quad \sum_{i=1}^{m} f_i(x)$$
$$\text{subject to} \quad x \in X,$$

where $f_i : \Re^n \mapsto \Re$, $i = 1, \ldots, m$, are convex real-valued functions, and X is a closed convex set. We have considered incremental gradient and subgradient methods in Sections 2.1.5 and 3.3.1, for this problem. We now consider an extension to proximal algorithms. The simplest one has the form

$$x_{k+1} \in \arg\min_{x \in X} \left\{ f_{i_k}(x) + \frac{1}{2\alpha_k}\|x - x_k\|^2 \right\}, \tag{6.67}$$

where i_k is an index from $\{1, \ldots, m\}$, selected in ways to be discussed shortly, and $\{\alpha_k\}$ is a positive scalar sequence. This method relates to the proximal algorithm in the same way that the incremental subgradient method of Section 3.3.1 relates to its nonincremental version of Section 3.2.

The motivation here is that with a favorable structure of the components, the proximal iteration (6.67) may be given in closed form or be relatively simple, in which case it may be preferable to a gradient or subgradient iteration, since it is generally more stable. For example in the nonincremental case, the proximal iteration converges essentially for any choice of α_k, while this is not so for gradient-type methods.

Unfortunately, while some cost function components may be well suited for a proximal iteration, others may not be because the minimization (6.67) is inconvenient. This leads us to consider combinations of gradient/subgradient and proximal iterations. In fact this was the motivation for the proximal gradient and related splitting algorithms that we discussed in the preceding section.

With similar motivation in mind, we adopt in this section a unified algorithmic framework that includes incremental gradient, subgradient, and proximal methods, and their combinations, and serves to highlight their common structure and behavior. We focus on problems of the form

$$\begin{aligned} &\text{minimize} \quad F(x) \stackrel{\text{def}}{=} \sum_{i=1}^{m} F_i(x) \\ &\text{subject to} \quad x \in X, \end{aligned} \tag{6.68}$$

where for all i,

$$F_i(x) = f_i(x) + h_i(x), \tag{6.69}$$

$f_i : \Re^n \mapsto \Re$ and $h_i : \Re^n \mapsto \Re$ are real-valued convex functions, and X is a nonempty closed convex set.

One of our algorithms has the form

$$z_k \in \arg\min_{x \in X} \left\{ f_{i_k}(x) + \frac{1}{2\alpha_k} \|x - x_k\|^2 \right\}, \tag{6.70}$$

$$x_{k+1} = P_X\big(z_k - \alpha_k \tilde{\nabla} h_{i_k}(z_k)\big), \tag{6.71}$$

where $\tilde{\nabla} h_{i_k}(z_k)$ is an arbitrary subgradient of h_{i_k} at z_k.†

Note that the iteration is well-defined because the minimum in Eq. (6.70) is uniquely attained, while the subdifferential $\partial h_{i_k}(z_k)$ is nonempty since h_{i_k} is real-valued. Note also that by choosing all the f_i or all the h_i to be identically zero, we obtain as special cases the subgradient and proximal iterations, respectively.

The iterations (6.70) and (6.71) maintain both sequences $\{z_k\}$ and $\{x_k\}$ within the constraint set X, but it may be convenient to relax this constraint for either the proximal or the subgradient iteration, thereby requiring a potentially simpler computation. This leads to the algorithm

$$z_k \in \arg\min_{x \in \Re^n} \left\{ f_{i_k}(x) + \frac{1}{2\alpha_k} \|x - x_k\|^2 \right\}, \tag{6.72}$$

$$x_{k+1} = P_X\big(z_k - \alpha_k \tilde{\nabla} h_{i_k}(z_k)\big), \tag{6.73}$$

where the restriction $x \in X$ has been omitted from the proximal iteration, and the algorithm

$$z_k = x_k - \alpha_k \tilde{\nabla} h_{i_k}(x_k), \tag{6.74}$$

$$x_{k+1} \in \arg\min_{x \in X} \left\{ f_{i_k}(x) + \frac{1}{2\alpha_k} \|x - z_k\|^2 \right\}, \tag{6.75}$$

† To facilitate notation when using both differentiable and nondifferentiable functions, we will use $\tilde{\nabla} h(x)$ to denote a subgradient of a convex function h at a point x. The method of choice of subgradient from within $\partial h(x)$ will be clear from the context.

where the projection onto X has been omitted from the subgradient iteration. It is also possible to use different stepsize sequences in the proximal and subgradient iterations, but for notational simplicity we will not discuss this type of algorithm.

Part (a) of the following proposition is a key fact about incremental proximal iterations. It shows that they are closely related to incremental subgradient iterations, with the only difference being that the subgradient is evaluated at the end point of the iteration rather than at the start point. Part (b) of the proposition is the three-term inequality, which was shown in Section 5.1 (cf. Prop. 5.1.2). It will be useful in our convergence analysis and is restated here for convenience.

Proposition 6.4.1: Let $f : \Re^n \mapsto (-\infty, \infty]$ be a closed proper convex function, and let X be a nonempty closed convex set such that $\text{ri}(X) \cap \text{ri}\big(\text{dom}(f)\big) \neq \emptyset$. For any $x_k \in \Re^n$ and $\alpha_k > 0$, consider the proximal iteration

$$x_{k+1} \in \arg\min_{x \in X} \left\{ f(x) + \frac{1}{2\alpha_k} \|x - x_k\|^2 \right\}. \tag{6.76}$$

(a) The iteration can be written as

$$x_{k+1} = P_X\big(x_k - \alpha_k \tilde{\nabla} f(x_{k+1})\big), \tag{6.77}$$

where $\tilde{\nabla} f(x_{k+1})$ is some subgradient of f at x_{k+1}.

(b) For all $y \in \Re^n$, we have

$$\begin{aligned} \|x_{k+1} - y\|^2 &\leq \|x_k - y\|^2 - 2\alpha_k\big(f(x_{k+1}) - f(y)\big) - \|x_k - x_{k+1}\|^2 \\ &\leq \|x_k - y\|^2 - 2\alpha_k\big(f(x_{k+1}) - f(y)\big). \end{aligned} \tag{6.78}$$

Proof: (a) We use the formula for the subdifferential of the sum of the three functions f, $(1/2\alpha_k)\|x - x_k\|^2$, and the indicator function of X (cf. Prop. 5.4.6 in Appendix B), together with the condition that 0 should belong to this subdifferential at the optimum x_{k+1}. We obtain that Eq. (6.76) holds if and only if

$$\frac{1}{\alpha_k}(x_k - x_{k+1}) \in \partial f(x_{k+1}) + N_X(x_{k+1}), \tag{6.79}$$

where $N_X(x_{k+1})$ is the normal cone of X at x_{k+1} [the set of vectors y such that $y'(x - x_{k+1}) \leq 0$ for all $x \in X$, and also the subdifferential of the indicator function of X at x_{k+1}; cf. Section 3.1]. This is true if and only if

$$x_k - x_{k+1} - \alpha_k \tilde{\nabla} f(x_{k+1}) \in N_X(x_{k+1}),$$

for some $\tilde{\nabla} f(x_{k+1}) \in \partial f(x_{k+1})$, which in turn is true if and only if Eq. (6.77) holds, by the Projection Theorem (Prop. 1.1.9 in Appendix B).

(b) See Prop. 5.1.2. **Q.E.D.**

Based on Prop. 6.4.1(a), we see that all the iterations (6.70)-(6.71), (6.72)-(6.73), and (6.74)-(6.75) can be written in an incremental subgradient format:

(a) Iteration (6.70)-(6.71) can be written as

$$z_k = P_X\big(x_k - \alpha_k \tilde{\nabla} f_{i_k}(z_k)\big), \qquad x_{k+1} = P_X\big(z_k - \alpha_k \tilde{\nabla} h_{i_k}(z_k)\big). \tag{6.80}$$

(b) Iteration (6.72)-(6.73) can be written as

$$z_k = x_k - \alpha_k \tilde{\nabla} f_{i_k}(z_k), \qquad x_{k+1} = P_X\big(z_k - \alpha_k \tilde{\nabla} h_{i_k}(z_k)\big). \tag{6.81}$$

(c) Iteration (6.74)-(6.75) can be written as

$$z_k = x_k - \alpha_k \tilde{\nabla} h_{i_k}(x_k), \qquad x_{k+1} = P_X\big(z_k - \alpha_k \tilde{\nabla} f_{i_k}(x_{k+1})\big). \tag{6.82}$$

Note that in all the preceding updates, the subgradient $\tilde{\nabla} h_{i_k}$ can be *any* vector in the subdifferential of h_{i_k}, while the subgradient $\tilde{\nabla} f_{i_k}$ must be a *specific* vector in the subdifferential of f_{i_k}, specified according to Prop. 6.4.1(a). Note also that iteration (6.81) can be written as

$$x_{k+1} = P_X\big(x_k - \alpha_k \tilde{\nabla} F_{i_k}(z_k)\big),$$

and resembles the incremental subgradient method for minimizing over X the cost function

$$F(x) = \sum_{i=1}^{m} F_i(x)$$

[cf. Eq. (6.68)], the only difference being that the subgradient of F_{i_k} is computed at z_k rather than x_k.

An important issue which affects the methods' effectiveness is the order in which the components $\{f_i, h_i\}$ are chosen for iteration. In this section we consider and analyze the convergence for two possibilities:

(1) A *cyclic order*, whereby $\{f_i, h_i\}$ are taken up in the fixed deterministic order $1, \ldots, m$, so that i_k is equal to (k modulo m) plus 1. A contiguous block of iterations involving

$$\{f_1, h_1\}, \ldots, \{f_m, h_m\}$$

in this order and exactly once is called a *cycle*. We assume that the stepsize α_k is constant within a cycle (for all k with $i_k = 1$ we have $\alpha_k = \alpha_{k+1} \cdots = \alpha_{k+m-1}$).

(2) A *randomized order based on uniform sampling*, whereby at each iteration a component pair $\{f_i, h_i\}$ is chosen randomly by sampling over all component pairs with a uniform distribution, independently of the past history of the algorithm.

It is essential to include all components in a cycle in the cyclic case, and to sample according to the uniform distribution in the randomized case. Otherwise some components will be sampled more often than others, leading to a bias in the convergence process, and convergence to an incorrect limit.

Another popular technique for incremental methods, which we discussed in Section 2.1.5, is to reshuffle randomly the order of the component functions after each cycle. This alternative order selection scheme leads to convergence, like the preceding two, and works well in practice. However, its convergence rate seems harder to analyze. In this section we will focus on the easier-to-analyze randomized uniform sampling order, and demonstrate its superiority over the cyclic order.

There are also irregular and possibly distributed incremental schemes where multiple subgradient iterations may be performed in between two proximal iterations, and reversely. Moreover the order of component selection may be made dependent on algorithmic progress, as long as all components are sampled with the same long-term frequency.

For the remainder of this section, we denote by F^* the optimal value of problem (6.68):

$$F^* = \inf_{x \in X} F(x),$$

and by X^* the set of optimal solutions (which could be empty):

$$X^* = \{x^* \mid x^* \in X,\ F(x^*) = F^*\}.$$

Also, for a nonempty closed convex set X, we denote by $\text{dist}(\cdot; X)$ the distance function given by

$$\text{dist}(x; X) = \min_{z \in X} \|x - z\|, \qquad x \in \Re^n.$$

6.4.1 Convergence for Methods with Cyclic Order

We first discuss convergence under the cyclic order. We focus on the sequence $\{x_k\}$ rather than $\{z_k\}$, which need not lie within X in the case of iterations (6.81) and (6.82) when $X \neq \Re^n$. In summary, the idea is to show that the effect of taking subgradients of f_i or h_i at points near x_k (e.g., at z_k rather than at x_k) is inconsequential, and diminishes as the stepsize α_k becomes smaller, as long as some subgradients relevant to the algorithms are uniformly bounded in norm by some constant. This is similar to the convergence mechanism of incremental gradient methods described informally in Section 3.3.1. We use the following assumptions throughout the present section.

Assumption 6.4.1: [For iterations (6.80) and (6.81)] There is a constant $c \in \Re$ such that for all k

$$\max\left\{\|\tilde{\nabla} f_{i_k}(z_k)\|, \|\tilde{\nabla} h_{i_k}(z_k)\|\right\} \le c. \tag{6.83}$$

Furthermore, for all k that mark the beginning of a cycle (i.e., all $k > 0$ with $i_k = 1$), we have for all $j = 1, \ldots, m$,

$$\max\left\{f_j(x_k) - f_j(z_{k+j-1}),\, h_j(x_k) - h_j(z_{k+j-1})\right\} \le c\,\|x_k - z_{k+j-1}\|. \tag{6.84}$$

Assumption 6.4.2: [For iteration (6.82)] There is a constant $c \in \Re$ such that for all k

$$\max\left\{\|\tilde{\nabla} f_{i_k}(x_{k+1})\|, \|\tilde{\nabla} h_{i_k}(x_k)\|\right\} \le c. \tag{6.85}$$

Furthermore, for all k that mark the beginning of a cycle (i.e., all $k > 0$ with $i_k = 1$), we have for all $j = 1, \ldots, m$,

$$\max\left\{f_j(x_k) - f_j(x_{k+j-1}),\, h_j(x_k) - h_j(x_{k+j-1})\right\} \le c\,\|x_k - x_{k+j-1}\|, \tag{6.86}$$

$$f_j(x_{k+j-1}) - f_j(x_{k+j}) \le c\,\|x_{k+j-1} - x_{k+j}\|. \tag{6.87}$$

Note that the condition (6.84) is satisfied if for each i and k, there is a subgradient of f_i at x_k and a subgradient of h_i at x_k, whose norms are bounded by c. Conditions that imply the preceding assumptions are:

(a) For algorithm (6.80): f_i and h_i are Lipschitz continuous over the set X.

(b) For algorithms (6.81) and (6.82): f_i and h_i are Lipschitz continuous over the entire space $\Re^n$.

(c) For all algorithms (6.80), (6.81), and (6.82): f_i and h_i are polyhedral [this is a special case of (a) and (b)].

(d) The sequences $\{x_k\}$ and $\{z_k\}$ are bounded (since then f_i and h_i, being real-valued and convex, are Lipschitz continuous over any bounded set that contains $\{x_k\}$ and $\{z_k\}$).

The following proposition provides a key estimate that reveals the convergence mechanism of the methods.

Proposition 6.4.2: Let $\{x_k\}$ be the sequence generated by any one of the algorithms (6.80)-(6.82), with a cyclic order of component selection. Then for all $y \in X$ and all k that mark the beginning of a cycle (i.e., all k with $i_k = 1$), we have

$$\|x_{k+m} - y\|^2 \leq \|x_k - y\|^2 - 2\alpha_k\big(F(x_k) - F(y)\big) + \alpha_k^2 \beta m^2 c^2, \quad (6.88)$$

where $\beta = \frac{1}{m} + 4$.

Proof: We first prove the result for algorithms (6.80) and (6.81), and then indicate the modifications necessary for algorithm (6.82). Using Prop. 6.4.1(b), we have for all $y \in X$ and k,

$$\|z_k - y\|^2 \leq \|x_k - y\|^2 - 2\alpha_k\big(f_{i_k}(z_k) - f_{i_k}(y)\big). \quad (6.89)$$

Also, using the nonexpansion property of the projection,

$$\big\|P_X(u) - P_X(v)\big\| \leq \|u - v\|, \qquad \forall\ u, v \in \Re^n,$$

the definition of subgradient, and Eq. (6.83), we obtain for all $y \in X$ and k,

$$\begin{aligned} \|x_{k+1} - y\|^2 &= \big\|P_X\big(z_k - \alpha_k \tilde{\nabla} h_{i_k}(z_k)\big) - y\big\|^2 \\ &\leq \|z_k - \alpha_k \tilde{\nabla} h_{i_k}(z_k) - y\|^2 \\ &= \|z_k - y\|^2 - 2\alpha_k \tilde{\nabla} h_{i_k}(z_k)'(z_k - y) + \alpha_k^2 \big\|\tilde{\nabla} h_{i_k}(z_k)\big\|^2 \\ &\leq \|z_k - y\|^2 - 2\alpha_k\big(h_{i_k}(z_k) - h_{i_k}(y)\big) + \alpha_k^2 c^2. \end{aligned} \quad (6.90)$$

Combining Eqs. (6.89) and (6.90), and using the definition $F_j = f_j + h_j$, we have

$$\begin{aligned} \|x_{k+1} - y\|^2 &\leq \|x_k - y\|^2 - 2\alpha_k\big(f_{i_k}(z_k) + h_{i_k}(z_k) - f_{i_k}(y) - h_{i_k}(y)\big) + \alpha_k^2 c^2 \\ &= \|x_k - y\|^2 - 2\alpha_k\big(F_{i_k}(z_k) - F_{i_k}(y)\big) + \alpha_k^2 c^2. \end{aligned} \quad (6.91)$$

Let now k mark the beginning of a cycle (i.e., $i_k = 1$). Then at iteration $k + j - 1$, $j = 1, \ldots, m$, the selected components are $\{f_j, h_j\}$, in view of the assumed cyclic order. We may thus replicate the preceding inequality with k replaced by $k + 1, \ldots, k + m - 1$, and add to obtain

$$\|x_{k+m} - y\|^2 \leq \|x_k - y\|^2 - 2\alpha_k \sum_{j=1}^{m} \big(F_j(z_{k+j-1}) - F_j(y)\big) + m\alpha_k^2 c^2,$$

or equivalently, since $F = \sum_{j=1}^{m} F_j$,

$$\|x_{k+m} - y\|^2 \leq \|x_k - y\|^2 - 2\alpha_k\big(F(x_k) - F(y)\big) + m\alpha_k^2 c^2 + 2\alpha_k \sum_{j=1}^{m} \big(F_j(x_k) - F_j(z_{k+j-1})\big). \tag{6.92}$$

The remainder of the proof deals with appropriately bounding the last term above.

From Eq. (6.84), we have for $j = 1, \ldots, m$,

$$F_j(x_k) - F_j(z_{k+j-1}) \leq 2c\,\|x_k - z_{k+j-1}\|. \tag{6.93}$$

We also have

$$\|x_k - z_{k+j-1}\| \leq \|x_k - x_{k+1}\| + \cdots + \|x_{k+j-2} - x_{k+j-1}\| + \|x_{k+j-1} - z_{k+j-1}\|, \tag{6.94}$$

and by the definition of the algorithms (6.80) and (6.81), the nonexpansion property of the projection, and Eq. (6.83), each of the terms in the right-hand side above is bounded by $2\alpha_k c$, except for the last, which is bounded by $\alpha_k c$. Thus Eq. (6.94) yields $\|x_k - z_{k+j-1}\| \leq \alpha_k(2j-1)c$, which together with Eq. (6.93), shows that

$$F_j(x_k) - F_j(z_{k+j-1}) \leq 2\alpha_k c^2(2j-1). \tag{6.95}$$

Combining Eqs. (6.92) and (6.95), we have

$$\|x_{k+m} - y\|^2 \leq \|x_k - y\|^2 - 2\alpha_k\big(F(x_k) - F(y)\big) + m\alpha_k^2 c^2 + 4\alpha_k^2 c^2 \sum_{j=1}^{m}(2j-1),$$

and finally

$$\|x_{k+m} - y\|^2 \leq \|x_k - y\|^2 - 2\alpha_k\big(F(x_k) - F(y)\big) + m\alpha_k^2 c^2 + 4\alpha_k^2 c^2 m^2,$$

which is of the form (6.88) with $\beta = \frac{1}{m} + 4$.

For the algorithm (6.82), a similar argument goes through using Assumption 6.4.2. In place of Eq. (6.89), using the nonexpansion property of the projection, the definition of subgradient, and Eq. (6.85), we obtain for all $y \in X$ and $k \geq 0$,

$$\|z_k - y\|^2 \leq \|x_k - y\|^2 - 2\alpha_k\big(h_{i_k}(x_k) - h_{i_k}(y)\big) + \alpha_k^2 c^2, \tag{6.96}$$

while in place of Eq. (6.90), using Prop. 6.4.1(b), we have

$$\|x_{k+1} - y\|^2 \leq \|z_k - y\|^2 - 2\alpha_k\big(f_{i_k}(x_{k+1}) - f_{i_k}(y)\big). \tag{6.97}$$

Combining these equations, in analogy with Eq. (6.91), we obtain

$$\begin{aligned}\|x_{k+1} - y\|^2 &\le \|x_k - y\|^2 - 2\alpha_k\big(f_{i_k}(x_{k+1}) + h_{i_k}(x_k) - f_{i_k}(y) - h_{i_k}(y)\big) \\ &\quad + \alpha_k^2 c^2 \\ &= \|x_k - y\|^2 - 2\alpha_k\big(F_{i_k}(x_k) - F_{i_k}(y)\big) + \alpha_k^2 c^2 \\ &\quad + 2\alpha_k\big(f_{i_k}(x_k) - f_{i_k}(x_{k+1})\big).\end{aligned} \tag{6.98}$$

As earlier, we let k mark the beginning of a cycle (i.e., $i_k = 1$). We replicate the preceding inequality with k replaced by $k+1, \ldots, k+m-1$, and add to obtain [in analogy with Eq. (6.92)]

$$\begin{aligned}\|x_{k+m} - y\|^2 &\le \|x_k - y\|^2 - 2\alpha_k\big(F(x_k) - F(y)\big) + m\alpha_k^2 c^2 \\ &\quad + 2\alpha_k \sum_{j=1}^m \big(F_j(x_k) - F_j(x_{k+j-1})\big) \\ &\quad + 2\alpha_k \sum_{j=1}^m \big(f_j(x_{k+j-1}) - f_j(x_{k+j})\big).\end{aligned} \tag{6.99}$$

We now bound the two sums in Eq. (6.99), using Assumption 6.4.2. From Eq. (6.86), we have

$$\begin{aligned}F_j(x_k) - F_j(x_{k+j-1}) &\le 2c\|x_k - x_{k+j-1}\| \\ &\le 2c\big(\|x_k - x_{k+1}\| + \cdots + \|x_{k+j-2} - x_{k+j-1}\|\big),\end{aligned}$$

and since by Eq. (6.85) and the definition of the algorithm, each of the norm terms in the right-hand side above is bounded by $2\alpha_k c$,

$$F_j(x_k) - F_j(x_{k+j-1}) \le 4\alpha_k c^2 (j-1).$$

Also from Eqs. (6.85) and (6.87), and the nonexpansion property of the projection, we have

$$f_j(x_{k+j-1}) - f_j(x_{k+j}) \le c\,\|x_{k+j-1} - x_{k+j}\| \le 2\alpha_k c^2.$$

Combining the preceding relations and adding, we obtain

$$\begin{aligned}2\alpha_k &\sum_{j=1}^m \big(F_j(x_k) - F_j(x_{k+j-1})\big) + 2\alpha_k \sum_{j=1}^m \big(f_j(x_{k+j-1}) - f_j(x_{k+j})\big) \\ &\le 8\alpha_k^2 c^2 \sum_{j=1}^m (j-1) + 4\alpha_k^2 c^2 m \\ &= 4\alpha_k^2 c^2 (m^2 - m) + 4\alpha_k^2 c^2 m \\ &= \left(4 + \frac{1}{m}\right) \alpha_k^2 c^2 m^2,\end{aligned}$$

which together with Eq. (6.99), yields Eq. (6.88). **Q.E.D.**

Among other things, Prop. 6.4.2 guarantees that with a cyclic order, given the iterate x_k at the start of a cycle and any point $y \in X$ having lower cost than x_k (for example an optimal point), the algorithm yields a point x_{k+m} at the end of the cycle that will be closer to y than x_k, provided the stepsize α_k satisfies

$$\alpha_k < \frac{2\big(F(x_k) - F(y)\big)}{\beta m^2 c^2}.$$

In particular, for any $\epsilon > 0$ and assuming that there exists an optimal solution x^*, either we are within $\frac{\alpha_k \beta m^2 c^2}{2} + \epsilon$ of the optimal value,

$$F(x_k) \leq F(x^*) + \frac{\alpha_k \beta m^2 c^2}{2} + \epsilon,$$

or else the squared distance to x^* will be strictly decreased by at least $2\alpha_k\epsilon$,

$$\|x_{k+m} - x^*\|^2 < \|x_k - x^*\|^2 - 2\alpha_k\epsilon.$$

Thus, using Prop. 6.4.2, we can provide various types of convergence results. As an example, for a constant stepsize ($\alpha_k \equiv \alpha$), convergence can be established to a neighborhood of the optimum, which shrinks to 0 as $\alpha \to 0$, as stated in the following proposition.

Proposition 6.4.3: Let $\{x_k\}$ be the sequence generated by any one of the algorithms (6.80)-(6.82), with a cyclic order of component selection, and let the stepsize α_k be fixed at some positive constant α.

(a) If $F^* = -\infty$, then

$$\liminf_{k\to\infty} F(x_k) = F^*.$$

(b) If $F^* > -\infty$, then

$$\liminf_{k\to\infty} F(x_k) \leq F^* + \frac{\alpha \beta m^2 c^2}{2},$$

where c and β are the constants of Prop. 6.4.2.

Proof: We prove (a) and (b) simultaneously. If the result does not hold, there must exist an $\epsilon > 0$ such that

$$\liminf_{k\to\infty} F(x_{km}) - \frac{\alpha \beta m^2 c^2}{2} - 2\epsilon > F^*.$$

Let $\hat{y} \in X$ be such that

$$\liminf_{k\to\infty} F(x_{km}) - \frac{\alpha\beta m^2c^2}{2} - 2\epsilon \geq F(\hat{y}),$$

and let k_0 be large enough so that for all $k \geq k_0$, we have

$$F(x_{km}) \geq \liminf_{k\to\infty} F(x_{km}) - \epsilon.$$

By combining the preceding two relations, we obtain for all $k \geq k_0$,

$$F(x_{km}) - F(\hat{y}) \geq \frac{\alpha\beta m^2c^2}{2} + \epsilon.$$

Using Prop. 6.4.2 for the case where $y = \hat{y}$ together with the above relation, we obtain for all $k \geq k_0$,

$$\begin{aligned}\|x_{(k+1)m} - \hat{y}\|^2 &\leq \|x_{km} - \hat{y}\|^2 - 2\alpha\big(F(x_{km}) - F(\hat{y})\big) + \beta\alpha^2m^2c^2\\ &\leq \|x_{km} - \hat{y}\|^2 - 2\alpha\epsilon.\end{aligned}$$

This relation implies that for all $k \geq k_0$,

$$\|x_{(k+1)m} - \hat{y}\|^2 \leq \|x_{(k-1)m} - \hat{y}\|^2 - 4\alpha\epsilon \leq \cdots \leq \|x_{k_0} - \hat{y}\|^2 - 2(k+1-k_0)\alpha\epsilon,$$

which cannot hold for k sufficiently large – a contradiction. **Q.E.D.**

The next proposition gives an estimate of the number of iterations needed to guarantee a given level of optimality up to the threshold tolerance $\alpha\beta m^2c^2/2$ of the preceding proposition.

Proposition 6.4.4: Assume that X^* is nonempty. Let $\{x_k\}$ be a sequence generated as in Prop. 6.4.3. Then for any $\epsilon > 0$, we have

$$\min_{0\leq k\leq N} F(x_k) \leq F^* + \frac{\alpha\beta m^2c^2 + \epsilon}{2}, \tag{6.100}$$

where N is given by

$$N = m\left\lfloor \frac{\text{dist}(x_0; X^*)^2}{\alpha\epsilon} \right\rfloor. \tag{6.101}$$

Proof: Assume, to arrive at a contradiction, that Eq. (6.100) does not hold, so that for all k with $0 \leq km \leq N$, we have

$$F(x_{km}) > F^* + \frac{\alpha\beta m^2c^2 + \epsilon}{2}.$$

By using this relation in Prop. 6.4.2 with α_k replaced by α and y equal to the vector of X^* that is at minimum distance from x_{km}, we obtain for all k with $0 \leq km \leq N$,

$$\begin{aligned} \text{dist}(x_{(k+1)m}; X^*)^2 &\leq \text{dist}(x_{km}; X^*)^2 - 2\alpha\big(F(x_{km}) - F^*\big) + \alpha^2\beta m^2 c^2 \\ &\leq \text{dist}(x_{km}; X^*)^2 - (\alpha^2\beta m^2 c^2 + \alpha\epsilon) + \alpha^2\beta m^2 c^2 \\ &= \text{dist}(x_{km}; X^*)^2 - \alpha\epsilon. \end{aligned}$$

Adding the above inequalities for $k = 0, \ldots, \frac{N}{m}$, we obtain

$$\text{dist}(x_{N+m}; X^*)^2 \leq \text{dist}(x_0; X^*)^2 - \left(\frac{N}{m} + 1\right)\alpha\epsilon,$$

so that

$$\left(\frac{N}{m} + 1\right)\alpha\epsilon \leq \text{dist}(x_0; X^*)^2,$$

which contradicts the definition of N. **Q.E.D.**

According to Prop. 6.4.4, to achieve a cost function value within $O(\epsilon)$ of the optimal, the term $\alpha\beta m^2 c^2$ must also be of order $O(\epsilon)$, so α must be of order $O(\epsilon/m^2c^2)$, and from Eq. (6.101), the number of necessary iterations N is $O(m^3c^2/\epsilon^2)$, and the number of necessary cycles is $O\big((mc)^2/\epsilon^2\big)$. This is the same type of estimate as for the nonincremental subgradient method [i.e., $O(1/\epsilon^2)$, counting a cycle as one iteration of the nonincremental method, and viewing mc as a Lipschitz constant for the entire cost function F], and does not reveal any advantage for the incremental methods given here. However, in the next section, we demonstrate a much more favorable iteration complexity estimate for the incremental methods that use a randomized order of component selection.

Exact Convergence for a Diminishing Stepsize

We can also obtain an exact convergence result for the case where the stepsize α_k diminishes to zero. The idea is that with a constant stepsize α we can get to within an $O(\alpha)$-neighborhood of the optimum, as shown above, so with a diminishing stepsize α_k, we should be able to reach an arbitrarily small neighborhood of the optimum. However, for this to happen, α_k should not be reduced too fast, and should satisfy $\sum_{k=0}^{\infty} \alpha_k = \infty$ (so that the method can "travel" infinitely far if necessary).

Proposition 6.4.5: Let $\{x_k\}$ be the sequence generated by any one of the algorithms (6.80)-(6.82), with a cyclic order of component selection, and let the stepsize α_k satisfy

$$\lim_{k\to\infty} \alpha_k = 0, \qquad \sum_{k=0}^{\infty} \alpha_k = \infty.$$

Then,

$$\liminf_{k\to\infty} F(x_k) = F^*.$$

Furthermore, if X^* is nonempty and

$$\sum_{k=0}^{\infty} \alpha_k^2 < \infty,$$

then $\{x_k\}$ converges to some $x^* \in X^*$.

Proof: For the first part, it will suffice to show that

$$\liminf_{k\to\infty} F(x_{km}) = F^*.$$

Assume, to arrive at a contradiction, that there exists an $\epsilon > 0$ such that

$$\liminf_{k\to\infty} F(x_{km}) - 2\epsilon > F^*.$$

Then there exists a point $\hat{y} \in X$ such that

$$\liminf_{k\to\infty} F(x_{km}) - 2\epsilon > F(\hat{y}).$$

Let k_0 be large enough so that for all $k \geq k_0$, we have

$$F(x_{km}) \geq \liminf_{k\to\infty} F(x_{km}) - \epsilon.$$

By combining the preceding two relations, we obtain for all $k \geq k_0$,

$$F(x_{km}) - F(\hat{y}) > \epsilon.$$

By setting $y = \hat{y}$ in Prop. 6.4.2, and by using the above relation, we have for all $k \geq k_0$,

$$\begin{aligned} \|x_{(k+1)m} - \hat{y}\|^2 &\leq \|x_{km} - \hat{y}\|^2 - 2\alpha_{km}\epsilon + \beta\alpha_{km}^2 m^2 c^2 \\ &= \|x_{km} - \hat{y}\|^2 - \alpha_{km}\left(2\epsilon - \beta\alpha_{km} m^2 c^2\right). \end{aligned}$$

Since $\alpha_k \to 0$, without loss of generality, we may assume that k_0 is large enough so that

$$2\epsilon - \beta\alpha_k m^2 c^2 \geq \epsilon, \qquad \forall\, k \geq k_0.$$

Therefore for all $k \geq k_0$, we have

$$\|x_{(k+1)m} - \hat{y}\|^2 \leq \|x_{km} - \hat{y}\|^2 - \alpha_{km}\epsilon \leq \cdots \leq \|x_{k_0m} - \hat{y}\|^2 - \epsilon \sum_{\ell=k_0}^{k} \alpha_{\ell m},$$

which cannot hold for k sufficiently large. Hence $\liminf_{k\to\infty} F(x_{km}) = F^*$.

To prove the second part of the proposition, note that from Prop. 6.4.2, for every $x^* \in X^*$ and $k \geq 0$ we have

$$\|x_{(k+1)m} - x^*\|^2 \leq \|x_{km} - x^*\|^2 - 2\alpha_{km}\big(F(x_{km}) - F(x^*)\big) + \alpha_{km}^2 \beta m^2 c^2. \tag{6.102}$$

If $\sum_{k=0}^{\infty} \alpha_k^2 < \infty$, Prop. A.4.4 of Appendix A implies that for each $x^* \in X^*$, $\|x_{(k+1)m} - x^*\|$ converges to some real number, and hence $\{x_{(k+1)m}\}$ is bounded. Consider a subsequence $\{x_{(k+1)m}\}_{\mathcal{K}}$ such that

$$\liminf_{k\to\infty} F(x_{km}) = F^*,$$

and let $\overline{x}$ be a limit point of $\{x_k\}_{\mathcal{K}}$. Since F is continuous, we must have $F(\overline{x}) = F^*$, so $\overline{x} \in X^*$. Using $x^* = \overline{x}$ in Eq. (6.102), it follows that $\|x_{(k+1)m} - \overline{x}\|$ converges to a real number, which must be equal to 0 since $\overline{x}$ is a limit point of $\{x_{(k+1)m}\}$. Thus $\overline{x}$ is the unique limit point of $\{x_{(k+1)m}\}$.

Finally, to show that the entire sequence $\{x_k\}$ also converges to $\overline{x}$, note that from Eqs. (6.83) and (6.85), and the form of the iterations (6.80)-(6.82), we have $\|x_{k+1} - x_k\| \leq 2\alpha_k c \to 0$. Since $\{x_{km}\}$ converges to $\overline{x}$, it follows that $\{x_k\}$ also converges to $\overline{x}$. **Q.E.D.**

6.4.2 Convergence for Methods with Randomized Order

In this section we discuss convergence for the randomized component selection order and a constant stepsize α. The randomized versions of iterations (6.80), (6.81), and (6.82), are

$$z_k = P_X\big(x_k - \alpha\tilde{\nabla} f_{\omega_k}(z_k)\big), \qquad x_{k+1} = P_X\big(z_k - \alpha\tilde{\nabla} h_{\omega_k}(z_k)\big), \tag{6.103}$$

$$z_k = x_k - \alpha\tilde{\nabla} f_{\omega_k}(z_k), \qquad x_{k+1} = P_X\big(z_k - \alpha\tilde{\nabla} h_{\omega_k}(z_k)\big), \tag{6.104}$$

$$z_k = x_k - \alpha\tilde{\nabla} h_{\omega_k}(x_k), \qquad x_{k+1} = P_X\big(z_k - \alpha\tilde{\nabla} f_{\omega_k}(x_{k+1})\big), \tag{6.105}$$

respectively, where $\{\omega_k\}$ is a sequence of random variables, taking values from the index set $\{1, \ldots, m\}$.

We assume the following throughout the present section.

Assumption 6.4.3: [For iterations (6.103) and (6.104)]

(a) $\{\omega_k\}$ is a sequence of random variables, each uniformly distributed over $\{1,\dots,m\}$, and such that for each k, ω_k is independent of the past history $\{x_k, z_{k-1}, x_{k-1}, \dots, z_0, x_0\}$.

(b) There is a constant $c \in \Re$ such that for all k, we have with probability 1, for all $i = 1,\dots,m$,

$$\max\big\{\|\tilde\nabla f_i(z_k^i)\|,\ \|\tilde\nabla h_i(z_k^i)\|\big\} \le c, \tag{6.106}$$

$$\max\big\{f_i(x_k) - f_i(z_k^i),\ h_i(x_k) - h_i(z_k^i)\big\} \le c\|x_k - z_k^i\|, \tag{6.107}$$

where z_k^i is the result of the proximal iteration, starting at x_k if ω_k would be i, i.e.,

$$z_k^i \in \arg\min_{x\in X}\left\{f_i(x) + \frac{1}{2\alpha_k}\|x - x_k\|^2\right\},$$

in the case of iteration (6.103), and

$$z_k^i \in \arg\min_{x\in\Re^n}\left\{f_i(x) + \frac{1}{2\alpha_k}\|x - x_k\|^2\right\},$$

in the case of iteration (6.104).

Assumption 6.4.4: [For iteration (6.105)]

(a) $\{\omega_k\}$ is a sequence of random variables, each uniformly distributed over $\{1,\dots,m\}$, and such that for each k, ω_k is independent of the past history $\{x_k, z_{k-1}, x_{k-1}, \dots, z_0, x_0\}$.

(b) There is a constant $c \in \Re$ such that for all k, we have with probability 1

$$\max\big\{\|\tilde\nabla f_i(x_{k+1}^i)\|,\ \|\tilde\nabla h_i(x_k)\|\big\} \le c, \qquad \forall\, i = 1,\dots,m, \tag{6.108}$$

$$f_i(x_k) - f_i(x_{k+1}^i) \le c\|x_k - x_{k+1}^i\|, \qquad \forall\, i = 1,\dots,m, \tag{6.109}$$

where x_{k+1}^i is the result of the iteration, starting at x_k if ω_k would be i, i.e.,

$$x_{k+1}^i = P_X\big(z_k^i - \alpha_k\tilde\nabla f_i(x_{k+1}^i)\big),$$

with

$$z_k^i = x_k - \alpha_k \tilde{\nabla} h_i(x_k).$$

Note that condition (6.107) is satisfied if there exist subgradients of f_i and h_i at x_k with norms less or equal to c. Thus the conditions (6.106) and (6.107) are similar, the main difference being that the first applies to "slopes" of f_i and h_i at z_k^i while the second applies to the "slopes" of f_i and h_i at x_k. As in the case of Assumption 6.4.1, these conditions are guaranteed by Lipschitz continuity assumptions on f_i and h_i. The convergence analysis of the randomized algorithms of this section is somewhat more complicated than the one of the cyclic order counterparts, and relies on the Supermartingale Convergence Theorem (Prop. A.4.5 in Appendix A). The following proposition deals with the case of a constant stepsize, and parallels Prop. 6.4.3 for the cyclic order case.

Proposition 6.4.6: Let $\{x_k\}$ be the sequence generated by one of the randomized incremental methods (6.103)-(6.105), and let the stepsize α_k be fixed at some positive constant α.

(a) If $F^* = -\infty$, then with probability 1

$$\inf_{k \geq 0} F(x_k) = F^*.$$

(b) If $F^* > -\infty$, then with probability 1

$$\inf_{k \geq 0} F(x_k) \leq F^* + \frac{\alpha \beta m c^2}{2},$$

where $\beta = 5$.

Proof: Consider first algorithms (6.103) and (6.104). By adapting the proof argument of Prop. 6.4.2 with F_{i_k} replaced by F_{ω_k} [cf. Eq. (6.91)], we have

$$\|x_{k+1} - y\|^2 \leq \|x_k - y\|^2 - 2\alpha\big(F_{\omega_k}(z_k) - F_{\omega_k}(y)\big) + \alpha^2 c^2, \quad \forall\, y \in X, \quad k \geq 0.$$

Taking conditional expectation of both sides given the set of random variables $\mathcal{F}_k = \{x_k, z_{k-1}, \ldots, z_0, x_0\}$, and using the fact that ω_k takes the values $i = 1, \ldots, m$ with equal probability $1/m$, we obtain for all $y \in X$

and k,

$$\begin{aligned} E\big\{\|x_{k+1}-y\|^2 \mid \mathcal{F}_k\big\} &\le \|x_k-y\|^2 - 2\alpha E\big\{F_{\omega_k}(z_k)-F_{\omega_k}(y) \mid \mathcal{F}_k\big\} + \alpha^2c^2 \\ &= \|x_k-y\|^2 - \frac{2\alpha}{m}\sum_{i=1}^m \big(F_i(z_k^i)-F_i(y)\big) + \alpha^2c^2 \\ &= \|x_k-y\|^2 - \frac{2\alpha}{m}\big(F(x_k)-F(y)\big) + \alpha^2c^2 \\ &\quad + \frac{2\alpha}{m}\sum_{i=1}^m \big(F_i(x_k)-F_i(z_k^i)\big). \end{aligned} \tag{6.110}$$

By using Eqs. (6.106) and (6.107),

$$\sum_{i=1}^m \big(F_i(x_k)-F_i(z_k^i)\big) \le 2c\sum_{i=1}^m \|x_k-z_k^i\| = 2c\alpha\sum_{i=1}^m \|\tilde{\nabla} f_i(z_k^i)\| \le 2m\alpha c^2.$$

By combining the preceding two relations, we obtain

$$\begin{aligned} E\big\{\|x_{k+1}-y\|^2 \mid \mathcal{F}_k\big\} &\le \|x_k-y\|^2 - \frac{2\alpha}{m}\big(F(x_k)-F(y)\big) + 4\alpha^2c^2 + \alpha^2c^2 \\ &= \|x_k-y\|^2 - \frac{2\alpha}{m}\big(F(x_k)-F(y)\big) + \beta\alpha^2c^2, \end{aligned} \tag{6.111}$$

where $\beta = 5$.

The preceding equation holds also for algorithm (6.105). To see this note that Eq. (6.98) yields for all $y \in X$

$$\begin{aligned} \|x_{k+1}-y\|^2 \le \|x_k-y\|^2 &- 2\alpha\big(F_{\omega_k}(x_k)-F_{\omega_k}(y)\big) + \alpha^2c^2 \\ &+ 2\alpha\big(f_{\omega_k}(x_k)-f_{\omega_k}(x_{k+1})\big), \end{aligned} \tag{6.112}$$

and similar to Eq. (6.110), we obtain

$$\begin{aligned} E\big\{\|x_{k+1}-y\|^2 \mid \mathcal{F}_k\big\} \le \|x_k-y\|^2 &- \frac{2\alpha}{m}\big(F(x_k)-F(y)\big) + \alpha^2c^2 \\ &+ \frac{2\alpha}{m}\sum_{i=1}^m \big(f_i(x_k)-f_i(x_{k+1}^i)\big). \end{aligned} \tag{6.113}$$

From Eq. (6.109), we have

$$f_i(x_k) - f_i(x_{k+1}^i) \le c\|x_k - x_{k+1}^i\|,$$

and from Eq. (6.108) and the nonexpansion property of the projection,

$$\begin{aligned} \|x_k - x_{k+1}^i\| &\le \big\|x_k - z_k^i + \alpha\tilde{\nabla} f_i(x_{k+1}^i)\big\| \\ &= \big\|x_k - x_k + \alpha\tilde{\nabla} h_i(x_k) + \alpha\tilde{\nabla} f_i(x_{k+1}^i)\big\| \\ &\le 2\alpha c. \end{aligned}$$

Combining the preceding inequalities, we obtain Eq. (6.111) with $\beta = 5$.

Let us fix a positive scalar γ, consider the level set L_γ defined by

$$L_\gamma = \begin{cases} \left\{x \in X \mid F(x) < -\gamma + 1 + \frac{\alpha\beta mc^2}{2}\right\} & \text{if } F^* = -\infty, \\ \left\{x \in X \mid F(x) < F^* + \frac{2}{\gamma} + \frac{\alpha\beta mc^2}{2}\right\} & \text{if } F^* > -\infty, \end{cases}$$

and let $y_\gamma \in X$ be such that

$$F(y_\gamma) = \begin{cases} -\gamma & \text{if } F^* = -\infty, \\ F^* + \frac{1}{\gamma} & \text{if } F^* > -\infty. \end{cases}$$

Note that $y_\gamma \in L_\gamma$ by construction. Define a new process $\{\hat{x}_k\}$ that is identical to $\{x_k\}$, except that once x_k enters the level set L_γ, the process terminates with $\hat{x}_k = y_\gamma$. We will now argue that for any fixed γ, $\{\hat{x}_k\}$ (and hence also $\{x_k\}$) will eventually enter L_γ, which will prove both parts (a) and (b).

Using Eq. (6.111) with $y = y_\gamma$, we have

$$E\{\|\hat{x}_{k+1} - y_\gamma\|^2 \mid \mathcal{F}_k\} \leq \|\hat{x}_k - y_\gamma\|^2 - \frac{2\alpha}{m}\big(F(\hat{x}_k) - F(y_\gamma)\big) + \beta\alpha^2c^2,$$

from which

$$E\{\|\hat{x}_{k+1} - y_\gamma\|^2 \mid \mathcal{F}_k\} \leq \|\hat{x}_k - y_\gamma\|^2 - v_k, \tag{6.114}$$

where

$$v_k = \begin{cases} \frac{2\alpha}{m}\big(F(\hat{x}_k) - F(y_\gamma)\big) - \beta\alpha^2c^2 & \text{if } \hat{x}_k \notin L_\gamma, \\ 0 & \text{if } \hat{x}_k = y_\gamma. \end{cases}$$

The idea of the subsequent argument is to show that as long as $\hat{x}_k \notin L_\gamma$, the scalar v_k (which is a measure of progress) is strictly positive and bounded away from 0.

(a) Let $F^* = -\infty$. Then if $\hat{x}_k \notin L_\gamma$, we have

$$\begin{aligned} v_k &= \frac{2\alpha}{m}\big(F(\hat{x}_k) - F(y_\gamma)\big) - \beta\alpha^2c^2 \\ &\geq \frac{2\alpha}{m}\left(-\gamma + 1 + \frac{\alpha\beta mc^2}{2} + \gamma\right) - \beta\alpha^2c^2 \\ &= \frac{2\alpha}{m}. \end{aligned}$$

Since $v_k = 0$ for $\hat{x}_k \in L_\gamma$, we have $v_k \geq 0$ for all k, and by Eq. (6.114) and the Supermartingale Convergence Theorem (Prop. A.4.5 in Appendix A), $\sum_{k=0}^\infty v_k < \infty$ implying that $\hat{x}_k \in L_\gamma$ for sufficiently large k, with probability 1. Therefore, in the original process we have

$$\inf_{k\geq 0} F(x_k) \leq -\gamma + 1 + \frac{\alpha\beta mc^2}{2}$$

with probability 1. Letting $\gamma \to \infty$, we obtain $\inf_{k\geq 0} F(x_k) = -\infty$ with probability 1.

(b) Let $F^* > -\infty$. Then if $\hat{x}_k \notin L_\gamma$, we have

$$\begin{aligned} v_k &= \frac{2\alpha}{m}\big(F(\hat{x}_k) - F(y_\gamma)\big) - \beta\alpha^2 c^2 \\ &\geq \frac{2\alpha}{m}\left(F^* + \frac{2}{\gamma} + \frac{\alpha\beta m c^2}{2} - F^* - \frac{1}{\gamma}\right) - \beta\alpha^2 c^2 \\ &= \frac{2\alpha}{m\gamma}. \end{aligned}$$

Hence, $v_k \geq 0$ for all k, and by the Supermartingale Convergence Theorem (Prop. A.4.5 in Appendix A), we have $\sum_{k=0}^{\infty} v_k < \infty$ implying that $\hat{x}_k \in L_\gamma$ for sufficiently large k, so that in the original process,

$$\inf_{k\geq 0} F(x_k) \leq F^* + \frac{2}{\gamma} + \frac{\alpha\beta m c^2}{2}$$

with probability 1. Letting $\gamma \to \infty$, we obtain $\inf_{k\geq 0} F(x_k) \leq F^* + \alpha\beta m c^2/2$. **Q.E.D.**

By comparing Prop. 6.4.6(b) with Prop. 6.4.3(b), we see that when $F^* > -\infty$ and the stepsize α is constant, the randomized methods (6.103), (6.104), and (6.105), have a better error bound (by a factor m) than their nonrandomized counterparts. In fact there is an example for the incremental subgradient method (see [BNO03], p. 514) that can be adapted to show that the bound of Prop. 6.4.3(b) is tight in the sense that for a bad problem/cyclic order we have $\liminf_{k\to\infty} F(x_k) - F^* = O(\alpha m^2 c^2)$. By contrast the randomized method will get to within $O(\alpha m c^2)$ with probability 1 for any problem, according to Prop. 6.4.6(b). Thus with the randomized algorithm we do not run the risk of choosing by accident a bad cyclic order. A related result is provided by the following proposition, which should be compared with Prop. 6.4.4 for the nonrandomized methods.

Proposition 6.4.7: Assume that X^* is nonempty. Let $\{x_k\}$ be a sequence generated as in Prop. 6.4.6. Then for any $\epsilon > 0$, we have with probability 1

$$\min_{0\leq k\leq N} F(x_k) \leq F^* + \frac{\alpha\beta m c^2 + \epsilon}{2}, \tag{6.115}$$

where N is a random variable with

$$E\{N\} \leq m\,\frac{\mathrm{dist}(x_0; X^*)^2}{\alpha\epsilon}. \tag{6.116}$$

Proof: Let $\hat{y}$ be some fixed vector in X^*. Define a new process $\{\hat{x}_k\}$ which is identical to $\{x_k\}$ except that once x_k enters the level set

$$L = \left\{ x \in X \;\middle|\; F(x) < F^* + \frac{\alpha\beta m c^2 + \epsilon}{2} \right\},$$

the process $\{\hat{x}_k\}$ terminates at $\hat{y}$. Similar to the proof of Prop. 6.4.6 [cf. Eq. (6.111) with y being the closest point of $\hat{x}_k$ in X^*], for the process $\{\hat{x}_k\}$ we obtain for all k,

$$\begin{aligned} E\{\text{dist}(\hat{x}_{k+1}; X^*)^2 \mid \mathcal{F}_k\} &\le E\{\|\hat{x}_{k+1} - y\|^2 \mid \mathcal{F}_k\} \\ &\le \text{dist}(\hat{x}_k; X^*)^2 - \frac{2\alpha}{m}\big(F(\hat{x}_k) - F^*\big) + \beta\alpha^2 c^2 \\ &= \text{dist}(\hat{x}_k; X^*)^2 - v_k, \end{aligned} \tag{6.117}$$

where $\mathcal{F}_k = \{x_k, z_{k-1}, \ldots, z_0, x_0\}$ and

$$v_k = \begin{cases} \frac{2\alpha}{m}\big(F(\hat{x}_k) - F^*\big) - \beta\alpha^2 c^2 & \text{if } \hat{x}_k \notin L, \\ 0 & \text{otherwise.} \end{cases}$$

In the case where $\hat{x}_k \notin L$, we have

$$v_k \ge \frac{2\alpha}{m}\left(F^* + \frac{\alpha\beta m c^2 + \epsilon}{2} - F^*\right) - \beta\alpha^2 c^2 = \frac{\alpha\epsilon}{m}. \tag{6.118}$$

By the Supermartingale Convergence Theorem (Prop. A.4.5 in Appendix A), from Eq. (6.117) we have $\sum_{k=0}^{\infty} v_k < \infty$ with probability 1, so that $v_k = 0$ for all $k \ge N$, where N is a random variable. Hence $\hat{x}_N \in L$ with probability 1, implying that in the original process we have

$$\min_{0 \le k \le N} F(x_k) \le F^* + \frac{\alpha\beta m c^2 + \epsilon}{2}$$

with probability 1. Furthermore, by taking the total expectation in Eq. (6.117), we obtain for all k,

$$\begin{aligned} E\{\text{dist}(\hat{x}_{k+1}; X^*)^2\} &\le E\{\text{dist}(\hat{x}_k; X^*)^2\} - E\{v_k\} \\ &\le \text{dist}(\hat{x}_0; X^*)^2 - E\left\{\sum_{j=0}^{k} v_j\right\}, \end{aligned}$$

where in the last inequality we use the facts $\hat{x}_0 = x_0$ and $E\{\text{dist}(\hat{x}_0; X^*)^2\} = \text{dist}(\hat{x}_0; X^*)^2$. Therefore, letting $k \to \infty$, and using the definition of v_k and Eq. (6.118),

$$\text{dist}(\hat{x}_0; X^*)^2 \ge E\left\{\sum_{k=0}^{\infty} v_k\right\} = E\left\{\sum_{k=0}^{N-1} v_k\right\} \ge E\left\{\frac{N\alpha\epsilon}{m}\right\} = \frac{\alpha\epsilon}{m} E\{N\}.$$

Q.E.D.

Like Prop. 6.4.6, a comparison of Props. 6.4.4 and 6.4.7 again suggests an advantage for the randomized methods: compared to their deterministic counterparts, they achieve a much smaller error tolerance (by a factor of m), in the same *expected* number of iterations. Note, however, that the preceding assessment is based on upper bound estimates, which may not be sharp on a given problem [although the bound of Prop. 6.4.3(b) is tight with a worst-case problem selection as mentioned earlier; see [BNO03], p. 514]. Moreover, the comparison based on worst-case values versus expected values may not be strictly valid. In particular, while Prop. 6.4.4 provides an upper bound estimate on N, Prop. 6.4.7 provides an upper bound estimate on $E\{N\}$, which is not quite the same. However, this comparison seems to be supported by the experimental results obtained so far.

Finally for the case of a diminishing stepsize, let us give the following proposition, which parallels Prop. 6.4.5 for the cyclic order.

Proposition 6.4.8: Let $\{x_k\}$ be the sequence generated by one of the randomized incremental methods (6.103)-(6.105), and let the stepsize α_k satisfy

$$\lim_{k\to\infty} \alpha_k = 0, \qquad \sum_{k=0}^{\infty} \alpha_k = \infty.$$

Then, with probability 1,

$$\liminf_{k\to\infty} F(x_k) = F^*.$$

Furthermore, if X^* is nonempty and

$$\sum_{k=0}^{\infty} \alpha_k^2 < \infty,$$

then $\{x_k\}$ converges to some $x^* \in X^*$ with probability 1.

Proof: The proof of the first part is nearly identical to the corresponding part of Prop. 6.4.5. To prove the second part, similar to the proof of Prop. 6.4.6, we obtain for all k and all $x^* \in X^*$,

$$E\{\|x_{k+1} - x^*\|^2 \mid \mathcal{F}_k\} \le \|x_k - x^*\|^2 - \frac{2\alpha_k}{m}\big(F(x_k) - F^*\big) + \beta\alpha_k^2 c^2 \quad (6.119)$$

[cf. Eq. (6.111) with α and y replaced with α_k and x^*, respectively], where $\mathcal{F}_k = \{x_k, z_{k-1}, \ldots, z_0, x_0\}$. According to the Supermartingale Convergence Theorem (Prop. A.4.5 in Appendix A), for each $x^* \in X^*$, there is a set Ω_{x^*} of sample paths of probability 1 such that for each sample path in Ω_{x^*}

$$\sum_{k=0}^{\infty} \frac{2\alpha_k}{m}\big(F(x_k) - F^*\big) < \infty, \tag{6.120}$$

and the sequence $\{\|x_k - x^*\|\}$ converges.

Let $\{v_i\}$ be a countable subset of the relative interior $\mathrm{ri}(X^*)$ that is dense in X^* [such a set exists since $\mathrm{ri}(X^*)$ is a relatively open subset of the affine hull of X^*; an example of such a set is the intersection of X^* with the set of vectors of the form $x^* + \sum_{i=1}^{p} r_i\xi_i$, where $\xi_1, \ldots, \xi_p$ are basis vectors for the affine hull of X^* and r_i are rational numbers]. Let also Ω_{v_i} be the set of sample paths defined earlier that corresponds to v_i. The intersection

$$\overline{\Omega} = \cap_{i=1}^{\infty} \Omega_{v_i}$$

has probability 1, since its complement $\overline{\Omega}^c$ is equal to $\cup_{i=1}^{\infty}\Omega_{v_i}^c$ and

$$\mathrm{Prob}\left(\cup_{i=1}^{\infty}\Omega_{v_i}^c\right) \le \sum_{i=1}^{\infty} \mathrm{Prob}\left(\Omega_{v_i}^c\right) = 0.$$

For each sample path in $\overline{\Omega}$, all the sequences $\{\|x_k - v_i\|\}$ converge so that $\{x_k\}$ is bounded, while by the first part of the proposition [or Eq. (6.120)] $\liminf_{k\to\infty} F(x_k) = F^*$. Therefore, $\{x_k\}$ has a limit point $\overline{x}$ in X^*. Since $\{v_i\}$ is dense in X^*, for every $\epsilon > 0$ there exists $v_{i(\epsilon)}$ such that $\|\overline{x} - v_{i(\epsilon)}\| < \epsilon$. Since the sequence $\{\|x_k - v_{i(\epsilon)}\|\}$ converges and $\overline{x}$ is a limit point of $\{x_k\}$, we have $\lim_{k\to\infty}\|x_k - v_{i(\epsilon)}\| < \epsilon$, so that

$$\limsup_{k\to\infty} \|x_k - \overline{x}\| \le \lim_{k\to\infty} \|x_k - v_{i(\epsilon)}\| + \|v_{i(\epsilon)} - \overline{x}\| < 2\epsilon.$$

By taking $\epsilon \to 0$, it follows that $x_k \to \overline{x}$. **Q.E.D.**

6.4.3 Application in Specially Structured Problems

We will now illustrate the application of our methods of this section to some specially structured problems.

ℓ_1-Regularization

Let us consider the ℓ_1-regularization problem

$$\begin{aligned} &\text{minimize} \quad \gamma\|x\|_1 + \frac{1}{2}\sum_{i=1}^{m}(c_i'x - b_i)^2 \\ &\text{subject to} \quad x \in \Re^n, \end{aligned} \tag{6.121}$$

where γ is a positive scalar and x^j is the jth coordinate of x (cf. Example 1.3.1). It is convenient to handle the regularization term with the proximal algorithm:

$$z_k \in \arg\min_{x\in\Re^n}\left\{\gamma\,\|x\|_1+\frac{1}{2\alpha_k}\|x-x_k\|^2\right\}.$$

This proximal iteration decomposes into the n one-dimensional minimizations

$$z_k^j \in \arg\min_{x^j\in\Re}\left\{\gamma\,|x^j|+\frac{1}{2\alpha_k}|x^j-x_k^j|^2\right\},\qquad j=1,\ldots,n.$$

and can be done in closed form

$$z_k^j=\begin{cases} x_k^j-\gamma\alpha_k & \text{if } \gamma\alpha_k\le x_k^j,\\ 0 & \text{if } -\gamma\alpha_k<x_k^j<\gamma\alpha_k,\\ x_k^j+\gamma\alpha_k & \text{if } x_k^j\le-\gamma\alpha_k,\end{cases}\qquad j=1,\ldots,n; \tag{6.122}$$

cf. the shrinkage operation discussed in Section 5.4.1.

Thus the incremental algorithms of this section are well-suited for solution of the ℓ_1-regularization problem. The kth incremental iteration may consist of selecting a pair (c_{i_k},b_{i_k}) and performing a proximal iteration of the form (6.122) to obtain z_k, followed by a gradient iteration on the component $\frac{1}{2}(c_{i_k}'x-b_{i_k})^2$, starting at z_k:

$$x_{k+1}=z_k-\alpha_k c_{i_k}(c_{i_k}'z_k-b_{i_k}).$$

This algorithm is the special case of the algorithms (6.80)-(6.82) (here $X=\Re^n$, and all three algorithms coincide), with $f_i(x)$ being $\gamma\|x\|_1$ (we use m copies of this function) and $h_i(x)=\frac{1}{2}(c_i'x-b_i)^2$.

Finally, let us note that as an alternative, the proximal iteration (6.122) could be replaced by a proximal iteration on $\gamma\,|x^j|$ for some index j, with all indexes selected cyclically in incremental iterations. Randomized selection of the data pair (c_{i_k},b_{i_k}) is also a possibility, particularly in contexts where the data has a natural stochastic interpretation.

Incremental and Distributed Augmented Lagrangian Methods

We will now revisit the augmented Lagrangian methodology of Section 5.2.1, in the context of large-scale separable problems and incremental proximal methods. Consider the separable constrained minimization problem

$$\begin{aligned} &\text{minimize} \quad \sum_{i=1}^m f_i(x^i)\\ &\text{subject to}\ \ x^i\in X_i,\ \ i=1,\ldots,m,\quad \sum_{i=1}^m (A_i x^i-b_i)=0,\end{aligned} \tag{6.123}$$

where $f_i : \Re^{n_i} \mapsto \Re$ are convex functions (n_i is a positive integer, which may depend on i), X_i are nonempty closed convex subsets of $\Re^{n_i}$, A_i are given $r \times n_i$ matrices, and $b_i \in \Re^r$ are given vectors. For simplicity, we focus on linear equality constraints, but the analysis can be extended to convex inequality constraints as well.

Similar to our discussion of separable problems in Section 1.1.1, the dual function is given by

$$q(\lambda) = \inf_{x^i \in X_i,\, i=1,\ldots,m} \left\{ \sum_{i=1}^{m} \big(f_i(x^i) + \lambda'(A_i x^i - b_i) \big) \right\},$$

and by decomposing the minimization over the components of x, it can be expressed in the additive form

$$q(\lambda) = \sum_{i=1}^{m} q_i(\lambda),$$

where

$$q_i(\lambda) = \inf_{x^i \in X_i} \big\{ f_i(x^i) + \lambda'(A_i x^i - b_i) \big\}. \tag{6.124}$$

A classical method for maximizing the dual function $q(\lambda)$, dating to [Eve63] [see [Las70] or [Ber99] (Section 6.2.2) for a discussion], is a gradient method, which, however, can be used only in the case where the functions q_i are differentiable. Assuming a constant stepsize $\alpha > 0$, the method takes the form

$$\lambda_{k+1} = \lambda_k + \alpha \sum_{i=1}^{m} \nabla q_i(\lambda_k),$$

with the gradients $\nabla q_i(\lambda_k)$ obtained as

$$\nabla q_i(\lambda_k) = A_i x_k^i - b_i, \qquad i = 1, \ldots, m,$$

where x_k^i attains the minimum of

$$f_i(x^i) + \lambda_k'(A_i x^i - b_i)$$

over $x^i \in X_i$; cf. Eq. (6.124) and Example 3.1.2. Thus this method exploits the separability of the problem, and is well suited for distributed computation, with the gradients $\nabla q_i(\lambda_k)$ computed in parallel at separate processors. However, the differentiability requirement on q_i is very strong [it is equivalent to the infimum being attained uniquely for all λ in Eq. (6.124)], and the convergence properties of this method tend to be fragile. We will consider instead an incremental proximal method that can be dually implemented (cf. Section 5.2) with decomposable augmented Lagrangian minimizations, and has more solid convergence properties.

In this connection, we observe that the dual problem,

$$\begin{aligned} &\text{maximize} \quad \sum_{i=1}^m q_i(\lambda) \\ &\text{subject to} \quad \lambda \in \Re^r, \end{aligned}$$

has a suitable form for application of the incremental proximal method [cf. Eq. (6.67)].† In particular, the incremental proximal algorithm updates the current vector λ_k to a new vector λ_{k+1} after a cycle of m subiterations:

$$\lambda_{k+1} = \psi_k^m, \tag{6.125}$$

where starting with $\psi_k^0 = \lambda_k$, we obtain ψ_k^m after the m proximal steps

$$\psi_k^i = \arg\max_{\lambda \in \Re^r} \left\{ q_i(\lambda) - \frac{1}{2\alpha_k}\|\lambda - \psi_k^{i-1}\|^2 \right\}, \qquad i = 1, \ldots, m, \tag{6.126}$$

where α_k is a positive parameter.

We now recall the Fenchel duality relation between proximal and augmented Lagrangian minimization, which was discussed in Section 5.2. Based on that relation, the proximal incremental update (6.126) can be written in terms of the data of the primal problem as

$$\psi_k^i = \psi_k^{i-1} + \alpha_k(A_k^i x_k^i - b_i), \tag{6.127}$$

where x_k^i is obtained from the minimization

$$x_k^i \in \arg\min_{x^i \in X_i} L_{\alpha_k,i}(x^i, \psi_k^{i-1}), \tag{6.128}$$

and $L_{\alpha_k,i}$ is the "incremental" augmented Lagrangian function

$$L_{\alpha_k,i}(x^i, \lambda) = f_i(x^i) + \lambda'(A_i x^i - b_i) + \frac{\alpha_k}{2}\|A_i x^i - b_i\|^2. \tag{6.129}$$

This algorithm allows decomposition within the augmented Lagrangian framework, which is not possible in the standard augmented Lagrangian method of Section 5.2.1, since the addition of the penalty term

$$\frac{c}{2}\left\|\sum_{i=1}^m (A_i x^i - b_i)\right\|^2$$

---

† The algorithm (6.67) requires that the functions $-q_i$ have a common effective domain, which is a closed convex set. This is true for example if q_i is real-valued, which occurs if X_i is compact. In unusual cases where $-q_i$ has an effective domain that depends on i and/or is not closed, the earlier convergence analysis does not apply and needs to be modified.

to the Lagrangian function destroys its separability.

We note that the algorithm has suitable structure for distributed asynchronous implementation, along the lines discussed in Section 2.1.6. In particular, the augmented Lagrangian minimizations (6.128) can be done at separate processors, each dedicated to updating a single component x^i, based on a multiplier vector λ that is updated at a central processor/coordinator, using incremental iterations of the form (6.127).

Finally, let us note that the algorithm bears similarity with the ADMM for separable problems of Section 5.4.2. The latter algorithm is not incremental, which can be a disadvantage, but uses a constant penalty parameter, which can be an advantage. Other differences are that the ADMM maintains iterates of variables z^i that approximate the constraint levels b_i of Eq. (6.123), and involves a form of averaging of the multiplier iterates.

6.4.4 Incremental Constraint Projection Methods

In this section we consider incremental approaches for problems involving complicated constraint sets that cannot be easily dealt with projection or proximal algorithms. In particular, we consider the following problem,

$$\begin{aligned} &\text{minimize} \quad f(x) \\ &\text{subject to} \quad x \in \cap_{i=1}^m X_i, \end{aligned} \tag{6.130}$$

where $f : \Re^n \mapsto \Re$ is a convex cost function, and X_i are closed convex sets that have a relatively simple form that is convenient for projection or proximal iterations.

While the problem (6.130) does not involve a sum of component functions, it may be converted into one that does by using a distance-based exact penalty function. In particular, consider the problem

$$\begin{aligned} &\text{minimize} \quad f(x) + c\sum_{i=1}^m \text{dist}(x; X_i) \\ &\text{subject to} \quad x \in \Re^n, \end{aligned} \tag{6.131}$$

where c is a positive penalty parameter. Then for f Lipschitz continuous and c sufficiently large, problems (6.130) and (6.131) are equivalent, as shown by the following proposition, which was proved in Section 1.5 (cf. Prop. 1.5.3).

Proposition 6.4.9: Let $f : \Re^n \mapsto \Re$ be a function, and let X_i, $i = 0, 1, \ldots, m$, be closed subsets of $\Re^n$ with nonempty intersection. Assume that f is Lipschitz continuous over X_0. Then there is a scalar $\bar{c} > 0$ such that for all $c \geq \bar{c}$, the set of minima of f over $\cap_{i=0}^m X_i$ coincides with the set of minima of

$$f(x) + c \sum_{i=1}^{m} \text{dist}(x; X_i)$$

over X_0.

From the preceding proposition, with $X_0 = \Re^n$, it follows that we can solve in place of the original problem (6.130) the additive cost problem (6.131) for which our incremental algorithms of the preceding sections apply (assuming a sufficiently large choice of c). In particular, let us consider the algorithms (6.80)-(6.82), with $X = \Re^n$, which involve a proximal iteration on one of the functions $c\,\text{dist}(x; X_i)$, $i = 1, \ldots, m$, followed by a subgradient iteration on f. A key fact here is that the proximal iteration

$$z_k \in \arg\min_{x \in \Re^n} \left\{ c\,\text{dist}(x; X_{i_k}) + \frac{1}{2\alpha_k}\|x - x_k\|^2 \right\} \tag{6.132}$$

involves a projection on X_{i_k} of x_k, followed by an interpolation; see Fig. 6.4.1. This is shown in the following proposition.

Proposition 6.4.10: Let z_k be the vector produced by the proximal iteration (6.132). If $x_k \in X_{i_k}$ then $z_k = x_k$, while if $x_k \notin X_{i_k}$,

$$z_k = \begin{cases} (1 - \beta_k)x_k + \beta_k P_{X_{i_k}}(x_k) & \text{if } \beta_k < 1, \\ P_{X_{i_k}}(x_k) & \text{if } \beta_k \geq 1, \end{cases} \tag{6.133}$$

where

$$\beta_k = \frac{\alpha_k c}{\text{dist}(x_k; X_{i_k})}.$$

Proof: The case $x_k \in X_{i_k}$ is evident, so assume that $x_k \notin X_{i_k}$. From the nature of the cost function in Eq. (6.132) we see that z_k is a vector that lies in the line segment between x_k and $P_{X_{i_k}}(x_k)$. Hence there are two possibilities: either

$$z_k = P_{X_{i_k}}(x_k), \tag{6.134}$$

or $z_k \notin X_{i_k}$ in which case by setting to 0 the gradient at z_k of the cost function in Eq. (6.132) yields

$$c\,\frac{z_k - P_{X_{i_k}}(z_k)}{\left\| z_k - P_{X_{i_k}}(z_k) \right\|} = \frac{1}{\alpha_k}(x_k - z_k).$$

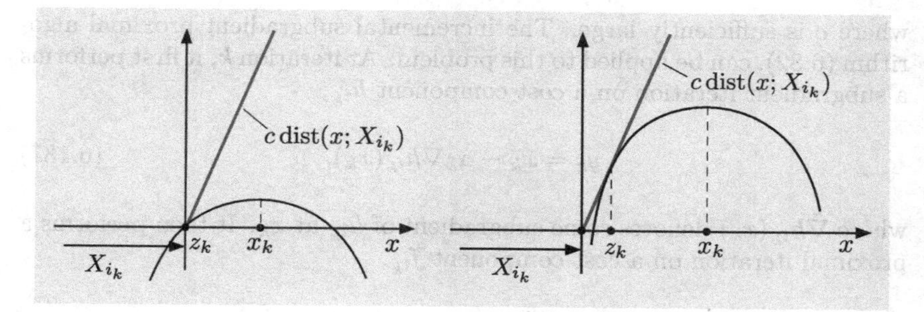

Figure 6.4.1. Illustration of a proximal iteration applied to the distance function $\mathrm{dist}(x; X_{i_k})$ of the set X_{i_k}:

$$z_k \in \arg\min_{x\in\Re^n}\left\{\mathrm{dist}(x; X_{i_k}) + \frac{1}{2\alpha_k}\|x - x_k\|^2\right\}.$$

In the left-hand side figure, we have

$$\alpha_k c \geq \mathrm{dist}(x_k; X_{i_k}),$$

and z_k is equal to the projection of x_k onto X_{i_k}. In the right-hand side figure, we have

$$\alpha_k c < \mathrm{dist}(x_k; X_{i_k}),$$

and z_k is obtained by calculating the projection of x_k onto X_{i_k}, and then interpolating according to Eq. (6.133).

This equation implies that x_k, z_k, and $P_{X_{i_k}}(z_k)$ lie on the same line, so that $P_{X_{i_k}}(z_k) = P_{X_{i_k}}(x_k)$ and

$$z_k = x_k - \frac{\alpha_k c}{\mathrm{dist}(x_k; X_{i_k})}\big(x_k - P_{X_{i_k}}(x_k)\big) = (1-\beta_k)x_k + \beta_k P_{X_{i_k}}(x_k). \quad (6.135)$$

By calculating and comparing the value of the cost function in Eq. (6.132) for each of the possibilities (6.134) and (6.135), we can verify that (6.135) gives a lower cost if and only if $\beta_k < 1$. **Q.E.D.**

Let us now consider the problem

$$\begin{aligned} &\text{minimize} \quad \sum_{i=1}^{m}\big(f_i(x) + h_i(x)\big) \\ &\text{subject to} \quad x \in \cap_{i=1}^{m} X_i. \end{aligned} \quad (6.136)$$

Based on the preceding analysis, we can convert this problem to the unconstrained minimization problem

$$\begin{aligned} &\text{minimize} \quad \sum_{i=1}^{m}\big(f_i(x) + h_i(x) + c\,\mathrm{dist}(x; X_i)\big) \\ &\text{subject to} \quad x \in \Re^n, \end{aligned}$$

where c is sufficiently large. The incremental subgradient proximal algorithm (6.82), can be applied to this problem. At iteration k, it first performs a subgradient iteration on a cost component h_{i_k},

$$y_k = x_k - \alpha_k \tilde{\nabla} h_{i_k}(x_k), \tag{6.137}$$

where $\tilde{\nabla} h_{i_k}(x_k)$ denotes some subgradient of h_{i_k} at x_k. It then performs a proximal iteration on a cost component f_{i_k},

$$z_k \in \arg\min_{x\in\Re^n} \left\{ f_{i_k}(x) + \frac{1}{2\alpha_k}\|x - y_k\|^2 \right\}. \tag{6.138}$$

Finally, it performs a proximal iteration on a constraint distance component $c\,\mathrm{dist}(\cdot; X_{i_k})$ according to Prop. 6.4.10 [cf. Eq. (6.133)],

$$x_{k+1} = \begin{cases} (1-\beta_k)z_k + \beta_k P_{X_{i_k}}(z_k) & \text{if } \beta_k < 1, \\ P_{X_{i_k}}(z_k) & \text{if } \beta_k \geq 1, \end{cases} \tag{6.139}$$

where

$$\beta_k = \frac{\alpha_k c}{\mathrm{dist}(z_k; X_{i_k})}, \tag{6.140}$$

with the convention that $\beta_k = \infty$ if $\mathrm{dist}(z_k; X_{i_k}) = 0$. The index i_k may be chosen either randomly or according to a cyclic rule. Our earlier convergence analysis extends straightforwardly to the case of three cost function components for each index i. Moreover the subgradient, proximal, and projection operations may be performed in any order.

Note that the penalty parameter c can be taken as large as desired, and it does not affect the algorithm as long as

$$\alpha_k c \geq \mathrm{dist}(z_k; X_{i_k}),$$

in which case $\beta_k \geq 1$ [cf. Eq. (6.140)]. Thus we may keep increasing c so that $\beta_k \geq 1$, up to the point where it reaches some "very large" threshold. It would thus appear that in practice we may be able to use a stepsize β_k that is always equal to 1 in Eq. (6.139), leading to the simpler algorithm

$$y_k = x_k - \alpha_k \tilde{\nabla} h_{i_k}(x_k),$$

$$z_k \in \arg\min_{x\in\Re^n} \left\{ f_{i_k}(x) + \frac{1}{2\alpha_k}\|x - y_k\|^2 \right\},$$

$$x_{k+1} = P_{X_{i_k}}(z_k),$$

which does not involve the parameter c. Indeed, this algorithm has been analyzed in the paper [WaB13a]. Convergence has been shown for a variety of randomized and cyclic sampling schemes for selecting the cost function components and the constraint components.

While this algorithm does not depend on the penalty parameter c, its currently available convergence proof requires an additional condition. This is the so-called *linear regularity* condition, namely that for some $\eta > 0$,

$$\left\| x - P_{\cap_{i=1}^m X_i}(x) \right\| \le \eta \max_{i=1,\ldots,m} \left\| x - P_{X_i}(x) \right\|, \qquad \forall\, x \in \Re^n,$$

where $P_Y(x)$ denotes projection of a vector x on the set Y. This property is satisfied in particular if all the sets X_i are polyhedral. By contrast, the algorithm (6.137)-(6.139) does not need this condition, but uses a (large) value of c to guard against the rare case where the scalar β_k of Eq. (6.140) gets smaller than 1, requiring an interpolation as per Eq. (6.139), which ensures convergence.

Finally, let us note a related problem for which incremental constraint projection methods are well-suited. This is the problem

$$\begin{aligned} &\text{minimize} \quad f(x) + c \sum_{j=1}^{r} \max\{0, g_j(x)\} \\ &\text{subject to} \quad x \in \cap_{i=1}^m X_i, \end{aligned}$$

which is obtained by replacing convex inequality constraints of the form $g_j(x) \le 0$ with the nondifferentiable penalty terms $c \max\{0, g_j(x)\}$, where $c > 0$ is a penalty parameter (cf. Section 1.5). Then a possible incremental method at each iteration, would either do a subgradient iteration on f, or select one of the violated constraints (if any) and perform a subgradient iteration on the corresponding function g_j, or select one of the sets X_i and do an interpolated projection on it.

6.5 COORDINATE DESCENT METHODS

In this section we will consider the block coordinate descent approach that we discussed briefly in Section 2.1.2. We focus on the problem

$$\begin{aligned} &\text{minimize} \quad f(x) \\ &\text{subject to} \quad x \in X, \end{aligned}$$

where $f : \Re^n \mapsto \Re$ is a differentiable convex function, and X is a Cartesian product of closed convex sets $X_1, \ldots, X_m$:

$$X = X_1 \times X_2 \times \cdots \times X_m,$$

where X_i is a subset of $\Re^{n_i}$ (we allow that $n_i > 1$, although the most common case is when $n_i = 1$ for all i). The vector x is partitioned as

$$x = (x^1, x^2, \ldots, x^m),$$

where each x^i is a "block component" of x that is constrained to be in X_i, so the constraint $x \in X$ is equivalent to $x^i \in X_i$ for all $i = 1, \ldots, m$.

Similar to the proximal gradient and incremental methods of the preceding two sections, the idea is to split the optimization process into a sequence of simpler optimizations, with the motivation to exploit the special structure of the problem. In the block coordinate descent method, however, the simpler optimizations revolve around the components x^i of x, rather than the components of f or the components of X, as in the proximal gradient and the incremental methods.

The most common form of block coordinate descent method is defined as follows: given the current iterate $x_k = (x_k^1, \ldots, x_k^m)$, we generate the next iterate $x_{k+1} = (x_{k+1}^1, \ldots, x_{k+1}^m)$, according to

$$x_{k+1}^i \in \arg\min_{\xi \in X_i} f(x_{k+1}^1, \ldots, x_{k+1}^{i-1}, \xi, x_k^{i+1}, \ldots, x_k^m), \quad i = 1, \ldots, m; \tag{6.141}$$

where we assume that the preceding minimization has at least one optimal solution. Thus, at each iteration, the cost is minimized with respect to each of the block components x_k^i, taken in cyclic order, with each minimization incorporating the results of the preceding minimizations. Naturally, the method makes practical sense if the minimization in Eq. (6.141) is fairly easy. This is frequently so when each x^i is a scalar, but there are also other cases of interest, where x^i is multidimensional.

The coordinate descent approach has a sound theoretical basis thanks to its iterative cost function descent character, and is often conveniently applicable. In particular, when the coordinate blocks are one-dimensional, the descent direction does not require a special calculation. Moreover if the cost function is a sum of functions with "loose coupling" between the block components (i.e., each block component appears in just a few of the functions in the sum), then the calculation of the minimum along each block component may be simplified. Another structure that favors the use of block coordinate descent is when the cost function involves terms of the form $h(Ax)$, where A is a matrix such that computing $y = Ax$ is far more expensive than computing $h(y)$; this simplifies the minimization over a block component.

The following proposition gives the basic convergence result for the method. It turns out that it is necessary to make an assumption implying that the minimum in Eq. (6.141) is uniquely attained. This assumption is satisfied if f is strictly convex in each block component when all other block components are held fixed. We will discuss later a version of the algorithm, which involves quadratic regularization and does not require this assumption. While the proposition as stated applies to convex optimization problems, consistently with the framework of this section, the proof can be adapted to use only the continuous differentiability of f and not its convexity (see the proposition after the next one).

Proposition 6.5.1: (Convergence of Block Coordinate Descent) Let $f : \Re^n \mapsto \Re$ be convex and differentiable, and let $X = X_1 \times X_2 \times \cdots \times X_m$, where X_i are closed and convex. Assume further that for each $x = (x^1, \ldots, x^m) \in X$ and i,

$$f(x^1, \ldots, x^{i-1}, \xi, x^{i+1}, \ldots, x^m)$$

viewed as a function of ξ, attains a unique minimum over X_i. Let $\{x_k\}$ be the sequence generated by the block coordinate descent method (6.141). Then, every limit point of $\{x_k\}$ minimizes f over X.

Proof: Denote

$$z_k^i = \left(x_{k+1}^1, \ldots, x_{k+1}^i, x_k^{i+1}, \ldots, x_k^m\right).$$

Using the definition (6.141) of the method, we obtain

$$f(x_k) \geq f(z_k^1) \geq f(z_k^2) \geq \cdots \geq f(z_k^{m-1}) \geq f(x_{k+1}), \qquad \forall\, k. \tag{6.142}$$

Let $\bar{x} = (\bar{x}^1, \ldots, \bar{x}^m)$ be a limit point of the sequence $\{x_k\}$, and note that $\bar{x} \in X$ since X is closed. Equation (6.142) implies that the sequence $\{f(x_k)\}$ converges to $f(\bar{x})$. We will show that $\bar{x}$ satisfies the optimality condition

$$\nabla f(\bar{x})'(x - \bar{x}) \geq 0, \qquad \forall\, x \in X;$$

cf. Prop. 1.1.8 in Appendix B.

Let $\{x_{k_j} \mid j = 0, 1, \ldots\}$ be a subsequence of $\{x_k\}$ that converges to $\bar{x}$. From the definition (6.141) of the algorithm and Eq. (6.142), we have

$$f(x_{k_{j+1}}) \leq f(z_{k_j}^1) \leq f(x^1, x_{k_j}^2, \ldots, x_{k_j}^m), \qquad \forall\, x^1 \in X_1.$$

Taking the limit as j tends to infinity, we obtain

$$f(\bar{x}) \leq f(x^1, \bar{x}^2, \ldots, \bar{x}^m), \qquad \forall\, x^1 \in X_1. \tag{6.143}$$

Using the optimality conditions of Prop. 1.1.8 in Appendix B, we conclude that

$$\nabla_1 f(\bar{x})'(x^1 - \bar{x}^1) \geq 0, \qquad \forall\, x^1 \in X_1,$$

where $\nabla_i f$ denotes the gradient of f with respect to the component x^i.

The idea of the proof is now to show that $\{z_{k_j}^1\}$ converges to $\bar{x}$ as $j \to \infty$, so that by repeating the preceding argument with $\{z_{k_j}^1\}$ in place of $\{x_{k_j}\}$, we will have

$$\nabla_2 f(\bar{x})'(x^2 - \bar{x}^2) \geq 0, \qquad \forall\, x^2 \in X_2.$$

We can then continue similarly to obtain

$$\nabla_i f(\bar{x})'(x^i - \bar{x}^i) \geq 0, \qquad \forall\, x^i \in X_i,$$

for all $i = 1, \ldots, m$. By adding these inequalities, and using the Cartesian product structure of the set X, it follows that $\nabla f(\bar{x})'(x - \bar{x}) \geq 0$ for all $x \in X$, thereby completing the proof.

To show that $\{z^1_{k_j}\}$ converges to $\bar{x}$ as $j \to \infty$, we assume the contrary, or equivalently that $\{z^1_{k_j} - x_{k_j}\}$ does not converge to zero. Let

$$\gamma_{k_j} = \|z^1_{k_j} - x_{k_j}\|.$$

By possibly restricting to a subsequence of $\{k_j\}$, we may assume that there exists some $\overline{\gamma} > 0$ such that $\gamma_{k_j} \geq \overline{\gamma}$ for all j. Let

$$s^1_{k_j} = (z^1_{k_j} - x_{k_j})/\gamma_{k_j}.$$

Thus, $z^1_{k_j} = x_{k_j} + \gamma_{k_j} s^1_{k_j}$, $\|s^1_{k_j}\| = 1$, and $s^1_{k_j}$ differs from zero only along the first block component. Notice that $s^1_{k_j}$ belongs to a compact set and therefore has a limit point $\bar{s}^1$. By restricting to a further subsequence of $\{k_j\}$, we assume that $s^1_{k_j}$ converges to $\bar{s}^1$.

Let us fix some $\epsilon \in [0,1]$. Since $0 \leq \epsilon\overline{\gamma} \leq \gamma_{k_j}$, the vector $x_{k_j} + \epsilon\overline{\gamma} s^1_{k_j}$ lies on the line segment joining x_{k_j} and $x_{k_j} + \gamma_{k_j} s^1_{k_j} = z^1_{k_j}$, and belongs to X since X is convex. Using the fact that f is monotonically nonincreasing on the interval from x_{k_j} to $z^1_{k_j}$ (by the convexity of f), we obtain

$$f(z^1_{k_j}) = f(x_{k_j} + \gamma_{k_j} s^1_{k_j}) \leq f(x_{k_j} + \epsilon\overline{\gamma} s^1_{k_j}) \leq f(x_{k_j}).$$

Since $f(x_k)$ converges to $f(\bar{x})$, Eq. (6.142) shows that $f(z^1_{k_j})$ also converges to $f(\bar{x})$. Taking the limit as j tends to infinity, we obtain

$$f(\bar{x}) \leq f(\bar{x} + \epsilon\overline{\gamma}\bar{s}^1) \leq f(\bar{x}).$$

We conclude that $f(\bar{x}) = f(\bar{x} + \epsilon\overline{\gamma}\bar{s}^1)$, for every $\epsilon \in [0,1]$. Since $\overline{\gamma}\bar{s}^1 \neq 0$ and by Eq. (6.143), $\bar{x}^1$ attains the minimum of $f(x^1, \bar{x}^2, \ldots, \bar{x}^m)$ over $x^1 \in X_1$, this contradicts the hypothesis that f is uniquely minimized when viewed as a function of the first block component. This contradiction establishes that $z^1_{k_j}$ converges to $\bar{x}$, which as noted earlier, shows that $\nabla_2 f(\bar{x})'(x^2 - \bar{x}^2) \geq 0$ for all $x^2 \in X_2$.

By using $\{z^1_{k_j}\}$ in place of $\{x_{k_j}\}$, and $\{z^2_{k_j}\}$ in place of $\{z^1_{k_j}\}$ in the preceding arguments, we can show that $\nabla_3 f(\bar{x})'(x^3 - \bar{x}^3) \geq 0$ for all $x^3 \in X_3$, and similarly $\nabla_i f(\bar{x})'(x^i - \bar{x}^i) \geq 0$ for all $x^i \in X_i$ and i. **Q.E.D.**

The preceding proposition applies to convex optimization problems. However, its proof can be simply adapted to use only the continuous differentiability of f and not its convexity. For this an extra assumption is needed (monotonic decrease to the minimum along each coordinate), as stated in the following proposition from [Ber99], Section 2.7.

Proposition 6.5.2: (Convergence of Block Coordinate Descent – Nonconvex Case) Let $f : \Re^n \mapsto \Re$ be continuously differentiable, and let $X = X_1 \times X_2 \times \cdots \times X_m$, where X_i are closed and convex. Assume further that for each $x = (x^1, \ldots, x^m) \in X$ and i,

$$f(x^1, \ldots, x^{i-1}, \xi, x^{i+1}, \ldots, x^m) \tag{6.144}$$

viewed as a function of ξ, attains a unique minimum $\bar{\xi}$ over X_i, and is monotonically nonincreasing in the interval from x^i to $\bar{\xi}$. Let $\{x_k\}$ be the sequence generated by the block coordinate descent method (6.141). Then, every limit point $\bar{x}$ of $\{x_k\}$ satisfies the optimality condition $\nabla f(\bar{x})'(x - \bar{x}) \geq 0$ for all $x \in X$.

The proof is nearly identical to the one of Prop. 6.5.1, using at the right point the monotonic nonincrease assumption in place of convexity of f. An alternative assumption, also discussed in [Ber99], Section 2.7, under which the conclusion of Prop. 6.5.2 can be shown with a similar proof is that the sets X_i are compact (as well as convex), and that for each i and $x \in X$, the function (6.144) of the ith block-component ξ attains a unique minimum over X_i, when all other block-components are held fixed.

The nonnegative matrix factorization problem, described in Section 1.3, is an important example where the cost function is convex as a function of each block component, but not convex as a function of the entire set of block components. For this problem, a special convergence result for the case of just two block components applies, which does not require the uniqueness of the minimum in the two block component minimizations. This result can be shown with a variation of the proofs of Props. 6.5.1 and 6.5.2; see [GrS99], [GrS00].

6.5.1 Variants of Coordinate Descent

There are many variations of coordinated descent, which are aimed at improved efficiency, application-specific structures, and distributed computing environments. We describe some of the possibilities here and in the exercises, and we refer to the literature for a more detailed analysis.

(a) We may apply *coordinate descent in the context of a dual problem.* This is often convenient because the dual constraint set often has the required Cartesian product structure (e.g., X may be the nonnegative orthant). An early algorithm of this type for quadratic programming was given in [Hil57]. For further discussion and development, we refer to the books [BeT89a], Section 3.4.1, [Ber99], Section 6.2.1, [CeZ97], and [Her09], and the references quoted there. For other applications in a dual context, including single commodity network flow optimization, we refer to the papers [BHT87], [TsB87], [TsB90], [TsB91], [Aus92].

(b) There is a *combination of coordinate descent with the proximal algorithm*, which aims to circumvent the need for the "uniqueness of minimum" assumption in Prop. 6.5.1; see [Tse91a], [Aus92], [GrS00]. This method is obtained by applying cyclically the coordinate iteration

$$x_{k+1}^i \in \arg\min_{\xi \in X_i} \left\{ f(x_{k+1}^1, \ldots, x_{k+1}^{i-1}, \xi, x_k^{i+1}, \ldots, x_k^m) + \frac{1}{2c}\|\xi - x_k^i\|^2 \right\},$$

where c is a positive scalar. Assuming that f is convex and differentiable, it can be seen that every limit point of the sequence $\{x_k\}$ is a global minimum. This is easily proved by applying the result of Prop. 6.5.1 to the cost function

$$F(x, y) = f(x) + \frac{1}{2c}\|x - y\|^2.$$

(c) The preceding combination of coordinate descent with the proximal algorithm is an example of a broader class of methods whereby instead of carrying out exactly the coordinate minimization

$$x_{k+1}^i \in \arg\min_{\xi \in X_i} f(x_{k+1}^1, \ldots, x_{k+1}^{i-1}, \xi, x_k^{i+1}, \ldots, x_k^m),$$

[cf. Eq. (6.141)], we perform one or more iterations of a descent algorithm aimed at solving this minimization. Aside from the proximal algorithm of (b) above, there are *combinations with other descent algorithms*, including conditional gradient, gradient projection, and two-metric projection methods. A key fact here is that there is guaranteed cost function descent after cycling through all the block components. See e.g., [Lin07], [LJS12], [Spa12], [Jag13], [BeT13], [RHL13], [RHZ14], [RiT14].

(d) When f is convex but nondifferentiable, the coordinate descent approach may fail because there may exist nonoptimal points starting from which it is impossible to obtain cost function descent along any one of the scalar coordinates. As an example consider minimization of $f(x^1, x^2) = x^1 + x^2 + 2|x^1 - x^2|$ over $x^1, x^2 \geq 0$. At points (x^1, x^2) with $x^1 = x^2 > 0$, coordinate descent makes no progress.

There is an important special case, however, where the nondifferentiability of f is inconsequential, and at each nonoptimal point, it is possible to find a direction of descent among the scalar coordinate directions. This is the case where *the nondifferentiable portion of* f *is separable*, i.e., f has the form

$$f(x) = F(x) + \sum_{i=1}^{n} G_i(x^i), \tag{6.145}$$

where F is convex and differentiable, and each G_i is a function of the ith block component that is convex but not necessarily differentiable; see Exercise 6.10 for a brief discussion, and [Aus76], [Tse01a], [TsY09], [FHT10], [Tse10], [Nes12], [RHL13], [SaT13], [BST14], [HCW14], [RiT14] for detailed analyses. The coordinate descent approach for this type of cost function may also be combined with the ideas of the proximal gradient method of Section 6.3; see [FeR12], [QSG13], [LLX14], [LiW14], [RiT14]. A case of special interest, which arises in machine learning problems involving ℓ_1-regularization, is when $\sum_{i=1}^{n} G_i(x^i)$ is a positive multiple of the ℓ_1 norm.

(e) For the case where f is convex and nondifferentiable but not of the separable form (6.145), a coordinate descent approach is possible but requires substantial modifications and special assumptions. In one possible approach, known as *auction and* ϵ*-relaxation*, the coordinate search is nonmonotonic, in the sense that a single coordinate may be allowed to change even if this deteriorates the cost function value. In particular, when a coordinate is increased (or decreased), it is set to $\epsilon > 0$ (or $-\epsilon$, respectively) plus the value that minimizes the cost function along that coordinate. For (single commodity) network special structures, including assignment, max-flow, linear cost transshipment, and convex separable network flow problems with and without gains, there are sound algorithms of this type, with excellent computational complexity properties, which can reach or approach an optimal solution, assuming that ϵ is sufficiently small; see the papers [Ber79], [Ber92], [BPT97a], [BPT97b], [Ber98], [TsB00], the books [BeT89a], [Ber91], [BeT97], [Ber98], and the references quoted there.

An alternative approach for nondifferentiable cost is to use (exact) coordinate descent when this leads to cost function improvement, and to compute a more complicated descent direction (a linear combination of multiple coordinate directions) otherwise. For dual problems of (single commodity) linear network optimization problems, the latter descent direction may be computed efficiently (it is the sum of appropriately chosen coordinate directions). This idea has led to fast algorithms, known as *relaxation methods*; see the papers [Ber81], [Ber85], [BeT94b], and the books [BeT89a], [Ber91], [Ber98]. In practice, most of the iterations use single coordinate directions, and the computationally more complex multiple-coordinate directions are needed infrequently. This approach has also been applied to network flow problems with gains [BeT88] and to convex network flow problems [BHT87], and may be applied more broadly to specially structured problems where multiple-coordinate descent directions can be efficiently computed.

(f) A possible way to improve the convergence properties of the coordinate descent method is to use *an irregular order instead of a fixed cyclic order* for coordinate selection. Convergence of such a method can usually be established simply: the result of Prop. 6.5.1 can be shown for a method where the order of iteration may be arbitrary as long as there is an integer M such that each block component is iterated at least once in every group of M contiguous iterations. The proof is similar to the one of Prop. 6.5.1.

An idea to accelerate convergence that has received a lot of attention is to use *randomization* in the choice of coordinate at each iteration; see [Spa03], [StV09], [LeL10], [Nee10], [Nes12], [LeS13], [NeC13]. An important question in this context, which has been addressed in various ways by these references, is whether the convergence rate of the method can be enhanced in this way. In particular, a complexity analysis in the paper [Nes12] of randomized coordinate selection has stimulated much interest.

An alternative possibility is to use a deterministic coordinate selection rule, which aims to identify promising coordinates that lead to large cost improvements. A classical approach of this type is the *Gauss-Southwell order*, which at each iterate chooses the coordinate direction of steepest descent (the one with minimum directional derivative). While this requires some overhead for coordinate selection, it is known to lead to faster convergence (fewer iterations) than cyclic or randomized selection, based on analysis and practical experience [DRT11], [ScF14] [see Exercise 6.11(b)].

(g) A more extreme type of irregular method is a *distributed asynchronous version*, which is executed concurrently at different processors with minimal coordination. The book [BeT89a] (Section 6.3.5) discusses the circumstances under which asynchrony may improve the convergence rate of the method. We discuss the convergence of asynchronous coordinate descent in the next section.

6.5.2 Distributed Asynchronous Coordinate Descent

We will now consider the distributed asynchronous implementation of fixed point algorithms, which applies as a special case to coordinate descent. The implementation is *totally asynchronous* in the sense that there is no bound on the size of the communication delays between the processors; cf. the terminology of Section 2.1.6. The analysis in this section is based on the author's paper [Ber83] (see [BeT89a], [BeT91], [FrS00] for broad surveys of totally asynchronous algorithms). A different line of analysis applies to *partially asynchronous* algorithms, for which it is necessary to have a bound on the size of the communication delays. Such algorithms will not be considered here; see [TBA86], [BeT89a], Chapter 7, for gradient-like methods, [BeT89a] for network flow algorithms, [TBT90] for nonexpansive iterations, and [LiW14] which focuses on coordinate descent methods, under less restrictive conditions than the ones of the present section (a sup-norm contraction property of the algorithmic map is not assumed).

Let us consider parallelizing a stationary fixed point algorithm by separating it into several local algorithms operating concurrently at different processors. As we discussed in Section 2.1.6, in an asynchronous algorithm, the local algorithms do not have to wait at predetermined points for predetermined information to become available. Thus some processors may execute more iterations than others, while the communication delays between processors may be unpredictable. Another practical setting that may be modeled well by a distributed asynchronous iteration is when all computation takes place at a single computer, but any number of coordinates may be simultaneously updated at a time, with the order of coordinate selection possibly being random.

With this context in mind, we introduce a model of asynchronous distributed solution of abstract fixed point problems of the form $x = F(x)$, where F is a given function. We represent x as $x = (x^1, \ldots, x^m)$, where $x^i \in R^{n_i}$ with n_i being some positive integer. Thus $x \in \Re^n$, where $n = n_1 + \cdots + n_m$, and F maps $\Re^n$ to $\Re^n$. We denote by $F_i : \Re^n \mapsto \Re^{n_i}$ the ith component of F, so $F(x) = \big(F_1(x), \ldots, F_m(x)\big)$. Our computation framework involves m interconnected processors, the ith of which updates the ith component x^i by applying the corresponding mapping F_i. Thus, in a (synchronous) distributed fixed point algorithm, processor i iterates at time t according to

$$x_{t+1}^i = F_i(x_t^1, \ldots, x_t^m), \qquad \forall\, i = 1, \ldots, m. \tag{6.146}$$

To accommodate the distributed algorithmic framework and its overloaded notation, we will use subscript t to denote iterations/times where some (but not all) processors update their corresponding components, reserving the index k for computation stages involving all processors, and also reserving superscript i to denote component/processor index.

In an asynchronous version of the algorithm, processor i updates x^i only for t in a selected subset $\mathcal{R}_i$ of iterations, and with components x^j, $j \neq i$, supplied by other processors with delays $t - \tau_{ij}(t)$,

$$x_{t+1}^i = \begin{cases} F_i\left(x_{\tau_{i1}(t)}^1, \ldots, x_{\tau_{im}(t)}^m\right) & \text{if } t \in \mathcal{R}_i, \\ x_t^i & \text{if } t \notin \mathcal{R}_i. \end{cases} \tag{6.147}$$

Here $\tau_{ij}(t)$ is the time at which the jth coordinate used in this update was computed, and the difference $t - \tau_{ij}(t)$ is referred to as the communication delay from j to i at time t.

We noted in Section 2.1.6 that an example of an algorithm of this type is a coordinate descent method, where we assume that the ith scalar coordinate is updated at a subset of times $\mathcal{R}_i \subset \{0, 1, \ldots\}$, according to

$$x_{t+1}^i \in \arg\min_{\xi \in \Re} f\left(x_{\tau_{i1}(t)}^1, \ldots, x_{\tau_{i,i-1}(t)}^{i-1}, \xi, x_{\tau_{i,i+1}(t)}^{i+1}, \ldots, x_{\tau_{im}(t)}^m\right),$$

and is left unchanged ($x_{k+1}^i = x_k^i$) if $t \notin \mathcal{R}_i$. Here we can assume without loss of generality that each scalar coordinate is assigned to a separate processor. The reason is that a physical processor that updates a block of scalar coordinates may be replaced by a block of fictitious processors, each assigned to a single scalar coordinate, and updating their coordinates simultaneously.

To discuss the convergence of the asynchronous algorithm (6.147), we introduce the following assumption.

Assumption 6.5.1: (Continuous Updating and Information Renewal)

(1) The set of times $\mathcal{R}_i$ at which processor i updates x^i is infinite, for each $i = 1, \ldots, m$.

(2) $\lim_{t\to\infty} \tau_{ij}(t) = \infty$ for all $i, j = 1, \ldots, m$.

Assumption 6.5.1 is natural, and is essential for any kind of convergence result about the algorithm. In particular, the condition $\tau_{ij}(t) \to \infty$ guarantees that outdated information about the processor updates will eventually be purged from the computation. It is also natural to assume that $\tau_{ij}(t)$ is monotonically increasing with t, but this assumption is not necessary for the subsequent analysis.

We wish to show that $\{x_t\}$ converges to a fixed point of F, and to this end we employ the following convergence theorem for totally asynchronous iterations from [Ber83]. The theorem has served as the basis for the treatment of totally asynchronous iterations in the book [BeT89a] (Chapter 6), including coordinate descent and asynchronous gradient-based optimization.

Proposition 6.5.3: (Asynchronous Convergence Theorem) Let F have a unique fixed point x^*, let Assumption 6.5.1 hold, and assume that there is a sequence of nonempty subsets $\{S(k)\} \subset \Re^n$ with

$$S(k+1) \subset S(k), \quad k = 0, 1, \ldots,$$

and is such that if $\{y_k\}$ is any sequence with $y_k \in S(k)$, for all $k \geq 0$, then $\{y_k\}$ converges to x^*. Assume further the following:

(1) *Synchronous Convergence Condition:* We have

$$F(x) \in S(k+1), \qquad \forall\, x \in S(k),\ k = 0, 1, \ldots.$$

(2) *Box Condition:* For all k, $S(k)$ is a Cartesian product of the form

$$S(k) = S_1(k) \times \cdots \times S_m(k),$$

where $S_i(k)$ is a subset of $\Re^{n_i}$, $i = 1, \ldots, m$.

Then for every initial vector $x_0 \in S(0)$, the sequence $\{x_t\}$ generated by the asynchronous algorithm (6.147) converges to x^*.

Proof: To explain the idea of the proof, let us note that the given conditions imply that updating any component x^i, by applying F to $x \in S(k)$, while leaving all other components unchanged, yields a vector in $S(k)$. Thus, once enough time passes so that the delays become "irrelevant," then after x enters $S(k)$, it stays within $S(k)$. Moreover, once a component x^i enters the subset $S_i(k)$ and the delays become "irrelevant," x^i gets permanently within the smaller subset $S_i(k+1)$ at the first time that x^i is iterated on with $x \in S(k)$. Once each component x^i, $i = 1, \ldots, m$, gets within $S_i(k+1)$, the entire vector x is within $S(k+1)$ by the box condition. Thus the iterates from $S(k)$ eventually get into $S(k+1)$ and so on, and converge pointwise to x^* in view of the assumed properties of $\{S(k)\}$.

With this idea in mind, we show by induction that for each $k \geq 0$, there is a time t_k such that:

(1) $x_t \in S(k)$ for all $t \geq t_k$.

(2) For all i and $t \in \mathcal{R}_i$ with $t \geq t_k$, we have

$$\left(x^1_{\tau_{i1}(t)}, \ldots, x^m_{\tau_{im}(t)}\right) \in S(k).$$

[In words, after some time, all fixed point estimates will be in $S(k)$ and all estimates used in iteration (6.147) will come from $S(k)$.]

The induction hypothesis is true for $k = 0$ since $x_0 \in S(0)$. Assuming it is true for a given k, we will show that there exists a time t_{k+1} with the required properties. For each $i = 1, \ldots, m$, let $t(i)$ be the first element of $\mathcal{R}_i$ such that $t(i) \geq t_k$. Then by the synchronous convergence condition, we have $F(x_{t(i)}) \in S(k+1)$, implying (in view of the box condition) that

$$x^i_{t(i)+1} \in S_i(k+1).$$

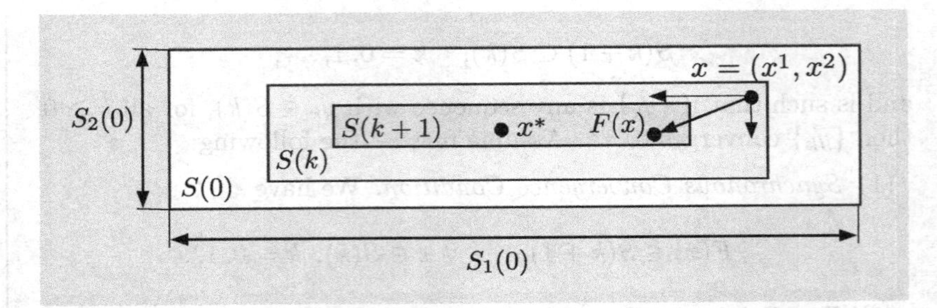

Figure 6.5.1. Geometric interpretation of the conditions of the asynchronous convergence theorem. We have a nested sequence of boxes $\{S(k)\}$ such that $F(x) \in S(k+1)$ for all $x \in S(k)$.

Similarly, for every $t \in \mathcal{R}_i$, $t \geq t(i)$, we have $x^i_{t+1} \in S_i(k+1)$. Between elements of $\mathcal{R}_i$, x^i_t does not change. Thus,

$$x^i_t \in S_i(k+1), \qquad \forall\, t \geq t(i) + 1.$$

Let $t'_k = \max_i\{t(i)\} + 1$. Then, using the box condition we have

$$x_t \in S(k+1), \qquad \forall\, t \geq t'_k.$$

Finally, since by Assumption 6.5.1, we have $\tau_{ij}(t) \to \infty$ as $t \to \infty$, $t \in \mathcal{R}_i$, we can choose a time $t_{k+1} \geq t'_k$ that is sufficiently large so that $\tau_{ij}(t) \geq t'_k$ for all i, j, and $t \in \mathcal{R}_i$ with $t \geq t_{k+1}$. We then have, for all $t \in \mathcal{R}_i$ with $t \geq t_{k+1}$ and $j = 1, \ldots, m$, $x^j_{\tau_{ji}(t)} \in S_j(k+1)$, which (by the box condition) implies that

$$\left(x^1_{\tau_{i1}(t)}, \ldots, x^m_{\tau_{im}(t)}\right) \in S(k+1).$$

The induction is complete. **Q.E.D.**

Figure 6.5.1 illustrates the assumptions of the preceding convergence theorem. The main issue in applying the theorem is to identify the set sequence $\{S(k)\}$ and to verify the assumptions of Prop. 6.5.3. These assumptions hold in two primary contexts of interest. The first is when $S(k)$ are spheres centered at x^* with respect to the weighted sup-norm

$$\|x\|^w_\infty = \max_{i=1,\ldots,n} \left|\frac{x^i}{w^i}\right|,$$

where $w = (w^1, \ldots, w^n)$ is a vector of positive weights (see the following proposition). The second context is based on monotonicity conditions, and is particularly useful in dynamic programming algorithms, for which we refer to the papers [Ber82c], [BeY10], and the books [Ber12], [Ber13].

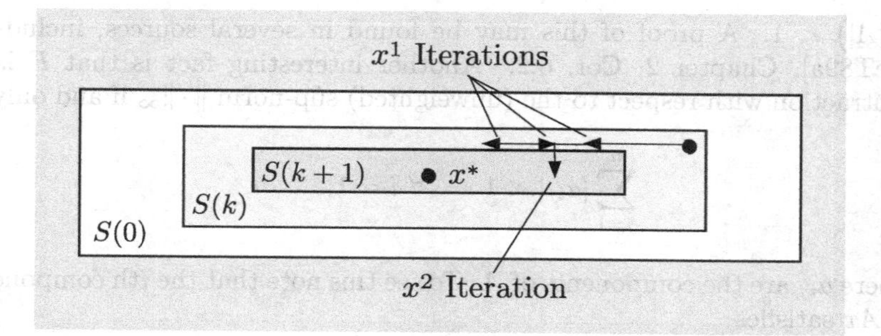

Figure 6.5.2. Geometric interpretation of the mechanism for asynchronous convergence. Iteration on a single component x^i, keeps x in $S(k)$, while it moves x^i into the corresponding component $S_i(k+1)$ of $S(k+1)$, where it remains throughout the subsequent iterations. Once all components x^i have been iterated on at least once, the iterate is guaranteed to be in $S(k+1)$.

Figure 6.5.2 illustrates the mechanism by which asynchronous convergence is achieved.

As an example, let us apply the preceding convergence theorem under a weighted sup-norm contraction assumption.

Proposition 6.5.4: Let F be a contraction with respect to a weighted sup-norm $\|\cdot\|_\infty^w$ and modulus $\rho < 1$, and let Assumption 6.5.1 hold. Then a sequence $\{x_t\}$ generated by the asynchronous algorithm (6.147) converges pointwise to x^*.

Proof: We apply Prop. 6.5.3 with

$$S(k) = \{x \in \Re^n \mid \|x_k - x^*\|_\infty^w \le \rho^k \|x_0 - x^*\|_\infty^w\}, \qquad k = 0, 1, \ldots.$$

Since F is a contraction with modulus ρ, the synchronous convergence condition is satisfied. Since F is a weighted sup-norm contraction, the box condition is also satisfied, and the result follows. **Q.E.D.**

The contraction property of the preceding proposition can be verified in a few interesting special cases. In particular, let F be linear of the form

$$F(x) = Ax + b,$$

where A and b are given $n \times n$ matrix and vector in $\Re^n$. Let us denote by $|A|$ the matrix whose components are the absolute values of the components of A and let $\sigma(|A|)$ denote the spectral radius of $|A|$ (the largest modulus among the moduli of the eigenvalues of $|A|$). Then it can be shown that F is a contraction with respect to some weighted sup-norm if and only if

$\sigma(|A|) < 1$. A proof of this may be found in several sources, including [BeT89a], Chapter 2, Cor. 6.2. Another interesting fact is that F is a contraction with respect to the (unweighted) sup-norm $\|\cdot\|_\infty$ if and only if

$$\sum_{j=1}^{n} |a_{ij}| < 1 \qquad \forall\, i = 1, \ldots, n,$$

where a_{ij} are the components of A. To see this note that the ith component of Ax satisfies

$$\big|(Ax)_i\big| \le \sum_{j=1}^{n} |a_{ij}|\,|x_j| \le \sum_{j=1}^{n} |a_{ij}|\,\|x\|_\infty,$$

so $\|Ax\|_\infty \le \rho \|x\|_\infty$, where $\rho = \max_i \sum_{j=1}^{n} |a_{ij}| < 1$. This shows that A (and hence also F) is a contraction with respect to $\|\cdot\|_\infty$. A similar argument shows also the reverse assertion.

We finally note a few extensions of the theorem. It is possible to allow F to be time-varying, so in place of F we operate with a sequence of mappings F_k, $k = 0, 1, \ldots$. Then if all F_k have a common fixed point, the conclusion of the theorem holds (see [BeT89a] for details). Another extension is to allow F to have multiple fixed points and introduce an assumption that roughly says that $\cap_{k=0}^{\infty} S(k)$ is the set of all fixed points. Then the conclusion is that any limit point of $\{x_t\}$ is a fixed point.

6.6 GENERALIZED PROXIMAL METHODS

The proximal algorithm admits several extensions, which may be particularly useful in specialized application domains, such as inference and signal processing. Moreover the algorithm, with some unavoidable limitations, applies to nonconvex problems as well. A general form of the algorithm for minimizing a function $f : \Re^n \mapsto (-\infty, \infty]$ is

$$x_{k+1} \in \arg\min_{x \in \Re^n} \big\{ f(x) + D_k(x, x_k) \big\}, \tag{6.148}$$

where $D_k : \Re^{2n} \mapsto (-\infty, \infty]$ is a regularization term that replaces the quadratic

$$\frac{1}{2c_k}\|x - x_k\|^2$$

in the proximal algorithm. We assume that $D_k(\cdot, x_k)$ is a closed proper (extended real-valued) convex function for each x_k.

The algorithm (6.148) can be graphically interpreted similar to the proximal algorithm, as illustrated in Fig. 6.6.1. This figure, and the convergence and rate of convergence results of Section 5.1 provide some qualitative

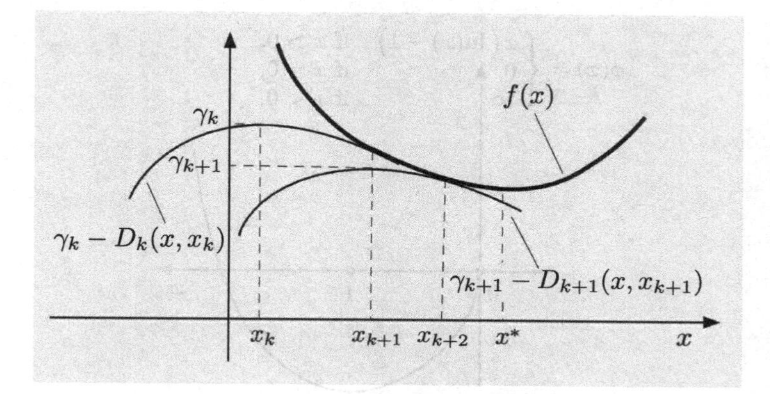

Figure 6.6.1. Illustration of the generalized proximal algorithm (6.148) for a convex cost function f. The regularization term is convex but need not be quadratic or real-valued. In this figure, γ_k is the scalar by which the graph of $-D_k(\cdot, x_k)$ must be raised so that it just touches the graph of f.

guidelines about the kind of behavior that may be expected from the algorithm when f is a closed proper convex function. In particular, under suitable assumptions on D_k, we expect to be able to show convergence to the optimal value and convergence to an optimal solution if one exists (cf. Prop. 5.1.3).

Example 6.6.1: (Entropy Minimization Algorithm)

Let us consider the case where

$$D_k(x, y) = \frac{1}{c_k} \sum_{i=1}^{n} x^i \left(\ln \left(\frac{x^i}{y^i} \right) - 1 \right),$$

where x^i and y^i denote the scalar components of x and y, respectively. This regularization term is based on the scalar function $\phi : \Re \mapsto (-\infty, \infty]$, given by

$$\phi(x) = \begin{cases} x\big(\ln(x) - 1\big) & \text{if } x > 0, \\ 0 & \text{if } x = 0, \\ \infty & \text{if } x < 0, \end{cases} \tag{6.149}$$

which is referred to as the *entropy function*, and is shown in Fig. 6.6.2. Note that ϕ is finite only for nonnegative arguments.

The proximal algorithm that uses the entropy function is given by

$$x_{k+1} \in \arg\min_{x \in \Re^n} \left\{ f(x) + \frac{1}{c_k} \sum_{i=1}^{n} x^i \left(\ln \left(\frac{x^i}{x_k^i} \right) - 1 \right) \right\}, \tag{6.150}$$

where x_k^i denotes the ith coordinate of x_k; see Fig. 6.6.3. Because the logarithm is finite only for positive arguments, the algorithm requires that $x_0^i > 0$

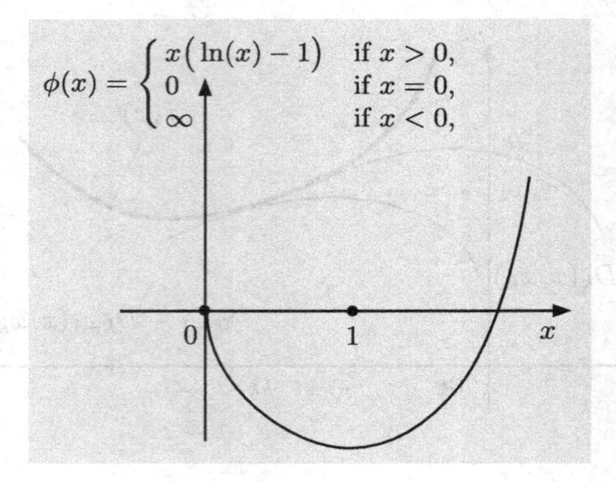

Figure 6.6.2. Illustration of the entropy function (6.149).

for all i, and must generate a sequence that lies strictly within the positive orthant. Thus the algorithm can be used only for functions f for which the minimum above is well-defined and guaranteed to be attained within the positive orthant.

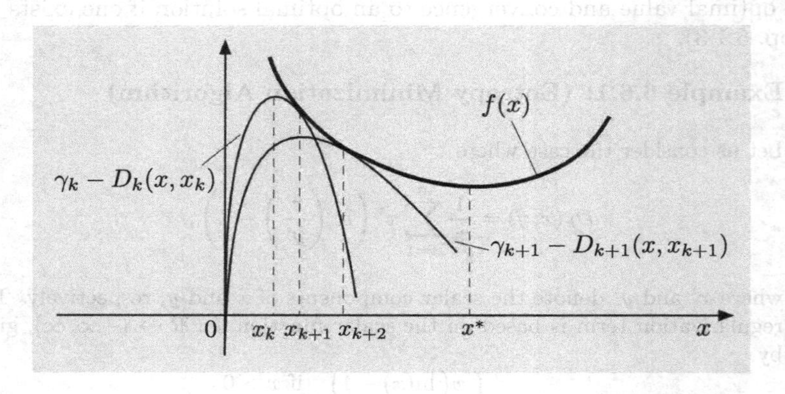

Figure 6.6.3. Illustration of the entropy minimization algorithm (6.150).

We may speculate on the properties and potential applications of the proximal approach with nonquadratic regularization, in various algorithmic contexts, based on the corresponding applications that we have considered so far. These are:

(a) *Dual proximal algorithms*, based on application of the Fenchel Duality Theorem (Prop. 1.2.1) to the minimization (6.148); cf. Section 5.2.

(b) *Augmented Lagrangian methods* with nonquadratic penalty functions.

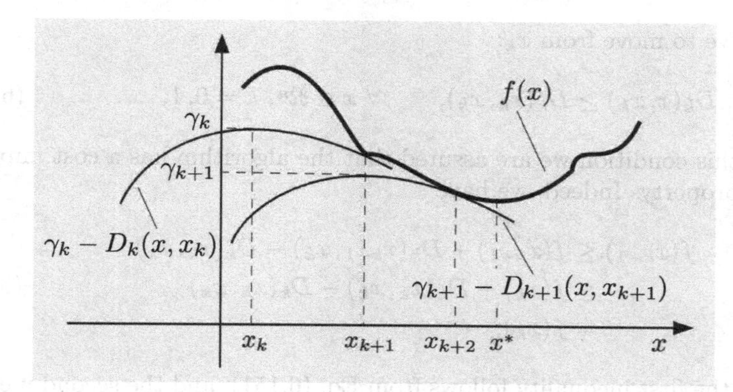

Figure 6.6.4. Illustration of the generalized proximal algorithm (6.148) for the case of a nonconvex cost function f.

Here the augmented Lagrangian function will be related to the conjugate of $D_k(\cdot, x_k)$; cf. Section 5.2.1. We will discuss later in this section an example involving an augmented Lagrangian with exponential penalty function (the conjugate of the entropy function).

(c) Combinations with *polyhedral approximations*. There are straightforward extensions of proximal cutting plane and bundle algorithms involving nonquadratic regularization terms; cf. Section 5.3.

(d) Extensions of *incremental subgradient proximal methods*; cf. Section 6.4. Again such extensions are straightforward.

(e) *Gradient projection algorithms with "nonquadratic metric."* We will discuss an example of such an algorithm, the so-called mirror descent method, later in this section.

We may also consider in principle the application of the approach to a nonconvex cost function f as well. Its behavior, however, in this case may be complicated and/or unreliable, as indicated in Fig. 6.6.4. An important question is whether the minimum in the proximal iteration (6.148) is attained for all k; if f is not assumed convex, this is not automatically guaranteed, even if D_k is quadratic [take for example $f(x) = -\|x\|^3$ and $D_k(x, x_k) = \|x - x_k\|^2$]. To simplify the presentation, we will implicitly assume the attainment of the minimum throughout our discussion; it is guaranteed for example if f is closed proper convex, and $D_k(\cdot, x_k)$ is closed and coercive [satisfies $D_k(x, x_k) \to \infty$ as $\|x\| \to \infty$], and its effective domain intersects with $\text{dom}(f)$ for all k (cf. Prop. 3.2.3 in Appendix B).

Let us now introduce two conditions on D_k that guarantee some sound behavior for the algorithm, even when f is not convex. The first is a "stabilization property," whereby adding D_k to f does not produce an incentive to move from x_k:

$$D_k(x, x_k) \geq D_k(x_k, x_k), \qquad \forall\, x \in \Re^n,\ k = 0, 1, \ldots. \tag{6.151}$$

With this condition we are assured that the algorithm has a cost improvement property. Indeed, we have

$$\begin{aligned} f(x_{k+1}) &\leq f(x_{k+1}) + D_k(x_{k+1}, x_k) - D_k(x_k, x_k) \\ &\leq f(x_k) + D_k(x_k, x_k) - D_k(x_k, x_k) \\ &= f(x_k), \end{aligned} \tag{6.152}$$

where the first inequality follows from Eq. (6.151), and the second inequality follows from the definition (6.148) of the algorithm. The condition (6.151) also guarantees that

$$x^* \in \arg\min_{x \in \Re^n} f(x) \qquad \Rightarrow \qquad x^* \in \arg\min_{x \in \Re^n} \big\{ f(x) + D_k(x, x^*) \big\},$$

in which case we assume that the algorithm stops.

However, for the algorithm to be reliable, an additional condition is required to guarantee that it produces strict cost improvement when outside of X^*, the set of (global) minima of f. One such condition is that the algorithm can stop only at points of X^*, i.e.,

$$x_k \in \arg\min_{x \in \Re^n} \big\{ f(x) + D_k(x, x_k) \big\} \qquad \Rightarrow \qquad x_k \in X^*, \tag{6.153}$$

in which case, the second inequality in the calculation of Eq. (6.152) is strict when $x_k \notin X^*$, implying that

$$f(x_{k+1}) < f(x_k), \qquad \text{if } x_k \notin X^*. \tag{6.154}$$

A set of assumptions guaranteeing the condition (6.153) are:

(a) f is convex.

(b) $D_k(\cdot, x_k)$ satisfies Eq. (6.151), and is convex and differentiable at x_k.

(c) We have

$$\operatorname{ri}\big(\operatorname{dom}(f)\big) \cap \operatorname{ri}\big(\operatorname{dom}(D_k(\cdot, x_k))\big) \neq \varnothing. \tag{6.155}$$

To see this, note that if

$$x_k \in \arg\min_{x \in \Re^n} \big\{ f(x) + D_k(x, x_k) \big\},$$

by the Fenchel Duality Theorem (Prop. 1.2.1), there exists a dual optimal solution λ^* such that $-\lambda^*$ is a subgradient of $D_k(\cdot, x_k)$ at x_k, so that $\lambda^* = 0$ [by Eq. (6.151)], and also λ^* is a subgradient of f at x_k, so that x_k

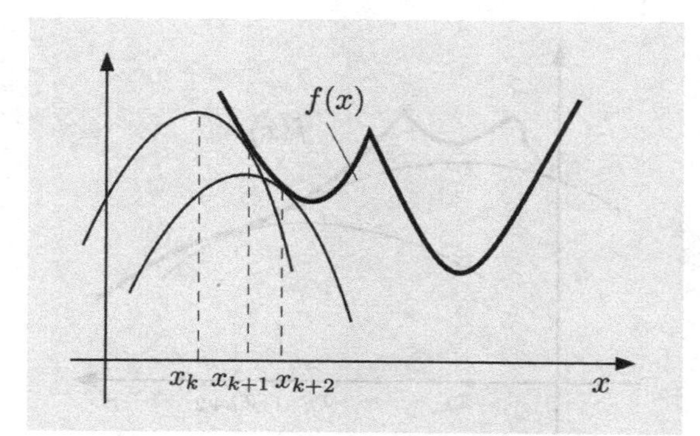

Figure 6.6.5. Illustration of a case where the generalized proximal algorithm (6.156) converges to a local minimum that is not global. In this example convergence to the global minimum would be attained if the regularization term $D_k(\cdot, x_k)$ were sufficiently "flat."

minimizes f. Note that the condition (6.153) may fail if $D_k(\cdot, x_k)$ is not differentiable. For example if $f(x) = \frac{1}{2}\|x\|^2$ and $D_k(x, x_k) = \frac{1}{c}\|x - x_k\|$, then for any $c > 0$, the points $x_k \in [-1/c, 1/c]$ minimize $f(\cdot) + D_k(\cdot, x_k)$. Simple examples can also be constructed to show that the relative interior condition is essential to guarantee the condition (6.153).

We summarize the preceding discussion in the following proposition.

Proposition 6.6.1: Under the conditions (6.151) and (6.153), and assuming that the minimum of $f(x) + D_k(x, x_k)$ over x is attained for every k, the algorithm

$$x_{k+1} \in \arg\min_{x \in \Re^n} \{f(x) + D_k(x, x_k)\} \tag{6.156}$$

improves strictly the value of f at each iteration where x_k is not a global minimum of f, and stops at a global minimum of f.

Of course, cost improvement is a reassuring property for the algorithm (6.156), but does not guarantee convergence to a global minimum, particularly when f is convex (see Fig. 6.6.5). Thus despite the descent property established in the preceding proposition, the convergence of the algorithm may be problematic. In fact this is true even if f is assumed convex and has a nonempty set of minima X^*. Some extra conditions are required, but we will not pursue this issue further; see e.g., [ChT93], [Teb97].

If f is not convex, the difficulties may be formidable. First, the global minimum in Eq. (6.156) may be hard to compute, since the cost function $f(\cdot) + D_k(\cdot, x_k)$ may not be convex. Second, the algorithm may converge to local minima of f that are not global; see Fig. 6.6.5. As this figure indi-

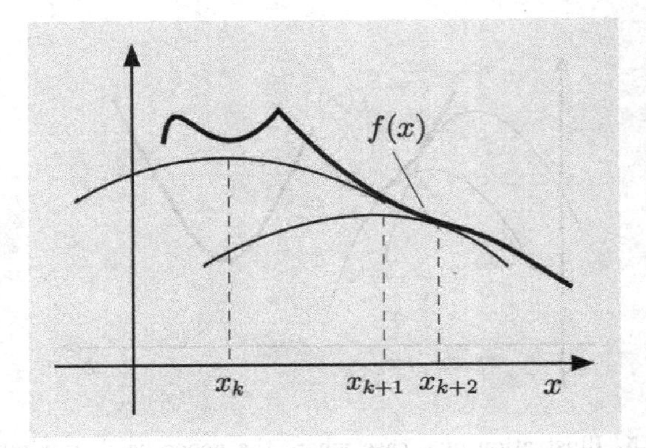

Figure 6.6.6. Illustration of a case where the generalized proximal algorithm (6.156) diverges even when started at a local minimum of f.

cates, convergence to a global minimum is facilitated if the regularization term $D_k(\cdot, x_k)$ is relatively "flat." The algorithm may also not converge at all, even if f has local minima and the algorithm is started near or at a local minimum of f (see Fig. 6.6.6). Still the algorithm has been used for nonconvex problems, often on a heuristic basis.

Some Examples

In what follows in this section we will give some examples of application of the generalized proximal algorithm (6.156) for the case where f is closed proper convex. We will not provide a convergence analysis, and refer instead to the literature.

Example 6.6.2: (Bregman Distance Function)

Let $\psi : \Re^n \mapsto (-\infty, \infty]$ be a convex function, which is differentiable within $\text{int}\big(\text{dom}(\psi)\big)$, and define for all $x, y \in \text{int}\big(\text{dom}(\psi)\big)$,

$$D_k(x, y) = \frac{1}{c_k}\big(\psi(x) - \psi(y) - \nabla\psi(y)'(x - y)\big), \tag{6.157}$$

where c_k is a positive penalty parameter. This is known as the *Bregman distance function*, and has been analyzed in connection with proximal-like algorithms in the paper [CeZ92] and the book [CeZ97]. Note that in the case where $\psi(x) = \frac{1}{2}\|x\|^2$, we have

$$D_k(x, y) = \frac{1}{c_k}\left(\frac{1}{2}\|x\|^2 - \frac{1}{2}\|y\|^2 - y'(x - y)\right) = \frac{1}{2c_k}\|x - y\|^2,$$

so the quadratic regularization term of the proximal algorithm is included as a special case.

Similarly, when $\psi(x) = \sum_{i=1}^{n} \psi_i(x^i)$, where

$$\psi_i(x^i) = \begin{cases} x^i \ln(x^i) & \text{if } x^i > 0, \\ 0 & \text{if } x^i = 0, \\ \infty & \text{if } x^i < 0, \end{cases}$$

with gradient $\nabla\psi_i(x^i) = \ln(x^i) + 1$, we obtain from Eq. (6.157) the function

$$D_k(x, y) = \frac{1}{c_k} \sum_{i=1}^{n} \left(x^i \left(\ln\left(\frac{x^i}{y^i}\right) - 1 \right) + y^i \right).$$

Except for the constant term $\frac{1}{c_k} \sum_{i=1}^{n} y^i$, which is inconsequential since it does not depend on x, this is the regularization function that is used in the entropy minimization algorithm of Example 6.6.1.

Note that because of the convexity of ψ, the condition (6.151) holds. Furthermore, because of the differentiability of $D_k(\cdot, x_k)$ (a consequence of the differentiability of ψ), the condition (6.153) holds as well when f is convex.

Example 6.6.3: (Exponential Augmented Lagrangian Method)

Consider the constrained minimization problem

$$\begin{aligned} &\text{minimize } f(x) \\ &\text{subject to } x \in X, \quad g_1(x) \le 0, \ldots, g_r(x) \le 0, \end{aligned}$$

where $f, g_1, \ldots, g_r : \Re^n \mapsto \Re$ are convex functions, and X is a closed convex set. Consider also the corresponding primal and dual functions

$$p(u) = \inf_{x \in X,\, g(x) \le u} f(x), \qquad q(\mu) = \inf_{x \in X} \big\{ f(x) + \mu' g(x) \big\}.$$

We assume that p is closed, so that there is no duality gap, and except for sign changes, q and p are conjugates of each other [i.e., $p(u)$ is equal to $(-q)^\star(-u)$; cf. Section 4.2 in Appendix B].

Let us consider the entropy minimization algorithm of Example 6.6.1, applied to maximization over $\mu \ge 0$ of the dual function. It is given by

$$\mu_{k+1} \in \arg\max_{\mu \ge 0} \left\{ q(\mu) - \frac{1}{c_k} \sum_{j=1}^{r} \mu^j \left(\ln\left(\frac{\mu^j}{\mu_k^j}\right) - 1 \right) \right\}, \tag{6.158}$$

where μ^j and μ_k^j denote the jth coordinates of μ and μ_k, respectively, and it corresponds to the case

$$D_k(\mu, \mu_k) = \frac{1}{c_k} \sum_{j=1}^{r} \mu^j \left(\ln\left(\frac{\mu^j}{\mu_k^j}\right) - 1 \right).$$

We now consider a dual implementation of the proximal iteration (6.158) (cf. Section 5.2). It is based on the Fenchel dual problem, which is to minimize over $u \in \Re^r$

$$(-q)^\star(-u) + D_k^\star(u, \mu_k),$$

where $(-q)^\star$ and $D_k^\star(\cdot, \mu_k)$ are the conjugates of $(-q)$ and $D_k(\cdot, \mu_k)$, respectively. Since $(-q)^\star(-u) = p(u)$, we see that the dual proximal iteration is

$$u_{k+1} \in \arg\min_{u \in \Re^n} \big\{p(u) + D_k^\star(u, \mu_k)\big\}, \tag{6.159}$$

and from the optimality condition of the Fenchel Duality Theorem (Prop. 1.2.1), the primal optimal solution of Eq. (6.158) is given by

$$\mu_{k+1} = \nabla D_k^\star(u_{k+1}, \mu_k). \tag{6.160}$$

To calculate $D_k^\star$, we first note that the conjugate of the entropy function

$$\phi(x) = \begin{cases} x\big(\ln(x) - 1\big) & \text{if } x > 0, \\ 0 & \text{if } x = 0, \\ \infty & \text{if } x < 0, \end{cases}$$

is the exponential function $\phi^\star(u) = e^u$, [to see this, simply calculate $\sup_x\{x'u - e^u\}$, the conjugate of the exponential e^u, and show that it is equal to $\phi(x)$]. Thus the conjugate of the function

$$D_k^j(\mu^j, \mu_k^j) = \frac{1}{c_k}\mu_k^j \phi\left(\frac{\mu^j}{\mu_k^j}\right),$$

which can be written as

$$\sup_{\mu^j}\left\{\mu^j u^j - \frac{1}{c_k}\mu_k^j \phi\left(\frac{\mu^j}{\mu_k^j}\right)\right\} = \frac{1}{c_k}\mu_k^j \sup_{\nu^j}\big\{\nu^j(c_k u^j) - \phi(\nu^j)\big\} = \frac{1}{c_k}\mu_k^j \phi^\star(c_k u^j),$$

is equal to

$$\frac{1}{c_k}\mu_k^j e^{c_k u^j}.$$

Since we have

$$D_k(\mu, \mu_k) = \sum_{j=1}^r \frac{1}{c_k}\mu_k^j \phi\left(\frac{\mu^j}{\mu_k^j}\right) = \sum_{j=1}^r D_k^j(\mu^j, \mu_k^j),$$

it follows that its conjugate is

$$D_k^\star(u, \mu_k) = \frac{1}{c_k}\sum_{j=1}^r \mu_k^j e^{c_k u^j}, \tag{6.161}$$

so the proximal minimization (6.159) is written as

$$u_{k+1} \in \arg\min_{u \in \Re^n}\left\{p(u) + \frac{1}{c_k}\sum_{j=1}^r \mu_k^j e^{c_k u^j}\right\}.$$

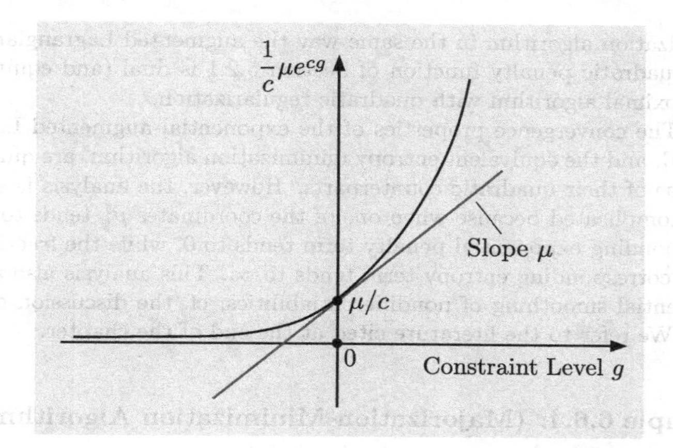

Figure 6.6.7. Illustration of the exponential penalty function.

Similar to the augmented Lagrangian method with quadratic penalty function of Section 5.2.1, the preceding minimization can be written as

$$u_{k+1} \in \arg\min_{u \in \Re^n} \left\{ \inf_{x \in X,\, g(x) \le u} \left\{ f(x) + \frac{1}{c_k} \sum_{j=1}^{r} \mu_k^j e^{c_k u^j} \right\} \right\}.$$

It can be seen that $u_{k+1} = g(x_k)$, where x_k is obtained through the minimization

$$\min_{x \in X} \left\{ f(x) + \frac{1}{c_k} \sum_{j=1}^{r} \mu_k^j e^{c_k g_j(x)} \right\},$$

or equivalently, by minimization of a corresponding augmented Lagrangian function:

$$x_k \in \arg\min_{x \in X} L_{c_k}(x, \mu_k) = \arg\min_{x \in X} \left\{ f(x) + \frac{1}{c_k} \sum_{j=1}^{r} \mu_k^j e^{c_k g_j(x)} \right\}. \tag{6.162}$$

From Eqs. (6.160) and (6.161), and the fact $u_{k+1} = g(x_k)$, it follows that the corresponding multiplier iteration is

$$\mu_{k+1}^j = \mu_k^j e^{c_k g_j(x_k)}, \qquad j = 1, \ldots, r. \tag{6.163}$$

The exponential penalty that is added to f to form the augmented Lagrangian in Eq. (6.162) is illustrated in Fig. 6.6.7. Contrary to its quadratic counterpart for inequality constraints, it is twice differentiable, which, depending on the problem at hand, can be a significant practical advantage when Newton-like methods are used for minimizing the augmented Lagrangian.

In summary, the exponential augmented Lagrangian method consists of sequential minimizations of the form (6.162) followed by multiplier iterations of the form (6.163). The method is dual (and equivalent) to the entropy minimization algorithm in the same way the augmented Lagrangian method with quadratic penalty function of Section 5.2.1 is dual (and equivalent) to the proximal algorithm with quadratic regularization.

The convergence properties of the exponential augmented Lagrangian method, and the equivalent entropy minimization algorithm, are quite similar to those of their quadratic counterparts. However, the analysis is somewhat more complicated because when one of the coordinates μ_k^j tends to zero, the corresponding exponential penalty term tends to 0, while the fraction μ^j/μ_k^j of the corresponding entropy term tends to ∞. This analysis also applies to exponential smoothing of nondifferentiabilities; cf. the discussion of Section 2.2.5. We refer to the literature cited at the end of the chapter.

Example 6.6.4: (Majorization-Minimization Algorithm)

An equivalent version of the generalized proximal algorithm (6.156) (known as *majorization-minimization* algorithm) is obtained by absorbing the cost function into the regularization term. This leads to the algorithm

$$x_{k+1} \in \arg\min_{x \in \Re^n} M_k(x, x_k), \tag{6.164}$$

where $M_k : \Re^{2n} \mapsto (-\infty, \infty]$ satisfies the conditions

$$M_k(x, x) = f(x), \qquad \forall\, x \in \Re^n,\ k = 0, 1, \ldots, \tag{6.165}$$

$$M_k(x, x_k) \geq f(x_k), \qquad \forall\, x \in \Re^n,\ k = 0, 1, \ldots \tag{6.166}$$

By defining

$$D_k(x, y) = M_k(x, y) - M_k(x, x),$$

we have

$$M_k(x, x_k) = f(x) + D_k(x, x_k),$$

so the algorithm (6.164) can be written in the generalized proximal format (6.156). Moreover the condition (6.166) is equivalent to the condition (6.151) that guarantees cost improvement, which is strict assuming also that

$$x_k \in \arg\min_{x \in \Re^n} M_k(x, x_k) \qquad \Rightarrow \qquad x_k \in X^*, \tag{6.167}$$

where X^* is the set of desirable points for convergence, cf. Eq. (6.153) and Prop. 6.6.1.

As an example, consider the problem of unconstrained minimization of the function

$$f(x) = R(x) + \|Ax - b\|^2,$$

where A is an $m \times n$ matrix, b is a vector in $\Re^m$, and $R : \Re^n \mapsto \Re$ is a nonnegative-valued convex regularization function. Let D be any symmetric matrix such that $D - A'A$ is positive definite (for example D may be a sufficiently large multiple of the identity). Let us define

$$M(x, y) = R(x) + \|Ay - b\|^2 + 2(x - y)'A'(Ay - b) + (x - y)'D(x - y),$$

and note that M satisfies the condition $M(x,x) = f(x)$ [cf. Eq. (6.165)], as well as the condition $M(x,x_k) \geq f(x_k)$ for all x and k [cf. Eq. (6.166)] in view of the calculation

$$\begin{aligned} M(x,y) - f(x) &= \|Ay-b\|^2 - \|Ax-b\|^2 \\ &\quad + 2(x-y)'A'(Ay-b) + (x-y)'D(x-y) \\ &= (x-y)'(D - A'A)(x-y). \end{aligned} \tag{6.168}$$

When D is the identity matrix I, by scaling A, we can make the matrix $I - A'A$ positive definite, and from Eq. (6.168), we have

$$M(x,y) = R(x) + \|Ax-b\|^2 - \|Ax-Ay\|^2 + \|x-y\|^2.$$

The majorization-minimization algorithm for this form of M has been used widely in signal processing applications.

Example 6.6.5: (Proximal Algorithm with Power Regularization – Superlinear Convergence)

Consider the generalized proximal algorithm

$$x_{k+1} \in \arg\min_{x\in\Re^n} \big\{ f(x) + D_k(x,x_k) \big\}, \tag{6.169}$$

where $D_k : \Re^{2n} \mapsto \Re$ is a regularization term that grows as the ρth power of the distance to the proximal center, where ρ is any scalar with $\rho > 1$ (instead of $\rho = 2$ as in the quadratic regularization case):

$$D_k(x,x_k) = \frac{1}{c_k}\sum_{i=1}^{n} \phi(x^i - x_k^i), \tag{6.170}$$

where c_k is a positive parameter, and ϕ is a scalar convex function with order of growth $\rho > 1$ around 0, such as for example

$$\phi(x^i - x_k^i) = \frac{1}{\rho}|x^i - x_k^i|^\rho. \tag{6.171}$$

We will aim to show that while the algorithm has satisfactory convergence properties for all $\rho > 1$, *it attains a superlinear convergence rate, provided ρ is larger than the order of growth of f around the optimum.* This occurs under natural conditions, even when c_k is kept constant - an old result, first obtained in [KoB76] (see also [Ber82a], Section 5.4, and [BeT94a], which we will follow in the subsequent derivation).

We assume that $f : \Re^n \mapsto (-\infty,\infty]$ is a closed convex function with a nonempty set of minima, denoted X^*. We also assume that for some scalars $\beta > 0$, $\delta > 0$, and $\gamma > 1$, we have

$$f^* + \beta\big(d(x)\big)^\gamma \leq f(x), \qquad \forall\, x \in \Re^n \text{ with } d(x) \leq \delta, \tag{6.172}$$

where

$$d(x) = \min_{x^* \in X^*} \|x - x^*\|$$

(cf. the assumption of Prop. 5.1.4). Moreover we require that $\phi : \Re \mapsto \Re$ is strictly convex, continuously differentiable, and satisfies the following conditions:

$$\phi(0) = 0, \quad \nabla\phi(0) = 0, \quad \lim_{z \to -\infty} \nabla\phi(z) = -\infty, \quad \lim_{z \to \infty} \nabla\phi(z) = \infty,$$

and for some scalar $M > 0$, we have

$$0 \le \phi(z) \le M\,|z|^\rho, \qquad \forall\, z \in [-\delta, \delta]. \tag{6.173}$$

An example is the order-ρ power function ϕ of Eq. (6.171). Since we want to focus on the rate of convergence, we will assume that the method converges in the sense that

$$d(x_k) \to 0, \qquad f(x_k) \to f^*, \qquad \|x_{k+1} - x_k\| \to 0, \tag{6.174}$$

where f^* is the optimal value (convergence proofs under mild conditions on f are given in [KoB76], [Ber82a], and [BeT94a]).

Let us denote by $\bar{x}_k$ the vector of X^*, which is at minimum distance from x_k. Let also η be a constant that bounds the ℓ_ρ norm in terms of the ℓ_2 norm:

$$\left(\sum_{i=1}^n |x^i|^\rho\right)^{1/\rho} = \|x\|_\rho \le \eta\,\|x\|_2, \qquad x \in \Re^n.$$

From the form of the proximal minimization (6.169)-(6.170), and using Eqs. (6.173), (6.174), we have for all k large enough so that $|x^i_{k+1} - x^i_k| \le \delta$,

$$\begin{aligned}
f(x_{k+1}) - f^* &\le f(x_{k+1}) + \frac{1}{c_k}\sum_{i=1}^n \phi(x^i_{k+1} - x^i_k) - f^* \\
&\le f(\bar{x}_k) + \frac{1}{c_k}\sum_{i=1}^n \phi(\bar{x}^i_k - x^i_k) - f^* \\
&= \frac{1}{c_k}\sum_{i=1}^n \phi(\bar{x}^i_k - x^i_k). \qquad (6.175) \\
&\le \frac{M}{c_k}\sum_{i=1}^n |\bar{x}^i_k - x^i_k|^\rho \\
&= \frac{M}{c_k}\|\bar{x}_k - x_k\|_\rho^\rho \\
&\le \frac{\eta^\rho M}{c_k} d(x_k)^\rho.
\end{aligned}$$

Also from the growth assumption (6.172), we have for all k large enough so that $d(x_k) \le \delta$,

$$d(x_k) \le \left(\frac{f(x_k) - f^*}{\beta}\right)^{1/\gamma}. \tag{6.176}$$

By combining Eq. (6.175) and (6.176), we obtain

$$f(x_{k+1}) - f^* \leq \frac{\eta^\rho M}{c_k} \left(\frac{f(x_k) - f^*}{\beta} \right)^{\rho/\gamma},$$

so if $\rho > \gamma$, the convergence rate is superlinear. In particular, in the special case where f is strongly convex so that $\gamma = 2$, $\{f(x_k)\}$ converges to f^* superlinearly when $\rho > 2$ [superlinear convergence is also attained if $\rho = 2$ and $c_k \to \infty$, cf. Prop. 5.1.4(c)].

The dual of the proximal algorithm (6.169)-(6.170) is an augmented Lagrangian method obtained via Fenchel duality. Computational experience with this method [KoB76], has shown that indeed its asymptotic convergence rate is very fast when $\rho > \gamma$. At the same time the use of order $\rho > 2$ regularization rather than $\rho = 2$ may lead to complications, because of the diminished regularization near 0. Such complications may be serious if a first order method is used for the proximal minimization. If on the other hand a Newton-like method can be used, much better results may be obtained with $\rho > 2$.

Example 6.6.6: (Mirror Descent Method)

Consider the general problem

$$\begin{aligned} &\text{minimize} \quad f(x) \\ &\text{subject to} \quad x \in X, \end{aligned}$$

where $f : \Re^n \mapsto \Re$ is a convex function, and X is a closed convex set. We noted earlier that the subgradient projection method

$$x_{k+1} = P_X\big(x_k - \alpha_k \tilde{\nabla} f(x_k)\big),$$

where $\tilde{\nabla} f(x_k)$ is a subgradient of f at x_k, can equivalently be written as

$$x_{k+1} \in \arg\min_{x \in X} \left\{ \tilde{\nabla} f(x_k)'(x - x_k) + \frac{1}{2\alpha_k} \|x - x_k\|^2 \right\};$$

cf. Prop. 6.1.4. In this form the method resembles the proximal algorithm, the difference being that $f(x)$ is replaced by its linearized version

$$f(x_k) + \tilde{\nabla} f(x_k)'(x - x_k),$$

and the stepsize α_k plays the role of the penalty parameter.

If we also replace the quadratic

$$\frac{1}{2\alpha_k} \|x - x_k\|^2$$

with a nonquadratic proximal term $D_k(x, x_k)$, we obtain a version of the subgradient projection method, called *mirror descent*. It has the form

$$x_{k+1} \in \arg\min_{x \in X} \big\{ \tilde{\nabla} f(x_k)'(x - x_k) + D_k(x, x_k) \big\}.$$

One advantage of this method is that using in place of f its linearization may simplify the minimization above for a problem with special structure.

As an example, consider the minimization of $f(x)$ over the unit simplex

$$X = \left\{ x \geq 0 \;\middle|\; \sum_{i=1}^{n} x^i = 1 \right\}.$$

A special case of the mirror descent method, called *entropic descent*, uses the entropy regularization function of Example 6.6.1 and has the form

$$x_{k+1} \in \arg\min_{x \in X} \sum_{i=1}^{n} \left(\tilde{\nabla}_i f(x_k) x^i + \frac{1}{\alpha_k} \left(x^i \ln\left(\frac{x^i}{x_k^i} \right) - 1 \right) \right),$$

where $\tilde{\nabla}_i f(x_k)$ are the components of $\tilde{\nabla} f(x_k)$. It can be verified that this minimization can be done in closed form as follows:

$$x_{k+1}^i = \frac{x_k^i e^{-\alpha_k \tilde{\nabla}_i f(x_k)}}{\sum_{j=1}^{n} x_k^j e^{-\alpha_k \tilde{\nabla}_j f(x_k)}}, \qquad i = 1, \ldots, n.$$

Thus it involves less overhead per iteration than the corresponding gradient projection iteration, which requires projection on the unit simplex, as well as the corresponding proximal iteration.

When f is differentiable, the convergence properties of mirror descent are similar to those of the gradient projection method, although depending on the problem at hand and the nature of $D_k(x, x_k)$ the analysis may be more complicated. When f is nondifferentiable, an analysis similar to the one for the subgradient projection method may be carried out; see [BeT03]. For extensions and further analysis of the method, we refer to the surveys [JuN11a], [JuN11b], and the references quoted there.

6.7 ϵ-DESCENT AND EXTENDED MONOTROPIC PROGRAMMING

In this section we return to the idea of cost function descent for nondifferentiable cost functions, which we discussed in Section 2.1.3. We noted there the theoretical difficulties around the use of the steepest descent direction, which is obtained by projection of the origin on the subdifferential. In this section we focus on the ϵ-subdifferential, aiming at theoretically more sound descent algorithms. We subsequently use these algorithms in an unusual way: to obtain a strong duality analysis for the extended monotropic programming problem that we discussed in Section 4.4, in connection with generalized polyhedral approximation.

6.7.1 ϵ-Subgradients

The discussion of Section 2.1.3 has indicated some of the deficiencies of subgradients and directional derivatives: anomalies may occur near points where the directional derivative is discontinuous, and at points of the relative boundary of the domain where the subdifferential may be empty. This motivates an attempt to rectify these deficiencies through the use of the ϵ-subdifferential, which turns out to have better continuity properties.

We recall from Section 3.3 that given a proper convex function $f : \Re^n \mapsto (-\infty, \infty]$ and a scalar $\epsilon > 0$, we say that a vector g is an *ϵ-subgradient* of f at a point $x \in \text{dom}(f)$ if

$$f(z) \geq f(x) + (z - x)'g - \epsilon, \qquad \forall\ z \in \Re^n. \tag{6.177}$$

The *ϵ-subdifferential* $\partial_\epsilon f(x)$ is the set of all ϵ-subgradients of f at x, and by convention, $\partial_\epsilon f(x) = \emptyset$ for $x \notin \text{dom}(f)$. It can be seen that

$$\partial_{\epsilon_1} f(x) \subset \partial_{\epsilon_2} f(x) \qquad \text{if } 0 < \epsilon_1 < \epsilon_2,$$

and that

$$\cap_{\epsilon \downarrow 0} \partial_\epsilon f(x) = \partial f(x).$$

We will now discuss in more detail the properties of ϵ-subgradients, with a view towards using them in cost function descent algorithms.

ϵ-Subgradients and Conjugate Functions

We first provide characterizations of the ϵ-subdifferential as a level set of a certain conjugate function. Consider a proper convex function $f : \Re^n \mapsto (-\infty, \infty]$, and for any $x \in \text{dom}(f)$, consider the x-translation of f, i.e., the function f_x given by

$$f_x(d) = f(x + d) - f(x), \qquad \forall\ d \in \Re^n.$$

The conjugate of f_x, is given by

$$f_x^\star(g) = \sup_{d \in \Re^n} \big\{d'g - f(x + d) + f(x)\big\}. \tag{6.178}$$

Since the definition of subgradient can be written as

$$g \in \partial f(x) \qquad \text{if and only if} \qquad \sup_{d \in \Re^n} \big\{g'd - f(x + d) + f(x)\big\} \leq 0,$$

we see from Eq. (6.178) that $\partial f(x)$ can be characterized as the 0-level set of $f_x^\star$:

$$\partial f(x) = \big\{g \mid f_x^\star(g) \leq 0\big\}. \tag{6.179}$$

Similarly, from Eq. (6.178), we see that

$$\partial_\epsilon f(x) = \{g \mid f_x^\star(g) \le \epsilon\}. \tag{6.180}$$

We will now use the preceding facts to discuss issues of nonemptiness and compactness of $\partial_\epsilon f(x)$.

We first observe that, viewed as a function of d, the conjugate of $f_x^\star$ is $(\mathrm{cl}\, f)(x+d) - f(x)$ (cf. Prop. 1.6.1 of Appendix B). Hence from the definition of conjugacy, for $d = 0$, we obtain

$$(\mathrm{cl}\, f)(x) - f(x) = \sup_{g \in \Re^n} \{-f_x^\star(g)\}.$$

Since $(\mathrm{cl}\, f)(x) \le f(x)$, we see that $0 \le \inf_{g \in \Re^n} f_x^\star(g)$ and

$$\inf_{g \in \Re^n} f_x^\star(g) = 0 \quad \text{if and only if} \quad (\mathrm{cl}\, f)(x) = f(x). \tag{6.181}$$

It follows from Eqs. (6.179) and (6.180) that for every $x \in \mathrm{dom}(f)$, there are two cases of interest:

(a) $(\mathrm{cl}\, f)(x) = f(x)$. Then, we have

$$\partial f(x) = \arg\min_{g \in \Re^n} f_x^\star(g) = \{g \mid f_x^\star(g) = 0\},$$

$$\partial_\epsilon f(x) = \{g \mid f_x^\star(g) \le \epsilon\}.$$

In this case, $\partial_\epsilon f(x)$ is nonempty, although $\partial f(x)$ may be empty.

(b) $(\mathrm{cl}\, f)(x) < f(x)$. In this case, $\partial f(x)$ is empty, and so is $\partial_\epsilon f(x)$ when

$$\epsilon < f(x) - (\mathrm{cl}\, f)(x).$$

We will now summarize the main properties of the ϵ-subdifferential in the following proposition. Part (b) is illustrated in Fig. 6.7.1 and will be the basis for the ϵ-descent method to be introduced shortly. Note the relation of this part with the formula for the support function of the subdifferential, given in Prop. 3.1.1(a) of Chapter 3 and Prop. 5.4.8 of Appendix B.

Proposition 6.7.1: Let $f : \Re^n \mapsto (-\infty, \infty]$ be a proper convex function and let ϵ be a positive scalar. For every $x \in \mathrm{dom}(f)$, the following hold:

(a) The ϵ-subdifferential $\partial_\epsilon f(x)$ is a closed convex set.

(b) If $(\mathrm{cl}\, f)(x) = f(x)$, then $\partial_\epsilon f(x)$ is nonempty and its support function is given by

$$\sigma_{\partial_\epsilon f(x)}(d) = \sup_{g \in \partial_\epsilon f(x)} d'g = \inf_{\alpha > 0} \frac{f(x + \alpha d) - f(x) + \epsilon}{\alpha}, \quad d \in \Re^n.$$

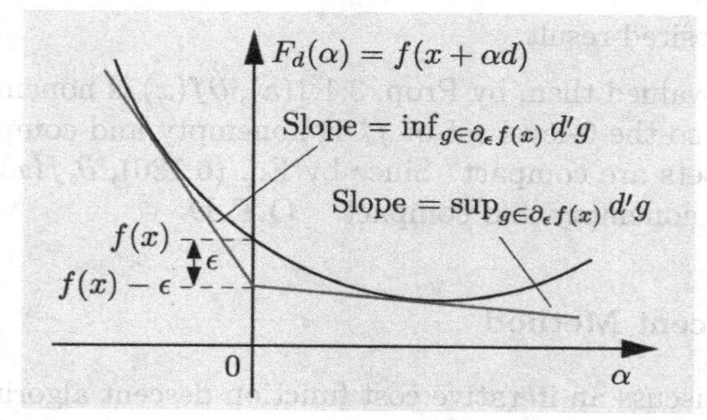

Figure 6.7.1. Illustration of the ϵ-subdifferential of a closed convex function $f : \Re \mapsto (-\infty, \infty]$, viewed along directions. The figure shows the function f along a direction d, starting at a point $x \in \text{dom}(f)$, i.e., the one-dimensional function

$$F_d(\alpha) = f(x + \alpha d).$$

As Prop. 6.7.1(b) shows, the minimal and maximal slopes of planes that support the graph of F_d and pass through $\big(0, f(x) - \epsilon\big)$ are

$$\inf_{g \in \partial_\epsilon f(x)} d'g \qquad \text{and} \qquad \sup_{g \in \partial_\epsilon f(x)} d'g.$$

These are also the two endpoints of the ϵ-subdifferential $\partial_\epsilon F_d(0)$.

(c) If f is real-valued, $\partial_\epsilon f(x)$ is nonempty and compact.

Proof: (a) We have shown that $\partial_\epsilon f(x)$ is the ϵ-level set of the function $f_x^\star$ [cf. Eq. (6.180)]. Since $f_x^\star$ is closed and convex, being a conjugate function, $\partial_\epsilon f(x)$ is closed and convex.

(b) By Eqs. (6.180) and (6.181), $\partial_\epsilon f(x)$ is the ϵ-level set of $f_x^\star$, while

$$\inf_{g \in \Re^n} f_x^\star(g) = 0.$$

It follows that $\partial_\epsilon f(x)$ is nonempty. Furthermore, by the discussion of support functions in Section 1.6 of Appendix B, the support function $\sigma_{\partial_\epsilon f(x)}$ of $\partial_\epsilon f(x)$ is the closed function generated by the conjugate of $f_x^\star - \epsilon$, which is $f_x + \epsilon$. Thus, $\text{epi}\big(\sigma_{\partial_\epsilon f(x)}\big)$ consists of the origin and the set

$$\cup_{\alpha > 0} \, \alpha^{-1} \, \text{epi}(f_x + \epsilon) = \cup_{\alpha > 0} \big\{ (\alpha^{-1} z, \alpha^{-1} w) \mid f_x(z) + \epsilon \le w \big\}.$$

Hence,

$$\sigma_{\partial_\epsilon f(x)}(d) = \inf_{\substack{\alpha^{-1} z = d, \; f_x(z) + \epsilon \le w \\ \alpha > 0}} \alpha^{-1} w = \inf_{\alpha > 0} \alpha^{-1} \big(f_x(\alpha d) + \epsilon \big),$$

which is the desired result.

(c) If f is real-valued then, by Prop. 3.1.1(a), $\partial f(x)$ is nonempty and compact for all x, so the 0-level set of $f_x^\star$ is nonempty and compact, implying that all level sets are compact. Since by Eq. (6.180), $\partial_\epsilon f(x)$ is the ϵ-level set of $f_x^\star$, it is nonempty and compact. **Q.E.D.**

6.7.2 ϵ-Descent Method

We will now discuss an iterative cost function descent algorithm that uses ϵ-subgradients. Let $f : \Re^n \mapsto (-\infty, \infty]$ be a proper convex function to be minimized.

We say that a direction d is an *ϵ-descent direction at $x \in \text{dom}(f)$*, where ϵ is a positive scalar, if

$$\inf_{\alpha>0} f(x + \alpha d) < f(x) - \epsilon,$$

or in words, there is a guarantee of a reduction of at least ϵ of the value of f along the direction d. Note that by Prop. 6.7.1(b), assuming

$$(\text{cl}\, f)(x) = f(x),$$

we have

$$\sigma_{\partial_\epsilon f(x)}(d) = \sup_{g\in\partial_\epsilon f(x)} d'g = \inf_{\alpha>0} \frac{f(x+\alpha d) - f(x) + \epsilon}{\alpha}, \qquad d \in \Re^n,$$

so

$$d \text{ is an } \epsilon\text{-descent direction} \quad \text{if and only if} \quad \sup_{g\in\partial_\epsilon f(x)} d'g < 0, \tag{6.182}$$

as illustrated in Fig. 6.7.2. Part (b) of the following proposition shows how to obtain an ϵ-descent direction if this is at all possible.

Proposition 6.7.2: Let $f : \Re^n \mapsto (-\infty, \infty]$ be a proper convex function, let ϵ be a positive scalar, and let $x \in \text{dom}(f)$ be such that $(\text{cl}\, f)(x) = f(x)$. Then:

(a) We have $0 \in \partial_\epsilon f(x)$ if and only if

$$f(x) \le \inf_{z\in\Re^n} f(z) + \epsilon.$$

(b) We have $0 \notin \partial_\epsilon f(x)$ if and only if there exists an ϵ-descent direction. In particular, if $0 \notin \partial_\epsilon f(x)$, the vector $-\overline{g}$ where

$$\overline{g} \in \arg \min_{g\in\partial_\epsilon f(x)} \|g\|,$$

is an ϵ-descent direction.

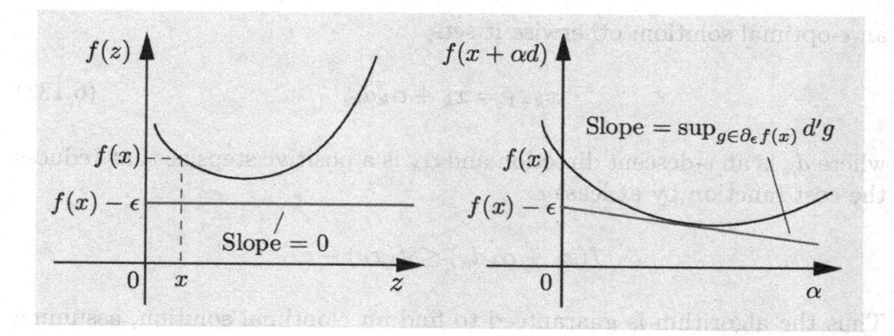

Figure 6.7.2. Illustration of the connection between $\partial_\epsilon f(x)$ and ϵ-descent directions [cf. Eq. (6.182)]. In the figure on the left, we have

$$f(x) \leq \inf_{z\in\Re^n} f(z) + \epsilon,$$

or equivalently, that the horizontal hyperplane [normal $(0, 1)$] that passes through $\big(x, f(x) - \epsilon\big)$ contains the epigraph of f in its upper halfspace, or equivalently, that $0 \in \partial_\epsilon f(x)$. In this case there is no ϵ-descent direction. In the figure on the right, d is an ϵ-descent direction because the slope shown is negative [cf. Prop. 6.7.1(b)].

Proof: (a) By definition, $0 \in \partial_\epsilon f(x)$ if and only if $f(z) \geq f(x) - \epsilon$ for all $z \in \Re^n$, which is equivalent to $\inf_{z\in\Re^n} f(z) + \epsilon \geq f(x)$.

(b) If there exists an ϵ-descent direction, then by part (a), we have $0 \notin \partial_\epsilon f(x)$. Conversely, assume that $0 \notin \partial_\epsilon f(x)$. The vector $\overline{g}$ is the projection of the origin on the closed convex set $\partial_\epsilon f(x)$, which is nonempty in view of the assumption $(\text{cl}\, f)(x) = f(x)$ [cf. Prop. 6.7.1(b)]. By the Projection Theorem (Prop. 1.1.9 in Appendix B),

$$0 \leq (g - \overline{g})'\overline{g}, \qquad \forall\, g \in \partial_\epsilon f(x),$$

or

$$\sup_{g\in\partial_\epsilon f(x)} (-\overline{g})'g \leq -\|\overline{g}\|^2 < 0,$$

where the last inequality follows from the hypothesis $0 \notin \partial_\epsilon f(x)$. By Eq. (6.182), this implies that $-\overline{g}$ is an ϵ-descent direction. **Q.E.D.**

The preceding proposition contains the elements of an iterative algorithm for minimizing f to within a tolerance of ϵ. This algorithm, called *ϵ-descent method*, is similar to the steepest descent method briefly discussed in Section 2.1.3, but is more general since it applies to extended real-valued functions, and has more sound theoretical convergence properties. At the kth iteration, it stops if there is no ϵ-descent direction, in which case x_k is an ϵ-optimal solution; otherwise it sets

$$x_{k+1} = x_k + \alpha_k d_k, \tag{6.183}$$

where d_k is an ϵ-descent direction and α_k is a positive stepsize that reduces the cost function by at least ϵ:

$$f(x_k + \alpha_k d_k) \leq f(x_k) - \epsilon.$$

Thus the algorithm is guaranteed to find an ϵ-optimal solution, assuming that f is bounded below, and to yield a sequence $\{x_k\}$ with $f(x_k) \to -\infty$, if f is unbounded below.

Note that by Prop. 6.7.2(a), the algorithm will stop if and only if $0 \in \partial_\epsilon f(x_k)$. This can be checked by finding the projection g_k of the origin onto $\partial_\epsilon f(x_k)$ to determine whether $g_k = 0$. If, however, $g_k \neq 0$, then by Prop. 6.7.2(b), $-g_k$ is an ϵ-descent direction, and can be used as the direction d_k in the iteration (6.183). In general, however, any ϵ-descent direction d_k can be used, and to verify the ϵ-descent property, it is sufficient to check that

$$\sup_{g \in \partial_\epsilon f(x_k)} d_k' g < 0$$

[cf. Eq. (6.182)].

A drawback of the ϵ-descent method is that at the typical iteration, an explicit representation of $\partial_\epsilon f(x_k)$ may be needed, which can be hard to obtain. This motivates a variant where $\partial_\epsilon f(x_k)$ is approximated by a set $A(x_k)$ that can be computed more easily than $\partial_\epsilon f(x_k)$. In this variant, given $A(x_k)$, the direction used in iteration (6.183) is $d_k = -g_k$, where g_k is the projection of the origin onto $A(x_k)$. One may consider two types of methods:

(a) *Outer approximation methods*: Here $\partial_\epsilon f(x_k)$ is approximated by a set $A(x)$ such that

$$\partial_\epsilon f(x_k) \subset A(x_k) \subset \partial_{\gamma\epsilon} f(x_k),$$

where γ is a scalar with $\gamma > 1$. If $g_k = 0$ [equivalently $0 \in A(x_k)$], the method stops, and from Prop. 6.7.2(a), it follows that x_k is within $\gamma\epsilon$ of being optimal. If $g_k \neq 0$, it follows from Prop. 6.7.2(b) that by suitable choice of the stepsize α_k, we can move along the direction $d_k = -g_k$ to decrease the cost function by at least ϵ. Thus for a fixed $\epsilon >$ and assuming that f is bounded below, the method is guaranteed to terminate in a finite number of iterations with a $\gamma\epsilon$-optimal solution. Aside from its computational value, the method will also be used for analytical purposes, to establish strong duality results for extended monotropic programming problems in the next section.

(b) *Inner approximation methods*: Here $A(x_k)$ is the convex hull of a finite number of ϵ-subgradients at x_k, and hence it is a subset of $\partial_\epsilon f(x_k)$. One method of this type, builds incrementally the approximation $A(x_k)$ of the ϵ-subdifferential, one element at a time, but does not need an explicit representation of the full ϵ-subdifferential $\partial_\epsilon f(x)$ or even the subdifferential $\partial f(x)$. Instead it requires that we be able to compute a single (arbitrary) element of $\partial f(x)$ at any x. This method was proposed in [Lem74] and is described in detail in [HiL93]. We will not consider it further in this book.

ϵ-Descent Based on Outer Approximation

We will now discuss outer approximation implementations of the ϵ-descent method. The idea is to replace $\partial_\epsilon f(x)$ by an outer approximation $A(x)$ that is easier to compute and/or easier to project on. To achieve this aim, it is typically necessary to exploit some special structure of the cost function. In the following, we restrict ourselves to an outer approximation method for the important case where f consists of a sum of functions $f = f_1 + \cdots + f_m$. The next proposition shows that we may use as approximation the closure of the vector sum of the ϵ-subdifferentials:

$$A(x) = \text{cl}\big(\partial_\epsilon f_1(x) + \cdots + \partial_\epsilon f_m(x)\big).$$

[Note here that the vector sum of the closed sets $\partial_\epsilon f_i(x)$ need not be closed; cf. the discussion of Section 1.4 of Appendix B.]

Proposition 6.7.3: Let f be the sum of m closed proper convex functions $f_i : \Re^n \mapsto (-\infty, \infty]$, $i = 1, \ldots, m$,

$$f(x) = f_1(x) + \cdots + f_m(x),$$

and let ϵ be a positive scalar. For any vector $x \in \text{dom}(f)$, we have

$$\partial_\epsilon f(x) \subset \text{cl}\big(\partial_\epsilon f_1(x) + \cdots + \partial_\epsilon f_m(x)\big) \subset \partial_{m\epsilon} f(x). \tag{6.184}$$

Proof: We first note that, by Prop. 6.7.1(b), the ϵ-subdifferentials $\partial_\epsilon f_i(x)$ are nonempty. Let $g_i \in \partial_\epsilon f_i(x)$ for $i = 1, \ldots, m$. Then we have

$$f_i(z) \geq f_i(x) + g_i'(z - x) - \epsilon, \qquad \forall\ z \in \Re^n,\ i = 1, \ldots, m.$$

By adding over all i, we obtain

$$f(z) \geq f(x) + (g_1 + \cdots + g_m)'(z - x) - m\epsilon, \qquad \forall\ z \in \Re^n.$$

Hence $g_1 + \cdots + g_m \in \partial_{m\epsilon} f(x)$, and it follows that

$$\partial_\epsilon f_1(x) + \cdots + \partial_\epsilon f_m(x) \subset \partial_{m\epsilon} f(x).$$

Since $\partial_{m\epsilon} f(x)$ is closed, this proves the right-hand side of Eq. (6.184).

To prove the left-hand side of Eq. (6.184), assume to arrive at a contradiction, that there exists a $g \in \partial_\epsilon f(x)$ such that

$$g \notin \mathrm{cl}\big(\partial_\epsilon f_1(x) + \cdots + \partial_\epsilon f_m(x)\big).$$

By the Strict Separation Theorem (Prop. 1.5.3 in Appendix B), there exists a hyperplane strictly separating g from the set $\mathrm{cl}\big(\partial_\epsilon f_1(x) + \cdots + \partial_\epsilon f_m(x)\big)$. Thus, there exist a vector d and a scalar b such that

$$d'(g_1 + \cdots + g_m) < b < d'g, \qquad \forall\ g_1 \in \partial_\epsilon f_1(x), \ldots, g_m \in \partial_\epsilon f_m(x).$$

From this we obtain

$$\sup_{g_1 \in \partial_\epsilon f_1(x)} d'g_1 + \cdots + \sup_{g_m \in \partial_\epsilon f_m(x)} d'g_m < d'g,$$

and by Prop. 6.7.1(b),

$$\inf_{\alpha>0} \frac{f_1(x+\alpha d) - f_1(x) + \epsilon}{\alpha} + \cdots + \inf_{\alpha>0} \frac{f_m(x+\alpha d) - f_m(x) + \epsilon}{\alpha} < d'g.$$

Let $\alpha_1, \ldots, \alpha_m$ be positive scalars such that

$$\frac{f_1(x+\alpha_1 d) - f_1(x) + \epsilon}{\alpha_1} + \cdots + \frac{f_m(x+\alpha_m d) - f_m(x) + \epsilon}{\alpha_m} < d'g, \quad (6.185)$$

and let

$$\overline{\alpha} = \frac{1}{1/\alpha_1 + \cdots + 1/\alpha_m}.$$

By the convexity of f_i, the ratio $\big(f_i(x+\alpha d) - f_i(x)\big)/\alpha$ is monotonically nondecreasing in α. Thus, since $\alpha_i \geq \overline{\alpha}$, we have

$$\frac{f_i(x+\alpha_i d) - f_i(x)}{\alpha_i} \geq \frac{f_i(x+\overline{\alpha} d) - f_i(x)}{\overline{\alpha}}, \qquad i = 1, \ldots, m,$$

and from Eq. (6.185) and the definition of $\overline{\alpha}$ we obtain

$$\begin{aligned} d'g &> \frac{f_1(x+\alpha_1 d) - f_1(x) + \epsilon}{\alpha_1} + \cdots + \frac{f_m(x+\alpha_m d) - f_m(x) + \epsilon}{\alpha_m} \\ &\geq \frac{f_1(x+\overline{\alpha} d) - f_1(x) + \epsilon}{\overline{\alpha}} + \cdots + \frac{f_m(x+\overline{\alpha} d) - f_m(x) + \epsilon}{\overline{\alpha}} \\ &= \frac{f(x+\overline{\alpha} d) - f(x) + \epsilon}{\overline{\alpha}} \\ &\geq \inf_{\alpha>0} \frac{f(x+\alpha d) - f(x) + \epsilon}{\alpha}. \end{aligned}$$

Since $g \in \partial_\epsilon f(x)$, this contradicts Prop. 6.7.1(b), and proves the left-hand side of Eq. (6.184). **Q.E.D.**

The potential lack of closure of the set $\partial_\epsilon f_1(x)+\cdots+\partial_\epsilon f_m(x)$ indicates a practical difficulty in implementing the method. In particular, in order to find an ϵ-descent direction one will ordinarily minimize $\|g_1 + \cdots + g_m\|$ over $g_i \in \partial_\epsilon f_i(x)$, $i = 1, \ldots, m$, but an optimal solution to this problem may not exist. Thus, it may be difficult to check computationally whether

$$0 \in \mathrm{cl}\big(\partial_\epsilon f_1(x) + \cdots + \partial_\epsilon f_m(x)\big),$$

which is the test for $m\epsilon$-optimality of x. The closure of the vector sum $\partial_\epsilon f_1(x)+\cdots+\partial_\epsilon f_m(x)$ may be guaranteed under various assumptions (e.g., the ones given in Section 1.4 of Appendix B; see also Section 6.7.4).

One may use Prop. 6.7.3 to approximate $\partial_\epsilon f(x)$ in cases where f is the sum of convex functions whose ϵ-subdifferential is easily computed or approximated. The following is an illustrative example.

Example 6.7.1: (ϵ-Descent for Separable Problems)

Let us consider the optimization problem

$$\begin{aligned} &\text{minimize} \quad \sum_{i=1}^{n} f_i(x_i) \\ &\text{subject to} \quad x \in P, \end{aligned}$$

where $x = (x_1, \ldots, x_n)$, each $f_i : \Re \mapsto \Re$ is a convex function of the scalar component x_i, and P is a polyhedral set of the form

$$P = P_1 \cap \cdots \cap P_r,$$

with

$$P_j = \{x \mid a_j'x \le b_j\}, \qquad j = 1, \ldots, r,$$

for some vectors a_j and scalars b_j. We can write this problem as

$$\begin{aligned} &\text{minimize} \quad \sum_{i=1}^{n} f_i(x_i) + \sum_{j=1}^{r} \delta_{P_j}(x) \\ &\text{subject to} \quad x \in \Re^n, \end{aligned}$$

where δ_{P_j} is the indicator function of P_j.

The ϵ-subdifferential of the cost function is not easily calculated, but can be approximated by a vector sum of intervals. In particular, it can be verified, using the definition, that the ϵ-subdifferential of δ_{P_j} is

$$\partial_\epsilon \delta_{P_j}(x) = \big\{\gamma a_j \mid 0 \le \gamma,\ \gamma(b_j - a_j'x) \le \epsilon\big\}, \qquad \forall\ x \in P_j,$$

which is an interval in $\Re^n$. Similarly, it can be seen that $\partial_\epsilon f_i(x_i)$ is a compact interval of the ith axis. Thus

$$\sum_{i=1}^{n} \partial_\epsilon f_i(x_i) + \sum_{i=1}^{m} \partial_\epsilon \delta_{P_i}(x),$$

is a vector sum of intervals, which is a polyhedral set and is therefore closed. Thus by Prop. 6.7.3, it can be used as an outer approximation of $\partial_\epsilon f(x)$. At each iteration of the corresponding ϵ-descent method, projection on this vector sum to obtain an ϵ-descent direction requires solution of a quadratic program, which depending on the problem, may be tractable.

6.7.3 Extended Monotropic Programming Duality

We now return to the Extended Monotropic Program of Section 4.4 (EMP for short):

$$\begin{aligned} &\text{minimize} \quad \sum_{i=1}^{m} f_i(x_i) \\ &\text{subject to} \quad x \in S, \end{aligned} \tag{6.186}$$

where

$$x \overset{\text{def}}{=} (x_1, \ldots, x_m),$$

is a vector in $\Re^{n_1+\cdots+n_m}$, with components $x_i \in \Re^{n_i}$, $i = 1, \ldots, m$, and

$f_i : \Re^{n_i} \mapsto (-\infty, \infty]$ is a closed proper convex function for each i,

S is a subspace of $\Re^{n_1+\cdots+n_m}$.

The dual problem was derived in Section 4.4. It has the form

$$\begin{aligned} &\text{minimize} \quad \sum_{i=1}^{m} f_i^\star(\lambda_i) \\ &\text{subject to} \quad \lambda \in S^\perp. \end{aligned} \tag{6.187}$$

In this section we will use the ϵ-descent method as an analytical tool to obtain conditions for strong duality.

Let f^* and q^* be the optimal values of the primal and dual problems (6.186) and (6.187), respectively, and note that by weak duality, we have $q^* \leq f^*$. Let us introduce the functions $\overline{f}_i : \Re^{n_1+\cdots+n_m} \mapsto (-\infty, \infty]$ of the vector $x = (x_1, \ldots, x_m)$, defined by

$$\overline{f}_i(x) = f_i(x_i), \qquad i = 1, \ldots, m.$$

Note that the ϵ-subdifferentials of $\overline{f}_i$ and f_i are related by

$$\partial_\epsilon \overline{f}_i(x) = \big\{(0, \ldots, 0, \lambda_i, 0, \ldots, 0) \mid \lambda_i \in \partial_\epsilon f_i(x_i)\big\}, \quad i = 1, \ldots, m, \tag{6.188}$$

where the nonzero element in $(0,\ldots,0,\lambda_i,0,\ldots,0)$ is in the ith position. The following proposition gives conditions for strong duality.

Proposition 6.7.4: (EMP Strong Duality) Assume that the EMP (6.186) is feasible, and that for all feasible solutions x and all $\epsilon > 0$, the set

$$T(x,\epsilon) = S^\perp + \partial_\epsilon \overline{f}_1(x) + \cdots + \partial_\epsilon \overline{f}_m(x)$$

is closed. Then $q^* = f^*$.

Proof: If $f^* = -\infty$, then $q^* = f^*$ by weak duality, so we may assume that $f^* > -\infty$. Let X denote the feasible region of the primal problem:

$$X = S \cap \left(\cap_{i=1}^m \mathrm{dom}(\overline{f}_i)\right).$$

We apply the ϵ-descent method based on outer approximation of the subdifferential to the minimization of the function

$$f = \delta_S + \sum_{i=1}^m \overline{f}_i = \delta_S + \sum_{i=1}^m f_i,$$

where δ_S is the indicator function of S, for which $\partial_\epsilon \delta_S(x) = S^\perp$ for all $x \in S$ and $\epsilon > 0$. In this method, we start with a vector $x^0 \in X$, and we generate a sequence $\{x^k\} \subset X$. At the kth iteration, given the current iterate x^k, we find the vector of minimum norm w^k on the set $T(x^k,\epsilon)$ (which is closed by assumption). If $w^k = 0$ the method stops, verifying that $0 \in \partial_{(m+1)\epsilon} f(x^k)$ [cf. the right side of Eq. (6.184)]. If $w^k \neq 0$, we generate a vector $x^{k+1} \in X$ of the form $x^{k+1} = x^k - \alpha_k w^k$, satisfying

$$f(x^{k+1}) < f(x^k) - \epsilon;$$

such a vector is guaranteed to exist, since $0 \notin T(x^k,\epsilon)$ and hence $0 \notin \partial_\epsilon f(x^k)$ by Prop. 6.7.3. Since $f(x^k) \geq f^*$ and at the current stage of the proof we have assumed that $f^* > -\infty$, the method must stop at some iteration with a vector $x = (x_1,\ldots,x_m)$ such that $0 \in T(x,\epsilon)$. Thus some vector in $\partial_\epsilon \overline{f}_1(x) + \cdots + \partial_\epsilon \overline{f}_m(x)$ must belong to $S^\perp$. In view of Eq. (6.188), it follows that there must exist vectors

$$\lambda_i \in \partial_\epsilon f_i(x_i), \qquad i = 1,\ldots,m,$$

such that

$$\lambda = (\lambda_1,\ldots,\lambda_m) \in S^\perp.$$

From the definition of an ϵ-subgradient we have

$$f_i(x_i) \leq -f_i^\star(\lambda_i) + \lambda_i' x_i + \epsilon, \qquad i = 1,\ldots,m,$$

and by adding over i, and using the fact $x \in S$ and $\lambda \in S^\perp$, we obtain

$$\sum_{i=1}^{m} f_i(x_i) \leq -\sum_{i=1}^{m} f_i^\star(\lambda_i) + m\epsilon.$$

Since x is primal feasible and $-\sum_{i=1}^{m} f_i^\star(\lambda_i)$ is the dual value at λ, it follows that

$$f^* \leq q^* + m\epsilon.$$

Taking the limit as $\epsilon \to 0$, we obtain $f^* \leq q^*$, and using also the weak duality relation $q^* \leq f^*$, we obtain $f^* = q^*$. **Q.E.D.**

6.7.4 Special Cases of Strong Duality

We now delineate some special cases where the assumptions for strong EMP duality of Prop. 6.7.4 are satisfied. We first note that in view of Eq. (6.188), the set $\partial_\epsilon \overline{f}_i(x)$ is compact if $\partial_\epsilon f_i(x_i)$ is compact, and it is polyhedral if $\partial_\epsilon f_i(x_i)$ is polyhedral. Since the vector sum of a compact set and a polyhedral set is closed (see the discussion at the end of Section 1.4 of Appendix B), it follows that if *each of the sets $\partial_\epsilon f_i(x_i)$ is either compact or polyhedral, then $T(x, \epsilon)$ is closed, and by Prop. 6.7.4, we have $q^* = f^*$*. Furthermore, from Prop. 5.4.1 of Appendix B, $\partial f_i(x_i)$ and hence also $\partial_\epsilon f_i(x_i)$ is compact if $x_i \in \text{int}(\text{dom}(f_i))$ (as in the case where f_i is real-valued). Moreover $\partial_\epsilon f_i(x_i)$ is polyhedral if f_i is polyhedral [being the level set of a polyhedral function, cf. Eq. (6.180)]. There are some other interesting special cases where $\partial_\epsilon f_i(x_i)$ is polyhedral, as we now describe.

One such special case is when f_i depends on a single scalar component of x, as in the case of a monotropic programming problem. The following definition introduces a more general case.

Definition 6.7.1: We say that a closed proper convex function $h : \Re^n \mapsto (-\infty, \infty]$ is *essentially one-dimensional* if it has the form

$$h(x) = \overline{h}(a'x),$$

where a is a vector in $\Re^n$ and $\overline{h} : \Re \mapsto (-\infty, \infty]$ is a scalar closed proper convex function.

The following proposition establishes the main associated property for our purposes.

Proposition 6.7.5: Let $h : \Re^n \mapsto (-\infty, \infty]$ be a closed proper convex essentially one-dimensional function. Then for all $x \in \text{dom}(h)$ and $\epsilon > 0$, the ϵ-subdifferential $\partial_\epsilon h(x)$ is nonempty and polyhedral.

Proof: Let $h(x) = \overline{h}(a'x)$, where a is a vector in $\Re^n$ and $\overline{h}$ is a scalar closed proper convex function. If $a = 0$, then h is a constant function, and $\partial_\epsilon h(x)$ is equal to $\{0\}$, a polyhedral set. Thus, we may assume that $a \neq 0$. We note that $\lambda \in \partial_\epsilon h(x)$ if and only if

$$\overline{h}(a'z) \geq \overline{h}(a'x) + (z - x)'\lambda - \epsilon, \qquad \forall\, z \in \Re^n.$$

Writing λ in the form $\lambda = \xi a + v$ with $\xi \in \Re$ and $v \perp a$, we have

$$\overline{h}(a'z) \geq \overline{h}(a'x) + (z - x)'(\xi a + v) - \epsilon, \qquad \forall\, z \in \Re^n,$$

and by taking $z = \gamma a + \delta v$ with $\gamma, \delta \in \Re$ and γ such that $\gamma\|a\|^2 \in \text{dom}(\overline{h})$, we obtain for all $\delta \in \Re$

$$\overline{h}(\gamma\|a\|^2) \geq \overline{h}(a'x) + (\gamma a + \delta v - x)'\lambda - \epsilon = \overline{h}(a'x) + (\gamma a - x)'\lambda - \epsilon + \delta v'\lambda.$$

Since $v'\lambda = \|v\|^2$ and δ can be arbitrarily large, this relation implies that $v = 0$, so it follows that every $\lambda \in \partial_\epsilon h(x)$ must be a scalar multiple of a. Since $\partial_\epsilon h(x)$ is also a closed convex set, it must be a nonempty closed interval in $\Re^n$, and hence is polyhedral. **Q.E.D.**

Another interesting special case is described in the following definition.

Definition 6.7.2: We say that a closed proper convex function $h : \Re^n \mapsto (-\infty, \infty]$ is *domain one-dimensional* if the affine hull of $\text{dom}(h)$ is either a single point or a line, i.e.,

$$\text{aff}\big(\text{dom}(h)\big) = \{\gamma a + b \mid \gamma \in \Re\},$$

where a and b are some vectors in $\Re^n$.

The following proposition parallels Prop. 6.7.5.

Proposition 6.7.6: Let $h : \Re^n \mapsto (-\infty, \infty]$ be a closed proper convex domain one-dimensional function. Then for all $x \in \text{dom}(h)$ and $\epsilon > 0$, the ϵ-subdifferential $\partial_\epsilon h(x)$ is nonempty and polyhedral.

Proof: Denote by a and b the vectors associated with the domain of h as per Definition 6.7.2. We note that for $\overline{\gamma}a + b \in \text{dom}(h)$, we have $\lambda \in \partial_\epsilon h(\overline{\gamma}a + b)$ if and only if

$$h(\gamma a + b) \geq h(\overline{\gamma}a + b) + (\gamma - \overline{\gamma})a'\lambda - \epsilon, \qquad \forall\, \gamma \in \Re,$$

or equivalently, if and only if $a'\lambda \in \partial_\epsilon \overline{h}(\overline{\gamma})$, where $\overline{h}$ is the one-dimensional convex function

$$\overline{h}(\gamma) = h(\gamma a + b), \qquad \gamma \in \Re.$$

Thus,

$$\partial_\epsilon h(\overline{\gamma}a + b) = \big\{\lambda \mid a'\lambda \in \partial_\epsilon \overline{h}(\overline{\gamma})\big\}.$$

Since $\partial_\epsilon \overline{h}(\overline{\gamma})$ is a nonempty closed interval ($\overline{h}$ is closed because h is), it follows that $\partial_\epsilon h(\overline{\gamma}a + b)$ is nonempty and polyhedral [if $a = 0$, it is equal to $\Re^n$, and if $a \neq 0$, it is the vector sum of two polyhedral sets: the interval $\big\{\gamma a \mid \gamma \|a\|^2 \in \partial_\epsilon \overline{h}(\overline{\gamma})\big\}$ and the subspace that is orthogonal to a]. **Q.E.D.**

By combining the preceding two propositions with Prop. 6.7.4, we obtain the following.

Proposition 6.7.7: Assume that the EMP (6.186) is feasible, and that each function f_i is real-valued, or is polyhedral, or is essentially one-dimensional, or is domain one-dimensional. Then $q^* = f^*$.

It turns out that there is a conjugacy relation between essentially one-dimensional functions and domain one-dimensional functions such that the affine hull of their domain is a subspace. This is shown in the following proposition, which establishes a somewhat more general connection, needed for our purposes.

Proposition 6.7.8:

(a) The conjugate of an essentially one-dimensional function is a domain one-dimensional function such that the affine hull of its domain is a subspace.

(b) The conjugate of a domain one-dimensional function is the sum of an essentially one-dimensional function and a linear function.

Proof: (a) Let $h : \Re^n \mapsto (-\infty, \infty]$ be essentially one-dimensional, so that

$$h(x) = \overline{h}(a'x),$$

where a is a vector in $\Re^n$ and $\overline{h} : \Re \mapsto (-\infty, \infty]$ is a scalar closed proper convex function. If $a = 0$, then h is a constant function, so its conjugate is domain one-dimensional, since its domain is $\{0\}$. We may thus assume that $a \neq 0$. We claim that the conjugate

$$h^\star(\lambda) = \sup_{x \in \Re^n} \{\lambda' x - \overline{h}(a'x)\}, \tag{6.189}$$

takes infinite values if λ is outside the one-dimensional subspace spanned by a, implying that $h^\star$ is domain one-dimensional with the desired property. Indeed, let λ be of the form $\lambda = \xi a + v$, where ξ is a scalar, and v is a nonzero vector with $v \perp a$. If we take $x = \gamma a + \delta v$ in Eq. (6.189), where γ is such that $\gamma \|a\|^2 \in \text{dom}(\overline{h})$, we obtain

$$\begin{aligned} h^\star(\lambda) &= \sup_{x \in \Re^n} \{\lambda' x - \overline{h}(a'x)\} \\ &\geq \sup_{\delta \in \Re} \big\{(\xi a + v)'(\gamma a + \delta v) - \overline{h}\big(\gamma \|a\|^2\big)\big\} \\ &= \xi \gamma \|a\|^2 - \overline{h}\big(\gamma \|a\|^2\big) + \sup_{\delta \in \Re} \{\delta \|v\|^2\}, \end{aligned}$$

so it follows that $h^\star(\lambda) = \infty$.

(b) Let $h : \Re^n \mapsto (-\infty, \infty]$ be domain one-dimensional, so that

$$\text{aff}\big(\text{dom}(h)\big) = \{\gamma a + b \mid \gamma \in \Re\},$$

for some vectors a and b. If $a = b = 0$, the domain of h is $\{0\}$, so its conjugate is the function taking the constant value $-h(0)$ and is essentially one-dimensional. If $b = 0$ and $a \neq 0$, then the conjugate is

$$h^\star(\lambda) = \sup_{x \in \Re^n} \{\lambda' x - h(x)\} = \sup_{\gamma \in \Re} \{\gamma a' \lambda - h(\gamma a)\},$$

so $h^\star(\lambda) = \overline{h}^\star(a'\lambda)$ where $\overline{h}^\star$ is the conjugate of the scalar function $\overline{h}(\gamma) = h(\gamma a)$. Since $\overline{h}$ is closed proper convex, the same is true for $\overline{h}^\star$, and it follows that $h^\star$ is essentially one-dimensional. Finally, consider the case where $b \neq 0$. Then we use a translation argument and write $h(x) = \hat{h}(x - b)$, where $\hat{h}$ is a function such that the affine hull of its domain is the subspace spanned by a. The conjugate of $\hat{h}$ is essentially one-dimensional (by the preceding argument), and the conjugate of h is obtained by adding $b'\lambda$ to it. **Q.E.D.**

We now turn to the dual problem, and derive a duality result that is analogous to the one of Prop. 6.7.7. We say that a function is *co-finite* if its conjugate is real-valued. If we apply Prop. 6.7.7 to the dual problem (6.187), we obtain the following.

Proposition 6.7.9: Assume that the dual EMP (6.187) is feasible. Assume further that each f_i is co-finite, or is polyhedral, or is essentially one-dimensional, or is domain one-dimensional. Then $q^* = f^*$.

In the special case of a monotropic programming problem, where the functions f_i are essentially one-dimensional (they depend on the single scalar component x_i), Props. 6.7.7 and 6.7.9 yield the following proposition. This is a central result for monotropic programming.

Proposition 6.7.10: (Monotropic Programming Strong Duality) Consider the monotropic programming problem, the special case of EMP where $n_i = 1$ for all i. Assume that either the problem is feasible, or else its dual problem is feasible. Then $q^* = f^*$.

Proof: This is a consequence of Props. 6.7.7 and 6.7.9, and the fact that when $n_i = 1$, the functions f_i and q_i are essentially one-dimensional. Applying Prop. 6.7.7 to the primal problem, shows that $q^* = f^*$ under the hypothesis that the primal problem is feasible. Applying Prop. 6.7.9 to the dual problem, shows that $q^* = f^*$ under the hypothesis that the dual problem is feasible. **Q.E.D.**

The preceding results can be used to establish conditions for $q^* = f^*$ in various specialized contexts, including multicommodity flow problems (cf. Example 1.4.5); see [Ber10a].

6.8 INTERIOR POINT METHODS

Let us consider inequality constrained problems of the form

$$\begin{aligned} &\text{minimize } f(x) \\ &\text{subject to } x \in X, \quad g_j(x) \le 0, \quad j = 1, \ldots, r, \end{aligned} \tag{6.190}$$

where f and g_j are real-valued convex functions, and X is a closed convex set. The interior (relative to X) of the set defined by the inequality constraints is

$$S = \big\{x \in X \mid g_j(x) < 0,\ j = 1, \ldots, r\big\},$$

and is assumed to be nonempty.

In interior point methods, we add to the cost a function $B(x)$ that is defined in the interior set S. This function, called the *barrier function*, is continuous and tends to ∞ as any one of the constraints $g_j(x)$ approaches

0 from negative values. A common example of barrier function is the *logarithmic*,

$$B(x) = -\sum_{j=1}^{r} \ln\{-g_j(x)\}.$$

Another example is the *inverse*,

$$B(x) = -\sum_{j=1}^{r} \frac{1}{g_j(x)}.$$

Note that both of these functions are convex since the constraints g_j are convex. Figure 6.8.1 illustrates the form of $B(x)$.

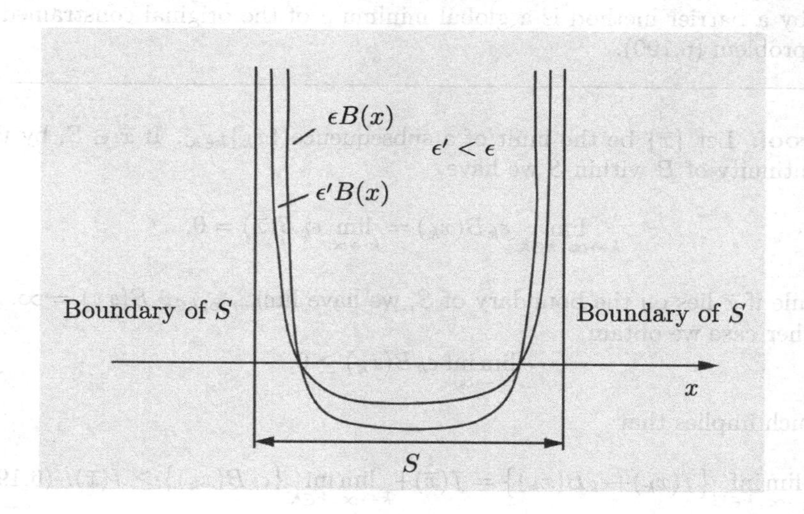

Figure 6.8.1. Form of a barrier function. The barrier term $\epsilon B(x)$ tends to zero for all interior points $x \in S$ as $\epsilon \to 0$.

The *barrier method* is defined by introducing a parameter sequence $\{\epsilon_k\}$ with

$$0 < \epsilon_{k+1} < \epsilon_k, \quad k = 0, 1, \ldots, \quad \epsilon_k \to 0.$$

It consists of finding

$$x_k \in \arg\min_{x \in S}\{f(x) + \epsilon_k B(x)\}, \qquad k = 0, 1, \ldots. \tag{6.191}$$

Since the barrier function is defined only on the interior set S, the successive iterates of any method used for this minimization must be interior points.

If $X = \Re^n$, one may use unconstrained methods such as Newton's method with the stepsize properly selected to ensure that all iterates lie in

S. Indeed, Newton's method is often recommended for reasons that have to do with *ill-conditioning*, a phenomenon that relates to the difficulty of carrying out the minimization (6.191) (see Fig. 6.8.2 and nonlinear programming sources such as [Ber99] for a discussion). Note that the barrier term $\epsilon_k B(x)$ goes to zero for all interior points $x \in S$ as $\epsilon_k \to 0$. Thus the barrier term becomes increasingly inconsequential as far as interior points are concerned, while progressively allowing x_k to get closer to the boundary of S (as it should if the solutions of the original constrained problem lie on the boundary of S). Figure 6.8.2 illustrates the convergence process, and the following proposition gives the main convergence result.

Proposition 6.8.1: Every limit point of a sequence $\{x_k\}$ generated by a barrier method is a global minimum of the original constrained problem (6.190).

Proof: Let $\{\overline{x}\}$ be the limit of a subsequence $\{x_k\}_{k\in\mathcal{K}}$. If $\overline{x} \in S$, by the continuity of B within S we have,

$$\lim_{k\to\infty,\, k\in\mathcal{K}} \epsilon_k B(x_k) = \lim_{k\to\infty} \epsilon_k B(\overline{x}) = 0,$$

while if $\overline{x}$ lies on the boundary of S, we have $\lim_{k\to\infty,\, k\in\mathcal{K}} B(x_k) = \infty$. In either case we obtain

$$\liminf_{k\to\infty} \epsilon_k B(x_k) \geq 0,$$

which implies that

$$\liminf_{k\to\infty,\, k\in\mathcal{K}} \big\{ f(x_k) + \epsilon_k B(x_k) \big\} = f(\overline{x}) + \liminf_{k\to\infty,\, k\in\mathcal{K}} \big\{ \epsilon_k B(x_k) \big\} \geq f(\overline{x}). \quad (6.192)$$

The vector $\overline{x}$ is a feasible point of the original problem (6.190), since $x_k \in S$ and X is a closed set. If $\overline{x}$ were not a global minimum, there would exist a feasible vector x^* such that $f(x^*) < f(\overline{x})$ and therefore also [since by the Line Segment Principle (Prop. 1.3.1 in Appendix B) x^* can be approached arbitrarily closely through the interior set S] an interior point $\tilde{x} \in S$ such that $f(\tilde{x}) < f(\overline{x})$. We now have by the definition of x_k,

$$f(x_k) + \epsilon_k B(x_k) \leq f(\tilde{x}) + \epsilon_k B(\tilde{x}), \qquad k = 0, 1, \ldots,$$

which by taking the limit as $k \to \infty$ and $k \in \mathcal{K}$, implies together with Eq. (6.192), that $f(\overline{x}) \leq f(\tilde{x})$. This is a contradiction, thereby proving that $\overline{x}$ is a global minimum of the original problem. **Q.E.D.**

The idea of using a barrier function as an approximation to constraints has been used in several different ways, in methods that generate

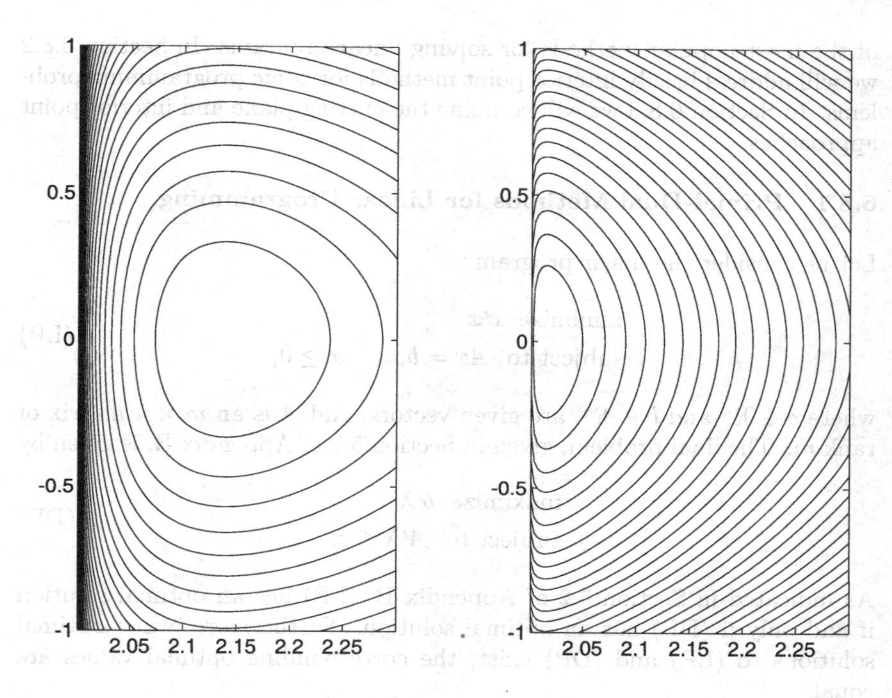

Figure 6.8.2. Illustration of the level sets of the barrier-augmented cost function, and the convergence process of the barrier method for the problem

$$\begin{aligned} &\text{minimize} \;\; f(x) = \tfrac{1}{2}\big((x^1)^2 + (x^2)^2\big) \\ &\text{subject to} \;\; 2 \le x^1, \end{aligned}$$

with optimal solution $x^* = (2,0)$. For the case of the logarithmic barrier function $B(x) = -\ln(x^1 - 2)$, we have

$$x_k \in \arg\min_{x^1 > 2} \left\{ \tfrac{1}{2}\big((x^1)^2 + (x^2)^2\big) - \epsilon_k \ln(x^1 - 2) \right\} = \left(1 + \sqrt{1 + \epsilon_k}\,, 0\right),$$

so as ϵ_k is decreased, the unconstrained minimum x_k approaches the constrained minimum $x^* = (2,0)$. The figure shows the level sets of $f(x) + \epsilon B(x)$ for $\epsilon = 0.3$ (left side) and $\epsilon = 0.03$ (right side). As $\epsilon_k \to 0$, computing x_k becomes more difficult due to ill-conditioning (the level sets become very elongated near x_k).

successive iterates lying in the interior of the constraint set. These methods are generically referred to as *interior point methods*, and have been extensively applied to linear, quadratic, and conic programming problems. The logarithmic barrier function has been central in many of these methods. In the next two sections we will discuss a few methods that are designed for problems with special structure. In particular, in Section 6.8.1 we will discuss in some detail primal-dual methods for linear programming, one of the most popular methods for solving linear programs. In Section 6.8.2 we will address briefly interior point methods for conic programming problems. In Section 6.8.3 we will combine the cutting plane and interior point approaches.

6.8.1 Primal-Dual Methods for Linear Programming

Let us consider the linear program

$$\begin{aligned} &\text{minimize} \;\; c'x \\ &\text{subject to} \;\; Ax = b, \qquad x \geq 0, \end{aligned} \tag{LP}$$

where $c \in \Re^n$ and $b \in \Re^m$ are given vectors, and A is an $m \times n$ matrix of rank m. The dual problem, given in Section 5.2 of Appendix B, is given by

$$\begin{aligned} &\text{maximize} \;\; b'\lambda \\ &\text{subject to} \;\; A'\lambda \leq c. \end{aligned} \tag{DP}$$

As indicated in Section 5.2 of Appendix B, (LP) has an optimal solution if and only if (DP) has an optimal solution. Furthermore, when optimal solutions to (LP) and (DP) exist, the corresponding optimal values are equal.

Recall that the logarithmic barrier method involves finding for various $\epsilon > 0$,

$$x(\epsilon) \in \arg\min_{x \in S} F_\epsilon(x), \tag{6.193}$$

where

$$F_\epsilon(x) = c'x - \epsilon \sum_{i=1}^{n} \ln x^i,$$

x^i is the ith component of x and S is the interior set

$$S = \{x \mid Ax = b,\, x > 0\}.$$

We assume that S is nonempty and bounded.

Rather than directly minimizing $F_\epsilon(x)$ for small values of ϵ [cf. Eq. (6.193)], we will apply Newton's method for solving the system of optimality conditions for the problem of minimizing $F_\epsilon(\cdot)$ over S. The salient features of this approach are:

(a) Only one Newton iteration is carried out for each value of ϵ_k.

(b) For every k, the pair (x_k, λ_k) is such that x_k is an interior point of the positive orthant, i.e., $x_k > 0$, while λ_k is an interior point of the dual feasible region, i.e.,

$$c - A'\lambda_k > 0.$$

(However, x_k need not be primal-feasible, that is, it need not satisfy the equation $Ax = b$.)

(c) Global convergence is enforced by ensuring that the expression

$$M_k = x_k{}'z_k + \|Ax_k - b\|, \tag{6.194}$$

is decreased to 0, where z_k is the vector

$$z_k = c - A'\lambda_k.$$

The expression (6.194) may be viewed as a *merit function*, and consists of two nonnegative terms: the first term is $x_k{}'z_k$, which is positive (since $x_k > 0$ and $z_k > 0$) and can be written as

$$x_k{}'z_k = x_k{}'(c - A'\lambda_k) = c'x_k - b'\lambda_k + (b - Ax_k)'\lambda_k.$$

Thus when x_k is primal-feasible ($Ax_k = b$), $x_k{}'z_k$ is equal to the duality gap, that is, the difference between the primal and the dual costs, $c'x_k - b'\lambda_k$. The second term is the norm of the primal constraint violation $\|Ax_k - b\|$. In the method to be described, neither of the terms $x_k{}'z_k$ and $\|Ax_k - b\|$ may increase at each iteration, so that $M_{k+1} \leq M_k$ (and typically $M_{k+1} < M_k$) for all k. If we can show that $M_k \to 0$, then asymptotically both the duality gap and the primal constraint violation will be driven to zero. Thus every limit point of $\{(x_k, \lambda_k)\}$ will be a pair of primal and dual optimal solutions, in view of the duality relation

$$\min_{Ax=b,\, x\geq 0} c'x = \max_{A'\lambda\leq c} b'\lambda,$$

given in Section 5.2 of Appendix B.

Let us write the necessary and sufficient conditions for (x, λ) to be a primal and dual optimal solution pair for the problem of minimizing the barrier function $F_\epsilon(x)$ subject to $Ax = b$. They are

$$c - \epsilon x^{-1} - A'\lambda = 0, \qquad Ax = b, \tag{6.195}$$

where x^{-1} denotes the vector with components $(x^i)^{-1}$. Let z be the vector of slack variables

$$z = c - A'\lambda.$$

Note that λ is dual feasible if and only if $z \geq 0$.

Using the vector z, we can write the first condition of Eq. (6.195) as $z - \epsilon x^{-1} = 0$ or, equivalently, $XZe = \epsilon e$, where X and Z are the diagonal matrices with the components of x and z, respectively, along the diagonal, and e is the vector with unit components,

$$X = \begin{pmatrix} x^1 & 0 & \cdots & 0 \\ 0 & x^2 & \cdots & 0 \\ \cdots & \cdots & \cdots & \cdots \\ 0 & 0 & \cdots & x^n \end{pmatrix}, \quad Z = \begin{pmatrix} z^1 & 0 & \cdots & 0 \\ 0 & z^2 & \cdots & 0 \\ \cdots & \cdots & \cdots & \cdots \\ 0 & 0 & \cdots & z^n \end{pmatrix}, \quad e = \begin{pmatrix} 1 \\ 1 \\ \vdots \\ 1 \end{pmatrix}.$$

Thus the optimality conditions (6.195) can be written in the equivalent form

$$XZe = \epsilon e, \tag{6.196}$$

$$Ax = b, \tag{6.197}$$

$$z + A'\lambda = c. \tag{6.198}$$

Given (x, λ, z) satisfying $z + A'\lambda = c$, and such that $x > 0$ and $z > 0$, a Newton iteration for solving this system is

$$x(\alpha, \epsilon) = x + \alpha \Delta x, \tag{6.199}$$

$$\lambda(\alpha, \epsilon) = \lambda + \alpha \Delta \lambda,$$

$$z(\alpha, \epsilon) = z + \alpha \Delta z,$$

where α is a stepsize such that $0 < \alpha \leq 1$ and

$$x(\alpha, \epsilon) > 0, \qquad z(\alpha, \epsilon) > 0,$$

and the Newton increment $(\Delta x, \Delta \lambda, \Delta z)$ solves the linearized version of the system (6.196)-(6.198)

$$X\Delta z + Z\Delta x = -v, \tag{6.200}$$

$$A\Delta x = b - Ax, \tag{6.201}$$

$$\Delta z + A'\Delta\lambda = 0, \tag{6.202}$$

with v defined by

$$v = XZe - \epsilon e. \tag{6.203}$$

After a straightforward calculation, the solution of the linearized system (6.200)-(6.202) can be written as

$$\Delta\lambda = \left(AZ^{-1}XA'\right)^{-1}\left(AZ^{-1}v + b - Ax\right), \tag{6.204}$$

$$\Delta z = -A'\Delta\lambda, \tag{6.205}$$

$$\Delta x = -Z^{-1}v - Z^{-1}X\Delta z.$$

Note that $\lambda(\alpha, \epsilon)$ is dual feasible, since from Eq. (6.202) and the condition $z + A'\lambda = c$, we see that

$$z(\alpha, \epsilon) + A'\lambda(\alpha, \epsilon) = c.$$

Note also that if $\alpha = 1$, i.e., a pure Newton step is used, $x(\alpha, \epsilon)$ is primal feasible, since from Eq. (6.201) we have $A(x + \Delta x) = b$.

Merit Function Improvement

We will now evaluate the changes in the constraint violation and the merit function (6.194) induced by the Newton iteration.

By using Eqs. (6.199) and (6.201), the new constraint violation is given by

$$Ax(\alpha,\epsilon)-b=Ax+\alpha A\Delta x-b=Ax+\alpha(b-Ax)-b=(1-\alpha)(Ax-b). \tag{6.206}$$

Thus, since $0<\alpha\le 1$, the new norm of constraint violation $\|Ax(\alpha,\epsilon)-b\|$ is always no larger than the old one. Furthermore, if x is primal-feasible ($Ax=b$), the new iterate $x(\alpha,\epsilon)$ is also primal-feasible.

The inner product

$$p=x'z \tag{6.207}$$

after the iteration becomes

$$\begin{aligned} p(\alpha,\epsilon)&=x(\alpha,\epsilon)'z(\alpha,\epsilon)\\ &=(x+\alpha\Delta x)'(z+\alpha\Delta z)\\ &=x'z+\alpha(x'\Delta z+z'\Delta x)+\alpha^2\Delta x'\Delta z. \end{aligned} \tag{6.208}$$

From Eqs. (6.201) and (6.205) we have

$$\Delta x'\Delta z=(Ax-b)'\Delta\lambda,$$

while by premultiplying Eq. (6.200) with e' and using the definition (6.203) for v, we obtain

$$x'\Delta z+z'\Delta x=-e'v=n\epsilon-x'z.$$

By substituting the last two relations in Eq. (6.208) and by using also the expression (6.207) for p, we see that

$$p(\alpha,\epsilon)=p-\alpha(p-n\epsilon)+\alpha^2(Ax-b)'\Delta\lambda. \tag{6.209}$$

Let us now denote by M and $M(\alpha,\epsilon)$ the value of the merit function (6.194) before and after the iteration, respectively. We have by using the expressions (6.206) and (6.209),

$$\begin{aligned} M(\alpha,\epsilon)&=p(\alpha,\epsilon)+\|Ax(\alpha,\epsilon)-b\|\\ &=p-\alpha(p-n\epsilon)+\alpha^2(Ax-b)'\Delta\lambda+(1-\alpha)\|Ax-b\|, \end{aligned}$$

or

$$M(\alpha,\epsilon)=M-\alpha\big(p-n\epsilon+\|Ax-b\|\big)+\alpha^2(Ax-b)'\Delta\lambda.$$

Thus if ϵ is chosen to satisfy

$$\epsilon<\frac{p}{n}$$

and α is chosen to be small enough so that the second order term $\alpha^2(Ax-b)'\Delta\lambda$ is dominated by the first order term $\alpha(p-n\epsilon)$, the merit function will be improved as a result of the iteration.

A General Class of Primal-Dual Algorithms

Let us consider now the general class of algorithms of the form

$$x_{k+1}=x(\alpha_k,\epsilon_k),\qquad \lambda_{k+1}=\lambda(\alpha_k,\epsilon_k),\qquad z_{k+1}=z(\alpha_k,\epsilon_k),$$

where α_k and ϵ_k are positive scalars such that

$$x_{k+1}>0,\qquad z_{k+1}>0,\qquad \epsilon_k<\frac{g_k}{n},$$

where

$$g_k=x_k{}'z_k+(Ax_k-b)'\lambda_k,$$

and α_k is such that the merit function M_k is reduced. Initially we must have $x_0>0$, and $z_0=c-A'\lambda_0>0$ (such a point can often be easily found; otherwise an appropriate reformulation of the problem is necessary for which we refer to the specialized literature). These methods are generally called *primal-dual*, in view of the fact that they operate simultaneously on the primal and dual variables.

It can be shown that it is possible to choose α_k and ϵ_k so that the merit function is not only reduced at each iteration, but also converges to zero. Furthermore, with suitable choices of α_k and ϵ_k, algorithms with good theoretical properties, such as polynomial complexity and superlinear convergence, can be derived.

Computational experience has shown that with properly chosen sequences α_k and ϵ_k, and appropriate implementation, the practical performance of the primal-dual methods is excellent. The choice

$$\epsilon_k=\frac{g_k}{n^2},$$

leading to the relation

$$g_{k+1}=(1-\alpha_k+\alpha_k/n)g_k$$

for feasible x_k, has been suggested as a good practical rule. Usually, when x_k has already become feasible, α_k is chosen as $\theta\tilde{\alpha}_k$, where θ is a factor very close to 1 (say 0.999), and $\tilde{\alpha}_k$ is the maximum stepsize α that guarantees that $x(\alpha,\epsilon_k)\geq 0$ and $z(\alpha,\epsilon_k)\geq 0$

$$\tilde{\alpha}_k=\min\left\{\min_{i=1,\ldots,n}\left\{\frac{x_k^i}{-\Delta x^i}\;\Big|\;\Delta x^i<0\right\},\ \min_{i=1,\ldots,n}\left\{\frac{z_k^i}{-\Delta z^i}\;\Big|\;\Delta z^i<0\right\}\right\}.$$

When x_k is not feasible, the choice of α_k must also be such that the merit function is improved. In some works, a different stepsize for the x update than for the (λ,z) update has been suggested. The stepsize for the x update is near the maximum stepsize α that guarantees $x(\alpha,\epsilon_k)\geq 0$, and the stepsize for the (λ, z) update is near the maximum stepsize α that guarantees $z(\alpha, \epsilon_k) \geq 0$.

There are a number of additional practical issues related to implementation, for which we refer to the specialized literature. We refer to the research monographs [Wri97], [Ye97], and other sources for a detailed discussion, as well as extensions to nonlinear/convex programming problems, such as quadratic programming. There are also more sophisticated implementations of the Newton/primal-dual idea, one of which we describe next.

Predictor-Corrector Variants

We will now discuss some modified versions of the preceding interior point methods, which are based on a variation of Newton's method where the Hessian is evaluated periodically every $q > 1$ iterations in order to economize in iteration overhead. When $q = 2$ and the problem is to solve the system $h(x) = 0$, where $g : \Re^n \mapsto \Re^n$, this variation of Newton's method takes the form

$$\hat{x}_k = x_k - \big(\nabla h(x_k)'\big)^{-1} h(x_k), \tag{6.210}$$

$$x_{k+1} = \hat{x}_k - \big(\nabla h(x_k)'\big)^{-1} h(\hat{x}_k). \tag{6.211}$$

Thus, given x_k, this iteration performs a regular Newton step to obtain $\hat{x}_k$, and then an approximate Newton step from $\hat{x}_k$, using, however, the already available Jacobian inverse $\big(\nabla h(x_k)'\big)^{-1}$. It can be shown that if $x_k \to x^*$, the order of convergence of the error $\|x_k - x^*\|$ is cubic, that is,

$$\limsup_{k\to\infty} \frac{\|x_{k+1} - x^*\|}{\|x_k - x^*\|^3} < \infty,$$

under the same assumptions that the ordinary Newton's method ($q = 1$) attains a quadratic order of convergence; see [OrR70], p. 315. Thus, the price for the 50% saving in Jacobian evaluations and inversions is a small degradation of the convergence rate over the ordinary Newton's method (which attains a quartic order of convergence when two successive ordinary Newton steps are counted as one).

Two-step Newton methods such as the iteration (6.210), (6.211), when applied to the system of optimality conditions (6.196)-(6.198) for the linear program (LP) are known as *predictor-corrector* methods (the name comes from their similarity with predictor-corrector methods for solving differential equations). They operate as follows:

Given (x, z, λ) with $x > 0$, and $z = c - A'\lambda > 0$, the *predictor iteration* [cf. Eq. (6.210)], solves for $\big(\Delta\hat{x}, \Delta\hat{z}, \Delta\hat{\lambda}\big)$ the system

$$X\Delta\hat{z} + Z\Delta\hat{x} = -\hat{v}, \tag{6.212}$$

$$A\Delta\hat{x} = b - Ax, \tag{6.213}$$

$$\Delta\hat{z} + A'\Delta\hat{\lambda} = 0, \tag{6.214}$$

with $\hat{v}$ defined by

$$\hat{v} = XZe - \hat{\epsilon}e, \tag{6.215}$$

[cf. Eqs. (6.200)-(6.203)].

The *corrector iteration* [cf. Eq. (6.211)], solves for $(\Delta\bar{x}, \Delta\bar{z}, \Delta\bar{\lambda})$ the system

$$X\Delta\bar{z} + Z\Delta\bar{x} = -\bar{v}, \tag{6.216}$$

$$A\Delta\bar{x} = b - A(x + \Delta\hat{x}), \tag{6.217}$$

$$\Delta\bar{z} + A'\Delta\bar{\lambda} = 0, \tag{6.218}$$

with $\bar{v}$ defined by

$$\bar{v} = (X + \Delta\hat{X})(Z + \Delta\hat{Z})e - \bar{\epsilon}e, \tag{6.219}$$

where $\Delta\hat{X}$ and $\Delta\hat{Z}$ are the diagonal matrices corresponding to $\Delta\hat{x}$ and $\Delta\hat{z}$, respectively. Here $\hat{\epsilon}$ and $\bar{\epsilon}$ are the barrier parameters corresponding to the two iterations.

The *composite Newton direction* is

$$\Delta x = \Delta\hat{x} + \Delta\bar{x},$$

$$\Delta z = \Delta\hat{z} + \Delta\bar{z},$$

$$\Delta\lambda = \Delta\hat{\lambda} + \Delta\bar{\lambda},$$

and the corresponding iteration is

$$x(\alpha, \epsilon) = x + \alpha\Delta x,$$

$$\lambda(\alpha, \epsilon) = \lambda + \alpha\Delta\lambda,$$

$$z(\alpha, \epsilon) = z + \alpha\Delta z,$$

where α is a stepsize such that $0 < \alpha \leq 1$ and

$$x(\alpha, \epsilon) > 0, \qquad z(\alpha, \epsilon) > 0.$$

We will now develop a system of equations that yields the composite Newton direction. By adding Eqs. (6.212)-(6.214) and Eqs. (6.216)-(6.218), we obtain

$$X(\Delta\hat{z} + \Delta\bar{z})z + Z(\Delta\hat{x} + \Delta\bar{x}) = -\hat{v} - \bar{v}, \tag{6.220}$$

$$A(\Delta\hat{x} + \Delta\bar{x})x = b - Ax + b - A(x + \Delta\hat{x}), \tag{6.221}$$

$$\Delta\hat{z} + \Delta\bar{z} + A'(\Delta\hat{\lambda} + \Delta\bar{\lambda}) = 0, \tag{6.222}$$

We use the fact

$$b - A(x + \Delta\hat{x}) = 0$$

[cf. Eq. (6.213)], and we also use Eqs. (6.219) and (6.212) to write

$$\begin{aligned}\bar{v} &= (X + \Delta\hat{X})(Z + \Delta\hat{Z})e - \bar{\epsilon}e \\ &= XZe + \Delta\hat{X}Ze + X\Delta\hat{Z}e + \Delta\hat{X}\Delta\hat{Z}e - \bar{\epsilon}e \\ &= XZe + Z\Delta\hat{x} + X\Delta\hat{z} + \Delta\hat{X}\Delta\hat{Z}e - \bar{\epsilon}e \\ &= XZe - \hat{v} + \Delta\hat{X}\Delta\hat{Z}e - \bar{\epsilon}e.\end{aligned}$$

Substituting in Eqs. (6.220)-(6.222) we obtain the following system of equations for the composite Newton direction $(\Delta x, \Delta z, \Delta\lambda) = (\Delta\hat{x} + \Delta\bar{x}, \Delta\hat{z} + \Delta\bar{z}, \Delta\hat{\lambda} + \Delta\bar{\lambda})$:

$$X\Delta z + Z\Delta x = -XZe - \Delta\hat{X}\Delta\hat{Z}e + \bar{\epsilon}e, \tag{6.223}$$

$$A\Delta x = b - Ax, \tag{6.224}$$

$$\Delta z + A'\Delta\lambda = 0. \tag{6.225}$$

To implement the predictor-corrector method, we need to solve the system (6.212)-(6.215) for some value of $\hat{\epsilon}$ to obtain $(\Delta\hat{X}, \Delta\hat{Z})$, and then to solve the system (6.223)-(6.225) for some value of $\bar{\epsilon}$ to obtain $(\Delta x, \Delta z, \Delta\lambda)$. It is important to note here that most of the work needed for the first system, namely the factorization of the matrix

$$AZ^{-1}XA'$$

in Eq. (6.204), need not be repeated when solving the second system, so that solving both systems requires relatively little extra work over solving the first one. We refer to the specialized literature for further details [LMS92], [Meh92], [Wri97], [Ye97].

6.8.2 Interior Point Methods for Conic Programming

We now discuss briefly interior point methods for the conic programming problems discussed in Section 1.2. Consider first the second order cone problem

$$\begin{aligned}&\text{minimize} \quad c'x \\ &\text{subject to} \quad A_i x - b_i \in C_i, \ i = 1, \ldots, m,\end{aligned} \tag{6.226}$$

where $x \in \Re^n$, c is a vector in $\Re^n$, and for $i = 1, \ldots, m$, A_i is an $n_i \times n$ matrix, b_i is a vector in $\Re^{n_i}$, and C_i is the second order cone of $\Re^{n_i}$. We approximate this problem with

$$\begin{aligned}&\text{minimize} \quad c'x + \epsilon_k \sum_{i=1}^{m} B_i(A_i x - b_i) \\ &\text{subject to} \quad A_i x - b_i \in \text{int}(C_i), \ i = 1, \ldots, m,\end{aligned} \tag{6.227}$$

where B_i is a function defined in the interior of the second order cone C_i, and given by

$$B_i(y) = -\ln\left(y_{n_i}^2 - (y_1^2 + \cdots + y_{n_i-1}^2)\right), \qquad y = (y_1, \ldots, y_{n_i}) \in \mathrm{int}(C_i),$$

and $\{\epsilon_k\}$ is a positive sequence that converges to 0. Note that as $A_i x - b_i$ approaches the boundary of C_i, the logarithmic penalty $B_i(A_i x - b_i)$ approaches ∞.

Similar to Prop. 6.8.1, it can be shown that if x_k is an optimal solution of the approximating problem (6.227), then every limit point of the sequence $\{x_k\}$ is an optimal solution of the original problem. For theoretical as well as practical reasons, the approximating problem (6.227) should not be solved exactly. In the most efficient methods, one or more Newton steps corresponding to a given value ϵ_k are performed, and then the value of ϵ_k is appropriately reduced. Similar, to the interior point methods for linear programming of the preceding section, Newton's method should be implemented with a stepsize to ensure that the iterates keep $A_i x - b_i$ within the second order cone.

If the aim is to achieve a favorable polynomial complexity result, a single Newton step should be performed between successive reductions of ϵ_k, and the subsequent reduction of ϵ_k must be correspondingly small, according to an appropriate formula, which is designed to enable a polynomial complexity proof. An alternative, which has proved more efficient in practice, is to allow multiple Newton steps until an appropriate termination criterion is satisfied, and then reduce ϵ_k substantially. When properly implemented, methods of this type seem to require a consistently small total number of Newton steps [a number typically no more than 50, regardless of dimension (!) is often reported]. This empirical observation is far more favorable than what is predicted by the theoretical complexity analysis.

There is a similar interior point method for the dual semidefinite cone problem involving the multiplier vector $\lambda = (\lambda_1, \ldots, \lambda_m)$:

$$\begin{aligned} &\text{maximize} \quad b'\lambda \\ &\text{subject to} \quad D - (\lambda_1 A_1 + \cdots + \lambda_m A_m) \in C, \end{aligned} \tag{6.228}$$

where $b \in \Re^m$, $D, A_1, \ldots, A_m$ are symmetric matrices, and C is the cone of positive semidefinite matrices. It consists of solving the problem

$$\begin{aligned} &\text{maximize} \quad b'\lambda + \epsilon_k \ln\left(\det(D - \lambda_1 A_1 - \cdots - \lambda_m A_m)\right) \\ &\text{subject to} \quad \lambda \in \Re^m, \end{aligned} \tag{6.229}$$

where $\{\epsilon_k\}$ is a positive sequence that converges to 0. Furthermore, a starting point such that $D - \lambda_1 A_1 - \cdots - \lambda_m A_m$ is positive definite should be used, and Newton's method should be implemented with a stepsize to ensure that the iterates keep $D - \lambda_1 A_1 - \cdots - \lambda_m A_m$ within the positive definite cone.

The properties of this method are similar to the ones of the preceding second order cone method. In particular, if x_k is an optimal solution of the approximating problem (6.229), then every limit point of $\{x_k\}$ is an optimal solution of the original problem (6.228).

We finally note that there are primal-dual interior point methods for conic programming problems, which bear similarity with the one given for linear programming in the preceding section. We refer to the literature for further discussion and a complexity analysis; see e.g., [NeN94], [BoV04].

6.8.3 Central Cutting Plane Methods

We now return to the general problem of minimizing a real-valued convex function f over a closed convex constraint set X. We will discuss a method that combines the interior point and cutting plane approaches. Like the cutting plane method of Section 4.1, it maintains a polyhedral approximation

$$F_k(x) = \max\{f(x_0) + (x - x_0)'g_0, \ldots, f(x_k) + (x - x_k)'g_k\}$$

to f, constructed using the points $x_0, \ldots, x_k$ generated so far, and associated subgradients $g_0, \ldots, g_k$, with $g_i \in \partial f(x_i)$ for all $i = 0, \ldots, k$. However, it generates the next vector x_{k+1} by using a different mechanism. In particular, instead of minimizing F_k over X, the method obtains x_{k+1} by finding a "central pair" (x_{k+1}, w_{k+1}) within the subset

$$S_k = \{(x, w) \mid x \in X,\, F_k(x) \le w \le \tilde{f}_k\},$$

where $\tilde{f}_k$ is the best upper bound to the optimal value that has been found so far,

$$\tilde{f}_k = \min_{i=0,\ldots,k} f(x_i)$$

(see Fig. 6.8.3).

There are several methods for finding the central pair (x_{k+1}, w_{k+1}). Roughly, the idea is that the central pair should be "somewhere in the middle" of S_k. For example, consider the case where S_k is polyhedral with nonempty interior. Then (x_{k+1}, w_{k+1}) could be the *analytic center* of S_k, where for any polyhedral set

$$P = \{y \mid a_p'y \le c_p,\, p = 1, \ldots, m\} \tag{6.230}$$

with nonempty interior, its analytic center is defined as the unique maximizer of

$$\sum_{p=1}^{m} \ln(c_p - a_p'y)$$

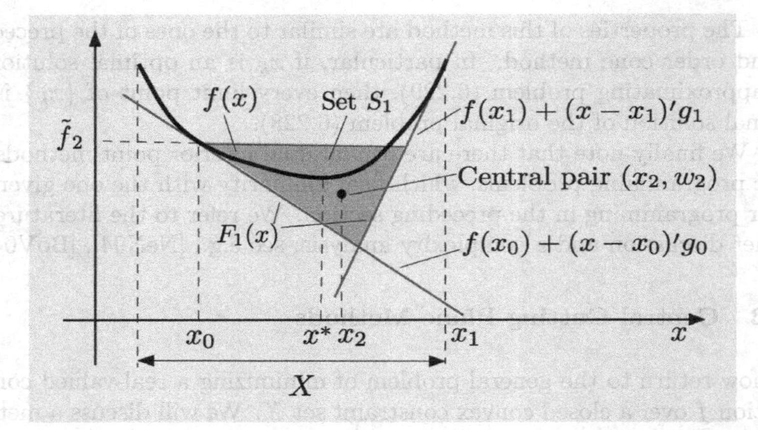

Figure 6.8.3. Illustration of the set

$$S_k = \{(x, w) \mid x \in X,\ F_k(x) \leq w \leq \tilde{f}_k\}$$

in the central cutting plane method.

over $y \in P$.

Another possibility is the *ball center* of S, i.e., the center of the largest inscribed sphere in S_k; for the generic polyhedral set P of the form (6.230), the ball center can be obtained by solving the following problem with optimization variables (y, σ):

$$\begin{aligned} &\text{maximize } \sigma \\ &\text{subject to } a_p'(y + d) \leq c_p, \quad \forall\ \|d\| \leq \sigma,\ p = 1, \ldots, m, \end{aligned}$$

assuming that P has nonempty interior. By maximizing over all d with $\|d\| \leq \sigma$, it can be seen that this problem is equivalent to the linear program

$$\begin{aligned} &\text{maximize } \sigma \\ &\text{subject to } a_p'y + \|a_p\|\sigma \leq c_p, \quad p = 1, \ldots, m. \end{aligned}$$

Central cutting plane methods have satisfactory convergence properties, even though they do not terminate finitely in the case of a polyhedral cost function f, as the ordinary cutting plane method does. Since they are closely related to the interior point methods, they have benefited from advances in the practical implementation methodology of these methods.

6.9 NOTES, SOURCES, AND EXERCISES

As we noted earlier, the discussion in this chapter is often not as detailed as in earlier chapters, as some of the methods described are under active development. Moreover the literature in the field has grown explosively in the decade preceding the writing of this book. As a result our presentation and references are not comprehensive. They tend to reflect to some extent the author's reading preferences and research orientation. The textbooks, research monographs, and surveys that we cite may provide the reader with a more comprehensive view of the many different research lines in the field.

Section 6.1: Convergence rate analysis of optimization methods (as well as related methods for solving systems of linear and nonlinear equations) has traditionally been local in nature, i.e., asymptotic estimates of the number of iterations, starting sufficiently close to the point of convergence, and an emphasis on issues of condition number and scaling (cf. our discussion in Section 2.1.1). Most of the books in nonlinear programming follow this approach. Over time and beginning in the late 70s, a more global approach has received attention, influenced in part by computational complexity ideas, and emphasizing first order methods. In this connection, we mention the books of Nemirovskii and Yudin [NeY83], and Nesterov and Nemirovskii [NeN94].

Section 6.2: The optimal complexity gradient projection/extrapolation method is due to Nesterov [Nes83], [Nes04], [Nes05]; see also Tseng [Tse08], Beck and Teboulle [BeT10], Lu, Monteiro, and Yuan [LMY12], and Gonzaga and Karas [GoK13], for proposals and analysis of variants and more general methods. Some of these variants also apply to important classes of nondifferentiable cost problems, similar to the ones treated by the proximal gradient methods of Section 6.3.

The idea of using smoothing in conjunction with a gradient method to construct optimal algorithms for nondifferentiabe convex problems is due to Nesterov [Nes05]. In his work he proves the Lipschitz property of Prop. 6.2.2 for the more general case, where p is convex but not necessarily differentiable, and analyzes several important special cases. The algorithm, and its complexity oriented analysis, have had a strong influence on convex optimization algorithms research. In our presentation, we follow the analysis of Tseng [Tse08].

Section 6.3: There have been several proposals of combinations of gradient and proximal methods for minimizing the sum of two convex functions (or more generally, finding a zero of the sum of two nonlinear monotone operators). These methods, commonly called *splitting algorithms*, have a long history, dating to the papers of Lions and Mercier [LiM79], and Passty [Pas79]. Like the ADMM, the basic proximal gradient algorithm, written in the form (6.46), admits an extension to the problem of finding a zero of the sum of two nonlinear maximal monotone operators, called the *forward-backward* algorithm (one of the two operators must be single-valued). The form of the forward-backward algorithm is illustrated in Fig. 6.3.1: we simply need to replace the subdifferential ∂h with a general multivalued maximal monotone operator, and the gradient ∇f with a general single-valued monotone operator.

The forward-backward algorithm was proposed and analyzed by Lions and Mercier [LiM79], and Passty [Pas79]. Additional convergence results for the algorithm and a discussion of its applications were given by Gabay [Gab83] and Tseng [Tse91b]. The convergence result of Prop. 6.3.3, in the case where the stepsize is constant, descends from the more general results of [Gab83] and [Tse91b]. A modification that converges under weaker assumptions is given by Tseng [Tse00]. The rate of convergence has been further discussed by Chen and Rockafellar [ChR97].

Variants of proximal gradient and Newton-like methods have been proposed and analyzed by several authors, including cases where the differentiable function is not convex; see e.g., Fukushima and Mine [FuM81], [MiF81], Patriksson [Pat93], [Pat98], [Pat99], Tseng and Yun [TsY09], and Schmidt [Sch10]. The methods have received renewed attention, as they are well-matched to the structure of some large-scale machine learning and signal processing problems; see Beck and Teboulle [BeT09a], [BeT09b], [BeT10], and the references they give to algorithms for problems with special structures.

There has been a lot of additional recent work in this area, which cannot be fully surveyed here. Methods (with and without extrapolation), which replace the gradient with an aggregated gradient that is calculated incrementally, are proposed and analyzed by Xiao [Xia10], and Xiao and Zhang [XiZ14]. Inexact variants that admit errors in the proximal minimization and the gradient calculation, in the spirit of the ϵ-subgradient methods of Section 3.3, have been discussed by Schmidt, Roux, and Bach [SRB11]. The convergence rate for some interesting special cases was investigated by Tseng [Tse10], Hou et al. [HZS13], and Zhang, Jiang, and Luo [ZJL13]. Algorithms where f has an additive form with components treated incrementally are discussed by Duchi and Singer [DuS09], and by Langford, Li, and Zhang [LLZ09]. For recent work on proximal Newton-like methods, see Becker and Fadili [BeF12], Lee, Sun, and Saunders [LLS12], [LLS14], and Chouzenoux, Pesquet, and Repetti [CPR14]. The finite and superlinear convergence rate results of Exercises 6.4 and 6.5 are new to the author's knowledge.

Section 6.4: Incremental subgradient methods were proposed by several authors in the 60s and 70s. Perhaps the earliest paper is by Litvakov [Lit66], which considered convex/nondifferentiable extensions of linear least squares problems. There were several other related subsequent proposals, including the paper by Kibardin [Kib80]. These works remained unnoticed in the Western literature, where incremental methods were reinvented often in different contexts and with different lines of analysis. We mention the papers by Solodov and Zavriev [SoZ98], Bertsekas [Ber99] (Section 6.3.2), Ben-Tal, Margalit, and Nemirovski [BMN01], Nedić and Bertsekas

[NeB00], [NeB01], [NeB10], Nedić, Bertsekas, and Borkar [NBB01], Kiwiel [Kiw04], Rabbat and Nowak [RaN04], [RaN05], Gaudioso, Giallombardo, and Miglionico [GGM06], Shalev-Shwartz et al. [SSS07], Helou and De Pierro [HeD09], Johansson, Rabi, and Johansson [JRJ09], Predd, Kulkarni, and Poor [PKP09], Ram, Nedić, and Veeravalli [RNV09], Agarwal and Duchi [AgD11], Duchi, Hazan, and Singer [DHS11], Nedić [Ned11], Duchi, Bartlett, and Wainwright [DBW12], Wang and Bertsekas [WaB13a], [WaB14], and Wang, Fang, and Liu [WFL14].

The advantage of using deliberate randomization in selecting components for deterministic additive cost functions was first established by Nedić and Bertsekas [NeB00], [NeB01]. Asynchronous incremental subgradient methods were proposed and analyzed by Nedić, Bertsekas, and Borkar [NBB01].

The incremental proximal methods and their combinations with subgradient methods (cf. Sections 6.4.1 and 6.4.2) were first proposed by the author in [Ber10b], [Ber11], which we follow in our development here. For recent work and applications, see Andersen and Hansen [AnH13], Couellan and Trafalis [CoT13], Weinmann, Demaret, and Storath [WDS13], Bacak [Bac14], Bergmann et al. [BSL14], Richard, Gaiffas, and Vayatis [RGV14], and You, Song, and Qiu [YSQ14].

The incremental augmented Lagrangian method of Section 6.4.3 is new. The idea of this method is simple: just as the proximal algorithm, when dualized, yields the augmented Lagrangian method, the incremental proximal algorithm, when dualized, should yield a form of incremental augmented Lagrangian method. While we have focused on the case of a particular type of separable problem, this idea applies more broadly to other contexts that involve additive cost functions.

Incremental constraint projection methods are related to classical feasibility methods, which have been discussed by many authors; see e.g., Gubin, Polyak, and Raik [GPR67], the survey by Bauschke and Borwein [BaB96], and recent papers such as Bauschke [Bau01], Bauschke, Combettes, and Kruk [BCK06], Cegielski and Suchocka [CeS08], and Nedić [Ned10], and their bibliographies.

Incremental constraint projection methods (with a nonzero convex cost function) were first proposed by Nedić [Ned11]. The algorithm of Section 6.4.4 was proposed by the author in [Ber10b], [Ber11], and is similar but differs in a few ways from the one of [Ned11]. The latter algorithm uses a stepsize $\beta_k \equiv 1$ and requires the linear regularity assumption, noted in Section 6.4.4, in order to prove convergence. Moreover, it does not consider cost functions that are sums of components that can be treated incrementally. We refer to the papers by Wang and Bertsekas [WaB13a], [WaB14] for extensions of the results of [Ber11] and [Ned11], which involve incremental treatment of the cost function and the constraints, and a unified convergence analysis of a variety of component selection rules, both deterministic and randomized.

Section 6.5: There is extensive literature on coordinate descent methods, and it has recently grown tremendously (a Google Scholar search produced many thousands of papers appearing in the two years preceding the publication of this book). Our references here and in Section 6.5 are consequently highly selective, with a primary focus on early contributions.

Coordinate descent algorithms have their origin in the classical Jacobi and Gauss-Seidel methods for solving systems of linear and nonlinear equations. These were among the first methods to receive extensive attention in the field of scientific computation in the 50s; see the books by Ortega and Rheinboldt [OrR70] for a numerical analysis point of view, and by Bertsekas and Tsitsiklis [BeT89a] for a distributed synchronous and asynchronous computation point of view.

In the field of optimization, Zangwill [Zan69] gave the first convergence proof of coordinate descent for unconstrained continuously differentiable problems, and focused attention on the need for a unique attainment assumption of the minimum along each coordinate direction. Powell [Pow73] gave a counterexample showing that for three or more block components this assumption is essential.

The convergence rate of coordinate descent has been discussed by Luenberger [Lue84], and more recently by Nesterov [Nes12], and by Schmidt and Friedlander [ScF14] (see Exercise 6.11). Convergence analyses for constrained problems have been given by a number of authors under various conditions. We have followed here the line of analysis of Bertsekas and Tsitsiklis [BeT89a] (Section 3.3.5), also used in Bertsekas [Ber99] (Section 2.7).

Coordinate descent methods are often well suited for solving dual problems, and within this specialized context there has been much analysis; see Hildreth [Hil57], Cryer [Cry71], Pang [Pan84], Censor and Herman [CeH87], Hearn and Lawphongpanich [HeL89], Lin and Pang [LiP87], Bertsekas, Hossein, and Tseng [BHT87], Tseng and Bertsekas [TsB87], [TsB90], [TsB91], Tseng [Tse91a], [Tse93], Hager and Hearn [HaH93], Luo and Tseng [LuT93a], [LuT93b], and Censor and Zenios [CeZ97]. Some of these papers also deal with the case of multiple dual optimal solutions.

The convergence to a unique limit and the rate of convergence of the coordinate descent method applied to quadratic programming problems were analyzed by Luo and Tseng [LuT91], [LuT92] (without an assumption of a unique minimum). These two papers marked the beginning of an important line of analysis of the convergence and rate of convergence of several types of constrained optimization algorithms, including coordinate descent, which is based on the Hoffman bound and other related error bounds; see [LuT93a], [LuT93b], [LuT93c], [LuT94a], [LuT94b]. For a survey of works in this area see Tseng [Tse10].

Recent attention has focused on variants of coordinate descent methods with inexact minimization along each block component, extensions to nondifferentiable cost problems with special structure, alternative orders for coordinate selection, and asynchronous distributed computation issues (see the discussion and references in Section 6.5). A survey of the field is given by Wright [Wri14].

Section 6.6: Nonquadratic augmented Lagrangians and their associated proximal minimization-type algorithms that use nonquadratic proximal terms were introduced by Kort and Bertsekas; see the paper [KoB72], which included as a special case the exponential augmented Lagrangian method, the thesis [Kor75], the paper [KoB76], and the augmented Lagrangian monograph [Ber82a]. Progressively stronger convergence results for the exponential method were established in [KoB72], [Ber82a], [TsB93].

There has been continuing interest in proximal algorithms with nonquadratic regularization, and related augmented Lagrangian algorithms, directed at obtaining additional classes of methods, sharper convergence results, understanding the properties that enhance computational performance, and specialized applications. For representative works, see Polyak [Pol88], [Pol92], Censor and Zenios [CeZ92], [CeZ97], Guler [Gul92], Teboulle [Teb92], Chen and Teboulle [ChT93], [ChT94], Tseng and Bertsekas [TsB93], Bertsekas and Tseng [BeT94a], Eckstein [Eck94], Iusem, Svaiter, and Teboulle [IST94], Iusem and Teboulle [IuT95], Auslender, Cominetti, and Haddou [ACH97], Ben-Tal and Zibulevsky [BeZ97], Polyak and Teboulle [PoT97], Iusem [Ius99], Facchinei and Pang [FaP03], Auslender and Teboulle [AuT03], and Tseng [Tse04].

Section 6.7: Descent methods for minimax problems were considered in several works during the 60s and early 70s, as extensions of feasible direction methods for constrained minimization (a minimax problem can be converted to a constrained optimization problem); see e.g., the book by Demjanov and Rubinov [DeR70], and the papers by Pshenichnyi [Psh65], Demjanov [Dem66], [Dem68], and Demjanov and Malozemov [DeM71]. These methods typically involved an ϵ parameter to deal with convergence difficulties due to proximity to the constraint boundaries.

The ϵ-descent method of Section 6.7 relates to these earlier methods, but applies more generally to any closed proper convex function. It was proposed by Bertsekas and Mitter [BeM71], [BeM73], together with its outer approximation variant for sums of functions. This variant is well-suited for application to convex network optimization problems; see Rockafellar [Roc84], Bertsekas, Hossein, and Tseng [BHT87], Bertsekas and Tsitsiklis [BeT89a], and Bertsekas [Ber98]. Inner approximation variants of the ϵ-descent method were proposed by Lemaréchal [Lem74], [Lem75], and a related algorithm was independently derived by Wolfe [Wol75]. These papers, together with other papers in the collection edited by Balinski and Wolfe [BaW75], were influential in focusing attention on nondifferentiable convex optimization as a field with broad applications.

The extended monotropic programming problem was introduced and studied by the author [Ber10a], as an extension of Rockafellar's monotropic programming framework, which was extensively developed in the book [Roc84]. The latter framework is the case where each function f_i is one-dimensional, and includes as special cases linear, quadratic, and convex separable single commodity network flow problems (see [Roc84], [Ber98]). Extended monotropic programming is more general in that it allows multi-dimensional component functions f_i, but requires constraint qualifications for strong duality, which we have discussed in Sections 6.7.3 and 6.7.4.

Monotropic programming is the largest class of convex optimization problems with linear constraints for which strong duality holds without any additional qualifications, such as the Slater or other relative interior point conditions (cf. Section 1.1), or the ϵ-subdifferential condition of Prop. 6.7.4. The EMP strong duality theorem (Prop. 6.7.4) extends a theorem proved for the case of a monotropic programming problem in [Roc84]. Related analyses and infinite dimensional extensions of extended monotropic programming are given by Borwein, Burachik, and Yao [BBY12], Burachik and Majeed [BuM13], and Burachik, Kaya, and Majeed [BKM14].

Section 6.8: Interior point methods date to the work of Frisch in the 50s [Fri56]. A textbook treatment was given by Fiacco and McCormick [FiM68]. The methods became popular in the middle 80s, when they were systematically applied to linear programming problems, following the intense interest in the complexity oriented work of Karmarkar [Kar84].

The textbooks by Bertsimas and Tsitsiklis [BeT97], and Vanderbei [Van01] provide accounts of interior point methods for linear programming. The research monographs by Wright [Wri97], Ye [Ye97], and Roos, Terlaky, and Vial [RTV06], the edited volume by Terlaky [Ter96], and the survey by Forsgren, Gill, and Wright [FGW02] are devoted to interior point methods for linear, quadratic, and convex programming. For a convergence rate analysis of primal-dual interior point methods, see Zhang, Tapia, and Dennis [ZTD92], and Ye et al. [YGT93]. The predictor-corrector variant was proposed by Mehrotra [Meh92]; see also Lustig, Marsten, and Shanno [LMS92]. We have followed closely here the development of the author's textbook [Ber99].

Interior point methods were adapted for conic programming, starting with the complexity-oriented monograph by Nesterov and Nemirovskii [NeN94]. The book by Boyd and Vanderbergue [BoV04] has a special focus on interior point methods. Furthermore, related convex optimization software have become popular (Grant, Boyd, and Ye [GBY12]).

Central cutting plane methods were introduced by Elzinga and Moore [ElM75]. More recent proposals, some of which relate to interior point methods, are discussed in Goffin and Vial [GoV90], Goffin, Haurie, and Vial [GHV92], Ye [Ye92], Kortanek and No [KoN93], Goffin, Luo, and Ye [GLY94], [GLY96], Atkinson and Vaidya [AtV95], den Hertog et al. [HKR95], Nesterov [Nes95]. For a textbook treatment, see Ye [Ye97], and for a survey, see Goffin and Vial [GoV02].

EXERCISES

6.1 (Equivalent Properties of Convex Functions with Lipschitz Gradient)

Given a differentiable convex function $f : \Re^n \mapsto \Re$ and a scalar $L > 0$, show that the following five properties are equivalent:

(i) $\big\|\nabla f(x) - \nabla f(y)\big\| \leq L\,\|x - y\|$, for all $x, y \in \Re^n$.

(ii) $f(y) \geq f(x) + \nabla f(x)'(y - x) + \frac{1}{2L}\big\|\nabla f(x) - \nabla f(y)\big\|^2$, for all $x, y \in \Re^n$.

(iii) $\big(\nabla f(x) - \nabla f(y)\big)'(x - y) \geq \frac{1}{L}\big\|\nabla f(x) - \nabla f(y)\big\|^2$, for all $x, y \in \Re^n$.

(iv) $f(y) \leq f(x) + \nabla f(x)'(y - x) + \frac{L}{2}\|y - x\|^2$, for all $x, y \in \Re^n$.

(v) $\big(\nabla f(x) - \nabla f(y)\big)'(x - y) \leq L\|x - y\|^2$, for all $x, y \in \Re^n$.

Note: This exercise proves among others the converse to Prop. 6.1.9(a). It is given as part of Th. 2.1.5 in [Nes04], and also in part as Prop. 12.60 of [RoW98].
Proof: In Prop. 6.1.9(a), we showed that (i) implies (ii). To show that (ii) implies (iii), we write (ii) twice, with the roles of x and y interchanged, and add. Also (iii) and the Schwarz inequality imply (i). Thus (i), (ii), and (iii) are equivalent.

To show that (iv) implies (v), we write (iv) twice, with the roles of x and y interchanged, and add. To show that (v) implies (iv), we argue similar to the descent lemma (Prop. 6.1.2). Let $g(t) = f\big(x + t(y - x)\big)$, $t \in \Re$. We have

$$f(y) - f(x) = g(1) - g(0) = \int_0^1 \frac{dg}{dt}(t)\,dt = \int_0^1 \nabla f\big(x + t(y - x)\big)'(y - x)\,dt,$$

so that

$$f(y)-f(x)-\nabla f(x)'(y-x) = \int_0^1 \Big(\nabla f\big(x+t(y-x)\big)-\nabla f(x)\Big)'(y-x)\,dt \leq \frac{L}{2}\|y-x\|^2$$

where the last inequality follows by using (v) with y replaced by $x + t(y - x)$ to obtain

$$\Big(\nabla f\big(x + t(y - x)\big) - \nabla f(x)\Big)'(y - x) \leq Lt\,\|y - x\|^2, \qquad \forall\, t \in [0, 1].$$

Thus (iv) and (v) are equivalent.

We complete the proof by showing that (i) is equivalent to (iv). The descent lemma (Prop. 6.1.2) states that (i) implies (iv). To show that (iv) implies (i), we use the following proof, given to us by Huizhen Yu (a close examination of the proof of Th. 2.1.5 of [Nes04] shows that it does not include a proof of this part).

Fix x and consider the following expression as a function of y,

$$\frac{1}{L}\big(f(y) - f(x) - \nabla f(x)'(y - x)\big).$$

Making a change of variable $z = y - x$, we define

$$h_x(z) = \frac{1}{L}\big(f(z+x) - f(x) - \nabla f(x)'z\big).$$

By definition

$$\nabla h_x(y-x) = \frac{1}{L}\big(\nabla f(y) - \nabla f(x)\big), \tag{6.231}$$

while by the assumption (iv), we have $\frac{1}{2}\|z\|^2 \geq h_x(z)$ for all z, implying that

$$h_x^\star(\theta) \geq \frac{1}{2}\|\theta\|^2, \qquad \forall\, \theta \in \Re^n, \tag{6.232}$$

where $h_x^\star$ is the conjugate of h_x.

Similarly, fix y and define

$$h_y(z) = \frac{1}{L}\big(f(z+y) - f(y) - \nabla f(y)'z\big),$$

to obtain

$$\nabla h_y(x-y) = \frac{1}{L}\big(\nabla f(x) - \nabla f(y)\big),$$

and

$$h_y^\star(\theta) \geq \frac{1}{2}\|\theta\|^2, \qquad \forall\, \theta \in \Re^n. \tag{6.233}$$

Let $\theta = \frac{1}{L}\big(\nabla f(y) - \nabla f(x)\big)$. Since by Eq. (6.231), θ is the gradient of h_x at $y - x$, by the Conjugate Subgradient Theorem (Prop. 5.4.3 in Appendix B), we have

$$h_x^\star(\theta) + h_x(y-x) = \theta'(y-x),$$

and similarly

$$h_y^\star(-\theta) + h_y(x-y) = -\theta'(x-y).$$

Also by the definition of h_x and h_y,

$$h_x(y-x) + h_y(x-y) = \theta'(y-x).$$

By combining the preceding three relations,

$$h_x^\star(\theta) + h_y^\star(-\theta) = \theta'(y-x),$$

and by using also Eqs. (6.232) and (6.233),

$$\|\theta\|^2 \leq h_x^\star(\theta) + h_y^\star(-\theta) = \theta'(y-x) \leq \|\theta\|\|y-x\|.$$

Thus we obtain $\|\theta\| \leq \|y - x\|$, i.e., that $\big\|\nabla f(y) - \nabla f(x)\big\| \leq L\|y-x\|$.

6.2 (Finite Convergence of Gradient Projection for Linear Programming)

Consider the gradient projection method, applied to minimization of a linear function f over a polyhedral set X, using a stepsize α_k such that $\sum_{k=0}^{\infty} \alpha_k = \infty$. Show that if the set X^* of optimal solutions is nonempty, the method converges to some $x^* \in X^*$ in a finite number of iterations. *Hint*: Show that for this problem the method is equivalent to the proximal algorithm and apply Prop. 5.1.5.

6.3 (Convergence Rate of Proximal Gradient Method Under Strong Convexity Conditions)

This exercise, based on an unpublished analysis in [Sch14b], derives conditions under which the proximal gradient method converges linearly. Consider the minimization of $f + h$, where f is strongly convex and differentiable with Lipschitz continuous gradient [cf. Eq. (6.44)], and h is closed proper convex (cf. Section 6.3). The proximal gradient method with constant stepsize $\alpha > 0$ is written as

$$z_k = x_k - \alpha \nabla f(x_k), \qquad x_{k+1} = P_{\alpha,h}(z_k),$$

where $P_{\alpha,h}$ is the proximal operator corresponding to α and h (cf. Section 5.1.4). Denote by x^* the optimal solution (which exists and is unique by the strong convexity of f), and let $z^* = x^* - \alpha \nabla f(x^*)$. We assume that $\alpha \leq 1/L$, where L is the Lipschitz constant of ∇f, so that $x_k \to x^*$ (cf. Prop. 6.3.3).

(a) Show that for some scalars $p \in (0,1)$ and $q \in (0,1]$, we have

$$\big\| x - \alpha \nabla f(x) - z^* \big\| \leq p \, \| x - x^* \|, \qquad \forall\, x \in \Re^n, \tag{6.234}$$

$$\big\| P_{\alpha,h}(z) - P_{\alpha,h}(z^*) \big\| \leq q \, \| z - z^* \|, \qquad \forall\, z \in \Re^n. \tag{6.235}$$

Show also that

$$\| x_{k+1} - x^* \| \leq pq \, \| x_k - x^* \|, \qquad \forall\, k. \tag{6.236}$$

Hint: See Prop. 6.1.8 and also [SRB11] for Eq. (6.234). Because $P_{\alpha,h}$ is nonexpansive (cf. Prop. 5.1.8), we can set $q = 1$ in Eq. (6.235). Finally, note that as shown in Section 6.3, the set of minima of $f + h$ coincides with the set of fixed points of the mapping $P_{\alpha,h}\big(x - \alpha \nabla f(x)\big)$ [the composition of the two mappings in Eqs. (6.234) and (6.235)]. *Note*: Conditions under which $P_{\alpha,h}$ is a contraction, so values $q < 1$ can be used in Eq. (6.235), are given in Prop. 5.1.4 and Exercise 5.2 (see also [Roc76a] and [Luq84]). A local version of Eq. (6.236) can also be proved, when Eqs. (6.234) and (6.235) hold locally, within spheres centered at x^* and z^*, respectively.

(b) Assume that f is positive definite quadratic with minimum and maximum eigenvalues denoted by m and M, respectively, and that h is positive semidefinite quadratic with minimum eigenvalue λ. Show that

$$\| x_{k+1} - x^* \| \leq \frac{\max\big\{ |1 - \alpha m|, \, |1 - \alpha M| \big\}}{1 + \alpha \lambda} \, \| x_k - x^* \|, \qquad \forall\, k.$$

Hint: Use Eq. (6.236), Exercise 2.1, and the linearity of the proximal operator.

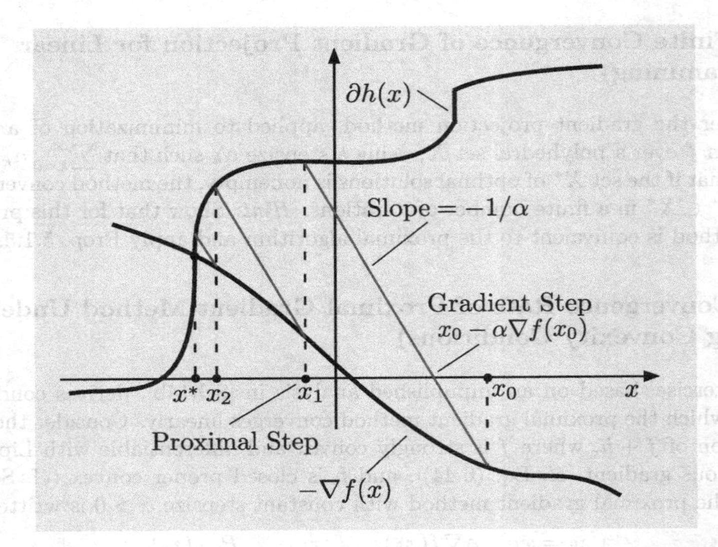

Figure 6.9.1. Illustration of the finite convergence process of the proximal gradient method for the case of a sharp minimum, where h is nondifferentiable at x^* and $-\nabla f(x^*) \in \text{int}\big(\partial h(x^*)\big)$ (cf. Exercise 6.4). The figure also illustrates how the method can attain superlinear convergence (cf. Exercise 6.5). These results should be compared with the convergence rate analysis of the proximal algorithm in Section 5.1.2.

6.4 (Finite Convergence of Proximal Gradient Method for Sharp Minima)

Consider the minimization of $f + h$, where f is convex and differentiable with Lipschitz continuous gradient [cf. Eq. (6.44)], and h is closed proper convex (cf. Section 6.3). Assume that there is a unique optimal solution, denoted x^*, and that $\alpha \le 1/L$, where L is the Lipschitz constant of ∇f, so that $x_k \to x^*$ (cf. Prop. 6.3.3). Show that if the interior point condition

$$0 \in \text{int}\big(\nabla f(x^*) + \partial h(x^*)\big) \tag{6.237}$$

holds, then the method converges to x^* in a finite number of iterations. *Note*: The condition (6.237) is illustrated in Fig. 6.9.1, and is known as a *sharp minimum* condition. It is equivalent to the existence of some $\beta > 0$ such that

$$f(x^*) + h(x^*) + \beta\|x - x^*\| \le f(x) + h(x), \qquad \forall\, x \in \Re^n;$$

cf. the finite convergence result of Prop. 5.1.5 for the proximal algorithm (which, however, does not assume uniqueness of the optimal solution). *Abbreviated proof*: By continuity of ∇f there is an open sphere S_{x^*} centered at x^* with

$$x - \alpha\nabla f(x) \in x^* + \alpha\partial h(x^*), \qquad \forall\, x \in S_{x^*}.$$

Once $x_k \in S_{x^*}$, the method terminates after one more iteration.

6.5 (Superlinear Convergence of Proximal Gradient Method)

Consider the minimization of $f + h$, where f is convex and differentiable with Lipschitz continuous gradient [cf. Eq. (6.44)], and h is closed proper convex (cf. Section 6.3). Assume that the optimal solution set X^* is nonempty, and that $\alpha \le 1/L$, where L is the Lipschitz constant of ∇f, so that x_k converges to some point in X^* (cf. Prop. 6.3.3). Assume also that for some scalars $\beta > 0$, $\delta > 0$, and $\gamma \in (1, 2)$, we have

$$F^* + \beta\big(d(x)\big)^\gamma \le f(x) + h(x), \qquad \forall\, x \in \Re^n \text{ with } d(x) \le \delta, \tag{6.238}$$

where F^* is the optimal value of the problem and

$$d(x) = \min_{x \in X^*} \|x - x^*\|.$$

Show that there exists $\bar{k} \ge 0$ such that for all $k \ge \bar{k}$ we have

$$f(x_{k+1}) + h(x_{k+1}) - F^* \le \frac{1}{2\alpha}\left(\frac{f(x_k) + h(x_k) - F^*}{\beta}\right)^{2/\gamma}.$$

Notes: If Eq. (6.238) holds with $\gamma = 1$ it holds for all $\gamma \in (1, 2)$, so this exercise shows arbitrarily fast superlinear convergence for the case of a sharp minimum, even when X^* contains multiple points (cf. Exercise 6.4). Figure 6.9.1 provides a geometric interpretation of the mechanism for superlinear convergence. The graph of ∂h changes almost "vertically" as we move away from the optimal solution set. *Abbreviated proof*: From the descent lemma (Prop. 6.1.2), we have the inequality

$$f(x_{k+1}) \le f(x_k) + \nabla f(x_k)'(x_{k+1} - x_k) + \frac{1}{2\alpha}\|x_{k+1} - x_k\|^2,$$

so that for all $x^* \in X^*$, we obtain

$$\begin{aligned}
f(x_{k+1}) + h(x_{k+1}) &\le f(x_k) + \nabla f(x_k)'(x_{k+1} - x_k) + h(x_{k+1}) + \frac{1}{2\alpha}\|x_{k+1} - x_k\|^2 \\
&= \min_{x \in \Re^n}\left\{f(x_k) + \nabla f(x_k)'(x - x_k) + h(x) + \frac{1}{2\alpha}\|x - x_k\|^2\right\} \\
&\le f(x_k) + \nabla f(x_k)'(x^* - x_k) + h(x^*) + \frac{1}{2\alpha}\|x^* - x_k\|^2 \\
&\le f(x^*) + h(x^*) + \frac{1}{2\alpha}\|x^* - x_k\|^2,
\end{aligned}$$

where the last step uses the gradient inequality $f(x_k) + \nabla f(x_k)'(x^* - x_k) \le f(x^*)$. Letting x^* be the vector of X^* that is at minimum distance from x_k, we obtain

$$f(x_{k+1}) + h(x_{k+1}) - F^* \le \frac{1}{2\alpha}d(x_k)^2.$$

Since x_k converges to some point of X^*, by using the hypothesis (6.238), we have for sufficiently large k,

$$d(x_k) \le \left(\frac{f(x_k) + h(x_k) - F^*}{\beta}\right)^{1/\gamma},$$

and the result follows by combining the last two relations.

6.6 (Proximal Gradient Method with Diagonal Scaling)

Consider the minimization of $f + h$, where f is convex and differentiable with Lipschitz continuous gradient [cf. Eq. (6.44)], and h is closed proper convex (cf. the framework of Section 6.3). Consider the algorithm

$$z_k^i = x_k - d_k^i \frac{\partial f(x_k)}{\partial x^i}, \qquad i = 1, \ldots, n,$$

$$x_{k+1} \in \arg \min_{x \in \Re^n} \left\{ h(x) + \sum_{i=1}^n \frac{1}{2d_k^i} \|x - z_k^i\|^2 \right\},$$

where d_k^i are positive scalars.

(a) Assume that $x_k \to x^*$ for some vector x^*, and $d_k^i \to d^i$ for some positive scalars d^i, $i = 1, \ldots, n$. Show that x^* minimizes $f + h$. *Hint*: Show that the vector g_k with components given by

$$g_k^i = \frac{1}{d_k^i}(z_k^i - x_{k+1}^i), \qquad i = 1, \ldots, n,$$

belongs to $\partial h(x_k)$, and that the limit of g_k belongs to $\partial h(x^*)$.

(b) Assuming that f is strongly convex, derive conditions under which $x_k \to x^*$ and $d_k^i \to d^i$.

6.7 (Parallel Projections Algorithm as a Block Coordinate Descent Method)

Let $X_1, \ldots, X_m$ be given closed convex sets in $\Re^n$. Consider the problem of finding a point in their intersection, and the equivalent problem

$$\begin{aligned} &\text{minimize} \quad \tfrac{1}{2} \textstyle\sum_{i=1}^m \|z^i - x\|^2 \\ &\text{subject to} \quad x \in \Re^n, \; z^i \in X_i, \; i = 1, \ldots, m. \end{aligned}$$

Show that a block coordinate descent algorithm is given by

$$x_{k+1} = \frac{1}{m} \sum_{i=1}^m z_k^i, \qquad z_{k+1}^i = P_{X_i}(x_{k+1}), \quad i = 1, \ldots, m.$$

Verify that the convergence result of Prop. 6.5.1 applies to this algorithm.

6.8 (Proximal Algorithm as a Block Coordinate Descent Method)

Let $f : \Re^n \mapsto \Re$ be a continuously differentiable convex function, let X be a closed convex set, and let c be a positive scalar. Show that the proximal algorithm

$$x_{k+1} \in \arg\min_{x \in X} \left\{ f(x) + \frac{1}{2c} \|x - x_k\|^2 \right\}$$

is a special case of the block coordinate descent method applied to the problem

$$\begin{aligned} &\text{minimize} \quad f(x) + \frac{1}{2c} \|x - y\|^2 \\ &\text{subject to} \quad x \in X, \quad y \in \Re^n, \end{aligned}$$

which is equivalent to the problem of minimizing f over X. *Hint*: Consider the cost function

$$g(x, y) = f(x) + \frac{1}{2c} \|x - y\|^2.$$

6.9 (Combination of Coordinate Descent and Proximal Algorithm [Tse91a])

Consider the minimization of a continuously differentiable function f of the vector $x = (x^1, \ldots, x^m)$ subject to $x^i \in X_i$, where X_i are closed convex sets. Consider the following variation of the block coordinate descent method (6.141):

$$x_{k+1}^i \in \arg\min_{\xi \in X_i} f(x_{k+1}^1, \ldots, x_{k+1}^{i-1}, \xi, x_k^{i+1}, \ldots, x_k^m) + \frac{1}{2c} \|\xi - x_k^i\|^2,$$

for some scalar $c > 0$. Assuming that f is convex, show that every limit point of the sequence of vectors $x_k = (x_k^1, \ldots, x_k^m)$ is a global minimum. *Hint*: Apply the result of Prop. 6.5.1 to the cost function

$$g(x, y) = f(x^1, \ldots, x^m) + \frac{1}{2c} \sum_{i=1}^m \|x^i - y^i\|^2.$$

For a related analysis of this type of algorithm see [Aus92], and for a recent analysis see [BST14].

6.10 (Coordinate Descent for Convex Nondifferentiable Problems)

Consider the minimization of $F + G$, where $F : \Re^n \mapsto \Re$ is a differentiable convex function and $G : \Re^n \mapsto (-\infty, \infty]$ is a closed proper convex function such that $\partial G(x) \neq \varnothing$ for all $x \in \text{dom}(G)$.

(a) Use the optimality conditions of Prop. 5.4.7 in Appendix B to show that

$$x^* \in \arg\min_{x \in \Re^n} \big\{F(x) + G(x)\big\}$$

if and only if

$$x^* \in \arg\min_{x \in \Re^n} \big\{\nabla F(x^*)'x + G(x)\big\}.$$

(b) Assume that G is separable of the form

$$G(x) = \sum_{i=1}^{n} G_i(x^i),$$

where x^i, $i = 1, \ldots, n$, are the one-dimensional coordinates of x, and $G_i : \Re \mapsto (-\infty, \infty]$ are closed proper convex functions. Use part (a) to show that x^* minimizes $F + G$ if and only if $(x^i)^*$ minimizes $F + G$ in the ith coordinate direction for each i.

6.11 (Convergence Rate of Coordinate Descent Under Strong Convexity Conditions)

This exercise compares the convergence rates associated with various orders of coordinate selection in the coordinate descent method, and illustrates how a good deterministic order can outperform a randomized order. We consider the minimization of a differentiable function $f : \Re^n \mapsto \Re$, with Lipschitz continuous gradient along the ith coordinate, i.e., for some $L > 0$, we have

$$\big|\nabla_i f(x + \alpha e_i) - \nabla_i f(x)\big| \leq L\,|\alpha|, \qquad \forall\, x \in \Re^n,\ \alpha \in \Re,\ i = 1, \ldots, n, \tag{6.239}$$

where e_i is the ith coordinate direction, $e_i = (0, \ldots, 0, 1, 0, \ldots, 0)$ with the 1 in the ith position, and $\nabla_i f$ is the ith component of the gradient. We also assume that f is strongly convex in the sense that for some $\sigma > 0$

$$f(y) \geq f(x) + \nabla f(x)'(y - x) + \frac{\sigma}{2}\|x - y\|^2, \qquad \forall\, x, y \in \Re^n, \tag{6.240}$$

and we denote by x^* the unique minimum of f. Consider the algorithm

$$x_{k+1} = x_k - \frac{1}{L}\nabla_{i_k} f(x_k)\, e_{i_k},$$

where the index i_k is chosen in some way to be specified shortly.

(a) (*Randomized Coordinate Selection* [LeL10], [Nes12]) Assume that i_k is selected by uniform randomization over the index set $\{1, \ldots, n\}$, independently of previous selections. Show that for all k,

$$E\big\{f(x_{k+1})\big\} - f(x^*) \leq \left(1 - \frac{\sigma}{Ln}\right)\big(f(x_k) - f(x^*)\big). \tag{6.241}$$

Abbreviated Solution: By the descent lemma (Prop. 6.1.2) and Eq. (6.239), we have

$$\begin{aligned} f(x_{k+1}) &\leq f(x_k) + \nabla_{i_k} f(x_k)(x_{k+1}^{i_k} - x_k^{i_k}) + \frac{L}{2}(x_{k+1}^{i_k} - x_k^{i_k})^2 \\ &= f(x_k) - \frac{1}{2L}\big|\nabla_{i_k} f(x_k)\big|^2, \end{aligned} \tag{6.242}$$

while by minimizing over y both sides of Eq. (6.240) with $x = x_k$, we have

$$f(x^*) \geq f(x_k) - \frac{1}{2\sigma}\big\|\nabla f(x_k)\big\|^2. \tag{6.243}$$

Taking conditional expected value, given x_k, in Eq. (6.242), we obtain

$$\begin{aligned} E\big\{f(x_{k+1})\big\} &\leq E\left\{f(x_k) - \frac{1}{2L}\big|\nabla_{i_k} f(x_k)\big|^2\right\} \\ &= f(x_k) - \frac{1}{2L}\sum_{i=1}^{n}\frac{1}{n}\big|\nabla_{i_k} f(x_k)\big|^2 \\ &= f(x_k) - \frac{1}{2Ln}\big\|\nabla f(x_k)\big\|^2. \end{aligned}$$

Subtracting $f(x^*)$ from both sides and using Eq. (6.243), we obtain Eq. (6.241).

(b) (*Gauss-Southwell Coordinate Selection* [ScF14]) Assume that i_k is selected according to

$$i_k \in \arg\max_{i=1,\ldots,n} \big|\nabla_i f(x_k)\big|,$$

and let the following strong convexity assumption hold for some $\sigma_1 > 0$,

$$f(y) \geq f(x) + \nabla f(x)'(y - x) + \frac{\sigma_1}{2}\|x - y\|_1^2, \qquad \forall\, x, y \in \Re^n, \tag{6.244}$$

where $\|\cdot\|_1$ denotes the ℓ_1 norm. Show that for all k,

$$f(x_{k+1}) - f(x^*) \leq \left(1 - \frac{\sigma_1}{L}\right)\big(f(x_k) - f(x^*)\big). \tag{6.245}$$

Abbreviated Solution: Minimizing with respect to y both sides of Eq. (6.244) with $x = x_k$, we obtain

$$f(x^*) \geq f(x_k) - \frac{1}{2\sigma_1}\big\|\nabla f(x_k)\big\|_\infty^2.$$

Combining this with Eq. (6.242) and using the definition of i_k, which implies that

$$\big|\nabla_{i_k} f(x_k)\big|^2 = \big\|\nabla f(x_k)\big\|_\infty^2,$$

we obtain Eq. (6.245). *Note*: It can be shown that $\frac{\sigma}{n} \leq \sigma_1 \leq \sigma$, so that the convergence rate estimate (6.245) is more favorable than the estimate (6.241) of part (a); see [ScF14].

(c) Consider instead the algorithm

$$x_{k+1} = x_k - \frac{1}{L_{i_k}} \nabla_{i_k} f(x_k)\, e_{i_k},$$

where L_i is a Lipschitz constant for the gradient along the ith coordinate, i.e.,

$$\big|\nabla_i f(x + \alpha e_i) - \nabla_i f(x)\big| \le L_i\, |\alpha|, \qquad \forall\, x \in \Re^n,\ \alpha \in \Re,\ i = 1, \ldots, n.$$

Here i_k is selected as in part (a), by uniform randomization over the index set $\{1, \ldots, n\}$, independently of previous selections. Show that for all k,

$$E\big\{f(x_{k+1})\big\} - f(x^*) \le \left(1 - \frac{\sigma}{\bar{L}}\right) \big(f(x_k) - f(x^*)\big),$$

where $\bar{L} = \sum_{i=1}^n L_i$. *Note*: This is a stronger convergence rate estimate than the one of part (a), which applies with $L = \min\{L_1, \ldots, L_n\}$. The use of different stepsizes for different coordinates may be viewed as a form of diagonal scaling. In practice, L_{i_k} may be approximated by a finite difference approximation of the second derivative of f along e_{i_k} or some other crude line search scheme.

APPENDIX A: Mathematical Background

In Sections A.1-A.3 of this appendix, we provide some basic definitions, notational conventions, and results from linear algebra and real analysis. We assume that the reader is familiar with these subjects, so no proofs are given. For additional related material, we refer to textbooks such as Hoffman and Kunze [HoK71], Lancaster and Tismenetsky [LaT85], and Strang [Str76] (linear algebra), and Ash [Ash72], Ortega and Rheinboldt [OrR70], and Rudin [Rud76] (real analysis).

In Section A.4, we provide a few convergence theorems for deterministic and random sequences, which we will use for various convergence analyses of algorithms in the text. Except for the Supermartingale Convergence Theorem for sequences of random variables (Prop. A.4.5), we provide complete proofs.

Set Notation

If X is a set and x is an element of X, we write $x \in X$. A set can be specified in the form $X = \{x \mid x \text{ satisfies } P\}$, as the set of all elements satisfying property P. The union of two sets X_1 and X_2 is denoted by $X_1 \cup X_2$, and their intersection by $X_1 \cap X_2$. The symbols $\exists$ and $\forall$ have the meanings "there exists" and "for all," respectively. The empty set is denoted by $\emptyset$.

The set of real numbers (also referred to as scalars) is denoted by $\Re$. The set $\Re$ augmented with $+\infty$ and $-\infty$ is called the *set of extended real numbers*. We write $-\infty < x < \infty$ for all real numbers x, and $-\infty \leq x \leq \infty$ for all extended real numbers x. We denote by $[a, b]$ the set of (possibly extended) real numbers x satisfying $a \leq x \leq b$. A rounded, instead of square, bracket denotes strict inequality in the definition. Thus $(a, b]$, $[a, b)$, and (a, b) denote the set of all x satisfying $a < x \leq b$, $a \leq x < b$, and

$a < x < b$, respectively. Furthermore, we use the natural extensions of the rules of arithmetic: $x \cdot 0 = 0$ for every extended real number x, $x \cdot \infty = \infty$ if $x > 0$, $x \cdot \infty = -\infty$ if $x < 0$, and $x + \infty = \infty$ and $x - \infty = -\infty$ for every scalar x. The expression $\infty - \infty$ is meaningless and is never allowed to occur.

Inf and Sup Notation

The *supremum* of a nonempty set X of scalars, denoted by $\sup X$, is defined as the smallest scalar y such that $y \geq x$ for all $x \in X$. If no such scalar exists, we say that the supremum of X is ∞. Similarly, the *infimum* of X, denoted by $\inf X$, is defined as the largest scalar y such that $y \leq x$ for all $x \in X$, and is equal to $-\infty$ if no such scalar exists. For the empty set, we use the convention

$$\sup \emptyset = -\infty, \qquad \inf \emptyset = \infty.$$

If $\sup X$ is equal to a scalar $\overline{x}$ that belongs to the set X, we say that $\overline{x}$ is the *maximum point* of X and we write $\overline{x} = \max X$. Similarly, if $\inf X$ is equal to a scalar $\overline{x}$ that belongs to the set X, we say that $\overline{x}$ is the *minimum point* of X and we write $\overline{x} = \min X$. Thus, when we write $\max X$ (or $\min X$) in place of $\sup X$ (or $\inf X$, respectively), we do so just for emphasis: we indicate that it is either evident, or it is known through earlier analysis, or it is about to be shown that the maximum (or minimum, respectively) of the set X is attained at one of its points.

Vector Notation

We denote by $\Re^n$ the set of n-dimensional real vectors. For any $x \in \Re^n$, we use x_i (or sometimes x^i) to indicate its ith *coordinate*, also called its ith *component*. Vectors in $\Re^n$ will be viewed as column vectors, unless the contrary is explicitly stated. For any $x \in \Re^n$, x' denotes the transpose of x, which is an n-dimensional row vector. The *inner product* of two vectors $x = (x_1, \ldots, x_n)$ and $y = (y_1, \ldots, y_n)$ is defined by $x'y = \sum_{i=1}^{n} x_i y_i$. Two vectors $x, y \in \Re^n$ satisfying $x'y = 0$ are called *orthogonal*.

If x is a vector in $\Re^n$, the notations $x > 0$ and $x \geq 0$ indicate that all components of x are positive and nonnegative, respectively. For any two vectors x and y, the notation $x > y$ means that $x - y > 0$. The notations $x \geq y$, $x < y$, etc., are to be interpreted accordingly.

Function Notation and Terminology

If f is a function, we use the notation $f : X \mapsto Y$ to indicate the fact that f is defined on a nonempty set X (its *domain*) and takes values in a set Y (its *range*). Thus when using the notation $f : X \mapsto Y$, we implicitly assume that X is nonempty. If $f : X \mapsto Y$ is a function, and U and V are subsets of X and Y, respectively, the set $\{f(x) \mid x \in U\}$ is called the *image* or *forward image of* U *under* f, and the set $\{x \in X \mid f(x) \in V\}$ is called the *inverse image of* V *under* f.

A function $f : \Re^n \mapsto \Re$ is said to be *affine* if it has the form $f(x) = a'x + b$ for some $a \in \Re^n$ and $b \in \Re$. Similarly, a function $f : \Re^n \mapsto \Re^m$ is said to be *affine* if it has the form $f(x) = Ax + b$ for some $m \times n$ matrix A and some $b \in \Re^m$. If $b = 0$, f is said to be a *linear function* or *linear transformation*. Sometimes, with slight abuse of terminology, an equation or inequality involving a linear function, such as $a'x = b$ or $a'x \leq b$, is referred to as a *linear equation or inequality*, respectively.

A.1 LINEAR ALGEBRA

If X is a subset of $\Re^n$ and λ is a scalar, we denote by λX the set $\{\lambda x \mid x \in X\}$. If X and Y are two subsets of $\Re^n$, we denote by $X + Y$ the set

$$\{x + y \mid x \in X,\, y \in Y\},$$

which is referred to as the *vector sum of* X *and* Y. We use a similar notation for the sum of any finite number of subsets. In the case where one of the subsets consists of a single vector $\overline{x}$, we simplify this notation as follows:

$$\overline{x} + X = \{\overline{x} + x \mid x \in X\}.$$

We also denote by $X - Y$ the set

$$\{x - y \mid x \in X,\, y \in Y\}.$$

Given sets $X_i \subset \Re^{n_i}$, $i = 1, \ldots, m$, the *Cartesian product* of the X_i, denoted by $X_1 \times \cdots \times X_m$, is the set

$$\{(x_1, \ldots, x_m) \mid x_i \in X_i,\, i = 1, \ldots, m\},$$

which is viewed as a subset of $\Re^{n_1 + \cdots + n_m}$.

Subspaces and Linear Independence

A nonempty subset S of $\Re^n$ is called a *subspace* if $ax + by \in S$ for every $x, y \in S$ and every $a, b \in \Re$. An *affine set* in $\Re^n$ is a translated subspace, i.e., a set X of the form $X = \overline{x} + S = \{\overline{x} + x \mid x \in S\}$, where $\overline{x}$ is a vector in $\Re^n$ and S is a subspace of $\Re^n$, called the *subspace parallel to* X. Note that there can be only one subspace S associated with an affine set in this manner. [To see this, let $X = x + S$ and $X = \overline{x} + \overline{S}$ be two representations of the affine set X. Then, we must have $x = \overline{x} + \overline{s}$ for some $\overline{s} \in \overline{S}$ (since

$x \in X$), so that $X = \overline{x} + \overline{s} + S$. Since we also have $X = \overline{x} + \overline{S}$, it follows that $S = \overline{S} - \overline{s} = \overline{S}$.] A nonempty set X is a subspace if and only if it contains the origin, and every line that passes through any pair of its points that are distinct, i.e., it contains 0 and all points $\alpha x + (1-\alpha)y$, where $\alpha \in \Re$ and $x, y \in X$ with $x \neq y$. Similarly X is affine if and only if it contains every line that passes through any pair of its points that are distinct. The *span* of a finite collection $\{x_1, \ldots, x_m\}$ of elements of $\Re^n$, denoted by $\text{span}(x_1, \ldots, x_m)$, is the subspace consisting of all vectors y of the form $y = \sum_{k=1}^{m} \alpha_k x_k$, where each α_k is a scalar.

The vectors $x_1, \ldots, x_m \in \Re^n$ are called *linearly independent* if there exists no set of scalars $\alpha_1, \ldots, \alpha_m$, at least one of which is nonzero, such that $\sum_{k=1}^{m} \alpha_k x_k = 0$. An equivalent definition is that $x_1 \neq 0$, and for every $k > 1$, the vector x_k does not belong to the span of $x_1, \ldots, x_{k-1}$.

If S is a subspace of $\Re^n$ containing at least one nonzero vector, a *basis* for S is a collection of vectors that are linearly independent and whose span is equal to S. Every basis of a given subspace has the same number of vectors. This number is called the *dimension* of S. By convention, the subspace $\{0\}$ is said to have dimension zero. Every subspace of nonzero dimension has a basis that is orthogonal (i.e., any pair of distinct vectors from the basis is orthogonal). The *dimension of an affine set* $\overline{x} + S$ is the dimension of the corresponding subspace S. An $(n-1)$-dimensional affine subset of $\Re^n$ is called a *hyperplane*, assuming $n \geq 2$. It is a set specified by a single linear equation, i.e., a set of the form $\{x \mid a'x = b\}$, where $a \neq 0$ and $b \in \Re$.

Given any subset X of $\Re^n$, the set of vectors that are orthogonal to all elements of X is a subspace denoted by $X^\perp$:

$$X^\perp = \{y \mid y'x = 0, \ \forall \ x \in X\}.$$

If S is a subspace, $S^\perp$ is called the *orthogonal complement* of S. Any vector x can be uniquely decomposed as the sum of a vector from S and a vector from $S^\perp$. Furthermore, we have $(S^\perp)^\perp = S$.

Matrices

For any matrix A, we use A_{ij}, $[A]_{ij}$, or a_{ij} to denote its ijth component. The *transpose* of A, denoted by A', is defined by $[A']_{ij} = a_{ji}$. For any two matrices A and B of compatible dimensions, the transpose of the product matrix AB satisfies $(AB)' = B'A'$. The inverse of a square and invertible A is denoted A^{-1}.

If X is a subset of $\Re^n$ and A is an $m \times n$ matrix, then the *image of X under A* is denoted by AX (or $A \cdot X$ if this enhances notational clarity):

$$AX = \{Ax \mid x \in X\}.$$

If Y is a subset of $\Re^m$, the *inverse image of Y under A* is denoted by $A^{-1}Y$:

$$A^{-1}Y = \{x \mid Ax \in Y\}.$$

Let A be an $m \times n$ matrix. The *range space* of A, denoted by $R(A)$, is the set of all vectors $y \in \Re^m$ such that $y = Ax$ for some $x \in \Re^n$. The *nullspace* of A, denoted by $N(A)$, is the set of all vectors $x \in \Re^n$ such that $Ax = 0$. It is seen that the range space and the nullspace of A are subspaces. The *rank* of A is the dimension of the range space of A. The rank of A is equal to the maximal number of linearly independent columns of A, and is also equal to the maximal number of linearly independent rows of A. The matrix A and its transpose A' have the same rank. We say that A has *full rank*, if its rank is equal to $\min\{m, n\}$. This is true if and only if either all the rows of A are linearly independent, or all the columns of A are linearly independent. The range space of an $m \times n$ matrix A is equal to the orthogonal complement of the nullspace of its transpose, i.e., $R(A) = N(A')^\perp$.

Square Matrices

By a square matrix we mean any $n \times n$ matrix, where $n \geq 1$. The determinant of a square matrix A is denoted by $\det(A)$.

Definition A.1.1: A square matrix A is called *singular* if its determinant is zero. Otherwise it is called *nonsingular* or *invertible*.

Definition A.1.2: The *characteristic polynomial* ϕ of an $n \times n$ matrix A is defined by $\phi(\lambda) = \det(\lambda I - A)$, where I is the identity matrix of the same size as A. The n (possibly repeated and complex) roots of ϕ are called the *eigenvalues* of A. A nonzero vector x (with possibly complex coordinates) such that $Ax = \lambda x$, where λ is an eigenvalue of A, is called an *eigenvector* of A associated with λ.

Note that the only use of complex numbers in this book is in relation to eigenvalues and eigenvectors. All other matrices or vectors are implicitly assumed to have real components.

Proposition A.1.1:

(a) Let A be an $n \times n$ matrix. The following are equivalent:

(i) The matrix A is nonsingular.

(ii) The matrix A' is nonsingular.

(iii) For every nonzero $x \in \Re^n$, we have $Ax \neq 0$.

(iv) For every $y \in \Re^n$, there is a unique $x \in \Re^n$ such that $Ax = y$.

(v) There is an $n \times n$ matrix B such that $AB = I = BA$.

(vi) The columns of A are linearly independent.

(vii) The rows of A are linearly independent.

(viii) All eigenvalues of A are nonzero.

(b) Assuming that A is nonsingular, the matrix B of statement (v) (called the *inverse* of A and denoted by A^{-1}) is unique.

(c) For any two square invertible matrices A and B of the same dimensions, we have $(AB)^{-1} = B^{-1}A^{-1}$.

Proposition A.1.2: Let A be an $n \times n$ matrix.

(a) If T is a nonsingular matrix and $B = TAT^{-1}$, then the eigenvalues of A and B coincide.

(b) For any scalar c, the eigenvalues of $cI + A$ are equal to $c + \lambda_1, \ldots, c + \lambda_n$, where $\lambda_1, \ldots, \lambda_n$ are the eigenvalues of A.

(c) The eigenvalues of A^k are equal to $\lambda_1^k, \ldots, \lambda_n^k$, where $\lambda_1, \ldots, \lambda_n$ are the eigenvalues of A.

(d) If A is nonsingular, then the eigenvalues of A^{-1} are the reciprocals of the eigenvalues of A.

(e) The eigenvalues of A and A' coincide.

Let A and B be square matrices, and let C be a matrix of appropriate dimension. Then we have

$$(A + CBC')^{-1} = A^{-1} - A^{-1}C(B^{-1} + C'A^{-1}C)^{-1}C'A^{-1},$$

provided all the inverses appearing above exist. For a proof, multiply the right-hand side by $A + CBC'$ and show that the product is the identity.

Another useful formula provides the inverse of the partitioned matrix

$$M = \begin{bmatrix} A & B \\ C & D \end{bmatrix}.$$

There holds

$$M^{-1} = \begin{bmatrix} Q & -QBD^{-1} \\ -D^{-1}CQ & D^{-1} + D^{-1}CQBD^{-1} \end{bmatrix},$$

where

$$Q = (A - BD^{-1}C)^{-1},$$

provided all the inverses appearing above exist. For a proof, multiply M with the given expression for M^{-1} and verify that the product is the identity.

Symmetric and Positive Definite Matrices

A square matrix A is said to be *symmetric* if $A = A'$. Symmetric matrices have several special properties, particularly regarding their eigenvalues and eigenvectors.

Proposition A.1.3: Let A be a symmetric $n \times n$ matrix. Then:

(a) The eigenvalues of A are real.

(b) The matrix A has a set of n mutually orthogonal, real, and nonzero eigenvectors $x_1, \ldots, x_n$.

(c) There holds

$$\underline{\lambda}\, x'x \leq x'Ax \leq \bar{\lambda}\, x'x, \qquad \forall\, x \in \Re^n,$$

where $\underline{\lambda}$ and $\bar{\lambda}$ are the smallest and largest eigenvalues of A, respectively.

Definition A.1.3: A symmetric $n \times n$ matrix A is called *positive definite* if $x'Ax > 0$ for all $x \in \Re^n$, $x \neq 0$. It is called *positive semidefinite* if $x'Ax \geq 0$ for all $x \in \Re^n$.

Throughout this book, the notion of positive definiteness applies exclusively to symmetric matrices. Thus *whenever we say that a matrix is positive (semi)definite, we implicitly assume that the matrix is symmetric*, although we usually add the term "symmetric" for clarity.

Proposition A.1.4:

(a) A square matrix is symmetric and positive definite if and only if it is invertible and its inverse is symmetric and positive definite.

(b) The sum of two symmetric positive semidefinite matrices is positive semidefinite. If one of the two matrices is positive definite, the sum is positive definite.

(c) If A is a symmetric positive semidefinite $n \times n$ matrix and T is an $m \times n$ matrix, then the matrix TAT' is positive semidefinite. If A is positive definite and T is invertible, then TAT' is positive definite.

(d) If A is a symmetric positive definite $n \times n$ matrix, there exists a unique symmetric positive definite matrix that yields A when multiplied with itself. This matrix is called the *square root of* A. It is denoted by $A^{1/2}$, and its inverse is denoted by $A^{-1/2}$.

A.2 TOPOLOGICAL PROPERTIES

Definition A.2.1: A *norm* $\|\cdot\|$ on $\Re^n$ is a function that assigns a scalar $\|x\|$ to every $x \in \Re^n$ and that has the following properties:

(a) $\|x\| \geq 0$ for all $x \in \Re^n$.

(b) $\|\alpha x\| = |\alpha| \cdot \|x\|$ for every scalar α and every $x \in \Re^n$.

(c) $\|x\| = 0$ if and only if $x = 0$.

(d) $\|x + y\| \leq \|x\| + \|y\|$ for all $x, y \in \Re^n$ (this is referred to as the *triangle inequality*).

The *Euclidean norm* of a vector $x = (x_1, \ldots, x_n)$ is defined by

$$\|x\| = (x'x)^{1/2} = \left(\sum_{i=1}^{n} |x_i|^2\right)^{1/2}.$$

Except for specialized contexts, we use this norm. In particular, *in the absence of a clear indication to the contrary,* $\|\cdot\|$ *will denote the Euclidean norm.* The *Schwarz inequality* states that for any two vectors x and y, we have

$$|x'y| \leq \|x\| \cdot \|y\|,$$

with equality holding if and only if $x = \alpha y$ for some scalar α. The *Pythagorean Theorem* states that for any two vectors x and y that are orthogonal, we have

$$\|x + y\|^2 = \|x\|^2 + \|y\|^2.$$

Two other important norms are the *maximum norm* $\|\cdot\|_\infty$ (also called *sup-norm* or ℓ_∞*-norm*), defined by

$$\|x\|_\infty = \max_{i=1,\ldots,n} |x_i|,$$

and the *ℓ_1-norm* $\|\cdot\|_1$, defined by

$$\|x\|_1 = \sum_{i=1}^{n} |x_i|.$$

Sequences

We use both subscripts and superscripts in sequence notation. Generally, we prefer subscripts, but sometimes we use superscripts whenever we need to reserve the subscript notation for indexing components of vectors and functions. The meaning of the subscripts and superscripts should be clear from the context in which they are used.

A scalar sequence $\{x_k \mid k = 1, 2, \ldots\}$ (or $\{x_k\}$ for short) is said to *converge* if there exists a scalar x such that for every $\epsilon > 0$ we have $|x_k - x| < \epsilon$ for every k greater than some integer K (that depends on ϵ). The scalar x is said to be the *limit* of $\{x_k\}$, and the sequence $\{x_k\}$ is said to *converge to* x; symbolically, $x_k \to x$ or $\lim_{k\to\infty} x_k = x$. If for every scalar b there exists some K (that depends on b) such that $x_k \geq b$ for all $k \geq K$, we write $x_k \to \infty$ and $\lim_{k\to\infty} x_k = \infty$. Similarly, if for every scalar b there exists some integer K such that $x_k \leq b$ for all $k \geq K$, we write $x_k \to -\infty$ and $\lim_{k\to\infty} x_k = -\infty$. Note, however, that implicit in any of the statements "$\{x_k\}$ converges" or "the limit of $\{x_k\}$ exists" or "$\{x_k\}$ has a limit" is that the limit of $\{x_k\}$ is a scalar.

A scalar sequence $\{x_k\}$ is said to be *bounded above* (respectively, *below*) if there exists some scalar b such that $x_k \leq b$ (respectively, $x_k \geq b$) for all k. It is said to be *bounded* if it is bounded above and bounded below. The sequence $\{x_k\}$ is said to be monotonically *nonincreasing* (respectively, *nondecreasing*) if $x_{k+1} \leq x_k$ (respectively, $x_{k+1} \geq x_k$) for all k. If $x_k \to x$ and $\{x_k\}$ is monotonically nonincreasing (nondecreasing), we also use the notation $x_k \downarrow x$ ($x_k \uparrow x$, respectively).

Proposition A.2.1: Every bounded and monotonically nonincreasing or nondecreasing scalar sequence converges.

Note that a monotonically nondecreasing sequence $\{x_k\}$ is either bounded, in which case it converges to some scalar x by the above proposition, or else it is unbounded, in which case $x_k \to \infty$. Similarly, a monotonically nonincreasing sequence $\{x_k\}$ is either bounded and converges, or it is unbounded, in which case $x_k \to -\infty$.

Given a scalar sequence $\{x_k\}$, let

$$y_m = \sup\{x_k \mid k \geq m\}, \qquad z_m = \inf\{x_k \mid k \geq m\}.$$

The sequences $\{y_m\}$ and $\{z_m\}$ are nonincreasing and nondecreasing, respectively, and therefore have a limit whenever $\{x_k\}$ is bounded above or is bounded below, respectively (Prop. A.2.1). The limit of y_m is denoted by $\limsup_{k\to\infty} x_k$, and is referred to as the *upper limit* of $\{x_k\}$. The limit of z_m is denoted by $\liminf_{k\to\infty} x_k$, and is referred to as the *lower limit* of $\{x_k\}$. If $\{x_k\}$ is unbounded above, we write $\limsup_{k\to\infty} x_k = \infty$, and if it is unbounded below, we write $\liminf_{k\to\infty} x_k = -\infty$.

Proposition A.2.2: Let $\{x_k\}$ and $\{y_k\}$ be scalar sequences.

(a) We have

$$\inf\{x_k \mid k \geq 0\} \leq \liminf_{k\to\infty} x_k \leq \limsup_{k\to\infty} x_k \leq \sup\{x_k \mid k \geq 0\}.$$

(b) $\{x_k\}$ converges if and only if

$$-\infty < \liminf_{k\to\infty} x_k = \limsup_{k\to\infty} x_k < \infty.$$

Furthermore, if $\{x_k\}$ converges, its limit is equal to the common scalar value of $\liminf_{k\to\infty} x_k$ and $\limsup_{k\to\infty} x_k$.

(c) If $x_k \leq y_k$ for all k, then

$$\liminf_{k\to\infty} x_k \leq \liminf_{k\to\infty} y_k, \qquad \limsup_{k\to\infty} x_k \leq \limsup_{k\to\infty} y_k.$$

(d) We have

$$\liminf_{k\to\infty} x_k + \liminf_{k\to\infty} y_k \leq \liminf_{k\to\infty}(x_k + y_k),$$

$$\limsup_{k\to\infty} x_k + \limsup_{k\to\infty} y_k \geq \limsup_{k\to\infty}(x_k + y_k).$$

A sequence $\{x_k\}$ of vectors in $\Re^n$ is said to converge to some $x \in \Re^n$ if the ith component of x_k converges to the ith component of x for every i. We use the notations $x_k \to x$ and $\lim_{k\to\infty} x_k = x$ to indicate convergence for vector sequences as well. A sequence $\{x_k\} \subset \Re^n$ is said to be a *Cauchy sequence* if $\|x_m - x_n\| \to 0$ as $m, n \to \infty$, i.e., given any $\epsilon > 0$, there exists N such that $\|x_m - x_n\| \leq \epsilon$ for all $m, n \geq N$. A sequence is Cauchy if and only if it converges to some vector. The sequence $\{x_k\}$ is called bounded if each of its corresponding component sequences is bounded. It can be seen that $\{x_k\}$ is bounded if and only if there exists a scalar c such that $\|x_k\| \leq c$ for all k. An infinite subset of a sequence $\{x_k\}$ is called a *subsequence* of $\{x_k\}$. Thus a subsequence can itself be viewed as a sequence, and can be represented as a set $\{x_k \mid k \in \mathcal{K}\}$, where $\mathcal{K}$ is an infinite subset of positive integers (the notation $\{x_k\}_\mathcal{K}$ will also be used).

A vector $x \in \Re^n$ is said to be a *limit point* of a sequence $\{x_k\}$ if there exists a subsequence of $\{x_k\}$ that converges to x. The following is a classical result that will be used often.

Proposition A.2.3: (Bolzano-Weierstrass Theorem) A bounded sequence in $\Re^n$ has at least one limit point.

$o(\cdot)$ Notation

For a function $h : \Re^n \mapsto \Re^m$ we write $h(x) = o(\|x\|^p)$, where p is a positive integer, if

$$\lim_{k\to\infty} \frac{h(x_k)}{\|x_k\|^p} = 0,$$

for all sequences $\{x_k\}$ such that $x_k \to 0$ and $x_k \neq 0$ for all k.

Closed and Open Sets

We say that x is a *closure point of a subset* X of $\Re^n$ if there exists a sequence $\{x_k\} \subset X$ that converges to x. The *closure* of X, denoted cl(X), is the set of all closure points of X.

Definition A.2.2: A subset X of $\Re^n$ is called *closed* if it is equal to its closure. It is called *open* if its complement, $\{x \mid x \notin X\}$, is closed. It is called *bounded* if there exists a scalar c such that $\|x\| \leq c$ for all $x \in X$. It is called *compact* if it is closed and bounded.

Given $x^* \in \Re^n$ and $\epsilon > 0$, the sets $\{x \mid \|x - x^*\| < \epsilon\}$ and $\{x \mid \|x - x^*\| \leq \epsilon\}$ are called an *open sphere* and a *closed sphere* centered at x^*, respectively. Sometimes the terms *open ball* and *closed ball* are used. A consequence of the definitions, is that a subset X of $\Re^n$ is open if and only if for every $x \in X$ there is an open sphere that is centered at x and is contained in X. A *neighborhood* of a vector x is an open set containing x.

Definition A.2.3: We say that x is an *interior point* of a subset X of $\Re^n$ if there exists a neighborhood of x that is contained in X. The set of all interior points of X is called the *interior* of X, and is denoted by int(X). A vector $x \in \text{cl}(X)$ which is not an interior point of X is said to be a *boundary point* of X. The set of all boundary points of X is called the *boundary* of X.

Proposition A.2.4:

(a) The union of a finite collection of closed sets is closed.

(b) The intersection of any collection of closed sets is closed.

(c) The union of any collection of open sets is open.

(d) The intersection of a finite collection of open sets is open.

(e) A set is open if and only if all of its elements are interior points.

(f) Every subspace of $\Re^n$ is closed.

(g) A set $X \subset \Re^n$ is compact if and only if every sequence of elements of X has a subsequence that converges to an element of X.

(h) If $\{X_k\}$ is a sequence of nonempty and compact subsets of $\Re^n$ such that $X_{k+1} \subset X_k$ for all k, then the intersection $\cap_{k=0}^{\infty} X_k$ is nonempty and compact.

The topological properties of sets in $\Re^n$, such as being open, closed, or compact, do not depend on the norm being used. This is a consequence of the following proposition.

Proposition A.2.5: (Norm Equivalence Property)

(a) For any two norms $\|\cdot\|$ and $\|\cdot\|'$ on $\Re^n$, there exists a scalar c such that

$$\|x\| \leq c\|x\|', \qquad \forall\, x \in \Re^n.$$

(b) If a subset of $\Re^n$ is open (respectively, closed, bounded, or compact) with respect to some norm, it is open (respectively, closed, bounded, or compact) with respect to all other norms.

Continuity

Let $f : X \mapsto \Re^m$ be a function, where X is a subset of $\Re^n$, and let x be a vector in X. If there exists a vector $y \in \Re^m$ such that the sequence $\{f(x_k)\}$ converges to y for every sequence $\{x_k\} \subset X$ such that $\lim_{k\to\infty} x_k = x$, we write $\lim_{z\to x} f(z) = y$. If there exists a vector $y \in \Re^m$ such that the sequence $\{f(x_k)\}$ converges to y for every sequence $\{x_k\} \subset X$ such that $\lim_{k\to\infty} x_k = x$ and $x_k \leq x$ (respectively, $x_k \geq x$) for all k, we write $\lim_{z\uparrow x} f(z) = y$ [respectively, $\lim_{z\downarrow x} f(z) = y$].

Definition A.2.4: Let X be a nonempty subset of $\Re^n$.

(a) A function $f : X \mapsto \Re^m$ is called *continuous* at a vector $x \in X$ if $\lim_{z\to x} f(z) = f(x)$.

(b) A function $f : X \mapsto \Re^m$ is called *right-continuous* (respectively, *left-continuous*) at a vector $x \in X$ if $\lim_{z\downarrow x} f(z) = f(x)$ [respectively, $\lim_{z\uparrow x} f(z) = f(x)$].

(c) A function $f : X \mapsto \Re^m$ is called *Lipschitz continuous* over X if there exists a scalar L such that

$$\|f(x) - f(y)\| \leq L\|x - y\|, \qquad \forall\, x, y \in X.$$

(d) A real-valued function $f : X \mapsto \Re$ is called *upper semicontinuous* (respectively, *lower semicontinuous*) at a vector $x \in X$ if $f(x) \geq \limsup_{k\to\infty} f(x_k)$ [respectively, $f(x) \leq \liminf_{k\to\infty} f(x_k)$] for every sequence $\{x_k\} \subset X$ that converges to x.

If $f : X \mapsto \Re^m$ is continuous at every vector in a subset of its domain X, we say that f *is continuous over that subset*. If $f : X \mapsto \Re^m$ is continuous at every vector in its domain X, we say that f *is continuous* (without qualification). We use similar terminology for right-continuous, left-continuous, Lipschitz continuous, upper semicontinuous, and lower semicontinuous functions.

Proposition A.2.6:

(a) Any vector norm on $\Re^n$ is a continuous function.

(b) Let $f : \Re^m \mapsto \Re^p$ and $g : \Re^n \mapsto \Re^m$ be continuous functions. The composition $f \cdot g : \Re^n \mapsto \Re^p$, defined by $(f \cdot g)(x) = f\big(g(x)\big)$, is a continuous function.

(c) Let $f : \Re^n \mapsto \Re^m$ be continuous, and let Y be an open (respectively, closed) subset of $\Re^m$. Then the inverse image of Y, $\big\{x \in \Re^n \mid f(x) \in Y\big\}$, is open (respectively, closed).

(d) Let $f : \Re^n \mapsto \Re^m$ be continuous, and let X be a compact subset of $\Re^n$. Then the image of X, $\big\{f(x) \mid x \in X\big\}$, is compact.

If $f : \Re^n \mapsto \Re$ is a continuous function and $X \subset \Re^n$ is compact, by Prop. A.2.6(c), the sets

$$V_\gamma = \{x \in X \mid f(x) \leq \gamma\}$$

are nonempty and compact for all $\gamma \in \Re$ with $\gamma > f^*$, where

$$f^* = \inf_{x \in X} f(x).$$

Since the set of minima of f is the intersection of the nonempty and compact sets V_{γ_k} for any sequence $\{\gamma_k\}$ with $\gamma_k \downarrow f^*$ and $\gamma_k > f^*$ for all k, it follows from Prop. A.2.4(h) that the set of minima is nonempty. This proves the following classical theorem of Weierstrass.

Proposition A.2.7: (Weierstrass' Theorem for Continuous Functions) A continuous function $f : \Re^n \mapsto \Re$ attains a minimum over any compact subset of $\Re^n$.

A.3 DERIVATIVES

Let $f : \Re^n \mapsto \Re$ be some function, fix $x \in \Re^n$, and consider the expression

$$\lim_{\alpha \to 0} \frac{f(x + \alpha e_i) - f(x)}{\alpha},$$

where e_i is the ith unit vector (all components are 0 except for the ith component which is 1). If the above limit exists, it is called the ith *partial derivative* of f at the vector x and it is denoted by $(\partial f / \partial x_i)(x)$ or $\partial f(x) / \partial x_i$ (x_i in this section will denote the ith component of the vector x). Assuming all of these partial derivatives exist, the *gradient* of f at x is defined as the column vector

$$\nabla f(x) = \begin{bmatrix} \frac{\partial f(x)}{\partial x_1} \\ \vdots \\ \frac{\partial f(x)}{\partial x_n} \end{bmatrix}.$$

For any $d \in \Re^n$, we define the one-sided *directional derivative* of f at a vector x in the direction d by

$$f'(x; d) = \lim_{\alpha \downarrow 0} \frac{f(x + \alpha d) - f(x)}{\alpha},$$

provided that the limit exists.

If the directional derivative of f at a vector x exists in all directions and $f'(x;d)$ is a linear function of d, we say that f is *differentiable* at x. It can be seen that f is differentiable at x if and only if the gradient $\nabla f(x)$ exists and satisfies $\nabla f(x)'d = f'(x;d)$ for all $d \in \Re^n$, or equivalently

$$f(x + \alpha d) = f(x) + \alpha \nabla f(x)'d + o(|\alpha|), \qquad \forall\, \alpha \in \Re.$$

The function f is called *differentiable over a subset S of $\Re^n$* if it is differentiable at every $x \in S$. The function f is called *differentiable* (without qualification) if it is differentiable at all $x \in \Re^n$.

If f is differentiable over an open set S and $\nabla f(\cdot)$ is continuous at all $x \in S$, f is said to be *continuously differentiable over S*. It can then be shown that for any $x \in S$ and norm $\|\cdot\|$,

$$f(x + d) = f(x) + \nabla f(x)'d + o(\|d\|), \qquad \forall\, d \in \Re^n.$$

The function f is called *continuously differentiable* (without qualification) if it is differentiable and $\nabla f(\cdot)$ is continuous at all $x \in \Re^n$. In our development, whenever we assume that f is differentiable, we also assume that it is continuously differentiable. Part of the reason is that a convex differentiable function is automatically continuously differentiable over $\Re^n$ (see Section 3.1).

If each one of the partial derivatives of a function $f : \Re^n \mapsto \Re$ is a continuously differentiable function of x over an open set S, we say that f is *twice continuously differentiable* over S. We then denote by

$$\frac{\partial^2 f(x)}{\partial x_i \partial x_j}$$

the ith partial derivative of $\partial f / \partial x_j$ at a vector $x \in \Re^n$. The *Hessian* of f at x, denoted by $\nabla^2 f(x)$, is the matrix whose components are the above second derivatives. The matrix $\nabla^2 f(x)$ is symmetric. In our development, whenever we assume that f is twice differentiable, we also assume that it is twice continuously differentiable.

We now state some theorems relating to differentiable functions.

Proposition A.3.1: (Mean Value Theorem) Let $f : \Re^n \mapsto \Re$ be continuously differentiable over an open sphere S, and let x be a vector in S. Then for all y such that $x + y \in S$, there exists an $\alpha \in [0, 1]$ such that

$$f(x + y) = f(x) + \nabla f(x + \alpha y)'y.$$

Proposition A.3.2: (Second Order Expansions) Let $f : \Re^n \mapsto \Re$ be twice continuously differentiable over an open sphere S, and let x be a vector in S. Then for all y such that $x + y \in S$:

(a) There exists an $\alpha \in [0, 1]$ such that

$$f(x+y) = f(x) + y'\nabla f(x) + \tfrac{1}{2}y'\nabla^2 f(x+\alpha y)y.$$

(b) We have

$$f(x+y) = f(x) + y'\nabla f(x) + \tfrac{1}{2}y'\nabla^2 f(x)y + o(\|y\|^2).$$

A.4 CONVERGENCE THEOREMS

We will now discuss a few convergence theorems relating to iterative algorithms. Given a mapping $T : \Re^n \mapsto \Re^n$, the iteration

$$x_{k+1} = T(x_k),$$

aims at finding a fixed point of T, i.e., a vector x^* such that $x^* = T(x^*)$. A common criterion for existence of a fixed point is that T is a *contraction mapping* (or contraction for short) with respect to some norm, i.e., for some $\beta < 1$, and some norm $\|\cdot\|$ (not necessarily the Euclidean norm), we have

$$\|T(x) - T(y)\| \leq \beta\|x - y\|, \qquad \forall\, x, y \in \Re^n.$$

When T is a contraction, it has a unique fixed point and the iteration $x_{k+1} = T(x_k)$ converges to the fixed point. This is shown in the following classical theorem.

Proposition A.4.1: (Contraction Mapping Theorem) Let $T : \Re^n \mapsto \Re^n$ be a contraction mapping. Then T has a unique fixed point x^*, and the sequence generated by the iteration $x_{k+1} = T(x_k)$ converges to x^*, starting from any $x_0 \in \Re^n$.

Proof: We first note that T can have at most one fixed point (if $\tilde{x}$ and $\hat{x}$ are two fixed points, we have

$$\|\tilde{x} - \hat{x}\| = \|T(\tilde{x}) - T(\hat{x})\| \leq \beta\|\tilde{x} - \hat{x}\|,$$

which implies that $\tilde{x} = \hat{x}$). Using the contraction property, we have for all $k, m > 0$

$$\|x_{k+m} - x_k\| \leq \beta^k \|x_m - x_0\| \leq \beta^k \sum_{\ell=1}^{m} \|x_\ell - x_{\ell-1}\| \leq \beta^k \sum_{\ell=0}^{m-1} \beta^\ell \|x_1 - x_0\|,$$

and finally,

$$\|x_{k+m} - x_k\| \leq \frac{\beta^k(1-\beta^m)}{1-\beta} \|x_1 - x_0\|.$$

Thus $\{x_k\}$ is a Cauchy sequence, and hence converges to some x^*. Taking the limit in the equation $x_{k+1} = T(x_k)$ and using the continuity of T (implied by the contraction property), we see that x^* must be a fixed point of T. **Q.E.D.**

In the case of a linear mapping

$$T(x) = Ax + b,$$

where A is an $n \times n$ matrix and $b \in \Re^n$, it can be shown that T is a contraction mapping with respect to some norm (but not necessarily all norms) if and only if all the eigenvalues of A lie strictly within the unit circle.

The next theorem applies to a mapping that is nonexpansive with respect to the Euclidean norm. It shows that a fixed point of such a mapping can be found by an interpolated iteration, provided at least one fixed point exists. The idea underlying the theorem is quite intuitive: if x^* is a fixed point of T, the distance $\|T(x_k) - x^*\|$ cannot be larger than the distance $\|x_k - x^*\|$ (by nonexpansiveness of T):

$$\|T(x_k) - x^*\| = \|T(x_k) - T(x^*)\| \leq \|x_k - x^*\|.$$

Hence, if $x_k \neq T(x_k)$, any point obtained by strict interpolation between x_k and $T(x_k)$ must be strictly closer to x^* than x_k (by Euclidean geometry). Note, however, that for this argument to work, we need to know that T has at least one fixed point. If T is a contraction, this is automatically guaranteed, but if T is just nonexpansive, there may not exist a fixed point [as an example, just let $T(x) = 1 + x$].

Proposition A.4.2: (Krasnosel'skii-Mann Theorem for Nonexpansive Iterations [Kra55], [Man53]) Consider a mapping $T : \Re^n \mapsto \Re^n$ that is nonexpansive with respect to the Euclidean norm $\|\cdot\|$, i.e.,

$$\|T(x) - T(y)\| \leq \|x - y\|, \qquad \forall\, x, y \in \Re^n,$$

and has at least one fixed point. Then the iteration

$$x_{k+1} = (1-\alpha_k)x_k + \alpha_k T(x_k), \tag{A.1}$$

where $\alpha_k \in [0,1]$ for all k and

$$\sum_{k=0}^{\infty} \alpha_k(1-\alpha_k) = \infty,$$

converges to a fixed point of T, starting from any $x_0 \in \Re^n$.

Proof: We will use the identity

$$\big\|\alpha x + (1-\alpha)y\big\|^2 = \alpha\|x\|^2 + (1-\alpha)\|y\|^2 - \alpha(1-\alpha)\|x-y\|^2, \tag{A.2}$$

which holds for all $x, y \in \Re^n$, and $\alpha \in [0,1]$, as can be verified by a straightforward calculation. For any fixed point x^* of T, we have

$$\begin{aligned} \|x_{k+1} - x^*\|^2 &= \big\|(1-\alpha_k)(x_k - x^*) + \alpha_k\big(T(x_k) - T(x^*)\big)\big\|^2 \\ &= (1-\alpha_k)\|x_k - x^*\|^2 + \alpha_k\big\|T(x_k) - T(x^*)\big\|^2 \\ &\quad - \alpha_k(1-\alpha_k)\big\|T(x_k) - x_k\big\|^2 \\ &\le \|x_k - x^*\|^2 - \alpha_k(1-\alpha_k)\big\|T(x_k) - x_k\big\|^2, \end{aligned} \tag{A.3}$$

where for the first equality we use iteration (A.1) and the fact $x^* = T(x^*)$, for the second equality we apply the identity (A.2), and for the inequality we use the nonexpansiveness of T. By adding Eq. (A.3) for all k, we obtain

$$\sum_{k=0}^{\infty} \alpha_k(1-\alpha_k)\big\|T(x_k) - x_k\big\|^2 \le \|x_0 - x^*\|^2.$$

In view of the hypothesis $\sum_{k=0}^{\infty} \alpha_k(1-\alpha_k) = \infty$, it follows that

$$\lim_{k\to\infty,\, k\in\mathcal{K}} \big\|T(x_k) - x_k\big\| = 0, \tag{A.4}$$

for some subsequence $\{x_k\}_{\mathcal{K}}$. Since from Eq. (A.3), $\{x_k\}_{\mathcal{K}}$ is bounded, it has at least one limit point, call it $\overline{x}$, so $\{x_k\}_{\overline{\mathcal{K}}} \to \overline{x}$ for an infinite index set $\overline{\mathcal{K}} \subset \mathcal{K}$. Since T is nonexpansive it is continuous, so $\big\{T(x_k)\big\}_{\overline{\mathcal{K}}} \to T(\overline{x})$, and in view of Eq. (A.4), it follows that $\overline{x}$ is a fixed point of T. Letting $x^* = \overline{x}$ in Eq. (A.3), we see that $\big\{\|x_k - \overline{x}\|\big\}$ is nonincreasing and hence converges, necessarily to 0, so the entire sequence $\{x_k\}$ converges to the fixed point $\overline{x}$. **Q.E.D.**

Nonstationary Iterations

For nonstationary iterations of the form $x_{k+1} = T_k(x_k)$, where the function T_k depends on k, the ideas of the preceding propositions may apply but with modifications. The following proposition is often useful in this respect.

Proposition A.4.3: Let $\{\alpha_k\}$ be a nonnegative sequence satisfying

$$\alpha_{k+1} \le (1-\gamma_k)\alpha_k + \beta_k, \qquad \forall\, k = 0, 1, \ldots,$$

where $0 \le \beta_k$, $0 < \gamma_k \le 1$ for all k, and

$$\sum_{k=0}^{\infty} \gamma_k = \infty, \qquad \frac{\beta_k}{\gamma_k} \to 0.$$

Then $\alpha_k \to 0$.

Proof: We first show that given any $\epsilon > 0$, we have $\alpha_k < \epsilon$ for infinitely many k. Indeed, if this were not so, by letting $\overline{k}$ be such that $\alpha_k \ge \epsilon$ and $\beta_k/\gamma_k \le \epsilon/2$ for all $k \ge \overline{k}$, we would have for all $k \ge \overline{k}$

$$\alpha_{k+1} \le \alpha_k - \gamma_k \alpha_k + \beta_k \le \alpha_k - \gamma_k \epsilon + \frac{\gamma_k \epsilon}{2} = \alpha_k - \frac{\gamma_k \epsilon}{2}.$$

Therefore, for all $m \ge \overline{k}$,

$$\alpha_{m+1} \le \alpha_{\overline{k}} - \frac{\epsilon}{2} \sum_{k=\overline{k}}^{m} \gamma_k.$$

Since $\{\alpha_k\}$ is nonnegative and $\sum_{k=0}^{\infty} \gamma_k = \infty$, we obtain a contradiction.

Thus, given any $\epsilon > 0$, there exists $\overline{k}$ such that $\beta_k/\gamma_k < \epsilon$ for all $k \ge \overline{k}$ and $\alpha_{\overline{k}} < \epsilon$. We then have

$$\alpha_{\overline{k}+1} \le (1-\gamma_k)\alpha_{\overline{k}} + \beta_k < (1-\gamma_k)\epsilon + \gamma_k \epsilon = \epsilon.$$

By repeating this argument, we obtain $\alpha_k < \epsilon$ for all $k \ge \overline{k}$. Since ϵ can be arbitrarily small, it follows that $\alpha_k \to 0$. **Q.E.D.**

As an example, consider a sequence of "approximate" contraction mappings $T_k : \Re^n \mapsto \Re^n$, satisfying

$$\|T_k(x) - T_k(y)\| \le (1-\gamma_k)\|x - y\| + \beta_k, \qquad \forall\, x, y \in \Re^n,\ k = 0, 1, \ldots,$$

where $\gamma_k \in (0,1]$, for all k, and

$$\sum_{k=0}^{\infty} \gamma_k = \infty, \qquad \frac{\beta_k}{\gamma_k} \to 0.$$

Assume also that all the mappings T_k have a common fixed point x^*. Then

$$\|x_{k+1} - x^*\| = \|T_k(x_k) - T_k(x^*)\| \le (1-\gamma_k)\|x_k - x^*\| + \beta_k,$$

and from Prop. A.4.3, it follows that the sequence $\{x_k\}$ generated by the iteration $x_{k+1} = T_k(x_k)$ converges to x^* starting from any $x_0 \in \Re^n$.

Supermartingale Convergence

We now give two theorems relating to *supermartingale convergence* analysis (the term refers to a collection of convergence theorems for sequences of nonnegative scalars or random variables, which satisfy certain inequalities implying that the sequences are "almost" nonincreasing). The first theorem relates to deterministic sequences, while the second theorem relates to sequences of random variables. We prove the first theorem, and we refer to the literature on stochastic processes and iterative methods for the proof of the second.

Proposition A.4.4: Let $\{Y_k\}$, $\{Z_k\}$, $\{W_k\}$, and $\{V_k\}$ be four scalar sequences such that

$$Y_{k+1} \le (1+V_k)Y_k - Z_k + W_k, \qquad k = 0,1,\ldots, \tag{A.5}$$

$\{Z_k\}$, $\{W_k\}$, and $\{V_k\}$ are nonnegative, and

$$\sum_{k=0}^{\infty} W_k < \infty, \qquad \sum_{k=0}^{\infty} V_k < \infty.$$

Then either $Y_k \to -\infty$, or else $\{Y_k\}$ converges to a finite value and $\sum_{k=0}^{\infty} Z_k < \infty$.

Proof: We first give the proof assuming that $V_k \equiv 0$, and then generalize. In this case, using the nonnegativity of $\{Z_k\}$, we have

$$Y_{k+1} \le Y_k + W_k.$$

By writing this relation for the index k set to $\bar{k}, \ldots, k$, where $k \ge \bar{k}$, and adding, we have

$$Y_{k+1} \le Y_{\bar{k}} + \sum_{\ell=\bar{k}}^{k} W_\ell \le Y_{\bar{k}} + \sum_{\ell=\bar{k}}^{\infty} W_\ell.$$

Since $\sum_{k=0}^{\infty} W_k < \infty$, it follows that $\{Y_k\}$ is bounded above, and by taking upper limit of the left hand side as $k \to \infty$ and lower limit of the right hand side as $\bar{k} \to \infty$, we have

$$\limsup_{k\to\infty} Y_k \le \liminf_{\bar{k}\to\infty} Y_{\bar{k}} < \infty.$$

This implies that either $Y_k \to -\infty$, or else $\{Y_k\}$ converges to a finite value. In the latter case, by writing Eq. (A.5) for the index k set to $0, \ldots, k$, and adding, we have

$$\sum_{\ell=0}^{k} Z_\ell \le Y_0 + \sum_{\ell=0}^{k} W_\ell - Y_{k+1}, \qquad \forall\, k = 0, 1, \ldots,$$

so by taking the limit as $k \to \infty$, we obtain $\sum_{\ell=0}^{\infty} Z_\ell < \infty$.

We now extend the proof to the case of a general nonnegative sequence $\{V_k\}$. We first note that

$$\log \prod_{\ell=0}^{k} (1 + V_\ell) = \sum_{\ell=0}^{k} \log(1 + V_\ell) \le \sum_{k=0}^{\infty} V_k,$$

since we generally have $(1+a) \le e^a$ and $\log(1+a) \le a$ for any $a \ge 0$. Thus the assumption $\sum_{k=0}^{\infty} V_k < \infty$ implies that

$$\prod_{\ell=0}^{\infty} (1 + V_\ell) < \infty. \tag{A.6}$$

Define

$$\overline{Y}_k = Y_k \prod_{\ell=0}^{k-1} (1+V_\ell)^{-1}, \quad \overline{Z}_k = Z_k \prod_{\ell=0}^{k} (1+V_\ell)^{-1}, \quad \overline{W}_k = W_k \prod_{\ell=0}^{k} (1+V_\ell)^{-1}.$$

Multiplying Eq. (A.5) with $\prod_{\ell=0}^{k} (1 + V_\ell)^{-1}$, we obtain

$$\overline{Y}_{k+1} \le \overline{Y}_k - \overline{Z}_k + \overline{W}_k.$$

Since $\overline{W}_k \le W_k$, the hypothesis $\sum_{k=0}^{\infty} W_k < \infty$ implies $\sum_{k=0}^{\infty} \overline{W}_k < \infty$, so from the special case of the result already shown, we have that either $\overline{Y}_k \to -\infty$ or else $\{\overline{Y}_k\}$ converges to a finite value and $\sum_{k=0}^{\infty} \overline{Z}_k < \infty$. Since

$$Y_k = \overline{Y}_k \prod_{\ell=0}^{k-1} (1 + V_\ell), \qquad Z_k = \overline{Z}_k \prod_{\ell=0}^{k} (1 + V_\ell),$$

and $\prod_{\ell=0}^{k-1} (1 + V_\ell)$ converges to a finite value by the nonnegativity of $\{V_k\}$ and Eq. (A.6), it follows that either $Y_k \to -\infty$ or else $\{Y_k\}$ converges to a finite value and $\sum_{k=0}^{\infty} Z_k < \infty$. **Q.E.D.**

The next theorem has a long history. The particular version we give here is due to Robbins and Sigmund [RoS71]. Their proof assumes the special case of the theorem where $V_k \equiv 0$ (see Neveu [Nev75], p. 33, for a proof of this special case), and then uses the line of proof of the preceding proposition. Note, however, that contrary to the preceding proposition, the following theorem requires nonnegativity of the sequence $\{Y_k\}$.

Proposition A.4.5: (Supermartingale Convergence Theorem) Let $\{Y_k\}$, $\{Z_k\}$, $\{W_k\}$, and $\{V_k\}$ be four nonnegative sequences of random variables, and let $\mathcal{F}_k$, $k = 0, 1, \ldots$, be sets of random variables such that $\mathcal{F}_k \subset \mathcal{F}_{k+1}$ for all k. Assume that:

(1) For each k, Y_k, Z_k, W_k, and V_k are functions of the random variables in $\mathcal{F}_k$.

(2) We have

$$E\{Y_{k+1} \mid \mathcal{F}_k\} \leq (1 + V_k)Y_k - Z_k + W_k, \qquad k = 0, 1, \ldots.$$

(3) There holds, with probability 1,

$$\sum_{k=0}^{\infty} W_k < \infty, \qquad \sum_{k=0}^{\infty} V_k < \infty.$$

Then $\{Y_k\}$ converges to a nonnegative random variable Y, and we have $\sum_{k=0}^{\infty} Z_k < \infty$, with probability 1.

Fejér Monotonicity

The supermartingale convergence theorems can be applied in a variety of contexts. One such context, the so called *Fejér monotonicity* theory, deals with iterations that "almost" decrease the distance to *every* element of some given set X^*. We may then often show that such iterations are convergent to a (unique) element of X^*. Applications of this idea arise when X^* is the set of optimal solutions of an optimization problem or the set of fixed points of a certain mapping. Examples are various gradient and subgradient projection methods with a diminishing stepsize that arise in various contexts in this book, as well as the Krasnosel'skii-Mann Theorem [Prop. A.4.2; see Eq. (A.3)].

The following theorem is appropriate for our purposes. There are several related but somewhat different theorems in the literature, and for complementary discussions, we refer to [BaB96], [Com01], [BaC11], [CoV13].

Proposition A.4.6: (Fejér Convergence Theorem) Let X^* be a nonempty subset of $\Re^n$, and let $\{x_k\} \subset \Re^n$ be a sequence satisfying for some $p > 0$ and for all k,

$$\|x_{k+1} - x^*\|^p \leq (1+\beta_k)\|x_k - x^*\|^p - \gamma_k\, \phi(x_k; x^*) + \delta_k, \qquad \forall\, x^* \in X^*,$$

where $\{\beta_k\}$, $\{\gamma_k\}$, and $\{\delta_k\}$ are nonnegative sequences satisfying

$$\sum_{k=0}^{\infty} \beta_k < \infty, \qquad \sum_{k=0}^{\infty} \gamma_k = \infty, \qquad \sum_{k=0}^{\infty} \delta_k < \infty,$$

$\phi : \Re^n \times X^* \mapsto [0, \infty)$ is some nonnegative function, and $\|\cdot\|$ is some norm. Then:

(a) The minimum distance sequence $\inf_{x^* \in X^*} \|x_k - x^*\|$ converges, and in particular, $\{x_k\}$ is bounded.

(b) If $\{x_k\}$ has a limit point $\overline{x}$ that belongs to X^*, then the entire sequence $\{x_k\}$ converges to $\overline{x}$.

(c) Suppose that for some $x^* \in X^*$, $\phi(\cdot; x^*)$ is lower semicontinuous and satisfies

$$\phi(x; x^*) = 0 \quad \text{if and only if} \quad x \in X^*. \tag{A.7}$$

Then $\{x_k\}$ converges to a point in X^*.

Proof: (a) Let $\{\epsilon_k\}$ be a positive sequence such that $\sum_{k=0}^{\infty}(1+\beta_k)\epsilon_k < \infty$, and let x_k^* be a point of X^* such that

$$\|x_k - x_k^*\|^p \leq \inf_{x^* \in X^*} \|x_k - x^*\|^p + \epsilon_k.$$

Then since ϕ is nonnegative, we have for all k,

$$\inf_{x^* \in X^*} \|x_{k+1} - x^*\|^p \leq \|x_{k+1} - x_k^*\|^p \leq (1+\beta_k)\|x_k - x_k^*\|^p + \delta_k,$$

and by combining the last two relations, we obtain

$$\inf_{x^* \in X^*} \|x_{k+1} - x^*\|^p \leq (1+\beta_k) \inf_{x^* \in X^*} \|x_k - x^*\|^p + (1+\beta_k)\epsilon_k + \delta_k.$$

The result follows by applying Prop. A.4.4 with

$$Y_k = \inf_{x^* \in X^*} \|x_k - x^*\|^p, \quad Z_k = 0, \quad W_k = (1+\beta_k)\epsilon_k + \delta_k, \quad V_k = \beta_k.$$

(b) Following the argument of the proof of Prop. A.4.4, define for all k,

$$\overline{Y}_k = \|x_k - \overline{x}\|^p \prod_{\ell=0}^{k-1}(1+\beta_\ell)^{-1}, \qquad \overline{\delta}_k = \delta_k \prod_{\ell=0}^{k}(1+\beta_\ell)^{-1}.$$

Then from our hypotheses, we have $\sum_{k=0}^{\infty} \overline{\delta}_k < \infty$ and

$$\overline{Y}_{k+1} \leq \overline{Y}_k + \overline{\delta}_k, \qquad \forall\, k = 0, 1, \ldots, \tag{A.8}$$

while $\{\overline{Y}_k\}$ has a limit point at 0, since $\overline{x}$ is a limit point of $\{x_k\}$. For any $\epsilon > 0$, let $\overline{k}$ be such that

$$\overline{Y}_{\overline{k}} \leq \epsilon, \qquad \sum_{\ell=\overline{k}}^{\infty} \overline{\delta}_\ell \leq \epsilon,$$

so that by adding Eq. (A.8), we obtain for all $k > \overline{k}$,

$$\overline{Y}_k \leq \overline{Y}_{\overline{k}} + \sum_{\ell=\overline{k}}^{\infty} \overline{\delta}_\ell \leq 2\epsilon.$$

Since ϵ is arbitrarily small, it follows that $\overline{Y}_k \to 0$. We now note that as in Eq. (A.6),

$$\prod_{\ell=0}^{\infty}(1+\beta_\ell)^{-1} < \infty,$$

so that $\overline{Y}_k \to 0$ implies that $\|x_k - \overline{x}\|^p \to 0$, and hence $x_k \to \overline{x}$.

(c) From Prop. A.4.4, it follows that

$$\sum_{k=0}^{\infty} \gamma_k\, \phi(x_k; x^*) < \infty.$$

Thus $\lim_{k\to\infty,\, k\in\mathcal{K}} \phi(x_k; x^*) = 0$ for some subsequence $\{x_k\}_\mathcal{K}$. By part (a), $\{x_k\}$ is bounded, so the subsequence $\{x_k\}_\mathcal{K}$ has a limit point $\overline{x}$, and by the lower semicontinuity of $\phi(\cdot; x^*)$, we must have

$$\phi(\overline{x}; x^*) \leq \lim_{k\to\infty,\, k\in\mathcal{K}} \phi(x_k; x^*) = 0,$$

which in view of the nonnegativity of ϕ, implies that $\phi(\overline{x}; x^*) = 0$. Using the hypothesis (A.7), it follows that $\overline{x} \in X^*$, so by part (b), the entire sequence $\{x_k\}$ converges to $\overline{x}$. **Q.E.D.**

APPENDIX B

Convex Optimization Theory: A Summary

In this appendix, we provide a summary of theoretical concepts and results relating to convex analysis, convex optimization, and duality theory. In particular, we list the relevant definitions and propositions (without proofs) of the author's book "Convex Optimization Theory," Athena Scientific, 2009. For ease of use, the chapter, section, definition, and proposition numbers of the latter book are identical to the ones of this appendix.

CHAPTER 1: Basic Concepts of Convex Analysis

Section 1.1. Convex Sets and Functions

Definition 1.1.1: A subset C of $\Re^n$ is called *convex* if

$$\alpha x + (1-\alpha)y \in C, \qquad \forall\, x, y \in C,\ \forall\, \alpha \in [0,1].$$

Proposition 1.1.1:

(a) The intersection $\cap_{i\in I} C_i$ of any collection $\{C_i \mid i \in I\}$ of convex sets is convex.

(b) The vector sum $C_1 + C_2$ of two convex sets C_1 and C_2 is convex.

(c) The set λC is convex for any convex set C and scalar λ. Furthermore, if C is a convex set and λ_1, λ_2 are positive scalars,

$$(\lambda_1 + \lambda_2)C = \lambda_1 C + \lambda_2 C.$$

(d) The closure and the interior of a convex set are convex.

(e) The image and the inverse image of a convex set under an affine function are convex.

A *hyperplane* is a set of the form $\{x \mid a'x = b\}$, where a is a nonzero vector and b is a scalar. A *halfspace* is a set specified by a single linear inequality, i.e., a set of the form $\{x \mid a'x \le b\}$, where a is a nonzero vector and b is a scalar. A set is said to be *polyhedral* if it is nonempty and it has the form $\{x \mid a_j'x \le b_j,\ j = 1,\ldots,r\}$, where $a_1,\ldots,a_r$ and $b_1,\ldots,b_r$ are some vectors in $\Re^n$ and scalars, respectively. A set C is said to be a *cone* if for all $x \in C$ and $\lambda > 0$, we have $\lambda x \in C$.

Definition 1.1.2: Let C be a convex subset of $\Re^n$. We say that a function $f : C \mapsto \Re$ is *convex* if

$$f\big(\alpha x + (1-\alpha)y\big) \le \alpha f(x) + (1-\alpha)f(y), \qquad \forall\, x, y \in C,\ \forall\, \alpha \in [0,1].$$

A convex function $f : C \mapsto \Re$ is called *strictly convex* if

$$f\big(\alpha x + (1-\alpha)y\big) < \alpha f(x) + (1-\alpha)f(y)$$

for all $x, y \in C$ with $x \neq y$, and all $\alpha \in (0,1)$. A function $f : C \mapsto \Re$, where C is a convex set, is called *concave* if the function $(-f)$ is convex.

The *epigraph* of a function $f : X \mapsto [-\infty, \infty]$, where $X \subset \Re^n$, is defined to be the subset of $\Re^{n+1}$ given by

$$\text{epi}(f) = \big\{(x, w) \mid x \in X,\ w \in \Re,\ f(x) \le w\big\}.$$

The *effective domain* of f is defined to be the set

$$\text{dom}(f) = \big\{x \in X \mid f(x) < \infty\big\}.$$

We say that f is *proper* if $f(x) < \infty$ for at least one $x \in X$ and $f(x) > -\infty$ for all $x \in X$, and we say that f *improper* if it is not proper. Thus f is proper if and only if epi(f) is nonempty and does not contain a vertical line.

Definition 1.1.3: Let C be a convex subset of $\Re^n$. We say that an extended real-valued function $f : C \mapsto [-\infty, \infty]$ is *convex* if epi(f) is a convex subset of $\Re^{n+1}$.

Definition 1.1.4: Let C and X be subsets of $\Re^n$ such that C is nonempty and convex, and $C \subset X$. We say that an extended real-valued function $f : X \mapsto [-\infty, \infty]$ is *convex over* C if f becomes convex when the domain of f is restricted to C, i.e., if the function $\tilde{f} : C \mapsto [-\infty, \infty]$, defined by $\tilde{f}(x) = f(x)$ for all $x \in C$, is convex.

We say that a function $f : X \mapsto [-\infty, \infty]$ is *closed* if $\text{epi}(f)$ is a closed set. We say that f is *lower semicontinuous* at a vector $x \in X$ if $f(x) \leq \liminf_{k\to\infty} f(x_k)$ for every sequence $\{x_k\} \subset X$ with $x_k \to x$. We say that f *is lower semicontinuous* if it is lower semicontinuous at each point x in its domain X. We say that f is *upper semicontinuous* if $-f$ is lower semicontinuous.

Proposition 1.1.2: For a function $f : \Re^n \mapsto [-\infty, \infty]$, the following are equivalent:

(i) The level set $V_\gamma = \{x \mid f(x) \leq \gamma\}$ is closed for every scalar γ.

(ii) f is lower semicontinuous.

(iii) $\text{epi}(f)$ is closed.

Proposition 1.1.3: Let $f : X \mapsto [-\infty, \infty]$ be a function. If $\text{dom}(f)$ is closed and f is lower semicontinuous at each $x \in \text{dom}(f)$, then f is closed.

Proposition 1.1.4: Let $f : \Re^m \mapsto (-\infty, \infty]$ be a given function, let A be an $m \times n$ matrix, and let $F : \Re^n \mapsto (-\infty, \infty]$ be the function

$$F(x) = f(Ax), \qquad x \in \Re^n.$$

If f is convex, then F is also convex, while if f is closed, then F is also closed.

Proposition 1.1.5: Let $f_i : \Re^n \mapsto (-\infty, \infty]$, $i = 1, \ldots, m$, be given functions, let $\gamma_1, \ldots, \gamma_m$ be positive scalars, and let $F : \Re^n \mapsto (-\infty, \infty]$

be the function

$$F(x) = \gamma_1 f_1(x) + \cdots + \gamma_m f_m(x), \qquad x \in \Re^n.$$

If $f_1, \ldots, f_m$ are convex, then F is also convex, while if $f_1, \ldots, f_m$ are closed, then F is also closed.

Proposition 1.1.6: Let $f_i : \Re^n \mapsto (-\infty, \infty]$ be given functions for $i \in I$, where I is an arbitrary index set, and let $f : \Re^n \mapsto (-\infty, \infty]$ be the function given by

$$f(x) = \sup_{i \in I} f_i(x).$$

If f_i, $i \in I$, are convex, then f is also convex, while if f_i, $i \in I$, are closed, then f is also closed.

Proposition 1.1.7: Let C be a nonempty convex subset of $\Re^n$ and let $f : \Re^n \mapsto \Re$ be differentiable over an open set that contains C.

(a) f is convex over C if and only if

$$f(z) \geq f(x) + \nabla f(x)'(z - x), \qquad \forall\, x, z \in C.$$

(b) f is strictly convex over C if and only if the above inequality is strict whenever $x \neq z$.

Proposition 1.1.8: Let C be a nonempty convex subset of $\Re^n$ and let $f : \Re^n \mapsto \Re$ be convex and differentiable over an open set that contains C. Then a vector $x^* \in C$ minimizes f over C if and only if

$$\nabla f(x^*)'(z - x^*) \geq 0, \qquad \forall\, z \in C.$$

When f is not convex but is differentiable over an open set that contains C, the condition of the above proposition is necessary but not sufficient for optimality of x^* (see e.g., [Ber99], Section 2.1).

Proposition 1.1.9: (Projection Theorem) Let C be a nonempty closed convex subset of $\Re^n$, and let z be a vector in $\Re^n$. There exists a unique vector that minimizes $\|z-x\|$ over $x \in C$, called the projection of z on C. Furthermore, a vector x^* is the projection of z on C if and only if

$$(z - x^*)'(x - x^*) \leq 0, \qquad \forall\, x \in C.$$

Proposition 1.1.10: Let C be a nonempty convex subset of $\Re^n$ and let $f : \Re^n \mapsto \Re$ be twice continuously differentiable over an open set that contains C.

(a) If $\nabla^2 f(x)$ is positive semidefinite for all $x \in C$, then f is convex over C.

(b) If $\nabla^2 f(x)$ is positive definite for all $x \in C$, then f is strictly convex over C.

(c) If C is open and f is convex over C, then $\nabla^2 f(x)$ is positive semidefinite for all $x \in C$.

Strong Convexity

If $f : \Re^n \mapsto \Re$ is a function that is continuous over a closed convex set $C \subset \Re^n$, and σ is a positive scalar, we say that f *is strongly convex over C with coefficient σ* if for all $x, y \in C$ and all $\alpha \in [0, 1]$, we have

$$f\big(\alpha x + (1-\alpha)y\big) + \frac{\sigma}{2}\alpha(1-\alpha)\|x-y\|^2 \leq \alpha f(x) + (1-\alpha)f(y).$$

Then f is strictly convex over C. Furthermore, there exists a unique $x^* \in C$ that minimizes f over C, and by applying the definition with $y = x^*$ and letting $\alpha \downarrow 0$, it can be seen that

$$f(x) \geq f(x^*) + \frac{\sigma}{2}\|x - x^*\|^2, \qquad \forall\, x \in C.$$

If $\text{int}(C)$, the interior of C, is nonempty, and f is continuously differentiable over $\text{int}(C)$, the following are equivalent:

(i) f is strongly convex with coefficient σ over C.

(ii) $\big(\nabla f(x) - \nabla f(y)\big)'(x-y) \geq \sigma\|x-y\|^2, \qquad \forall\, x, y \in \text{int}(C)$.

(iii) $f(y) \geq f(x) + \nabla f(x)'(y-x) + \frac{\sigma}{2}\|x-y\|^2, \qquad \forall\, x, y \in \text{int}(C)$.

Furthermore, if f is twice continuously differentiable over $\text{int}(C)$, the above three properties are equivalent to:

(iv) The matrix $\nabla^2 f(x) - \sigma I$ is positive semidefinite for every $x \in \text{int}(C)$, where I is the identity matrix.

A proof may be found in the on-line exercises of Chapter 1 of [Ber09].

Section 1.2. Convex and Affine Hulls

The *convex hull* of a set X, denoted $\text{conv}(X)$, is the intersection of all convex sets containing X. A *convex combination* of elements of X is a vector of the form $\sum_{i=1}^m \alpha_i x_i$, where m is a positive integer, $x_1, \ldots, x_m$ belong to X, and $\alpha_1, \ldots, \alpha_m$ are scalars such that

$$\alpha_i \geq 0, \quad i = 1, \ldots, m, \qquad \sum_{i=1}^m \alpha_i = 1.$$

The convex hull $\text{conv}(X)$ is equal to the set of all convex combinations of elements of X. Also, for any set S and linear transformation A, we have $\text{conv}(AS) = A\,\text{conv}(S)$. From this it follows that for any sets $S_1, \ldots, S_m$, we have $\text{conv}(S_1 + \cdots + S_m) = \text{conv}(S_1) + \cdots + \text{conv}(S_m)$.

If X is a subset of $\Re^n$, the *affine hull* of X, denoted $\text{aff}(X)$, is the intersection of all affine sets containing X. Note that $\text{aff}(X)$ is itself an affine set and that it contains $\text{conv}(X)$. The dimension of $\text{aff}(X)$ is defined to be the dimension of the subspace parallel to $\text{aff}(X)$. It can be shown that $\text{aff}(X) = \text{aff}\big(\text{conv}(X)\big) = \text{aff}\big(\text{cl}(X)\big)$. For a convex set C, the *dimension* of C is defined to be the dimension of $\text{aff}(C)$.

Given a nonempty subset X of $\Re^n$, a *nonnegative combination* of elements of X is a vector of the form $\sum_{i=1}^m \alpha_i x_i$, where m is a positive integer, $x_1, \ldots, x_m$ belong to X, and $\alpha_1, \ldots, \alpha_m$ are nonnegative scalars. If the scalars α_i are all positive, $\sum_{i=1}^m \alpha_i x_i$ is said to be a *positive combination*. The *cone generated by* X, denoted $\text{cone}(X)$, is the set of all nonnegative combinations of elements of X.

Proposition 1.2.1: (Caratheodory's Theorem) Let X be a nonempty subset of $\Re^n$.

(a) Every nonzero vector from $\text{cone}(X)$ can be represented as a positive combination of linearly independent vectors from X.

(b) Every vector from $\text{conv}(X)$ can be represented as a convex combination of no more than $n+1$ vectors from X.

Proposition 1.2.2: The convex hull of a compact set is compact.

Section 1.3. Relative Interior and Closure

Let C be a nonempty convex set. We say that x is a *relative interior point* of C if $x \in C$ and there exists an open sphere S centered at x such that

$$S \cap \text{aff}(C) \subset C,$$

i.e., x is an interior point of C relative to the affine hull of C. The set of relative interior points of C is called the *relative interior of* C, and is denoted by $\text{ri}(C)$. The set C is said to be *relatively open* if $\text{ri}(C) = C$. The vectors in $\text{cl}(C)$ that are not relative interior points are said to be *relative boundary points* of C, and their collection is called the *relative boundary* of C.

Proposition 1.3.1: (Line Segment Principle) Let C be a nonempty convex set. If $x \in \text{ri}(C)$ and $\overline{x} \in \text{cl}(C)$, then all points on the line segment connecting x and $\overline{x}$, except possibly $\overline{x}$, belong to $\text{ri}(C)$.

Proposition 1.3.2: (Nonemptiness of Relative Interior) Let C be a nonempty convex set. Then:

(a) $\text{ri}(C)$ is a nonempty convex set, and has the same affine hull as C.

(b) If m is the dimension of $\text{aff}(C)$ and $m > 0$, there exist vectors $x_0, x_1, \ldots, x_m \in \text{ri}(C)$ such that $x_1 - x_0, \ldots, x_m - x_0$ span the subspace parallel to $\text{aff}(C)$.

Proposition 1.3.3: (Prolongation Lemma) Let C be a nonempty convex set. A vector x is a relative interior point of C if and only if every line segment in C having x as one endpoint can be prolonged beyond x without leaving C [i.e., for every $\overline{x} \in C$, there exists a $\gamma > 0$ such that $x + \gamma(x - \overline{x}) \in C$].

Proposition 1.3.4: Let X be a nonempty convex subset of $\Re^n$, let $f : X \mapsto \Re$ be a concave function, and let X^* be the set of vectors where f attains a minimum over X, i.e.,

$$X^* = \left\{ x^* \in X \;\middle|\; f(x^*) = \inf_{x \in X} f(x) \right\}.$$

If X^* contains a relative interior point of X, then f must be constant over X, i.e., $X^* = X$.

Proposition 1.3.5: Let C be a nonempty convex set.

(a) We have $\text{cl}(C) = \text{cl}\big(\text{ri}(C)\big)$.

(b) We have $\text{ri}(C) = \text{ri}\big(\text{cl}(C)\big)$.

(c) Let $\overline{C}$ be another nonempty convex set. Then the following three conditions are equivalent:

  (i) C and $\overline{C}$ have the same relative interior.

  (ii) C and $\overline{C}$ have the same closure.

  (iii) $\text{ri}(C) \subset \overline{C} \subset \text{cl}(C)$.

Proposition 1.3.6: Let C be a nonempty convex subset of $\Re^n$ and let A be an $m \times n$ matrix.

(a) We have $A \cdot \text{ri}(C) = \text{ri}(A \cdot C)$.

(b) We have $A \cdot \text{cl}(C) \subset \text{cl}(A \cdot C)$. Furthermore, if C is bounded, then $A \cdot \text{cl}(C) = \text{cl}(A \cdot C)$.

Proposition 1.3.7: Let C_1 and C_2 be nonempty convex sets. We have

$$\text{ri}(C_1 + C_2) = \text{ri}(C_1) + \text{ri}(C_2), \qquad \text{cl}(C_1) + \text{cl}(C_2) \subset \text{cl}(C_1 + C_2).$$

Furthermore, if at least one of the sets C_1 and C_2 is bounded, then

$$\text{cl}(C_1) + \text{cl}(C_2) = \text{cl}(C_1 + C_2).$$

Proposition 1.3.8: Let C_1 and C_2 be nonempty convex sets. We have

$$\text{ri}(C_1) \cap \text{ri}(C_2) \subset \text{ri}(C_1 \cap C_2), \qquad \text{cl}(C_1 \cap C_2) \subset \text{cl}(C_1) \cap \text{cl}(C_2).$$

Furthermore, if the sets $\text{ri}(C_1)$ and $\text{ri}(C_2)$ have a nonempty intersection, then

$$\text{ri}(C_1 \cap C_2) = \text{ri}(C_1) \cap \text{ri}(C_2), \qquad \text{cl}(C_1 \cap C_2) = \text{cl}(C_1) \cap \text{cl}(C_2).$$

Proposition 1.3.9: Let C be a nonempty convex subset of $\Re^m$, and let A be an $m \times n$ matrix. If $A^{-1} \cdot \text{ri}(C)$ is nonempty, then

$$\text{ri}(A^{-1} \cdot C) = A^{-1} \cdot \text{ri}(C), \qquad \text{cl}(A^{-1} \cdot C) = A^{-1} \cdot \text{cl}(C),$$

where A^{-1} denotes inverse image of the corresponding set under A.

Proposition 1.3.10: Let C be a convex subset of $\Re^{n+m}$. For $x \in \Re^n$, denote

$$C_x = \{y \mid (x, y) \in C\},$$

and let

$$D = \{x \mid C_x \neq \emptyset\}.$$

Then

$$\text{ri}(C) = \{(x, y) \mid x \in \text{ri}(D),\ y \in \text{ri}(C_x)\}.$$

Continuity of Convex Functions

Proposition 1.3.11: If $f : \Re^n \mapsto \Re$ is convex, then it is continuous. More generally, if $f : \Re^n \mapsto (-\infty, \infty]$ is a proper convex function, then f, restricted to $\text{dom}(f)$, is continuous over the relative interior of $\text{dom}(f)$.

Proposition 1.3.12: If C is a closed interval of the real line, and $f : C \mapsto \Re$ is closed and convex, then f is continuous over C.

Closures of Functions

The closure of the epigraph of a function $f : X \mapsto [-\infty, \infty]$ can be seen to be a legitimate epigraph of another function. This function, called the *closure of* f and denoted $\mathrm{cl}\, f : \Re^n \mapsto [-\infty, \infty]$, is given by

$$(\mathrm{cl}\, f)(x) = \inf\big\{w \mid (x, w) \in \mathrm{cl}\big(\mathrm{epi}(f)\big)\big\}, \qquad x \in \Re^n.$$

The closure of the convex hull of the epigraph of f is the epigraph of some function, denoted $\check{\mathrm{cl}}\, f$ called the *convex closure of* f. It can be seen that $\check{\mathrm{cl}}\, f$ is the closure of the function $F : \Re^n \mapsto [-\infty, \infty]$ given by

$$F(x) = \inf\big\{w \mid (x, w) \in \mathrm{conv}\big(\mathrm{epi}(f)\big)\big\}, \qquad x \in \Re^n. \tag{B.1}$$

It is easily shown that F is convex, but it need not be closed and its domain may be strictly contained in $\mathrm{dom}(\check{\mathrm{cl}}\, f)$ (it can be seen though that the closures of the domains of F and $\check{\mathrm{cl}}\, f$ coincide).

Proposition 1.3.13: Let $f : X \mapsto [-\infty, \infty]$ be a function. Then

$$\inf_{x \in X} f(x) = \inf_{x \in X} (\mathrm{cl}\, f)(x) = \inf_{x \in \Re^n} (\mathrm{cl}\, f)(x) = \inf_{x \in \Re^n} F(x) = \inf_{x \in \Re^n} (\check{\mathrm{cl}}\, f)(x),$$

where F is given by Eq. (B.1). Furthermore, any vector that attains the infimum of f over X also attains the infimum of $\mathrm{cl}\, f$, F, and $\check{\mathrm{cl}}\, f$.

Proposition 1.3.14: Let $f : \Re^n \mapsto [-\infty, \infty]$ be a function.

(a) $\mathrm{cl}\, f$ is the greatest closed function majorized by f, i.e., if $g : \Re^n \mapsto [-\infty, \infty]$ is closed and satisfies $g(x) \le f(x)$ for all $x \in \Re^n$, then $g(x) \le (\mathrm{cl}\, f)(x)$ for all $x \in \Re^n$.

(b) $\check{\mathrm{cl}}\, f$ is the greatest closed and convex function majorized by f, i.e., if $g : \Re^n \mapsto [-\infty, \infty]$ is closed and convex, and satisfies $g(x) \le f(x)$ for all $x \in \Re^n$, then $g(x) \le (\check{\mathrm{cl}}\, f)(x)$ for all $x \in \Re^n$.

Proposition 1.3.15: Let $f : \Re^n \mapsto [-\infty, \infty]$ be a convex function. Then:

(a) We have

$$\mathrm{cl}\big(\mathrm{dom}(f)\big) = \mathrm{cl}\big(\mathrm{dom}(\mathrm{cl}\, f)\big), \qquad \mathrm{ri}\big(\mathrm{dom}(f)\big) = \mathrm{ri}\big(\mathrm{dom}(\mathrm{cl}\, f)\big),$$

$$(\mathrm{cl}\, f)(x) = f(x), \qquad \forall\, x \in \mathrm{ri}\big(\mathrm{dom}(f)\big).$$

Furthermore, $\mathrm{cl}\, f$ is proper if and only if f is proper.

(b) If $x \in \mathrm{ri}\big(\mathrm{dom}(f)\big)$, we have

$$(\mathrm{cl}\, f)(y) = \lim_{\alpha \downarrow 0} f\big(y + \alpha(x - y)\big), \qquad \forall\, y \in \Re^n.$$

Proposition 1.3.16: Let $f : \Re^m \mapsto [-\infty, \infty]$ be a convex function and A be an $m \times n$ matrix such that the range of A contains a point in $\mathrm{ri}\big(\mathrm{dom}(f)\big)$. The function F defined by

$$F(x) = f(Ax),$$

is convex and

$$(\mathrm{cl}\, F)(x) = (\mathrm{cl}\, f)(Ax), \qquad \forall\, x \in \Re^n.$$

Proposition 1.3.17: Let $f_i : \Re^n \mapsto [-\infty, \infty]$, $i = 1, \ldots, m$, be convex functions such that

$$\cap_{i=1}^m \mathrm{ri}\big(\mathrm{dom}(f_i)\big) \neq \emptyset.$$

The function F defined by

$$F(x) = f_1(x) + \cdots + f_m(x),$$

is convex and

$$(\mathrm{cl}\, F)(x) = (\mathrm{cl}\, f_1)(x) + \cdots + (\mathrm{cl}\, f_m)(x), \qquad \forall\, x \in \Re^n.$$

Section 1.4. Recession Cones

Given a nonempty convex set C, we say that a vector d is a *direction of recession* of C if $x + \alpha d \in C$ for all $x \in C$ and $\alpha \geq 0$. The set of all directions of recession is a cone containing the origin, called the *recession cone* of C, and denoted by R_C.

Proposition 1.4.1: (Recession Cone Theorem) Let C be a nonempty closed convex set.

(a) The recession cone R_C is closed and convex.

(b) A vector d belongs to R_C if and only if there exists a vector $x \in C$ such that $x + \alpha d \in C$ for all $\alpha \geq 0$.

Proposition 1.4.2: (Properties of Recession Cones) Let C be a nonempty closed convex set.

(a) R_C contains a nonzero direction if and only if C is unbounded.

(b) $R_C = R_{\mathrm{ri}(C)}$.

(c) For any collection of closed convex sets C_i, $i \in I$, where I is an arbitrary index set and $\cap_{i \in I} C_i \neq \emptyset$, we have

$$R_{\cap_{i \in I} C_i} = \cap_{i \in I} R_{C_i}.$$

(d) Let W be a compact and convex subset of $\Re^m$, and let A be an $m \times n$ matrix. The recession cone of the set

$$V = \{x \in C \mid Ax \in W\}$$

(assuming this set is nonempty) is $R_C \cap N(A)$, where $N(A)$ is the nullspace of A.

Given a convex set C the *lineality space* of C, denoted by L_C, is the set of directions of recession d whose opposite, $-d$, are also directions of recession:

$$L_C = R_C \cap (-R_C).$$

Proposition 1.4.3: (Properties of Lineality Space) Let C be a nonempty closed convex subset of $\Re^n$.

(a) L_C is a subspace of $\Re^n$.

(b) $L_C = L_{\mathrm{ri}(C)}$.

(c) For any collection of closed convex sets C_i, $i \in I$, where I is an arbitrary index set and $\cap_{i\in I} C_i \neq \emptyset$, we have

$$L_{\cap_{i\in I} C_i} = \cap_{i\in I} L_{C_i}.$$

(d) Let W be a compact and convex subset of $\Re^m$, and let A be an $m \times n$ matrix. The lineality space of the set

$$V = \{x \in C \mid Ax \in W\}$$

(assuming it is nonempty) is $L_C \cap N(A)$, where $N(A)$ is the nullspace of A.

Proposition 1.4.4: (Decomposition of a Convex Set) Let C be a nonempty convex subset of $\Re^n$. Then, for every subspace S that is contained in the lineality space L_C, we have

$$C = S + (C \cap S^\perp).$$

The notion of direction of recession of a convex function f can be described in terms of its epigraph via the following proposition.

Proposition 1.4.5: Let $f : \Re^n \mapsto (-\infty, \infty]$ be a closed proper convex function and consider the level sets

$$V_\gamma = \{x \mid f(x) \le \gamma\}, \qquad \gamma \in \Re.$$

Then:

(a) All the nonempty level sets V_γ have the same recession cone, denoted R_f, and given by

$$R_f = \{d \mid (d, 0) \in R_{\text{epi}(f)}\},$$

where $R_{\text{epi}(f)}$ is the recession cone of the epigraph of f.

(b) If one nonempty level set V_γ is compact, then all of these level sets are compact.

For a closed proper convex function $f : \Re^n \mapsto (-\infty, \infty]$, the (common) recession cone R_f of the nonempty level sets is called the *recession cone of* f. A vector $d \in R_f$ is called a *direction of recession of* f. The *recession function of* f, denoted r_f, is the closed proper convex function whose epigraph is R_f.

The lineality space of the recession cone R_f of a closed proper convex function f is denoted by L_f, and is the subspace of all $d \in \Re^n$ such that both d and $-d$ are directions of recession of f, i.e.,

$$L_f = R_f \cap (-R_f).$$

We have that $d \in L_f$ if and only if

$$f(x + \alpha d) = f(x), \qquad \forall\, x \in \text{dom}(f),\ \forall\, \alpha \in \Re.$$

Consequently, any $d \in L_f$ is called a *direction in which f is constant*, and L_f is called the *constancy space of* f.

Proposition 1.4.6: Let $f : \Re^n \mapsto (-\infty, \infty]$ be a closed proper convex function. Then the recession cone and constancy space of f are given in terms of its recession function by

$$R_f = \{d \mid r_f(d) \le 0\}, \qquad L_f = \{d \mid r_f(d) = r_f(-d) = 0\}.$$

Proposition 1.4.7: Let $f : \Re^n \mapsto (-\infty, \infty]$ be a closed proper convex function. Then, for all $x \in \text{dom}(f)$ and $d \in \Re^n$,

$$r_f(d) = \sup_{\alpha > 0} \frac{f(x + \alpha d) - f(x)}{\alpha} = \lim_{\alpha \to \infty} \frac{f(x + \alpha d) - f(x)}{\alpha}.$$

Proposition 1.4.8: (Recession Function of a Sum) Let $f_i : \Re^n \mapsto (-\infty, \infty]$, $i = 1, \ldots, m$, be closed proper convex functions such that the function $f = f_1 + \cdots + f_m$ is proper. Then

$$r_f(d) = r_{f_1}(d) + \cdots + r_{f_m}(d), \qquad \forall\, d \in \Re^n.$$

Nonemptiness of Set Intersections

Let $\{C_k\}$ be a sequence of nonempty closed sets in $\Re^n$ with $C_{k+1} \subset C_k$ for all k (such a sequence is said to be *nested*). We are concerned with the question whether $\cap_{k=0}^{\infty} C_k$ is nonempty.

Definition 1.4.1: Let $\{C_k\}$ be a nested sequence of nonempty closed convex sets. We say that $\{x_k\}$ is an *asymptotic sequence of* $\{C_k\}$ if $x_k \neq 0$, $x_k \in C_k$ for all k, and

$$\|x_k\| \to \infty, \qquad \frac{x_k}{\|x_k\|} \to \frac{d}{\|d\|},$$

where d is some nonzero common direction of recession of the sets C_k,

$$d \neq 0, \qquad d \in \cap_{k=0}^{\infty} R_{C_k}.$$

A special case is when all the sets C_k are equal. In particular, for a nonempty closed convex set C, and a sequence $\{x_k\} \subset C$, we say that $\{x_k\}$ is an asymptotic sequence of C if $\{x_k\}$ is asymptotic (as per the preceding definition) for the sequence $\{C_k\}$, where $C_k \equiv C$.

Given any unbounded sequence $\{x_k\}$ such that $x_k \in C_k$ for each k, there exists a subsequence $\{x_k\}_{k\in\mathcal{K}}$ that is asymptotic for the corresponding subsequence $\{C_k\}_{k\in\mathcal{K}}$. In fact, any limit point of $\{x_k/\|x_k\|\}$ is a common direction of recession of the sets C_k.

Definition 1.4.2: Let $\{C_k\}$ be a nested sequence of nonempty closed convex sets. We say that *an asymptotic sequence* $\{x_k\}$ *is retractive* if for the direction d corresponding to $\{x_k\}$ as per Definition 1.4.1, there exists an index $\overline{k}$ such that

$$x_k - d \in C_k, \qquad \forall\ k \geq \overline{k}.$$

We say that *the sequence* $\{C_k\}$ *is retractive* if all its asymptotic sequences are retractive. In the special case $C_k \equiv C$, we say that *the set* C *is retractive* if all its asymptotic sequences are retractive.

A closed halfspace is retractive. Intersections and Cartesian products, involving a finite number of sets, preserve retractiveness. In particular, if $\{C_k^1\}, \ldots, \{C_k^r\}$ are retractive nested sequences of nonempty closed convex sets, the sequences $\{N_k\}$ and $\{T_k\}$ are retractive, where

$$N_k = C_k^1 \cap C_k^2 \cap \cdots \cap C_k^r, \qquad T_k = C_k^1 \times C_k^2 \times \cdots \times C_k^r, \quad \forall\ k,$$

and we assume that all the sets N_k are nonempty. A simple consequence is that a polyhedral set is retractive, since it is the nonempty intersection of a finite number of closed halfspaces.

Proposition 1.4.9: A polyhedral set is retractive.

The importance of retractive sequences is motivated by the following proposition.

Proposition 1.4.10: A retractive nested sequence of nonempty closed convex sets has nonempty intersection.

Proposition 1.4.11: Let $\{C_k\}$ be a nested sequence of nonempty closed convex sets. Denote

$$R = \cap_{k=0}^{\infty} R_{C_k}, \qquad L = \cap_{k=0}^{\infty} L_{C_k}.$$

(a) If $R = L$, then $\{C_k\}$ is retractive, and $\cap_{k=0}^{\infty} C_k$ is nonempty. Furthermore,

$$\cap_{k=0}^{\infty} C_k = L + \tilde{C},$$

where $\tilde{C}$ is some nonempty and compact set.

(b) Let X be a retractive closed convex set. Assume that all the sets $\overline{C}_k = X \cap C_k$ are nonempty, and that

$$R_X \cap R \subset L.$$

Then, $\{\overline{C}_k\}$ is retractive, and $\cap_{k=0}^{\infty} \overline{C}_k$ is nonempty.

Proposition 1.4.12: (Existence of Solutions of Convex Quadratic Programs) Let Q be a symmetric positive semidefinite $n \times n$ matrix, let c and $a_1, \ldots, a_r$ be vectors in $\Re^n$, and let $b_1, \ldots, b_r$ be scalars. Assume that the optimal value of the problem

$$\begin{aligned} &\text{minimize} \quad x'Qx + c'x \\ &\text{subject to} \quad a_j'x \le b_j, \quad j = 1, \ldots, r, \end{aligned}$$

is finite. Then the problem has at least one optimal solution.

Closedness under Linear Transformation and Vector Sum

The conditions of Prop. 1.4.11 can be translated to conditions guaranteeing the closedness of the image, AC, of a closed convex set C under a linear transformation A.

Proposition 1.4.13: Let X and C be nonempty closed convex sets in $\Re^n$, and let A be an $m \times n$ matrix with nullspace denoted by $N(A)$. If X is a retractive closed convex set and

$$R_X \cap R_C \cap N(A) \subset L_C,$$

then $A(X \cap C)$ is a closed set.

A special case relates to vector sums.

Proposition 1.4.14: Let $C_1, \ldots, C_m$ be nonempty closed convex subsets of $\Re^n$ such that the equality $d_1 + \cdots + d_m = 0$ for some vectors $d_i \in R_{C_i}$ implies that $d_i \in L_{C_i}$ for all $i = 1, \ldots, m$. Then $C_1 + \cdots + C_m$ is a closed set.

When specialized to just two sets, the above proposition implies that if C_1 and $-C_2$ are closed convex sets, then $C_1 - C_2$ is closed if there is no common nonzero direction of recession of C_1 and C_2, i.e.

$$R_{C_1} \cap R_{C_2} = \{0\}.$$

This is true in particular if either C_1 or C_2 is bounded, in which case either $R_{C_1} = \{0\}$ or $R_{C_2} = \{0\}$, respectively. For an example of two unbounded closed convex sets in the plane whose vector sum is not closed, let

$$C_1 = \{(x_1, x_2) \mid x_1 x_2 \geq 0\}, \qquad C_2 = \{(x_1, x_2) \mid x_1 = 0\}.$$

Some other conditions asserting the closedness of vector sums can be derived from Prop. 1.4.13. For example, we can show that the vector sum of a finite number of polyhedral sets is closed, since it can be viewed as the image of their Cartesian product (clearly a polyhedral set) under a linear transformation. Another useful result is that if X is a polyhedral set, and C is a closed convex set, then $X + C$ is closed if every direction of recession of X whose opposite is a direction of recession of C lies also in the lineality space of C. In particular, $X + C$ is closed if X is polyhedral, and C is closed.

Section 1.5. Hyperplanes

A *hyperplane* in $\Re^n$ is a set of the form

$$\{x \mid a'x = b\},$$

where a is nonzero vector in $\Re^n$ (called the *normal* of the hyperplane), and b is a scalar. The sets

$$\{x \mid a'x \geq b\}, \qquad \{x \mid a'x \leq b\},$$

are called the *closed halfspaces* associated with the hyperplane (also referred to as the *positive and negative halfspaces*, respectively). The sets

$$\{x \mid a'x > b\}, \qquad \{x \mid a'x < b\},$$

are called the *open halfspaces* associated with the hyperplane.

Proposition 1.5.1: (Supporting Hyperplane Theorem) Let C be a nonempty convex subset of $\Re^n$ and let $\overline{x}$ be a vector in $\Re^n$. If $\overline{x}$ is not an interior point of C, there exists a hyperplane that passes through $\overline{x}$ and contains C in one of its closed halfspaces, i.e., there exists a vector $a \neq 0$ such that

$$a'\overline{x} \leq a'x, \qquad \forall\, x \in C.$$

Proposition 1.5.2: (Separating Hyperplane Theorem) Let C_1 and C_2 be two nonempty convex subsets of $\Re^n$. If C_1 and C_2 are disjoint, there exists a hyperplane that separates C_1 and C_2, i.e., there exists a vector $a \neq 0$ such that

$$a'x_1 \leq a'x_2, \qquad \forall\, x_1 \in C_1,\ \forall\, x_2 \in C_2.$$

Proposition 1.5.3: (Strict Separation Theorem) Let C_1 and C_2 be two disjoint nonempty convex sets. Then under any one of the following five conditions, there exists a hyperplane that strictly separates C_1 and C_2, i.e., a vector $a \neq 0$ and a scalar b such that

$$a'x_1 < b < a'x_2, \qquad \forall\, x_1 \in C_1,\ \forall\, x_2 \in C_2.$$

(1) $C_2 - C_1$ is closed.

(2) C_1 is closed and C_2 is compact.

(3) C_1 and C_2 are polyhedral.

(4) C_1 and C_2 are closed, and

$$R_{C_1} \cap R_{C_2} = L_{C_1} \cap L_{C_2},$$

where R_{C_i} and L_{C_i} denote the recession cone and the lineality space of C_i, $i = 1, 2$.

(5) C_1 is closed, C_2 is polyhedral, and $R_{C_1} \cap R_{C_2} \subset L_{C_1}$.

Proposition 1.5.4: The closure of the convex hull of a set C is the intersection of the closed halfspaces that contain C. In particular, a closed convex set is the intersection of the closed halfspaces that contain it.

Let C_1 and C_2 be two subsets of $\Re^n$. We say that a hyperplane *properly separates* C_1 *and* C_2 if it separates C_1 and C_2, and does not fully contain both C_1 and C_2. If C is a subset of $\Re^n$ and $\overline{x}$ is a vector in $\Re^n$, we say that a hyperplane *properly separates* C *and* $\overline{x}$ if it properly separates

C and the singleton set $\{\overline{x}\}$.

Proposition 1.5.5: (Proper Separation Theorem) Let C be a nonempty convex subset of $\Re^n$ and let $\overline{x}$ be a vector in $\Re^n$. There exists a hyperplane that properly separates C and $\overline{x}$ if and only if $\overline{x} \notin \mathrm{ri}(C)$.

Proposition 1.5.6: (Proper Separation of Two Convex Sets) Let C_1 and C_2 be two nonempty convex subsets of $\Re^n$. There exists a hyperplane that properly separates C_1 and C_2 if and only if

$$\mathrm{ri}(C_1) \cap \mathrm{ri}(C_2) = \emptyset.$$

Proposition 1.5.7: (Polyhedral Proper Separation Theorem) Let C and P be two nonempty convex subsets of $\Re^n$ such that P is polyhedral. There exists a hyperplane that separates C and P, and does not contain C if and only if

$$\mathrm{ri}(C) \cap P = \emptyset.$$

Consider a hyperplane in $\Re^{n+1}$ with a normal of the form (μ, β), where $\mu \in \Re^n$ and $\beta \in \Re$. We say that such a hyperplane is *vertical* if $\beta = 0$, and *nonvertical* if $\beta \neq 0$.

Proposition 1.5.8: (Nonvertical Hyperplane Theorem) Let C be a nonempty convex subset of $\Re^{n+1}$ that contains no vertical lines. Let the vectors in $\Re^{n+1}$ be denoted by (u, w), where $u \in \Re^n$ and $w \in \Re$. Then:

(a) C is contained in a closed halfspace corresponding to a nonvertical hyperplane, i.e., there exist a vector $\mu \in \Re^n$, a scalar $\beta \neq 0$, and a scalar γ such that

$$\mu' u + \beta w \geq \gamma, \qquad \forall\ (u, w) \in C.$$

(b) If $(\overline{u}, \overline{w})$ does not belong to $\text{cl}(C)$, there exists a nonvertical hyperplane strictly separating $(\overline{u}, \overline{w})$ and C.

Section 1.6. Conjugate Functions

Consider an extended real-valued function $f : \Re^n \mapsto [-\infty, \infty]$. The *conjugate function* of f is the function $f^\star : \Re^n \mapsto [-\infty, \infty]$ defined by

$$f^\star(y) = \sup_{x \in \Re^n} \big\{x'y - f(x)\big\}, \qquad y \in \Re^n. \tag{B.2}$$

Proposition 1.6.1: (Conjugacy Theorem) Let $f : \Re^n \mapsto [-\infty, \infty]$ be a function, let $f^\star$ be its conjugate, and consider the double conjugate $f^{\star\star} = (f^\star)^\star$. Then:

(a) We have

$$f(x) \geq f^{\star\star}(x), \qquad \forall\ x \in \Re^n.$$

(b) If f is convex, then properness of any one of the functions f, $f^\star$, and $f^{\star\star}$ implies properness of the other two.

(c) If f is closed proper convex, then

$$f(x) = f^{\star\star}(x), \qquad \forall\ x \in \Re^n.$$

(d) The conjugates of f and its convex closure $\check{\text{cl}}\, f$ are equal. Furthermore, if $\check{\text{cl}}\, f$ is proper, then

$$(\check{\text{cl}}\, f)(x) = f^{\star\star}(x), \qquad \forall\ x \in \Re^n.$$

Positively Homogeneous Functions and Support Functions

Given a nonempty set X, consider the *indicator function* of X, defined by

$$\delta_X(x) = \begin{cases} 0 & \text{if } x \in X, \\ \infty & \text{if } x \notin X. \end{cases}$$

The conjugate of δ_X is given by

$$\sigma_X(y) = \sup_{x \in X} y'x$$

and is called the *support function of* X.

Let C be a convex cone. The conjugate of its indicator function δ_C is its support function,

$$\sigma_C(y) = \sup_{x \in C} y'x.$$

The support/conjugate function σ_C is the indicator function δ_{C^*} of the cone

$$C^* = \{y \mid y'x \leq 0, \ \forall \ x \in C\},$$

called the *polar cone of* C. By the Conjugacy Theorem [Prop. 1.6.1(d)], the polar cone of C^* is $\text{cl}(C)$. In particular, if C is closed, the polar of its polar is equal to the original. This is a special case of the *Polar Cone Theorem*, given in Section 2.2.

A function $f : \Re^n \mapsto [-\infty, \infty]$ is called *positively homogeneous* if its epigraph is a cone in $\Re^{n+1}$. Equivalently, f is positively homogeneous if and only if

$$f(\gamma x) = \gamma f(x), \qquad \forall \ \gamma > 0, \ \forall \ x \in \Re^n.$$

Positively homogeneous functions are closely connected with support functions. Clearly, the support function σ_X of a set X is closed convex and positively homogeneous. Moreover, if $\sigma : \Re^n \mapsto (-\infty, \infty]$ is a proper convex positively homogeneous function, then we claim that the conjugate of σ is the indicator function of the closed convex set

$$X = \big\{x \mid y'x \leq \sigma(y), \ \forall \ y \in \Re^n\big\},$$

and that $\text{cl}\,\sigma$ is the support function of X. For a proof, let δ be the conjugate of σ:

$$\delta(x) = \sup_{y \in \Re^n} \big\{y'x - \sigma(y)\big\}.$$

Since σ is positively homogeneous, we have for any $\gamma > 0$,

$$\gamma\, \delta(x) = \sup_{y \in \Re^n} \big\{\gamma y'x - \gamma\, \sigma(y)\big\} = \sup_{y \in \Re^n} \big\{(\gamma y)'x - \sigma(\gamma y)\big\}.$$

The right-hand sides of the preceding two relations are equal, so we obtain

$$\delta(x) = \gamma\, \delta(x), \qquad \forall \ \gamma > 0,$$

which implies that δ takes only the values 0 and ∞ (since σ and hence also its conjugate δ is proper). Thus, δ is the indicator function of a set, call it X, and we have

$$\begin{aligned} X &= \big\{x \mid \delta(x) \leq 0\big\} \\ &= \left\{x \;\Big|\; \sup_{y \in \Re^n} \big\{y'x - \sigma(y)\big\} \leq 0\right\} \\ &= \big\{x \mid y'x \leq \sigma(y), \ \forall \ y \in \Re^n\big\}. \end{aligned}$$

Finally, since δ is the conjugate of σ, we see that $\text{cl}\,\sigma$ is the conjugate of δ; cf. the Conjugacy Theorem [Prop. 1.6.1(c)]. Since δ is the indicator function of X, it follows that $\text{cl}\,\sigma$ is the support function of X.

We now discuss a characterization of the support function of the 0-level set of a closed proper convex function $f : \Re^n \mapsto (-\infty, \infty]$. The closure of the cone generated by $\text{epi}(f)$, is the epigraph of a closed convex positively homogeneous function, called the *closed function generated by* f, and denoted by $\text{gen}\, f$. The epigraph of $\text{gen}\, f$ is the intersection of all the closed cones that contain $\text{epi}(f)$. Moreover, if $\text{gen}\, f$ is proper, then $\text{epi}(\text{gen}\, f)$ is the intersection of all the halfspaces that contain $\text{epi}(f)$ and contain 0 in their boundary.

Consider the conjugate $f^\star$ of a closed proper convex function $f : \Re^n \mapsto (-\infty, \infty]$. We claim that if the level set $\{y \mid f^\star(y) \le 0\}$ [or the level set $\{x \mid f(x) \le 0\}$] is nonempty, its support function is $\text{gen}\, f$ (or respectively $\text{gen}\, f^\star$). Indeed, if the level set $\{y \mid f^\star(y) \le 0\}$ is nonempty, any y such that $f^\star(y) \le 0$, or equivalently $y'x \le f(x)$ for all x, defines a nonvertical hyperplane that separates the origin from $\text{epi}(f)$, implying that the epigraph of $\text{gen}\, f$ does not contain a line, so $\text{gen}\, f$ is proper. Since $\text{gen}\, f$ is also closed, convex, and positively homogeneous, by our earlier analysis it follows that $\text{gen}\, f$ is the support function of the set

$$Y = \{y \mid y'x \le (\text{gen}\, f)(x),\ \forall\, x \in \Re^n\}.$$

Since $\text{epi}(\text{gen}\, f)$ is the intersection of all the halfspaces that contain $\text{epi}(f)$ and contain 0 in their boundary, the set Y can be written as

$$Y = \{y \mid y'x \le f(x),\ \forall\, x \in \Re^n\} = \left\{ y \;\middle|\; \sup_{x \in \Re^n} \{y'x - f(x)\} \le 0 \right\}.$$

We thus obtain that $\text{gen}\, f$ is the support function of the set

$$Y = \{y \mid f^\star(y) \le 0\},$$

assuming this set is nonempty.

Note that the method used to characterize the 0-level sets of f and $f^\star$ can be applied to any level set. In particular, a nonempty level set $L_\gamma = \{x \mid f(x) \le \gamma\}$ is the 0-level set of the function f_γ defined by $f_\gamma(x) = f(x) - \gamma$, and its support function is the closed function generated by $f_\gamma^\star$, the conjugate of f_γ, which is given by $f_\gamma^\star(y) = f^\star(y) + \gamma$.

CHAPTER 2: Basic Concepts of Polyhedral Convexity

Section 2.1. Extreme Points

In this chapter, we discuss polyhedral sets, i.e., nonempty sets specified by systems of a finite number of affine inequalities

$$a_j'x \le b_j, \qquad j = 1, \ldots, r,$$

where $a_1, \ldots, a_r$ are vectors in $\Re^n$, and $b_1, \ldots, b_r$ are scalars.

Given a nonempty convex set C, a vector $x \in C$ is said to be an *extreme point* of C if it does not lie strictly between the endpoints of any line segment contained in the set, i.e., if there do not exist vectors $y \in C$ and $z \in C$, with $y \neq x$ and $z \neq x$, and a scalar $\alpha \in (0, 1)$ such that $x = \alpha y + (1 - \alpha)z$.

Proposition 2.1.1: Let C be a convex subset of $\Re^n$, and let H be a hyperplane that contains C in one of its closed halfspaces. Then the extreme points of $C \cap H$ are precisely the extreme points of C that belong to H.

Proposition 2.1.2: A nonempty closed convex subset of $\Re^n$ has at least one extreme point if and only if it does not contain a line, i.e., a set of the form $\{x + \alpha d \mid \alpha \in \Re\}$, where x and d are vectors in $\Re^n$ with $d \neq 0$.

Proposition 2.1.3: Let C be a nonempty closed convex subset of $\Re^n$. Assume that for some $m \times n$ matrix A of rank n and some $b \in \Re^m$, we have

$$Ax \geq b, \qquad \forall\, x \in C.$$

Then C has at least one extreme point.

Proposition 2.1.4: Let P be a polyhedral subset of $\Re^n$.

(a) If P has the form

$$P = \{x \mid a_j'x \leq b_j,\ j = 1, \ldots, r\},$$

where $a_j \in \Re^n$, $b_j \in \Re$, $j = 1, \ldots, r$, then a vector $v \in P$ is an extreme point of P if and only if the set

$$A_v = \{a_j \mid a_j'v = b_j,\ j \in \{1, \ldots, r\}\}$$

contains n linearly independent vectors.

(b) If P has the form

$$P = \{x \mid Ax = b,\ x \geq 0\},$$

where A is an $m \times n$ matrix and b is a vector in $\Re^m$, then a vector $v \in P$ is an extreme point of P if and only if the columns of A corresponding to the nonzero coordinates of v are linearly independent.

(c) If P has the form

$$P = \{x \mid Ax = b,\ c \leq x \leq d\},$$

where A is an $m \times n$ matrix, b is a vector in $\Re^m$, and c, d are vectors in $\Re^n$, then a vector $v \in P$ is an extreme point of P if and only if the columns of A corresponding to the coordinates of v that lie strictly between the corresponding coordinates of c and d are linearly independent.

Proposition 2.1.5: A polyhedral set in $\Re^n$ of the form

$$\{x \mid a_j'x \leq b_j,\ j = 1, \ldots, r\}$$

has an extreme point if and only if the set $\{a_j \mid j = 1, \ldots, r\}$ contains n linearly independent vectors.

Section 2.2. Polar Cones

We return to the notion of *polar cone* of nonempty set C, denoted by C^*, and given by $C^* = \{y \mid y'x \leq 0,\ \forall\ x \in C\}$.

Proposition 2.2.1:

(a) For any nonempty set C, we have

$$C^* = \big(\text{cl}(C)\big)^* = \big(\text{conv}(C)\big)^* = \big(\text{cone}(C)\big)^*.$$

(b) (*Polar Cone Theorem*) For any nonempty cone C, we have

$$(C^*)^* = \mathrm{cl}\big(\mathrm{conv}(C)\big).$$

In particular, if C is closed and convex, we have $(C^*)^* = C$.

Section 2.3. Polyhedral Sets and Functions

We recall that a polyhedral cone $C \subset \Re^n$ is a polyhedral set of the form

$$C = \{x \mid a_j'x \leq 0,\ j = 1, \ldots, r\},$$

where $a_1, \ldots, a_r$ are some vectors in $\Re^n$, and r is a positive integer. We say that a cone $C \subset \Re^n$ is *finitely generated*, if it is generated by a finite set of vectors, i.e., if it has the form

$$C = \mathrm{cone}\big(\{a_1, \ldots, a_r\}\big) = \left\{ x \;\middle|\; x = \sum_{j=1}^{r} \mu_j a_j,\ \mu_j \geq 0,\ j = 1, \ldots, r \right\},$$

where $a_1, \ldots, a_r$ are some vectors in $\Re^n$, and r is a positive integer.

Proposition 2.3.1: (Farkas' Lemma) Let $a_1, \ldots, a_r$ be vectors in $\Re^n$. Then, $\{x \mid a_j'x \leq 0,\ j = 1, \ldots, r\}$ and $\mathrm{cone}\big(\{a_1, \ldots, a_r\}\big)$ are closed cones that are polar to each other.

Proposition 2.3.2: (Minkowski-Weyl Theorem) A cone is polyhedral if and only if it is finitely generated.

Proposition 2.3.3: (Minkowski-Weyl Representation) A set P is polyhedral if and only if there is a nonempty finite set $\{v_1, \ldots, v_m\}$ and a finitely generated cone C such that $P = \mathrm{conv}\big(\{v_1, \ldots, v_m\}\big) + C$, i.e.,

$$P = \left\{ x \,\middle|\, x = \sum_{j=1}^{m} \mu_j v_j + y, \ \sum_{j=1}^{m} \mu_j = 1, \ \mu_j \geq 0, \ j = 1, \ldots, m, \ y \in C \right\}.$$

Proposition 2.3.4: (Algebraic Operations on Polyhedral Sets)

(a) The intersection of polyhedral sets is polyhedral, if it is nonempty.

(b) The Cartesian product of polyhedral sets is polyhedral.

(c) The image of a polyhedral set under a linear transformation is a polyhedral set.

(d) The vector sum of two polyhedral sets is polyhedral.

(e) The inverse image of a polyhedral set under a linear transformation is polyhedral.

We say that a function $f : \Re^n \mapsto (-\infty, \infty]$ is *polyhedral* if its epigraph is a polyhedral set in $\Re^{n+1}$. Note that a polyhedral function f is, by definition, closed, convex, and also proper [since f cannot take the value $-\infty$, and $\text{epi}(f)$ is closed, convex, and nonempty (based on our convention that only nonempty sets can be polyhedral)].

Proposition 2.3.5: Let $f : \Re^n \mapsto (-\infty, \infty]$ be a convex function. Then f is polyhedral if and only if $\text{dom}(f)$ is a polyhedral set and

$$f(x) = \max_{j=1,\ldots,m} \{a_j' x + b_j\}, \qquad \forall \, x \in \text{dom}(f),$$

where a_j are vectors in $\Re^n$, b_j are scalars, and m is a positive integer.

Some common operations on polyhedral functions, such as sum and linear composition preserve their polyhedral character as shown by the following two propositions.

Proposition 2.3.6: The sum of two polyhedral functions f_1 and f_2, such that $\text{dom}(f_1) \cap \text{dom}(f_2) \neq \varnothing$, is a polyhedral function.

Proposition 2.3.7: If A is a matrix and g is a polyhedral function such that $\text{dom}(g)$ contains a point in the range of A, the function f given by $f(x) = g(Ax)$ is polyhedral.

Section 2.4. Polyhedral Aspects of Optimization

Polyhedral convexity plays a very important role in optimization. The following are two basic results related to linear programming, the minimization of a linear function over a polyhedral set.

Proposition 2.4.1: Let C be a closed convex subset of $\Re^n$ that has at least one extreme point. A concave function $f : C \mapsto \Re$ that attains a minimum over C attains the minimum at some extreme point of C.

Proposition 2.4.2: (Fundamental Theorem of Linear Programming) Let P be a polyhedral set that has at least one extreme point. A linear function that is bounded below over P attains a minimum at some extreme point of P.

CHAPTER 3: Basic Concepts of Convex Optimization

Section 3.1. Constrained Optimization

Let us consider the problem

$$\begin{aligned} &\text{minimize} \quad f(x) \\ &\text{subject to} \quad x \in X, \end{aligned}$$

where $f : \Re^n \mapsto (-\infty, \infty]$ is a function and X is a nonempty subset of $\Re^n$. Any vector $x \in X \cap \text{dom}(f)$ is said to be a *feasible solution* of the problem (we also use the terms *feasible vector* or *feasible point*). If there is at least one feasible solution, i.e., $X \cap \text{dom}(f) \neq \varnothing$, we say that the problem is *feasible*; otherwise we say that the problem is *infeasible*. Thus, when f is extended real-valued, we view only the points in $X \cap \text{dom}(f)$ as candidates for optimality, and we view $\text{dom}(f)$ as an implicit constraint set. Furthermore, feasibility of the problem is equivalent to $\inf_{x \in X} f(x) < \infty$.

We say that a vector x^* is a *minimum of f over X* if

$$x^* \in X \cap \text{dom}(f), \qquad \text{and} \qquad f(x^*) = \inf_{x \in X} f(x).$$

We also call x^* a *minimizing point* or *minimizer* or *global minimum of f over X*. Alternatively, we say that *f attains a minimum over X at x^**, and we indicate this by writing

$$x^* \in \arg\min_{x \in X} f(x).$$

If x^* is known to be the unique minimizer of f over X, with slight abuse of notation, we also occasionally write

$$x^* = \arg\min_{x \in X} f(x).$$

We use similar terminology for maxima.

Given a subset X of $\Re^n$ and a function $f : \Re^n \mapsto (-\infty, \infty]$, we say that a vector x^* is a *local minimum of f over X* if $x^* \in X \cap \text{dom}(f)$ and there exists some $\epsilon > 0$ such that

$$f(x^*) \le f(x), \qquad \forall\, x \in X \text{ with } \|x - x^*\| < \epsilon.$$

A local minimum x^* is said to be *strict* if there is no other local minimum within some open sphere centered at x^*. Local maxima are defined similarly.

Proposition 3.1.1: If X is a convex subset of $\Re^n$ and $f : \Re^n \mapsto (-\infty, \infty]$ is a convex function, then a local minimum of f over X is also a global minimum. If in addition f is strictly convex, then there exists at most one global minimum of f over X.

Section 3.2. Existence of Optimal Solutions

Proposition 3.2.1: (Weierstrass' Theorem) Consider a closed proper function $f : \Re^n \mapsto (-\infty, \infty]$, and assume that any one of the following three conditions holds:

(1) $\text{dom}(f)$ is bounded.

(2) There exists a scalar $\overline{\gamma}$ such that the level set

$$\{x \mid f(x) \le \overline{\gamma}\}$$

is nonempty and bounded.

(3) f is coercive, i.e., if for every sequence $\{x_k\}$ such that $\|x_k\| \to \infty$, we have $\lim_{k\to\infty} f(x_k) = \infty$.

Then the set of minima of f over $\Re^n$ is nonempty and compact.

Proposition 3.2.2: Let X be a closed convex subset of $\Re^n$, and let $f : \Re^n \mapsto (-\infty, \infty]$ be a closed convex function with $X \cap \text{dom}(f) \neq \emptyset$. The set of minima of f over X is nonempty and compact if and only if X and f have no common nonzero direction of recession.

Proposition 3.2.3: (Existence of Solution, Sum of Functions) Let $f_i : \Re^n \mapsto (-\infty, \infty]$, $i = 1, \ldots, m$, be closed proper convex functions such that the function $f = f_1 + \cdots + f_m$ is proper. Assume that the recession function of a single function f_i satisfies $r_{f_i}(d) = \infty$ for all $d \neq 0$. Then the set of minima of f is nonempty and compact.

Section 3.3. Partial Minimization of Convex Functions

Functions obtained by minimizing other functions partially, i.e., with respect to some of their variables, arise prominently in the treatment of duality and minimax theory. It is then useful to be able to deduce properties of the function obtained, such as convexity and closedness, from corresponding properties of the original.

Proposition 3.3.1: Consider a function $F : \Re^{n+m} \mapsto (-\infty, \infty]$ and the function $f : \Re^n \mapsto [-\infty, \infty]$ defined by

$$f(x) = \inf_{z \in \Re^m} F(x, z).$$

Then:

(a) If F is convex, then f is also convex.

(b) We have

$$P\big(\text{epi}(F)\big) \subset \text{epi}(f) \subset \text{cl}\Big(P\big(\text{epi}(F)\big)\Big),$$

where $P(\cdot)$ denotes projection on the space of (x, w), i.e., for any subset S of $\Re^{n+m+1}$, $P(S) = \{(x, w) \mid (x, z, w) \in S\}$.

Proposition 3.3.2: Let $F : \Re^{n+m} \mapsto (-\infty, \infty]$ be a closed proper convex function, and consider the function f given by

$$f(x) = \inf_{z \in \Re^m} F(x, z), \qquad x \in \Re^n.$$

Assume that for some $\overline{x} \in \Re^n$ and $\overline{\gamma} \in \Re$ the set

$$\{z \mid F(\overline{x}, z) \leq \overline{\gamma}\}$$

is nonempty and compact. Then f is closed proper convex. Furthermore, for each $x \in \text{dom}(f)$, the set of minima in the definition of $f(x)$ is nonempty and compact.

Proposition 3.3.3: Let X and Z be nonempty convex sets of $\Re^n$ and $\Re^m$, respectively, let $F : X \times Z \mapsto \Re$ be a closed convex function, and assume that Z is compact. Then the function f given by

$$f(x) = \inf_{z \in Z} F(x, z), \qquad x \in X,$$

is a real-valued convex function over X.

Proposition 3.3.4: Let $F : \Re^{n+m} \mapsto (-\infty, \infty]$ be a closed proper convex function, and consider the function f given by

$$f(x) = \inf_{z \in \Re^m} F(x, z), \qquad x \in \Re^n.$$

Assume that for some $\overline{x} \in \Re^n$ and $\overline{\gamma} \in \Re$ the set

$$\{z \mid F(\overline{x}, z) \leq \overline{\gamma}\}$$

is nonempty and its recession cone is equal to its lineality space. Then

f is closed proper convex. Furthermore, for each $x \in \text{dom}(f)$, the set of minima in the definition of $f(x)$ is nonempty.

Section 3.4. Saddle Point and Minimax Theory

Let us consider a function $\phi : X \times Z \mapsto \Re$, where X and Z are nonempty subsets of $\Re^n$ and $\Re^m$, respectively. An issue of interest is to derive conditions guaranteeing that

$$\sup_{z \in Z} \inf_{x \in X} \phi(x, z) = \inf_{x \in X} \sup_{z \in Z} \phi(x, z), \tag{B.3}$$

and that the infima and the suprema above are attained.

Definition 3.4.1: A pair of vectors $x^* \in X$ and $z^* \in Z$ is called a *saddle point* of ϕ if

$$\phi(x^*, z) \leq \phi(x^*, z^*) \leq \phi(x, z^*), \qquad \forall\, x \in X,\ \forall\, z \in Z.$$

Proposition 3.4.1: A pair (x^*, z^*) is a saddle point of ϕ if and only if the minimax equality (B.3) holds, and x^* is an optimal solution of the problem

$$\begin{aligned} &\text{minimize} && \sup_{z \in Z} \phi(x, z) \\ &\text{subject to} && x \in X, \end{aligned}$$

while z^* is an optimal solution of the problem

$$\begin{aligned} &\text{maximize} && \inf_{x \in X} \phi(x, z) \\ &\text{subject to} && z \in Z. \end{aligned}$$

CHAPTER 4: Geometric Duality Framework

Section 4.1. Min Common/Max Crossing Duality

We introduce a geometric framework for duality analysis, which aims to capture the most essential characteristics of duality in two simple geometrical problems, defined by a nonempty subset M of $\Re^{n+1}$.

(a) *Min Common Point Problem*: Consider all vectors that are common to M and the $(n+1)$st axis. We want to find one whose $(n+1)$st component is minimum.

(b) *Max Crossing Point Problem*: Consider nonvertical hyperplanes that contain M in their corresponding "upper" closed halfspace, i.e., the closed halfspace whose recession cone contains the vertical halfline $\{(0,w) \mid w \geq 0\}$. We want to find the maximum crossing point of the $(n+1)$st axis with such a hyperplane.

We refer to the two problems as the *min common/max crossing (MC/MC) framework*, and we will show that it can be used to develop much of the core theory of convex optimization in a unified way.

Mathematically, the min common problem is

$$\begin{aligned}&\text{minimize} \quad w\\&\text{subject to} \quad (0,w) \in M.\end{aligned}$$

We also refer to this as the *primal problem*, and we denote by w^* its optimal value,

$$w^* = \inf_{(0,w)\in M} w.$$

The max crossing problem is to maximize over all $\mu \in \Re^n$ the maximum crossing level corresponding to μ, i.e.,

$$\begin{aligned}&\text{maximize} \quad \inf_{(u,w)\in M} \{w + \mu' u\}\\&\text{subject to} \quad \mu \in \Re^n.\end{aligned} \tag{B.4}$$

We also refer to this as the *dual problem*, we denote by q^* its optimal value,

$$q^* = \sup_{\mu\in\Re^n} q(\mu),$$

and we refer to $q(\mu)$ as the *crossing* or *dual* function.

Proposition 4.1.1: The dual function q is concave and upper semicontinuous.

The following proposition states that we always have $q^* \leq w^*$; we refer to this as *weak duality*. When $q^* = w^*$, we say that *strong duality holds* or that *there is no duality gap*.

Proposition 4.1.2: (Weak Duality Theorem) We have $q^* \leq w^*$.

The feasible solutions of the max crossing problem are restricted by the horizontal directions of recession of $\overline{M}$. This is the essence of the following proposition.

Proposition 4.1.3: Assume that the set

$$\overline{M} = M + \{(0, w) \mid w \geq 0\}$$

is convex. Then the set of feasible solutions of the max crossing problem, $\{\mu \mid q(\mu) > -\infty\}$, is contained in the cone

$$\{\mu \mid \mu' d \geq 0 \text{ for all } d \text{ with } (d, 0) \in R_{\overline{M}}\},$$

where $R_{\overline{M}}$ is the recession cone of $\overline{M}$.

Section 4.2. Some Special Cases

There are several interesting special cases where the set M is the epigraph of some function. For example, consider the problem of minimizing a function $f : \Re^n \mapsto [-\infty, \infty]$. We introduce a function $F : \Re^{n+r} \mapsto [-\infty, \infty]$ of the pair (x, u), which satisfies

$$f(x) = F(x, 0), \qquad \forall\, x \in \Re^n. \tag{B.5}$$

Let the function $p : \Re^r \mapsto [-\infty, \infty]$ be defined by

$$p(u) = \inf_{x \in \Re^n} F(x, u), \tag{B.6}$$

and consider the MC/MC framework with

$$M = \text{epi}(p).$$

The min common value w^* is the minimal value of f, since

$$w^* = p(0) = \inf_{x \in \Re^n} F(x, 0) = \inf_{x \in \Re^n} f(x).$$

The max crossing problem (B.4) can be written as

$$\begin{aligned} &\text{maximize} \quad q(\mu) \\ &\text{subject to} \quad \mu \in \Re^r, \end{aligned}$$

where the dual function is

$$q(\mu) = \inf_{(u,w) \in M} \{w + \mu' u\} = \inf_{u \in \Re^r} \{p(u) + \mu' u\} = \inf_{(x,u) \in \Re^{n+r}} \{F(x, u) + \mu' u\}. \tag{B.7}$$

Note that from Eq. (B.7), an alternative expression for q is

$$q(\mu) = -\sup_{(x,u)\in\Re^{n+r}} \big\{-\mu' u - F(x,u)\big\} = -F^\star(0,-\mu),$$

where $F^\star$ is the conjugate of F, viewed as a function of (x,u). Since

$$q^* = \sup_{\mu\in\Re^r} q(\mu) = -\inf_{\mu\in\Re^r} F^\star(0,-\mu) = -\inf_{\mu\in\Re^r} F^\star(0,\mu),$$

the strong duality relation $w^* = q^*$ can be written as

$$\inf_{x\in\Re^n} F(x,0) = -\inf_{\mu\in\Re^r} F^\star(0,\mu).$$

Different choices of function F, as in Eqs. (B.5) and (B.6), yield corresponding MC/MC frameworks and dual problems. An example of this type is minimization with inequality constraints:

$$\begin{aligned} &\text{minimize} \quad f(x) \\ &\text{subject to} \quad x\in X, \quad g(x)\le 0, \end{aligned} \tag{B.8}$$

where X is a nonempty subset of $\Re^n$, $f : X \mapsto \Re$ is a given function, and $g(x) = \big(g_1(x),\ldots,g_r(x)\big)$ with $g_j : X \mapsto \Re$ being given functions. We introduce a "perturbed constraint set" of the form

$$C_u = \big\{x\in X \mid g(x)\le u\big\}, \qquad u\in\Re^r, \tag{B.9}$$

and the function

$$F(x,u) = \begin{cases} f(x) & \text{if } x\in C_u, \\ \infty & \text{otherwise,} \end{cases}$$

which satisfies the condition $F(x,0) = f(x)$ for all $x\in C_0$ [cf. Eq. (B.5)].

The function p of Eq. (B.6) is given by

$$p(u) = \inf_{x\in\Re^n} F(x,u) = \inf_{x\in X,\, g(x)\le u} f(x), \tag{B.10}$$

and is known as the *primal function* or *perturbation function*. It captures the essential structure of the constrained minimization problem, relating to duality and other properties, such as sensitivity. Consider now the MC/MC framework corresponding to $M = \text{epi}(p)$. From Eq. (B.7), we obtain with some calculation

$$q(\mu) = \begin{cases} \inf_{x\in X}\{f(x) + \mu' g(x)\} & \text{if } \mu\ge 0, \\ -\infty & \text{otherwise.} \end{cases}$$

The following proposition derives the primal and dual functions in the minimax framework. In this proposition, for a given x, we denote by

$(\hat{\text{cl}}\,\phi)(x,\cdot)$ the concave closure of $\phi(x,\cdot)$ [the smallest concave and upper semicontinuous function that majorizes $\phi(x,\cdot)$].

Proposition 4.2.1: Let X and Z be nonempty subsets of $\Re^n$ and $\Re^m$, respectively, and let $\phi : X \times Z \mapsto \Re$ be a function. Assume that $(-\hat{\text{cl}}\,\phi)(x,\cdot)$ is proper for all $x \in X$, and consider the MC/MC framework corresponding to $M = \text{epi}(p)$, where p is given by

$$p(u) = \inf_{x \in X} \sup_{z \in Z} \big\{ \phi(x,z) - u'z \big\}, \qquad u \in \Re^m.$$

Then the dual function is given by

$$q(\mu) = \inf_{x \in X} (\hat{\text{cl}}\,\phi)(x,\mu), \qquad \forall\, \mu \in \Re^m.$$

Section 4.3. Strong Duality Theorem

The following propositions give general results for strong duality.

Proposition 4.3.1: (MC/MC Strong Duality) Consider the min common and max crossing problems, and assume the following:

(1) Either $w^* < \infty$, or else $w^* = \infty$ and M contains no vertical lines.

(2) The set

$$\overline{M} = M + \big\{(0,w) \mid w \geq 0\big\}$$

is convex.

Then, we have $q^* = w^*$ if and only if for every sequence $\big\{(u_k, w_k)\big\} \subset M$ with $u_k \to 0$, there holds $w^* \leq \liminf_{k \to \infty} w_k$.

Proposition 4.3.2: Consider the MC/MC framework, assuming that $w^* < \infty$.

(a) Let M be closed and convex. Then $q^* = w^*$. Furthermore, the function

$$p(u) = \inf\{w \mid (u, w) \in M\}, \qquad u \in \Re^n,$$

is convex and its epigraph is the set

$$\overline{M} = M + \{(0, w) \mid w \geq 0\}.$$

If in addition $-\infty < w^*$, then p is closed and proper.

(b) q^* is equal to the optimal value of the min common problem corresponding to $\text{cl}\big(\text{conv}(M)\big)$.

(c) If M is of the form

$$M = \tilde{M} + \{(u, 0) \mid u \in C\},$$

where $\tilde{M}$ is a compact set and C is a closed convex set, then q^* is equal to the optimal value of the min common problem corresponding to $\text{conv}(M)$.

Section 4.4. Existence of Dual Optimal Solutions

The following propositions give general results for strong duality, as well existence of dual optimal solutions.

Proposition 4.4.1: (MC/MC Existence of Max Crossing Solutions) Consider the MC/MC framework and assume the following:

(1) $-\infty < w^*$.

(2) The set

$$\overline{M} = M + \{(0, w) \mid w \geq 0\}$$

is convex.

(3) The origin is a relative interior point of the set

$$D = \{u \mid \text{there exists } w \in \Re \text{ with } (u, w) \in \overline{M}\}.$$

Then $q^* = w^*$, and there exists at least one optimal solution of the max crossing problem.

Proposition 4.4.2: Let the assumptions of Prop. 4.4.1 hold. Then Q^*, the set of optimal solutions of the max crossing problem, has the form

$$Q^* = \big(\text{aff}(D)\big)^{\perp} + \tilde{Q},$$

where $\tilde{Q}$ is a nonempty, convex, and compact set. In particular, Q^* is compact if and only if the origin is an interior point of D.

Section 4.5. Duality and Polyhedral Convexity

The following propositions address special cases where the set M has partially polyhedral structure.

Proposition 4.5.1: Consider the MC/MC framework, and assume the following:

(1) $-\infty < w^*$.

(2) The set $\overline{M}$ has the form

$$\overline{M} = \tilde{M} - \{(u, 0) \mid u \in P\},$$

where $\tilde{M}$ and P are convex sets.

(3) Either $\text{ri}(\tilde{D}) \cap \text{ri}(P) \neq \emptyset$, or P is polyhedral and $\text{ri}(\tilde{D}) \cap P \neq \emptyset$, where $\tilde{D}$ is the set given by

$$\tilde{D} = \{u \mid \text{there exists } w \in \Re \text{ with } (u, w) \in \tilde{M}\}.$$

Then $q^* = w^*$, and Q^*, the set of optimal solutions of the max crossing problem, is a nonempty subset of R_P^*, the polar cone of the recession cone of P. Furthermore, Q^* is compact if $\text{int}(\tilde{D}) \cap P \neq \emptyset$.

Proposition 4.5.2: Consider the MC/MC framework, and assume that:

(1) $-\infty < w^*$.

(2) The set $\overline{M}$ is defined in terms of a polyhedral set P, an $r \times n$ matrix A, a vector $b \in \Re^r$, and a convex function $f : \Re^n \mapsto (-\infty, \infty]$ as follows:

$$\overline{M} = \{(u, w) \mid Ax - b - u \in P \text{ for some } (x, w) \in \text{epi}(f)\}.$$

(3) There is a vector $\overline{x} \in \text{ri}(\text{dom}(f))$ such that $A\overline{x} - b \in P$.

Then $q^* = w^*$ and Q^*, the set of optimal solutions of the max crossing problem, is a nonempty subset of R_P^*, the polar cone of the recession cone of P. Furthermore, Q^* is compact if the matrix A has rank r and there is a vector $\overline{x} \in \text{int}(\text{dom}(f))$ such that $A\overline{x} - b \in P$.

CHAPTER 5: Duality and Optimization

Section 5.1. Nonlinear Farkas' Lemma

A nonlinear version of Farkas' Lemma captures the essence of convex programming duality. The lemma involves a nonempty convex set $X \subset \Re^n$, and functions $f : X \mapsto \Re$ and $g_j : X \mapsto \Re$, $j = 1, \ldots, r$. We denote $g(x) = (g_1(x), \ldots, g_r(x))'$, and use the following assumption.

Assumption 5.1: The functions f and g_j, $j = 1, \ldots, r$, are convex, and

$$f(x) \geq 0, \qquad \forall\, x \in X \text{ with } g(x) \leq 0.$$

Proposition 5.1.1: (Nonlinear Farkas' Lemma) Let Assumption 5.1 hold and let Q^* be the subset of $\Re^r$ given by

$$Q^* = \{\mu \mid \mu \geq 0,\ f(x) + \mu' g(x) \geq 0,\ \forall\, x \in X\}.$$

Assume that one of the following two conditions holds:

(1) There exists $\overline{x} \in X$ such that $g_j(\overline{x}) < 0$ for all $j = 1, \ldots, r$.

(2) The functions g_j, $j = 1, \ldots, r$, are affine, and there exists $\overline{x} \in \text{ri}(X)$ such that $g(\overline{x}) \leq 0$.

Then Q^* is nonempty, and under condition (1) it is also compact.

The interior point condition (1) in the above proposition, and other propositions that follow, is known as the *Slater condition*. By selecting f and g_j to be linear, and X to be the entire space in the above proposition, we obtain a version of Farkas' Lemma (cf. Section 2.3) as a special case.

Proposition 5.1.2: (Linear Farkas' Lemma) Let A be an $m \times n$ matrix and c be a vector in $\Re^m$.

(a) The system $Ay = c$, $y \geq 0$ has a solution if and only if

$$A'x \leq 0 \quad \Rightarrow \quad c'x \leq 0.$$

(b) The system $Ay \geq c$ has a solution if and only if

$$A'x = 0,\ x \geq 0 \quad \Rightarrow \quad c'x \leq 0.$$

Section 5.2. Linear Programming Duality

One of the most important results in optimization is the linear programming duality theorem. Consider the problem

$$\begin{aligned} &\text{minimize } c'x \\ &\text{subject to } a_j'x \geq b_j, \quad j = 1, \ldots, r, \end{aligned}$$

where $c \in \Re^n$, $a_j \in \Re^n$, and $b_j \in \Re$, $j = 1, \ldots, r$. In the following proposition, we refer to this as the *primal problem*. We consider the *dual problem*

$$\begin{aligned} &\text{maximize } b'\mu \\ &\text{subject to } \sum_{j=1}^{r} a_j \mu_j = c, \quad \mu \geq 0, \end{aligned}$$

which can be derived from the MC/MC duality framework in Section 4.2. We denote the primal and dual optimal values by f^* and q^*, respectively.

Proposition 5.2.1: (Linear Programming Duality Theorem)

(a) If either f^* or q^* is finite, then $f^* = q^*$ and both the primal and the dual problem have optimal solutions.

(b) If $f^* = -\infty$, then $q^* = -\infty$.

(c) If $q^* = \infty$, then $f^* = \infty$.

Note that the theorem allows the possibility $f^* = \infty$ and $q^* = -\infty$. Another related result is the following necessary and sufficient condition for primal and dual optimality.

Proposition 5.2.2: (Linear Programming Optimality Conditions) A pair of vectors (x^*, μ^*) form a primal and dual optimal solution pair if and only if x^* is primal-feasible, μ^* is dual-feasible, and

$$\mu_j^*(b_j - a_j'x^*) = 0, \qquad \forall\, j = 1, \ldots, r.$$

Section 5.3. Convex Programming Duality

We first focus on the problem

$$\begin{aligned} &\text{minimize} \quad f(x) \\ &\text{subject to} \quad x \in X, \quad g(x) \leq 0, \end{aligned} \tag{B.11}$$

where X is a convex set in $\Re^n$, $g(x) = \big(g_1(x), \ldots, g_r(x)\big)'$, $f : X \mapsto \Re$ and $g_j : X \mapsto \Re$, $j = 1, \ldots, r$, are convex functions. The dual problem is

$$\begin{aligned} &\text{maximize} \quad \inf_{x \in X} L(x, \mu) \\ &\text{subject to} \quad \mu \geq 0, \end{aligned}$$

where L is the Lagrangian function

$$L(x, \mu) = f(x) + \mu' g(x), \qquad x \in X,\ \mu \in \Re^r.$$

For this and other similar problems, we denote the primal and dual optimal values by f^* and q^*, respectively. We always have the weak duality relation $q^* \leq f^*$; cf. Prop. 4.1.2. When strong duality holds, dual optimal solutions are also referred to as *Lagrange multipliers*. The following eight propositions are the main results relating to strong duality in a variety of contexts. They provide conditions (often called *constraint qualifications*), which guarantee that $q^* = f^*$.

Proposition 5.3.1: (Convex Programming Duality – Existence of Dual Optimal Solutions) Consider problem (B.11). Assume that f^* is finite, and that one of the following two conditions holds:

(1) There exists $\overline{x} \in X$ such that $g_j(\overline{x}) < 0$ for all $j = 1, \ldots, r$.

(2) The functions g_j, $j = 1, \ldots, r$, are affine, and there exists $\overline{x} \in \text{ri}(X)$ such that $g(\overline{x}) \leq 0$.

Then $q^* = f^*$ and the set of optimal solutions of the dual problem is nonempty. Under condition (1) this set is also compact.

Proposition 5.3.2: (Optimality Conditions) Consider problem (B.11). There holds $q^* = f^*$, and (x^*, μ^*) are a primal and dual optimal solution pair if and only if x^* is feasible, $\mu^* \geq 0$, and

$$x^* \in \arg\min_{x \in X} L(x, \mu^*), \qquad \mu_j^* g_j(x^*) = 0, \quad j = 1, \ldots, r.$$

The condition $\mu_j^* g_j(x^*) = 0$ is known as *complementary slackness*, and generalizes the corresponding condition for linear programming, given in Prop. 5.2.2. The preceding proposition actually can be proved without the convexity assumptions of X, f, and g, although this fact will not be useful to us.

The analysis for problem (B.11) can be refined by making more specific assumptions regarding available polyhedral structure in the constraint functions and the abstract constraint set X. Here is an extension of problem (B.11) with additional linear equality constraints:

$$\begin{aligned} &\text{minimize} \quad f(x) \\ &\text{subject to} \quad x \in X, \quad g(x) \leq 0, \quad Ax = b, \end{aligned} \tag{B.12}$$

where X is a convex set, $g(x) = \big(g_1(x), \ldots, g_r(x)\big)'$, $f : X \mapsto \Re$ and $g_j : X \mapsto \Re$, $j = 1, \ldots, r$, are convex functions, A is an $m \times n$ matrix, and $b \in \Re^m$. The corresponding Lagrangian function is

$$L(x, \mu, \lambda) = f(x) + \mu' g(x) + \lambda'(Ax - b),$$

and the dual problem is

$$\begin{aligned} &\text{maximize} \quad \inf_{x \in X} L(x, \mu, \lambda) \\ &\text{subject to} \quad \mu \geq 0, \ \lambda \in \Re^m. \end{aligned}$$

In the special case of a problem with just linear equality constraints:

$$\begin{aligned} &\text{minimize} \quad f(x) \\ &\text{subject to} \quad x \in X, \quad Ax = b, \end{aligned} \tag{B.13}$$

the Lagrangian function is

$$L(x, \lambda) = f(x) + \lambda'(Ax - b),$$

and the dual problem is

$$\begin{aligned} &\text{maximize} \quad \inf_{x \in X} L(x, \lambda) \\ &\text{subject to} \quad \lambda \in \Re^m. \end{aligned}$$

Proposition 5.3.3: (Convex Programming – Linear Equality Constraints) Consider problem (B.13).

(a) Assume that f^* is finite and that there exists $\overline{x} \in \text{ri}(X)$ such that $A\overline{x} = b$. Then $f^* = q^*$ and there exists at least one dual optimal solution.

(b) There holds $f^* = q^*$, and (x^*, λ^*) are a primal and dual optimal solution pair if and only if x^* is feasible and

$$x^* \in \arg\min_{x \in X} L(x, \lambda^*).$$

Proposition 5.3.4: (Convex Programming – Linear Equality and Inequality Constraints) Consider problem (B.12).

(a) Assume that f^* is finite, that the functions g_j are linear, and that there exists $\overline{x} \in \text{ri}(X)$ such that $A\overline{x} = b$ and $g(\overline{x}) \leq 0$. Then $q^* = f^*$ and there exists at least one dual optimal solution.

(b) There holds $f^* = q^*$, and (x^*, μ^*, λ^*) are a primal and dual optimal solution pair if and only if x^* is feasible, $\mu^* \geq 0$, and

$$x^* \in \arg\min_{x \in X} L(x, \mu^*, \lambda^*), \qquad \mu_j^* g_j(x^*) = 0, \quad j = 1, \ldots, r.$$

Proposition 5.3.5: (Convex Programming – Linear Equality and Nonlinear Inequality Constraints) Consider problem (B.12). Assume that f^* is finite, that there exists $\overline{x} \in X$ such that $A\overline{x} = b$ and $g(\overline{x}) < 0$, and that there exists $\tilde{x} \in \text{ri}(X)$ such that $A\tilde{x} = b$. Then $q^* = f^*$ and there exists at least one dual optimal solution.

Proposition 5.3.6: (Convex Programming – Mixed Polyhedral and Nonpolyhedral Constraints) Consider problem (B.12), where X is the intersection of a polyhedral set P and a convex set C,

$$X = P \cap C,$$

$g(x) = \big(g_1(x), \ldots, g_r(x)\big)'$, the functions $f : \Re^n \mapsto \Re$ and $g_j : \Re^n \mapsto \Re$, $j = 1, \ldots, r$, are defined over $\Re^n$, A is an $m \times n$ matrix, and $b \in \Re^m$. Assume that f^* is finite and that for some $\overline{r}$ with $1 \leq \overline{r} \leq r$, the functions g_j, $j = 1, \ldots, \overline{r}$, are polyhedral, and the functions f and g_j, $j = \overline{r}+1, \ldots, r$, are convex over C. Assume further that:

(1) There exists a vector $\tilde{x} \in \text{ri}(C)$ in the set

$$\tilde{P} = P \cap \big\{x \mid Ax = b,\ g_j(x) \leq 0,\ j = 1, \ldots, \overline{r}\big\}.$$

(2) There exists $\overline{x} \in \tilde{P} \cap C$ such that $g_j(\overline{x}) < 0$ for all $j = \overline{r}+1, \ldots, r$.

Then $q^* = f^*$ and there exists at least one dual optimal solution.

We will now give a different type of result, which under some compactness assumptions, guarantees strong duality and that there exists an optimal primal solution (even if there may be no dual optimal solution).

Proposition 5.3.7: (Convex Programming Duality – Existence of Primal Optimal Solutions) Assume that problem (B.11) is feasible, that the convex functions f and g_j are closed, and that the function

$$F(x, 0) = \begin{cases} f(x) & \text{if } g(x) \leq 0,\ x \in X, \\ \infty & \text{otherwise,} \end{cases}$$

has compact level sets. Then $f^* = q^*$ and the set of optimal solutions of the primal problem is nonempty and compact.

We now consider another important optimization framework, the problem

$$\begin{aligned} &\text{minimize} \quad f_1(x) + f_2(Ax) \\ &\text{subject to} \quad x \in \Re^n, \end{aligned} \tag{B.14}$$

where A is an $m \times n$ matrix, $f_1 : \Re^n \mapsto (-\infty, \infty]$ and $f_2 : \Re^m \mapsto (-\infty, \infty]$ are closed proper convex functions. We assume that there exists a feasible solution.

Proposition 5.3.8: (Fenchel Duality)

(a) If f^* is finite and $\big(A \cdot \text{ri}(\text{dom}(f_1))\big) \cap \text{ri}\big(\text{dom}(f_2)\big) \neq \emptyset$, then $f^* = q^*$ and there exists at least one dual optimal solution.

(b) There holds $f^* = q^*$, and (x^*, λ^*) is a primal and dual optimal solution pair if and only if

$$x^* \in \arg\min_{x\in\Re^n} \{f_1(x) - x'A'\lambda^*\} \text{ and } Ax^* \in \arg\min_{z\in\Re^m} \{f_2(z) + z'\lambda^*\}. \tag{B.15}$$

An important special case of Fenchel duality involves the problem

$$\begin{aligned} &\text{minimize} \quad f(x) \\ &\text{subject to} \quad x \in C, \end{aligned} \tag{B.16}$$

where $f : \Re^n \mapsto (-\infty, \infty]$ is a closed proper convex function and C is a closed convex cone in $\Re^n$. This is known as a *conic program*, and some of its special cases (semidefinite programming, second order cone programming) have many practical applications.

Proposition 5.3.9: (Conic Duality Theorem) Assume that the optimal value of the primal conic problem (B.16) is finite, and that $\text{ri}\big(\text{dom}(f)\big) \cap \text{ri}(C) \neq \emptyset$. Consider the dual problem

$$\begin{aligned} &\text{minimize} \quad f^\star(\lambda) \\ &\text{subject to} \quad \lambda \in \hat{C}, \end{aligned}$$

where $f^\star$ is the conjugate of f and $\hat{C}$ is the dual cone,

$$\hat{C} = -C^* = \{\lambda \mid \lambda' x \geq 0,\ \forall\, x \in C\}.$$

Then there is no duality gap and the dual problem has an optimal solution.

Section 5.4. Subgradients and Optimality Conditions

Let $f : \Re^n \mapsto (-\infty, \infty]$ be a proper convex function. We say that a vector $g \in \Re^n$ is a *subgradient* of f at a point $x \in \text{dom}(f)$ if

$$f(z) \geq f(x) + g'(z - x), \qquad \forall\, z \in \Re^n. \tag{B.17}$$

The set of all subgradients of f at x is called the *subdifferential of f at x* and is denoted by $\partial f(x)$. By convention, $\partial f(x)$ is considered empty for all $x \notin \text{dom}(f)$. Generally, $\partial f(x)$ is closed and convex, since based on the subgradient inequality (B.17), it is the intersection of a collection of closed halfspaces. Note that we restrict attention to proper functions (subgradients are not useful and make no sense for improper functions).

Proposition 5.4.1: Let $f : \Re^n \mapsto (-\infty, \infty]$ be a proper convex function. For every $x \in \text{ri}\big(\text{dom}(f)\big)$,

$$\partial f(x) = S^\perp + G,$$

where S is the subspace that is parallel to the affine hull of $\text{dom}(f)$, and G is a nonempty convex and compact set. In particular, if $x \in \text{int}\big(\text{dom}(f)\big)$, then $\partial f(x)$ is nonempty and compact.

It follows from the preceding proposition that *if f is real-valued, then $\partial f(x)$ is nonempty and compact for all $x \in \Re^n$*. An important property is that *if f is differentiable at some $x \in \text{int}\big(\text{dom}(f)\big)$, its gradient $\nabla f(x)$ is the unique subgradient at x*. We give a proof of these facts, together with the following proposition, in Section 3.1.

Proposition 5.4.2: (Subdifferential Boundedness and Lipschitz Continuity) Let $f : \Re^n \mapsto \Re$ be a real-valued convex function, and let X be a nonempty bounded subset of $\Re^n$.

(a) The set $\cup_{x \in X} \partial f(x)$ is nonempty and bounded.

(b) The function f is Lipschitz continuous over X, i.e., there exists a scalar L such that

$$\big|f(x) - f(z)\big| \leq L\,\|x - z\|, \qquad \forall\, x, z \in X.$$

Section 5.4.1. Subgradients of Conjugate Functions

We will now derive an important relation between the subdifferentials of a proper convex function $f : \Re^n \mapsto (-\infty, \infty]$ and its conjugate $f^\star$. Using the definition of conjugacy, we have

$$x'y \leq f(x) + f^\star(y), \qquad \forall\, x \in \Re^n,\ y \in \Re^n.$$

This is known as the *Fenchel inequality*. A pair (x, y) satisfies this inequality as an equation if and only if x attains the supremum in the definition

$$f^\star(y) = \sup_{z \in \Re^n} \big\{y'z - f(z)\big\}.$$

Pairs of this type are connected with the subdifferentials of f and $f^\star$, as shown in the following proposition.

Proposition 5.4.3: (Conjugate Subgradient Theorem) Let $f : \Re^n \mapsto (-\infty, \infty]$ be a proper convex function and let $f^\star$ be its conjugate. The following two relations are equivalent for a pair of vectors (x, y):

(i) $x'y = f(x) + f^\star(y)$.

(ii) $y \in \partial f(x)$.

If in addition f is closed, the relations (i) and (ii) are equivalent to

(iii) $x \in \partial f^\star(y)$.

For an application of the Conjugate Subgradient Theorem, note that the necessary and sufficient optimality condition (B.15) in the Fenchel Duality Theorem can be equivalently written as

$$A'\lambda^* \in \partial f_1(x^*), \qquad \lambda^* \in -\partial f_2(Ax^*).$$

The following proposition gives some useful corollaries of the Conjugate Subgradient Theorem:

Proposition 5.4.4: Let $f : \Re^n \mapsto (-\infty, \infty]$ be a closed proper convex function and let $f^\star$ be its conjugate.

(a) $f^\star$ is differentiable at a vector $y \in \operatorname{int}\big(\operatorname{dom}(f^\star)\big)$ if and only if the supremum of $x'y - f(x)$ over $x \in \Re^n$ is uniquely attained.

(b) The set of minima of f is given by

$$\arg\min_{x \in \Re^n} f(x) = \partial f^\star(0).$$

Section 5.4.2. Subdifferential Calculus

We now generalize some of the basic theorems of ordinary differentiation (Section 3.1 gives proofs for the case of real-valued functions).

Proposition 5.4.5: (Chain Rule) Let $f : \Re^m \mapsto (-\infty, \infty]$ be a convex function, let A be an $m \times n$ matrix, and assume that the function F given by

$$F(x) = f(Ax)$$

is proper. Then

$$\partial F(x) \supset A'\partial f(Ax), \qquad \forall\, x \in \Re^n.$$

Furthermore, if either f is polyhedral or else the range of A contains a point in the relative interior of $\text{dom}(f)$, we have

$$\partial F(x) = A'\partial f(Ax), \qquad \forall\, x \in \Re^n.$$

We also have the following proposition, which is a special case of the preceding one [cf. the proof of Prop. 3.1.3(b)].

Proposition 5.4.6: (Subdifferential of Sum of Functions) Let $f_i : \Re^n \mapsto (-\infty, \infty]$, $i = 1, \ldots, m$, be convex functions, and assume that the function $F = f_1 + \cdots + f_m$ is proper. Then

$$\partial F(x) \supset \partial f_1(x) + \cdots + \partial f_m(x), \qquad \forall\, x \in \Re^n.$$

Furthermore, if $\cap_{i=1}^m \text{ri}\big(\text{dom}(f_i)\big) \neq \emptyset$, we have

$$\partial F(x) = \partial f_1(x) + \cdots + \partial f_m(x), \qquad \forall\, x \in \Re^n.$$

More generally, the same is true if for some $\overline{m}$ with $1 \leq \overline{m} \leq m$, the functions f_i, $i = 1, \ldots, \overline{m}$, are polyhedral and

$$\Big(\cap_{i=1}^{\overline{m}} \text{dom}(f_i)\Big) \cap \Big(\cap_{i=\overline{m}+1}^{m} \text{ri}\big(\text{dom}(f_i)\big)\Big) \neq \emptyset.$$

Section 5.4.3. Optimality Conditions

It can be seen from the definition of subgradient that a vector x^* minimizes f over $\Re^n$ if and only if $0 \in \partial f(x^*)$. We give the following generalization of this condition to constrained problems.

Proposition 5.4.7: Let $f : \Re^n \mapsto (-\infty, \infty]$ be a proper convex function, let X be a nonempty convex subset of $\Re^n$, and assume that one of the following four conditions holds:

(1) $\mathrm{ri}\big(\mathrm{dom}(f)\big) \cap \mathrm{ri}(X) \neq \emptyset$.

(2) f is polyhedral and $\mathrm{dom}(f) \cap \mathrm{ri}(X) \neq \emptyset$.

(3) X is polyhedral and $\mathrm{ri}\big(\mathrm{dom}(f)\big) \cap X \neq \emptyset$.

(4) f and X are polyhedral, and $\mathrm{dom}(f) \cap X \neq \emptyset$.

Then, a vector x^* minimizes f over X if and only if there exists $g \in \partial f(x^*)$ such that

$$g'(x - x^*) \geq 0, \qquad \forall\, x \in X. \tag{B.18}$$

The relative interior condition (1) of the preceding proposition is automatically satisfied when f is real-valued [we have $\mathrm{dom}(f) = \Re^n$]; Section 3.1 gives a proof of the proposition for this case. If in addition, f is differentiable, the optimality condition (B.18) reduces to the one of Prop. 1.1.8 of this appendix:

$$\nabla f(x^*)'(x - x^*) \geq 0, \qquad \forall\, x \in X.$$

Section 5.4.4. Directional Derivatives

For a proper convex function $f : \Re^n \mapsto (-\infty, \infty]$, the directional derivative at any $x \in \mathrm{dom}(f)$ in a direction $d \in \Re^n$, is defined by

$$f'(x; d) = \lim_{\alpha \downarrow 0} \frac{f(x + \alpha d) - f(x)}{\alpha}. \tag{B.19}$$

An important fact here is that the ratio in Eq. (B.19) is monotonically nonincreasing as $\alpha \downarrow 0$, so that the limit above is well-defined. To verify this, note that for any $\overline{\alpha} > 0$, the convexity of f implies that for all $\alpha \in (0, \overline{\alpha})$,

$$f(x + \alpha d) \leq \frac{\alpha}{\overline{\alpha}} f(x + \overline{\alpha} d) + \left(1 - \frac{\alpha}{\overline{\alpha}}\right) f(x) = f(x) + \frac{\alpha}{\overline{\alpha}}\big(f(x + \overline{\alpha} d) - f(x)\big),$$

so that

$$\frac{f(x + \alpha d) - f(x)}{\alpha} \leq \frac{f(x + \overline{\alpha} d) - f(x)}{\overline{\alpha}}, \qquad \forall\, \alpha \in (0, \overline{\alpha}). \tag{B.20}$$

Thus the limit in Eq. (B.19) is well-defined (as a real number, or ∞, or $-\infty$) and an alternative definition of $f'(x; d)$ is

$$f'(x; d) = \inf_{\alpha > 0} \frac{f(x + \alpha d) - f(x)}{\alpha}, \qquad d \in \Re^n. \tag{B.21}$$

The directional derivative is related to the support function of the subdifferential $\partial f(x)$, as indicated in the following proposition.

Proposition 5.4.8: (Support Function of the Subdifferential) Let $f : \Re^n \mapsto (-\infty, \infty]$ be a proper convex function, and let $(\text{cl}\, f')(x; \cdot)$ be the closure of the directional derivative $f'(x; \cdot)$.

(a) For all $x \in \text{dom}(f)$ such that $\partial f(x)$ is nonempty, $(\text{cl}\, f')(x; \cdot)$ is the support function of $\partial f(x)$.

(b) For all $x \in \text{ri}\big(\text{dom}(f)\big)$, $f'(x; \cdot)$ is closed and it is the support function of $\partial f(x)$.

Section 5.5. Minimax Theory

We will now provide theorems regarding the validity of the minimax equality and the existence of saddle points. These theorems are obtained by specializing the MC/MC theorems of Chapter 4. We will assume throughout this section the following:

(a) X and Z are nonempty convex subsets of $\Re^n$ and $\Re^m$, respectively.

(b) $\phi : X \times Z \mapsto \Re$ is a function such that $\phi(\cdot, z) : X \mapsto \Re$ is convex and closed for each $z \in Z$, and $-\phi(x, \cdot) : Z \mapsto \Re$ is convex and closed for each $x \in X$.

Proposition 5.5.1: Assume that the function p given by

$$p(u) = \inf_{x \in X} \sup_{z \in Z} \{\phi(x, z) - u'z\}, \qquad u \in \Re^m,$$

satisfies either $p(0) < \infty$, or else $p(0) = \infty$ and $p(u) > -\infty$ for all $u \in \Re^m$. Then

$$\sup_{z \in Z} \inf_{x \in X} \phi(x, z) = \inf_{x \in X} \sup_{z \in Z} \phi(x, z)$$

if and only if p is lower semicontinuous at $u = 0$.

Proposition 5.5.2: Assume that $0 \in \text{ri}\big(\text{dom}(p)\big)$ and $p(0) > -\infty$. Then

$$\sup_{z \in Z} \inf_{x \in X} \phi(x, z) = \inf_{x \in X} \sup_{z \in Z} \phi(x, z),$$

and the supremum over Z in the left-hand side is finite and is attained. Furthermore, the set of $z \in Z$ attaining this supremum is compact if and only if 0 lies in the interior of $\text{dom}(p)$.

Proposition 5.5.3: (Classical Saddle Point Theorem) Let the sets X and Z be compact. Then the set of saddle points of ϕ is nonempty and compact.

To formulate more general saddle point theorems, we consider the convex functions $t : \Re^n \mapsto (-\infty, \infty]$ and $r : \Re^m \mapsto (-\infty, \infty]$ given by

$$t(x) = \begin{cases} \sup_{z \in Z} \phi(x, z) & \text{if } x \in X, \\ \infty & \text{if } x \notin X, \end{cases}$$

and

$$r(z) = \begin{cases} -\inf_{x \in X} \phi(x, z) & \text{if } z \in Z, \\ \infty & \text{if } z \notin Z. \end{cases}$$

Thus, by Prop. 3.4.1, (x^*, z^*) is a saddle point if and only if

$$\sup_{z \in Z} \inf_{x \in X} \phi(x, z) = \inf_{x \in X} \sup_{z \in Z} \phi(x, z),$$

and x^* minimizes t while z^* minimizes r.

The next two propositions provide conditions for the minimax equality to hold. These propositions are used to prove results about nonemptiness and compactness of the set of saddle points.

Proposition 5.5.4: Assume that t is proper and that the level sets $\{x \mid t(x) \leq \gamma\}$, $\gamma \in \Re$, are compact. Then

$$\sup_{z \in Z} \inf_{x \in X} \phi(x, z) = \inf_{x \in X} \sup_{z \in Z} \phi(x, z),$$

and the infimum over X in the right-hand side above is attained at a set of points that is nonempty and compact.

Proposition 5.5.5: Assume that t is proper, and that the recession cone and the constancy space of t are equal. Then

$$\sup_{z\in Z}\inf_{x\in X}\phi(x,z) = \inf_{x\in X}\sup_{z\in Z}\phi(x,z),$$

and the infimum over X in the right-hand side above is attained.

Proposition 5.5.6: Assume that either t is proper or r is proper.

(a) If the level sets $\{x \mid t(x) \leq \gamma\}$ and $\{z \mid r(z) \leq \gamma\}$, $\gamma \in \Re$, of t and r are compact, the set of saddle points of ϕ is nonempty and compact.

(b) If the recession cones of t and r are equal to the constancy spaces of t and r, respectively, the set of saddle points of ϕ is nonempty.

Proposition 5.5.7: (Saddle Point Theorem) The set of saddle points of ϕ is nonempty and compact under any one of the following conditions:

(1) X and Z are compact.

(2) Z is compact, and for some $\overline{z} \in Z$, $\overline{\gamma} \in \Re$, the level set

$$\{x \in X \mid \phi(x,\overline{z}) \leq \overline{\gamma}\}$$

is nonempty and compact.

(3) X is compact, and for some $\overline{x} \in X$, $\overline{\gamma} \in \Re$, the level set

$$\{z \in Z \mid \phi(\overline{x},z) \geq \overline{\gamma}\}$$

is nonempty and compact.

(4) For some $\overline{x} \in X$, $\overline{z} \in Z$, $\overline{\gamma} \in \Re$, the level sets

$$\{x \in X \mid \phi(x,\overline{z}) \leq \overline{\gamma}\}, \qquad \{z \in Z \mid \phi(\overline{x},z) \geq \overline{\gamma}\},$$

are nonempty and compact.

References

[ACH97] Auslender, A., Cominetti, R., and Haddou, M., 1997. "Asymptotic Analysis for Penalty and Barrier Methods in Convex and Linear Programming," Math. of Operations Research, Vol. 22, pp. 43-62.

[ALS14] Abernethy, J., Lee, C., Sinha, A., and Tewari, A., 2014. "Online Linear Optimization via Smoothing," arXiv preprint arXiv:1405.6076.

[AgB14] Agarwal, A., and Bottou, L., 2014. "A Lower Bound for the Optimization of Finite Sums," arXiv preprint arXiv:1410.0723.

[AgD11] Agarwal, A., and Duchi, J. C., 2011. "Distributed Delayed Stochastic Optimization," In Advances in Neural Information Processing Systems (NIPS 2011), pp. 873-881.

[AlG03] Alizadeh, F., and Goldfarb, D., 2003. "Second-Order Cone Programming," Math. Programming, Vol. 95, pp. 3-51.

[AnH13] Andersen, M. S., and Hansen, P. C., 2013. "Generalized Row-Action Methods for Tomographic Imaging," Numerical Algorithms, Vol. 67, pp. 1-24.

[Arm66] Armijo, L., 1966. "Minimization of Functions Having Continuous Partial Derivatives," Pacific J. Math., Vol. 16, pp. 1-3.

[Ash72] Ash, R. B., 1972. Real Analysis and Probability, Academic Press, NY.

[AtV95] Atkinson, D. S., and Vaidya, P. M., 1995. "A Cutting Plane Algorithm for Convex Programming that Uses Analytic Centers," Math. Programming, Vol. 69, pp. 1-44.

[AuE76] Aubin, J. P., and Ekeland, I., 1976. "Estimates of the Duality Gap in Nonconvex Optimization," Math. of Operations Research, Vol. 1, pp. 255- 245.

[AuT03] Auslender, A., and Teboulle, M., 2003. Asymptotic Cones and Functions in Optimization and Variational Inequalities, Springer, NY.

[AuT04] Auslender, A., and Teboulle, M., 2004. "Interior Gradient and Epsilon-Subgradient Descent Methods for Constrained Convex Minimization," Math. of Operations Research, Vol. 29, pp. 1-26.

[Aus76] Auslender, A., 1976. Optimization: Methodes Numeriques, Mason, Paris.

[Aus92] Auslender, A., 1992. "Asymptotic Properties of the Fenchel Dual

Functional and Applications to Decomposition Problems," J. of Optimization Theory and Applications, Vol. 73, pp. 427-449.

[BBC11] Bertsimas, D., Brown, D. B., and Caramanis, C., 2011. "Theory and Applications of Robust Optimization," SIAM Review, Vol. 53, pp. 464-501.

[BBG09] Bordes, A., Bottou, L., and Gallinari, P., 2009. "SGD-QN: Careful Quasi-Newton Stochastic Gradient Descent," J. of Machine Learning Research, Vol. 10, pp. 1737-1754.

[BBL07] Berry, M. W., Browne, M., Langville, A. N., Pauca, V. P., and Plemmons, R. J., 2007. "Algorithms and Applications for Approximate Nonnegative Matrix Factorization," Computational Statistics and Data Analysis, Vol. 52, pp. 155-173.

[BBY12] Borwein, J. M., Burachik, R. S., and Yao, L., 2012. "Conditions for Zero Duality Gap in Convex Programming," arXiv preprint arXiv:-1211.4953.

[BCK06] Bauschke, H. H., Combettes, P. L., and Kruk, S. G., 2006. "Extrapolation Algorithm for Affine-Convex Feasibility Problems," Numer. Algorithms, Vol. 41, pp. 239-274.

[BGI95] Burachik, R., Grana Drummond, L. M., Iusem, A. N., and Svaiter, B. F., 1995. "Full Convergence of the Steepest Descent Method with Inexact Line Searches," Optimization, Vol. 32, pp. 137-146.

[BGL06] Bonnans, J. F., Gilbert, J. C., Lemaréchal, C., and Sagastizábal, S. C., 2006. Numerical Optimization: Theoretical and Practical Aspects, Springer, NY.

[BGN09] Ben-Tal, A., El Ghaoui, L., and Nemirovski, A., 2009. Robust Optimization, Princeton Univ. Press, Princeton, NJ.

[BHG08] Blatt, D., Hero, A. O., Gauchman, H., 2008. "A Convergent Incremental Gradient Method with a Constant Step Size," SIAM J. Optimization, Vol. 18, pp. 29-51.

[BHT87] Bertsekas, D. P., Hossein, P., and Tseng, P., 1987. "Relaxation Methods for Network Flow Problems with Convex Arc Costs," SIAM J. on Control and Optimization, Vol. 25, pp. 1219-1243.

[BJM12] Bach, F., Jenatton, R., Mairal, J., and Obozinski, G., 2012. "Optimization with Sparsity-Inducing Penalties," Foundations and Trends in Machine Learning, Vol. 4, pp. 1-106.

[BKM14] Burachik, R. S., Kaya, C. Y., and Majeed, S. N., 2014. "A Duality Approach for Solving Control-Constrained Linear-Quadratic Optimal Control Problems," SIAM J. on Control and Optimization, Vol. 52, pp. 1423-1456.

[BLY14] Bragin, M. A., Luh, P. B., Yan, J. H., Yu, N., and Stern, G. A., 2014. "Convergence of the Surrogate Lagrangian Relaxation Method," J. of Optimization Theory and Applications, on line.

[BMN01] Ben-Tal, A., Margalit, T., and Nemirovski, A., 2001. "The Ordered Subsets Mirror Descent Optimization Method and its Use for the Positron Emission Tomography Reconstruction," in Inherently Parallel Algorithms in Feasibility and Optimization and their Applications (D. Butnariu, Y. Censor, and S. Reich, eds.), Elsevier, Amsterdam, Netherlands.

[BMR00] Birgin, E. G., Martinez, J. M., and Raydan, M., 2000. "Nonmonotone Spectral Projected Gradient Methods on Convex Sets," SIAM J. on Optimization, Vol. 10, pp. 1196-1211.

[BMS99] Boltyanski, V., Martini, H., and Soltan, V., 1999. Geometric Methods and Optimization Problems, Kluwer, Boston.

[BNO03] Bertsekas, D. P., Nedić, A., and Ozdaglar, A. E., 2003. Convex Analysis and Optimization, Athena Scientific, Belmont, MA.

[BOT06] Bertsekas, D. P., Ozdaglar, A. E., and Tseng, P., 2006 "Enhanced Fritz John Optimality Conditions for Convex Programming," SIAM J. on Optimization, Vol. 16, pp. 766-797.

[BPC11] Boyd, S., Parikh, N., Chu, E., Peleato, B., and Eckstein, J., 2011. Distributed Optimization and Statistical Learning via the Alternating Direction Method of Multipliers, Now Publishers Inc, Boston, MA.

[BPP13] Bhatnagar, S., Prasad, H., and Prashanth, L. A., 2013. Stochastic Recursive Algorithms for Optimization, Lecture Notes in Control and Information Sciences, Springer, NY.

[BPT97a] Bertsekas, D. P., Polymenakos, L. C., and Tseng, P., 1997. "An ϵ-Relaxation Method for Separable Convex Cost Network Flow Problems," SIAM J. on Optimization, Vol. 7, pp. 853-870.

[BPT97b] Bertsekas, D. P., Polymenakos, L. C., and Tseng, P., 1997. "Epsilon-Relaxation and Auction Methods for Separable Convex Cost Network Flow Problems," in Network Optimization, Pardalos, P. M., Hearn, D. W., and Hager, W. W., (Eds.), Lecture Notes in Economics and Mathematical Systems, Springer-Verlag, NY, pp. 103-126.

[BSL14] Bergmann, R., Steidl, G., Laus, F., and Weinmann, A., 2014. "Second Order Differences of Cyclic Data and Applications in Variational Denoising," arXiv preprint arXiv:1405.5349.

[BSS06] Bazaraa, M. S., Sherali, H. D., and Shetty, C. M., 2006. Nonlinear Programming: Theory and Algorithms, 3rd Edition, Wiley, NY.

[BST14] Bolte, J., Sabach, S., and Teboulle, M., 2014. "Proximal Alternating Linearized Minimization for Nonconvex and Nonsmooth Problems," Math. Programming, Vol. 146, pp. 1-36.

[BaB88] Barzilai, J., and Borwein, J. M., 1988. "Two Point Step Size Gradient Methods," IMA J. of Numerical Analysis, Vol. 8, pp. 141-148.

[BaB96] Bauschke, H. H., and Borwein, J. M., 1996. "On Projection Algorithms for Solving Convex Feasibility Problems," SIAM Review, Vol. 38, pp. 367-426.

[BaC11] Bauschke, H. H., and Combettes, P. L., 2011. Convex Analysis and Monotone Operator Theory in Hilbert Spaces, Springer, NY.

[BaM11] Bach, F., and E. Moulines, E., 2011. "Non-Asymptotic Analysis of Stochastic Approximation Algorithms for Machine Learning," Advances in Neural Information Processing Systems (NIPS 2011).

[BaT85] Balas, E., and Toth, P., 1985. "Branch and Bound Methods," in The Traveling Salesman Problem, Lawler, E., Lenstra, J. K., Rinnoy Kan, A. H. G., and Shmoys, D. B., (Eds.), Wiley, NY, pp. 361-401.

[BaW75] Balinski, M., and Wolfe, P., (Eds.), 1975. Nondifferentiable Optimization, Math. Programming Study 3, North-Holland, Amsterdam.

[Bac14] Bacak, M., 2014. "Computing Medians and Means in Hadamard Spaces," arXiv preprint arXiv:1210.2145v3.

[Bau01] Bauschke, H. H., 2001. "Projection Algorithms: Results and Open Problems," in Inherently Parallel Algorithms in Feasibility and Optimization and their Applications (D. Butnariu, Y. Censor, and S. Reich, eds.), Elsevier, Amsterdam, Netherlands.

[BeF12] Becker, S., and Fadili, J., 2012. "A Quasi-Newton Proximal Splitting Method," in Advances in Neural Information Processing Systems (NIPS 2012), pp. 2618-2626.

[BeG83] Bertsekas, D. P., and Gafni, E., 1983. "Projected Newton Methods and Optimization of Multicommodity Flows," IEEE Trans. Automat. Control, Vol. AC-28, pp. 1090-1096.

[BeG92] Bertsekas, D. P., and Gallager, R. G., 1992. Data Networks, 2nd Ed., Prentice-Hall, Englewood Cliffs, NJ.
On line at http://web.mit.edu/dimitrib/www/datanets.html.

[BeL88] Becker, S., and LeCun, Y., 1988. "Improving the Convergence of Back-Propagation Learning with Second Order Methods," in Proceedings of the 1988 Connectionist Models Summer School, San Matteo, CA.

[BeL07] Bengio, Y., and LeCun, Y., 2007. "Scaling Learning Algorithms Towards AI," Large-Scale Kernel Machines, Vol. 34, pp. 1-41.

[BeM71] Bertsekas, D. P, and Mitter, S. K., 1971. "Steepest Descent for Optimization Problems with Nondifferentiable Cost Functionals," Proc. 5th Annual Princeton Confer. Inform. Sci. Systems, Princeton, NJ, pp. 347-351.

[BeM73] Bertsekas, D. P., and Mitter, S. K., 1973. "A Descent Numerical Method for Optimization Problems with Nondifferentiable Cost Functionals," SIAM J. on Control, Vol. 11, pp. 637-652.

[BeN01] Ben-Tal, A., and Nemirovskii, A., 2001. Lectures on Modern Convex Optimization: Analysis, Algorithms, and Engineering Applications, SIAM, Philadelphia, PA.

[BeO02] Bertsekas, D. P., Ozdaglar, A. E., 2002. "Pseudonormality and a Lagrange Multiplier Theory for Constrained Optimization," J. of Optimization Theory and Applications, Vol. 114, pp. 287-343.

[BeS82] Bertsekas, D. P., and Sandell, N. R., 1982. "Estimates of the Duality Gap for Large-Scale Separable Nonconvex Optimization Problems," Proc. 1982 IEEE Conf. Decision and Control, pp. 782-785.

[BeT88] Bertsekas, D. P., and Tseng, P., 1988. "Relaxation Methods for Minimum Cost Ordinary and Generalized Network Flow Problems," Operations Research, Vol. 36, pp. 93-114.

[BeT89a] Bertsekas, D. P., and Tsitsiklis, J. N., 1989. Parallel and Distributed Computation: Numerical Methods, Prentice-Hall, Englewood Cliffs, NJ; republished by Athena Scientific, Belmont, MA, 1997. On line at http://web.mit.edu/dimitrib/www/pdc.html.

[BeT89b] Ben-Tal, A., and Teboulle, M., 1989. "A Smoothing Technique for Nondifferentiable Optimization Problems," Optimization, Lecture Notes in Mathematics, Vol. 1405, pp. 1-11.

[BeT91] Bertsekas, D. P., and Tsitsiklis, J. N., 1991. "Some Aspects of Parallel and Distributed Iterative Algorithms - A Survey," Automatica, Vol. 27, pp. 3-21.

[BeT94a] Bertsekas, D. P., and Tseng, P., 1994. "Partial Proximal Minimization Algorithms for Convex Programming," SIAM J. on Optimization, Vol. 4, pp. 551-572.

[BeT94b] Bertsekas, D. P., and Tseng, P., 1994. "RELAX-IV: A Faster Version of the RELAX Code for Solving Minimum Cost Flow Problems," Massachusetts Institute of Technology, Laboratory for Information and Decision Systems Report LIDS-P-2276, Cambridge, MA.

[BeT96] Bertsekas, D. P., and Tsitsiklis, J. N., 1996. Neuro-Dynamic Programming, Athena Scientific, Belmont, MA.

[BeT97] Bertsimas, D., and Tsitsiklis, J. N., 1997. Introduction to Linear Optimization, Athena Scientific, Belmont, MA.

[BeT00] Bertsekas, D. P., and Tsitsiklis, J. N., 2000. "Gradient Convergence of Gradient Methods with Errors," SIAM J. on Optimization, Vol. 36, pp. 627-642.

[BeT03] Beck, A., and Teboulle, M., 2003. "Mirror Descent and Nonlinear Projected Subgradient Methods for Convex Optimization," Operations Research Letters, Vol. 31, pp. 167-175.

[BeT09a] Beck, A., and Teboulle, M., 2009. "Fast Gradient-Based Algorithms for Constrained Total Variation Image Denoising and Deblurring Problems," IEEE Trans. on Image Processing, Vol. 18, pp. 2419-2434.

[BeT09b] Beck, A., and Teboulle, M., 2009. "A Fast Iterative Shrinkage-Thresholding Algorithm for Linear Inverse Problems," SIAM J. on Imaging Sciences, Vol. 2, pp. 183-202.

[BeT10] Beck, A., and Teboulle, M., 2010. "Gradient-Based Algorithms with Applications to Signal-Recovery Problems," in Convex Optimization in Signal Processing and Communications (Y. Eldar and D. Palomar, eds.), Cambridge Univ. Press, pp. 42-88.

[BeT13] Beck, A., and Tetruashvili, L., 2013. "On the Convergence of Block Coordinate Descent Type Methods," SIAM J. on Optimization, Vol. 23, pp. 2037-2060.

[BeY09] Bertsekas, D. P., and Yu, H., 2009. "Projected Equation Methods for Approximate Solution of Large Linear Systems," J. of Computational and Applied Mathematics, Vol. 227, pp. 27-50.

[BeY10] Bertsekas, D. P., and Yu, H., 2010. "Asynchronous Distributed Policy Iteration in Dynamic Programming," Proc. of Allerton Conf. on Communication, Control and Computing, Allerton Park, Ill, pp. 1368-1374.

[BeY11] Bertsekas, D. P., and Yu, H., 2011. "A Unifying Polyhedral Approximation Framework for Convex Optimization," SIAM J. on Optimization, Vol. 21, pp. 333-360.

[BeZ97] Ben-Tal, A., and Zibulevsky, M., 1997. "Penalty/Barrier Multiplier Methods for Convex Programming Problems," SIAM J. on Optimization, Vol. 7, pp. 347-366.

[Ben09] Bengio, Y., 2009. "Learning Deep Architectures for AI," Foundations and Trends in Machine Learning, Vol. 2, pp. 1-127.

[Ber71] Bertsekas, D. P., 1971. "Control of Uncertain Systems With a Set-Membership Description of the Uncertainty," Ph.D. Thesis, Dept. of EECS, MIT; may be downloaded from
http://web.mit.edu/dimitrib/www/publ.html.

[Ber72] Bertsekas, D. P., 1972. "Stochastic Optimization Problems with Nondifferentiable Cost Functionals with an Application in Stochastic Programming," Proc. 1972 IEEE Conf. Decision and Control, pp. 555-559.

[Ber73] Bertsekas, D. P., 1973. "Stochastic Optimization Problems with Nondifferentiable Cost Functionals," J. of Optimization Theory and Applications, Vol. 12, pp. 218-231.

[Ber75a] Bertsekas, D. P., 1975. "Necessary and Sufficient Conditions for a Penalty Method to be Exact," Math. Programming, Vol. 9, pp. 87-99.

[Ber75b] Bertsekas, D. P., 1975. "Nondifferentiable Optimization via Approximation," Math. Programming Study 3, Balinski, M., and Wolfe, P., (Eds.), North-Holland, Amsterdam, pp. 1-25.

[Ber75c] Bertsekas, D. P., 1975. "Combined Primal-Dual and Penalty Methods for Constrained Optimization," SIAM J. on Control, Vol. 13, pp. 521-544.

[Ber75d] Bertsekas, D. P., 1975. "On the Method of Multipliers for Convex Programming," IEEE Transactions on Aut. Control, Vol. 20, pp. 385-388.

[Ber76a] Bertsekas, D. P., 1976. "On the Goldstein-Levitin-Poljak Gradient Projection Method," IEEE Trans. Automat. Control, Vol. 21, pp. 174-184.

[Ber76b] Bertsekas, D. P., 1976. "Multiplier Methods: A Survey," Automatica, Vol. 12, pp. 133-145.

[Ber76c] Bertsekas, D. P., 1976. "On Penalty and Multiplier Methods for Constrained Optimization," SIAM J. on Control and Optimization, Vol. 14, pp. 216-235.

[Ber77] Bertsekas, D. P., 1977. "Approximation Procedures Based on the Method of Multipliers," J. Optimization Theory and Applications, Vol. 23, pp. 487-510.

[Ber79] Bertsekas, D. P., 1979. "A Distributed Algorithm for the Assignment Problem," Lab. for Information and Decision Systems Working Paper, MIT, Cambridge, MA.

[Ber81] Bertsekas, D. P., 1981. "A New Algorithm for the Assignment Problem," Mathematical Programming, Vol. 21, pp. 152-171.

[Ber82a] Bertsekas, D. P., 1982. Constrained Optimization and Lagrange Multiplier Methods, Academic Press, NY; republished in 1996 by Athena Scientific, Belmont, MA. On line at http://web.mit.edu/dimitrib/www/-lagrmult.html.

[Ber82b] Bertsekas, D. P., 1982. "Projected Newton Methods for Optimization Problems with Simple Constraints," SIAM J. on Control and Optimization, Vol. 20, pp. 221-246.

[Ber82c] Bertsekas, D. P., 1982. "Distributed Dynamic Programming," IEEE Trans. Aut. Control, Vol. AC-27, pp. 610-616.

[Ber83] Bertsekas, D. P., 1983. "Distributed Asynchronous Computation of Fixed Points," Math. Programming, Vol. 27, pp. 107-120.

[Ber85] Bertsekas, D. P., 1985. "A Unified Framework for Primal-Dual Methods in Minimum Cost Network Flow Problems," Mathematical Programming, Vol. 32, pp. 125-145.

[Ber91] Bertsekas, D. P., 1991. Linear Network Optimization: Algorithms and Codes, MIT Press, Cambridge, MA.

[Ber92] Bertsekas, D. P., 1992. "Auction Algorithms for Network Problems: A Tutorial Introduction," Computational Optimization and Applications, Vol. 1, pp. 7-66.

[Ber96] Bertsekas, D. P., 1996. "Incremental Least Squares Methods and the Extended Kalman Filter," SIAM J. on Optimization, Vol. 6, pp. 807-822.

[Ber97] Bertsekas, D. P., 1997. "A New Class of Incremental Gradient Methods for Least Squares Problems," SIAM J. on Optimization, Vol. 7, pp. 913-926.

[Ber98] Bertsekas, D. P., 1998. Network Optimization: Continuous and Discrete Models, Athena Scientific, Belmont, MA.

[Ber99] Bertsekas, D. P., 1999. Nonlinear Programming: 2nd Edition, Athena Scientific, Belmont, MA.

[Ber07] Bertsekas, D. P., 2007. Dynamic Programming and Optimal Control, Vol. I, 3rd Edition, Athena Scientific, Belmont, MA.

[Ber09] Bertsekas, D. P., 2009. Convex Optimization Theory, Athena Scientific, Belmont, MA.

[Ber10a] Bertsekas, D. P., 2010. "Extended Monotropic Programming and Duality," Lab. for Information and Decision Systems Report LIDS-P-2692, MIT, March 2006, Revised in Feb. 2010; a version appeared in J. of Optimization Theory and Applications, Vol. 139, pp. 209-225.

[Ber10b] Bertsekas, D. P., 2010. "Incremental Gradient, Subgradient, and Proximal Methods for Convex Optimization: A Survey," Lab. for Information and Decision Systems Report LIDS-P-2848, MIT.

[Ber11] Bertsekas, D. P., 2011. "Incremental Proximal Methods for Large Scale Convex Optimization," Math. Programming, Vol. 129, pp. 163-195.

[Ber12] Bertsekas, D. P., 2012. Dynamic Programming and Optimal Control, Vol. II, 4th Edition: Approximate Dynamic Programming, Athena Scientific, Belmont, MA.

[Ber13] Bertsekas, D. P., 2013. Abstract Dynamic Programming, Athena Scientific, Belmont, MA.

[BiF07] Bioucas-Dias, J., and Figueiredo, M. A. T., 2007. "A New TwIST: Two-Step Iterative Shrinkage/Thresholding Algorithms for Image Restoration," IEEE Trans. Image Processing, Vol. 16, pp. 2992-3004.

[BiL97] Birge, J. R., and Louveaux, 1997. Introduction to Stochastic Programming, Springer-Verlag, New York, NY.

[Bis95] Bishop, C. M, 1995. Neural Networks for Pattern Recognition, Oxford Univ. Press, NY.

[BoL00] Borwein, J. M., and Lewis, A. S., 2000. Convex Analysis and Nonlinear Optimization, Springer-Verlag, NY.

[BoL05] Bottou, L., and LeCun, Y., 2005. "On-Line Learning for Very Large Datasets," Applied Stochastic Models in Business and Industry, Vol. 21, pp. 137-151.

[BoS00] Bonnans, J. F., and Shapiro, A., 2000. Perturbation Analysis of Optimization Problems, Springer-Verlag, NY.

[BoV04] Boyd, S., and Vanderbergue, L., 2004. Convex Optimization, Cambridge Univ. Press, Cambridge, UK.

[Bor08] Borkar, V. S., 2008. Stochastic Approximation: A Dynamical Systems Viewpoint, Cambridge Univ. Press.

[Bot05] Bottou, L., 2005. "SGD: Stochastic Gradient Descent," http://leon.bottou.org/projects/sgd.

[Bot09] Bottou, L., 2009. "Curiously Fast Convergence of Some Stochastic Gradient Descent Algorithms," unpublished open problem offered to the attendance of the SLDS 2009 conference.

[Bot10] Bottou, L., 2010. "Large-Scale Machine Learning with Stochastic Gradient Descent," In Proc. of COMPSTAT 2010, pp. 177-186.

[BrL78] Brezis, H., and Lions, P. L., 1978. "Produits Infinis de Resolvantes," Israel J. of Mathematics, Vol. 29, pp. 329-345.

[BrS13] Brown, D. B., and Smith, J. E., 2013. "Information Relaxations, Duality, and Convex Stochastic Dynamic Programs," Working Paper, Fuqua School of Business, Durham, NC, USA.

[Bre73] Brezis, H.,1973. "Operateurs Maximaux Monotones et Semi-Groupes de Contractions Dans les Espaces de Hilbert," North-Holland, Amsterdam.

[BuM13] Burachik, R. S., and Majeed, S. N., 2013. "Strong Duality for Generalized Monotropic Programming in Infinite Dimensions," J. of Mathematical Analysis and Applications, Vol. 400, pp. 541-557.

[BuQ98] Burke, J. V., and Qian, M., 1998. "A Variable Metric Proximal Point Algorithm for Monotone Operators," SIAM J. on Control and Optimization, Vol. 37, pp. 353-375.

[Bur91] Burke, J. V., 1991. "An Exact Penalization Viewpoint of Constrained Optimization," SIAM J. on Control and Optimization, Vol. 29, pp. 968-998.

[CDS01] Chen, S. S., Donoho, D. L., and Saunders, M. A., 2001. "Atomic Decomposition by Basis Pursuit," SIAM Review, Vol. 43, pp. 129-159.

[CFM75] Camerini, P. M., Fratta, L., and Maffioli, F., 1975. "On Improving Relaxation Methods by Modified Gradient Techniques," Math. Programming Studies, Vol. 3, pp. 26-34.

[CGT00] Conn, A. R., Gould, N. I., and Toint, P. L., 2000. Trust Region Methods, SIAM, Philadelphia, PA.

[CHY13] Chen, C., He, B., Ye, Y., and Yuan, X., 2013. "The Direct Extension of ADMM for Multi-Block Convex Minimization Problems is not Necessarily Convergent," Optimization Online.

[CPR14] Chouzenoux, E., Pesquet, J. C., and Repetti, A., 2014. "Variable Metric Forward-Backward Algorithm for Minimizing the Sum of a Differentiable Function and a Convex Function," J. of Optimization Theory and Applications, Vol. 162, pp. 107-132.

[CPS92] Cottle, R. W., Pang, J.-S., and Stone, R. E., 1992. The Linear Complementarity Problem, Academic Press, NY.

[CRP12] Chandrasekaran, V., Recht, B., Parrilo, P. A., and Willsky, A. S., 2012. "The Convex Geometry of Linear Inverse Problems," Foundations of Computational Mathematics, Vol. 12, pp. 805-849.

[CaC68] Canon, M. D., and Cullum, C. D., 1968. "A Tight Upper Bound on the Rate of Convergence of the Frank-Wolfe Algorithm," SIAM J. on Control, Vol. 6, pp. 509-516.

[CaG74] Cantor, D. G., Gerla, M., 1974. "Optimal Routing in Packet

Switched Computer Networks," IEEE Trans. on Computing, Vol. C-23, pp. 1062-1068.

[CaR09] Candés, E. J., and Recht, B., 2009. "Exact Matrix Completion via Convex Optimization," Foundations of Computational Math., Vol. 9, pp. 717-772.

[CaT10] Candés, E. J., and Tao, T., 2010. "The Power of Convex Relaxation: Near-Optimal Matrix Completion," IEEE Trans. on Information Theory, Vol. 56, pp. 2053-2080.

[CeH87] Censor, Y., and Herman, G. T., 1987. "On Some Optimization Techniques in Image Reconstruction from Projections," Applied Numer. Math., Vol. 3, pp. 365-391.

[CeS08] Cegielski, A., and Suchocka, A., 2008. "Relaxed Alternating Projection Methods," SIAM J. Optimization, Vol. 19, pp. 1093-1106.

[CeZ92] Censor, Y., and Zenios, S. A., 1992. "The Proximal Minimization Algorithm with D-Functions," J. Opt. Theory and Appl., Vol. 73, pp. 451-464.

[CeZ97] Censor, Y., and Zenios, S. A., 1997. Parallel Optimization: Theory, Algorithms, and Applications, Oxford Univ. Press, NY.

[ChG59] Cheney, E. W., and Goldstein, A. A., 1959. "Newton's Method for Convex Programming and Tchebycheff Approximation," Numer. Math., Vol. I, pp. 253-268.

[ChR97] Chen, G. H., and Rockafellar, R. T., 1997. "Convergence Rates in Forward–Backward Splitting," SIAM J. on Optimization, Vol. 7, pp. 421-444.

[ChT93] Chen, G., and Teboulle, M., 1993. "Convergence Analysis of a Proximal-Like Minimization Algorithm Using Bregman Functions," SIAM J. on Optimization, Vol. 3, pp. 538-543.

[ChT94] Chen, G., and Teboulle, M., 1994. "A Proximal-Based Decomposition Method for Convex Minimization Problems," Mathematical Programming, Vol. 64, pp. 81-101.

[Cha04] Chambolle, A., 2004. "An Algorithm for Total Variation Minimization and Applications," J. of Mathematical Imaging and Vision, Vol. 20, pp. 89-97.

[Che07] Chen, Y., 2007. "A Smoothing Inexact Newton Method for Minimax Problems," Advances in Theoretical and Applied Mathematics, Vol. 2, pp. 137-143.

[Cla10] Clarkson, K. L., 2010. "Coresets, Sparse Greedy Approximation, and the Frank-Wolfe Algorithm," ACM Transactions on Algorithms, Vol. 6, pp. 63.

[CoT13] Couellan, N. P., and Trafalis, T. B., 2013. "On-line SVM Learning via an Incremental Primal-Dual Technique," Optimization Methods and Software, Vol. 28, pp. 256-275.

[CoV13] Combettes, P. L., and Vu, B. C., 2013. "Variable Metric Quasi-Fejér Monotonicity," Nonlinear Analysis: Theory, Methods and Applications, Vol. 78, pp. 17-31.

[Com01] Combettes, P. L., 2001. "Quasi-Fejérian Analysis of Some Optimization Algorithms," Studies in Computational Mathematics, Vol. 8, pp. 115-152.

[Cry71] Cryer, C. W., 1971. "The Solution of a Quadratic Programming Problem Using Systematic Overrelaxation," SIAM J. on Control, Vol. 9, pp. 385-392.

[DBW12] Duchi, J. C., Bartlett, P. L., and Wainwright, M. J., 2012. "Randomized Smoothing for Stochastic Optimization," SIAM J. on Optimization, Vol. 22, pp. 674-701.

[DCD14] Defazio, A. J., Caetano, T. S., and Domke, J., 2014. "Finito: A Faster, Permutable Incremental Gradient Method for Big Data Problems," Proceedings of the 31st ICML, Beijing.

[DHS06] Dai, Y. H., Hager, W. W., Schittkowski, K., and Zhang, H., 2006. "The Cyclic Barzilai-Borwein Method for Unconstrained Optimization," IMA J. of Numerical Analysis, Vol. 26, pp. 604-627.

[DHS11] Duchi, J., Hazan, E., and Singer, Y., 2011. "Adaptive Subgradient Methods for Online Learning and Stochastic Optimization," J. of Machine Learning Research, Vol. 12, pp. 2121-2159.

[DMM06] Drineas, P., Mahoney, M. W., and Muthukrishnan, S., 2006. "Sampling Algorithms for L2 Regression and Applications," Proc. 17th Annual SODA, pp. 1127-1136.

[DMM11] Drineas, P., Mahoney, M. W., Muthukrishnan, S., and Sarlos, T., 2011. "Faster Least Squares Approximation," Numerische Mathematik, Vol. 117, pp. 219-249.

[DRT11] Dhillon, I. S., Ravikumar, P., and Tewari, A., 2011. "Nearest Neighbor Based Greedy Coordinate Descent," in Advances in Neural Information Processing Systems 24, (NIPS 2011), pp. 2160-2168.

[DaW60] Dantzig, G. B., and Wolfe, P., 1960. "Decomposition Principle for Linear Programs," Operations Research, Vol. 8, pp. 101-111.

[DaY14a] Davis, D., and Yin, W., 2014. "Convergence Rate Analysis of Several Splitting Schemes," arXiv preprint arXiv:1406.4834.

[DaY14b] Davis, D., and Yin, W., 2014. "Convergence Rates of Relaxed Peaceman-Rachford and ADMM Under Regularity Assumptions, arXiv preprint arXiv:1407.5210.

[Dan67] Danskin, J. M., 1967. The Theory of Max-Min and its Application to Weapons Allocation Problems, Springer, NY.

[Dav76] Davidon, W. C., 1976. "New Least Squares Algorithms," J. Optimization Theory and Applications, Vol. 18, pp. 187-197.

[DeM71] Demjanov, V. F., and Malozemov, V. N., 1971. "On the Theory of Non-Linear Minimax Problems," Russian Math. Surveys, Vol. 26, p. 57.

[DeR70] Demjanov, V. F., and Rubinov, A. M., 1970. Approximate Methods in Optimization Problems, American Elsevier, NY.

[DeS96] Dennis, J. E., and Schnabel, R. B., 1996. Numerical Methods for Unconstrained Optimization and Nonlinear Equations, SIAM, Philadelphia, PA.

[DeT91] Dennis, J. E., and Torczon, V., 1991. "Direct Search Methods on Parallel Machines," SIAM J. on Optimization, Vol. 1, pp. 448-474.

[Dem66] Demjanov, V. F., 1966. "The Solution of Several Minimax Problems," Kibernetika, Vol. 2, pp. 58-66.

[Dem68] Demjanov, V. F., 1968. "Algorithms for Some Minimax Problems," J. of Computer and Systems Science, Vol. 2, pp. 342-380.

[DoE03] Donoho, D. L., Elad, M., 2003. "Optimally Sparse Representation in General (Nonorthogonal) Dictionaries via ℓ_1 Minimization," Proc. of the National Academy of Sciences, Vol. 100, pp. 2197-2202.

[DrH04] Drezner, Z., and Hamacher, H. W., 2004. Facility Location: Applications and Theory, Springer, NY.

[DuS83] Dunn, J. C., and Sachs, E., 1983. "The Effect of Perturbations on the Convergence Rates of Optimization Algorithms," Appl. Math. Optim., Vol. 10, pp. 143-157.

[DuS09] Duchi, J., and Singer, Y., 2009. "Efficient Online and Batch Learning Using Forward Backward Splitting, J. of Machine Learning Research, Vol. 10, pp. 2899-2934.

[Dun79] Dunn, J. C., 1979. "Rates of Convergence for Conditional Gradient Algorithms Near Singular and Nonsingular Extremals," SIAM J. on Control and Optimization, Vol. 17, pp. 187-211.

[Dun80] Dunn, J. C., 1980. "Convergence Rates for Conditional Gradient Sequences Generated by Implicit Step Length Rules," SIAM J. on Control and Optimization, Vol. 18, pp. 473-487.

[Dun81] Dunn, J. C., 1981. "Global and Asymptotic Convergence Rate Estimates for a Class of Projected Gradient Processes," SIAM J. on Control and Optimization, Vol. 19, pp. 368-400.

[Dun87] Dunn, J. C., 1987. "On the Convergence of Projected Gradient Processes to Singular Critical Points," J. of Optimization Theory and Applications, Vol. 55, pp. 203-216.

[Dun91] Dunn, J. C., 1991. "A Subspace Decomposition Principle for Scaled Gradient Projection Methods: Global Theory," SIAM J. on Control and Optimization, Vol. 29, pp. 219-246.

[EcB92] Eckstein, J., and Bertsekas, D. P., 1992. "On the Douglas-Rachford Splitting Method and the Proximal Point Algorithm for Maximal Monotone Operators," Math. Programming, Vol. 55, pp. 293-318.

[EcS13] Eckstein, J., and Silva, P. J. S., 2013. "A Practical Relative Error Criterion for Augmented Lagrangians," Math. Programming, Vol. 141, Ser. A, pp. 319-348.

[Eck94] Eckstein, J., 1994. "Nonlinear Proximal Point Algorithms Using Bregman Functions, with Applications to Convex Programming," Math. of Operations Research, Vol. 18, pp. 202-226.

[Eck03] Eckstein, J., 2003. "A Practical General Approximation Criterion for Methods of Multipliers Based on Bregman Distances," Math. Programming, Vol. 96, Ser. A, pp. 61-86.

[Eck12] Eckstein, J., 2012. "Augmented Lagrangian and Alternating Direction Methods for Convex Optimization: A Tutorial and Some Illustrative Computational Results," RUTCOR Research Report RRR 32-2012, Rutgers, Univ.

[EkT76] Ekeland, I., and Temam, R., 1976. Convex Analysis and Variational Problems, North-Holland Publ., Amsterdam.

[ElM75] Elzinga, J., and Moore, T. G., 1975. "A Central Cutting Plane Algorithm for the Convex Programming Problem," Math. Programming, Vol. 8, pp. 134-145.

[Erm76] Ermoliev, Yu. M., 1976. Stochastic Programming Methods, Nauka, Moscow.

[Eve63] Everett, H., 1963. "Generalized Lagrange Multiplier Method for Solving Problems of Optimal Allocation of Resources," Operations Research, Vol. 11, pp. 399-417.

[FGW02] Forsgren, A., Gill, P. E., and Wright, M. H., 2002. "Interior Methods for Nonlinear Optimization," SIAM Review, Vol. 44, pp. 525-597.

[FHT10] Friedman, J., Hastie, T., and Tibshirani, R., 2010. "Regularization Paths for Generalized Linear Models via Coordinate Descent," J. of Statistical Software, Vol. 33, pp. 1-22.

[FJS98] Facchinei, F., Judice, J., and Soares, J., 1998. "An Active Set Newton Algorithm for Large-Scale Nonlinear Programs with Box Constraints," SIAM J. on Optimization, Vol. 8, pp. 158-186.

[FLP02] Facchinei, F., Lucidi, S., and Palagi, L., 2002. "A Truncated Newton Algorithm for Large Scale Box Constrained Optimization," SIAM J. on Optimization, Vol. 12, pp. 1100-1125.

[FLT02] Fukushima, M., Luo, Z.-Q., and Tseng, P., 2002. "Smoothing Functions for Second-Order-Cone Complementarity Problems," SIAM J. Optimization, Vol. 12, pp. 436-460.

[FaP03] Facchinei, F., and Pang, J.-S., 2003. Finite-Dimensional Variational Inequalities and Complementarity Problems, Springer Verlag, NY.

[Fab73] Fabian, V., 1973. "Asymptotically Efficient Stochastic Approximation: The RM Case," Ann. Statist., Vol. 1, pp. 486-495.

[FeM91] Ferris, M. C., and Mangasarian, O. L., 1991. "Finite Perturbation of Convex Programs," Appl. Math. Optim., Vol. 23, pp. 263-273.

[FeM02] Ferris, M. C., and Munson, T. S., 2002. "Interior-Point Methods for Massive Support Vector Machines," SIAM J. on Optimization, Vol. 13, pp. 783-804.

[FeR12] Fercoq, O., and Richtarik, P., 2012. "Accelerated, Parallel, and Proximal Coordinate Descent," arXiv preprint arXiv:1312.5799.

[Fen51] Fenchel, W., 1951. Convex Cones, Sets, and Functions, Mimeographed Notes, Princeton Univ.

[FlH95] Florian, M. S., and Hearn, D., 1995. "Network Equilibrium Models and Algorithms," Handbooks in OR and MS, Ball, M. O., Magnanti, T. L., Monma, C. L., and Nemhauser, G. L., (Eds.), Vol. 8, North-Holland, Amsterdam, pp. 485-550.

[FiM68] Fiacco, A. V., and McCormick, G. P., 1968. Nonlinear Programming: Sequential Unconstrained Minimization Techniques, Wiley, NY.

[FiN03] Figueiredo, M. A. T., and Nowak, R. D., 2003. "An EM Algorithm for Wavelet-Based Image Restoration," IEEE Trans. Image Processing, Vol. 12, pp. 906-916.

[Fle00] Fletcher, R., 2000. Practical Methods of Optimization, 2nd edition, Wiley, NY.

[FoG83] Fortin, M., and Glowinski, R., 1983. "On Decomposition-Coordination Methods Using an Augmented Lagrangian," in: M. Fortin and R. Glowinski, eds., Augmented Lagrangian Methods: Applications to the Solution of Boundary-Value Problems, North-Holland, Amsterdam.

[FrG13] Friedlander, M. P., and Goh, G., 2013. "Tail Bounds for Stochastic Approximation," arXiv preprint arXiv:1304.5586.

[FrG14] Freund, R. M., and Grigas, P., 2014. "New Analysis and Results for the Frank-Wolfe Method," arXiv preprint arXiv:1307.0873, to appear in Math. Programming.

[FrS00] Frommer, A., and Szyld, D. B., 2000. "On Asynchronous Iterations," J. of Computational and Applied Mathematics, Vol. 123, pp. 201-216.

[FrS12] Friedlander, M. P., and Schmidt, M., 2012. "Hybrid Deterministic-Stochastic Methods for Data Fitting," SIAM J. Sci. Comput., Vol. 34, pp. A1380-A1405.

[FrT07] Friedlander, M. P., and Tseng, P., 2007. "Exact Regularization of Convex Programs," SIAM J. on Optimization, Vol. 18, pp. 1326-1350.

[FrW56] Frank, M., and Wolfe, P., 1956. "An Algorithm for Quadratic Programming," Naval Research Logistics Quarterly, Vol. 3, pp. 95-110.

[Fra02] Frangioni, A., 2002. "Generalized Bundle Methods," SIAM J. on Optimization, Vol. 13, pp. 117-156.

[Fri56] Frisch, M. R., 1956. "La Resolution des Problemes de Programme

Lineaire par la Methode du Potential Logarithmique," Cahiers du Seminaire D'Econometrie, Vol. 4, pp. 7-20.

[FuM81] Fukushima, M., and Mine, H., 1981. "A Generalized Proximal Point Algorithm for Certain Non-Convex Minimization Problems," Internat. J. Systems Sci., Vol. 12, pp. 989-1000.

[Fuk92] Fukushima, M., 1992. "Application of the Alternating Direction Method of Multipliers to Separable Convex Programming Problems," Computational Optimization and Applications, Vol. 1, pp. 93-111.

[GBY12] Grant, M., Boyd, S., and Ye, Y., 2012. "CVX: Matlab Software for Disciplined Convex Programming, Version 2.0 Beta," Recent Advances in Learning and Control, cvx.com.

[GGM06] Gaudioso, M., Giallombardo, G., and Miglionico, G., 2006. "An Incremental Method for Solving Convex Finite Min-Max Problems," Math. of Operations Research, Vol. 31, pp. 173-187.

[GHV92] Goffin, J. L., Haurie, A., and Vial, J. P., 1992. "Decomposition and Nondifferentiable Optimization with the Projective Algorithm," Management Science, Vol. 38, pp. 284-302.

[GKX10] Gupta, M. D., Kumar, S., and Xiao, J. 2010. "L1 Projections with Box Constraints," arXiv preprint arXiv:1010.0141.

[GLL86] Grippo, L., Lampariello, F., and Lucidi, S., 1986. "A Nonmonotone Line Search Technique for Newton's Method," SIAM J. on Numerical Analysis, Vol. 23, pp. 707-716.

[GLY94] Goffin, J. L., Luo, Z.-Q., and Ye, Y., 1994. "On the Complexity of a Column Generation Algorithm for Convex or Quasiconvex Feasibility Problems," in Large Scale Optimization: State of the Art, Hager, W. W., Hearn, D. W., and Pardalos, P. M., (Eds.), Kluwer, Boston.

[GLY96] Goffin, J. L., Luo, Z.-Q., and Ye, Y., 1996. "Complexity Analysis of an Interior Cutting Plane Method for Convex Feasibility Problems," SIAM J. on Optimization, Vol. 6, pp. 638-652.

[GMW81] Gill, P. E., Murray, W., and Wright, M. H., 1981. Practical Optimization, Academic Press, NY.

[GNS08] Griva, I., Nash, S. G., and Sofer, A., 2008. Linear and Nonlinear Optimization, 2nd Edition, SIAM, Philadelphia, PA.

[GOP14] Gurbuzbalaban, M., Ozdaglar, A., and Parrilo, P., 2014. "A Globally Convergent Incremental Newton Method," arXiv preprint arXiv:1410.-5284.

[GPR67] Gubin, L. G., Polyak, B. T., and Raik, E. V., 1967. "The Method of Projection for Finding the Common Point in Convex Sets," USSR Comput. Math. Phys., Vol. 7, pp. 1-24.

[GSW12] Gamarnik, D., Shah, D., and Wei, Y., 2012. "Belief Propagation for Min-Cost Network Flow: Convergence and Correctness," Operations Research, Vol. 60, pp. 410-428.

[GaB84] Gafni, E. M., and Bertsekas, D. P., 1984. "Two-Metric Projection Methods for Constrained Optimization," SIAM J. on Control and Optimization, Vol. 22, pp. 936-964.

[GaM76] Gabay, D., and Mercier, B., 1976. "A Dual Algorithm for the Solution of Nonlinear Variational Problems via Finite-Element Approximations," Comp. Math. Appl., Vol. 2, pp. 17-40.

[Gab79] Gabay, D., 1979. Methodes Numeriques pour l'Optimization Non Lineaire, These de Doctorat d'Etat et Sciences Mathematiques, Univ. Pierre at Marie Curie (Paris VI).

[Gab83] Gabay, D., 1983. "Applications of the Method of Multipliers to Variational Inequalities," in M. Fortin and R. Glowinski, eds., Augmented Lagrangian Methods: Applications to the Solution of Boundary-Value Problems, North-Holland, Amsterdam.

[GeM05] Gerlach, S., and Matzenmiller, A., 2005. "Comparison of Numerical Methods for Identification of Viscoelastic Line Spectra from Static Test Data," International J. for Numerical Methods in Engineering, Vol. 63, pp. 428-454.

[Geo72] Geoffrion, A. M., 1972. "Generalized Benders Decomposition," J. of Optimization Theory and Applications, Vol. 10, pp. 237-260.

[Geo77] Geoffrion, A. M., 1977. "Objective Function Approximations in Mathematical Programming," Math. Programming, Vol. 13, pp. 23-27.

[GiB14] Giselsson, P., and Boyd, S., 2014. "Metric Selection in Douglas-Rachford Splitting and ADMM," arXiv preprint arXiv:1410.8479.

[GiM74] Gill, P. E., and Murray, W., (Eds.), 1974. Numerical Methods for Constrained Optimization, Academic Press, NY.

[GlM75] Glowinski, R. and Marrocco, A., 1975. "Sur l' Approximation par Elements Finis d' Ordre un et la Resolution par Penalisation-Dualite d'une Classe de Problemes de Dirichlet Non Lineaires" Revue Francaise d'Automatique Informatique Recherche Operationnelle, Analyse Numerique, R-2, pp. 41-76.

[GoK13] Gonzaga, C. C., and Karas, E. W., 2013. "Fine Tuning Nesterov's Steepest Descent Algorithm for Differentiable Convex Programming," Math. Programming, Vol. 138, pp. 141-166.

[GoO09] Goldstein, T., and Osher, S., 2009. "The Split Bregman Method for L1-Regularized Problems," SIAM J. on Imaging Sciences, Vol. 2, pp. 323-343.

[GoS10] Goldstein, T., and Setzer, S., 2010. "High-Order Methods for Basis Pursuit," UCLA CAM Report, 10-41.

[GoV90] Goffin, J. L., and Vial, J. P., 1990. "Cutting Planes and Column Generation Techniques with the Projective Algorithm," J. Opt. Th. and Appl., Vol. 65, pp. 409-429.

[GoV02] Goffin, J. L., and Vial, J. P., 2002. "Convex Nondifferentiable

Optimization: A Survey Focussed on the Analytic Center Cutting Plane Method," Optimization Methods and Software, Vol. 17, pp. 805-867.

[GoZ12] Gong, P., and Zhang, C., 2012. "Efficient Nonnegative Matrix Factorization via Projected Newton Method," Pattern Recognition, Vol. 45, pp. 3557-3565.

[Gol64] Goldstein, A. A., 1964. "Convex Programming in Hibert Space," Bull. Amer. Math. Soc., Vol. 70, pp. 709-710.

[Gol85] Golshtein, E. G., 1985. "A Decomposition Method for Linear and Convex Programming Problems," Matecon, Vol. 21, pp. 1077-1091.

[Gon00] Gonzaga, C. C., 2000. "Two Facts on the Convergence of the Cauchy Algorithm," J. of Optimization Theory and Applications, Vol. 107, pp. 591-600.

[GrS99] Grippo, L., and Sciandrone, M., 1999. "Globally Convergent Block-Coordinate Techniques for Unconstrained Optimization," Optimization Methods and Software, Vol. 10, pp. 587-637.

[GrS00] Grippo, L., and Sciandrone, M., 2000. "On the Convergence of the Block Nonlinear Gauss-Seidel Method Under Convex Constraints," Operations Research Letters, Vol. 26, pp. 127-136.

[Gri94] Grippo, L., 1994. "A Class of Unconstrained Minimization Methods for Neural Network Training," Optim. Methods and Software, Vol. 4, pp. 135-150.

[Gri00] Grippo, L., 2000. "Convergent On-Line Algorithms for Supervised Learning in Neural Networks," IEEE Trans. Neural Networks, Vol. 11, pp. 1284-1299.

[Gul92] Guler, O., 1992. "New Proximal Point Algorithms for Convex Minimization," SIAM J. on Optimization, Vol. 2, pp. 649-664.

[HCW14] Hong, M., Chang, T. H., Wang, X., Razaviyayn, M., Ma, S., and Luo, Z. Q., 2014. "A Block Successive Upper Bound Minimization Method of Multipliers for Linearly Constrained Convex Optimization," arXiv preprint arXiv:1401.7079.

[HJN14] Harchaoui, Z., Juditsky, A., and Nemirovski, A., 2014. "Conditional Gradient Algorithms for Norm-Regularized Smooth Convex Optimization," Math. Programming, pp. 1-38.

[HKR95] den Hertog, D., Kaliski, J., Roos, C., and Terlaky, T., 1995. "A Path-Following Cutting Plane Method for Convex Programming," Annals of Operations Research, Vol. 58, pp. 69-98.

[HLV87] Hearn, D. W., Lawphongpanich, S., and Ventura, J. A., 1987. "Restricted Simplicial Decomposition: Computation and Extensions," Math. Programming Studies, Vol. 31, pp. 119-136.

[HMT10] Halko, N., Martinsson, P.-G., and Tropp, J. A., 2010. "Finding Structure with Randomness: Probabilistic Algorithms for Constructing Approximate Matrix Decompositions," arXiv preprint arXiv:0909.4061.

[HTF09] Hastie, T., Tibshirani, R., and Friedman, J., 2009. The Elements of Statistical Learning: Data Mining, Inference, and Prediction, 2nd Edition, Springer, NY. On line at http://statweb.stanford.edu/ tibs/ElemStatLearn/

[HYS15] Hu, Y., Yang, X., and Sim, C. K., 2015. "Inexact Subgradient Methods for Quasi-Convex Optimization Problems," European Journal of Operational Research, Vol. 240, pp. 315-327.

[HZS13] Hou, K., Zhou, Z., So, A. M. C., and Luo, Z. Q., 2013. "On the Linear Convergence of the Proximal Gradient Method for Trace Norm Regularization," in Advances in Neural Information Processing Systems (NIPS 2013), pp. 710-718.

[Ha90] Ha, C. D., 1990. "A Generalization of the Proximal Point Algorithm," SIAM J. on Control and Optimization, Vol. 28, pp. 503-512.

[HaB70] Haarhoff, P. C., and Buys, J. D, 1970. "A New Method for the Optimization of a Nonlinear Function Subject to Nonlinear Constraints," Computer J., Vol. 13, pp. 178-184.

[HaH93] Hager, W. W., and Hearn, D. W., 1993. "Application of the Dual Active Set Algorithm to Quadratic Network Optimization," Computational Optimization and Applications, Vol. 1, pp. 349-373.

[HaM79] Han, S. P., and Mangasarian, O. L., 1979. "Exact Penalty Functions in Nonlinear Programming," Math. Programming, Vol. 17, pp. 251-269.

[Hay08] Haykin, S., 2008. Neural Networks and Learning Machines, (3rd Ed.), Prentice Hall, Englewood Cliffs, NJ.

[HeD09] Helou, E. S., and De Pierro, A. R., 2009. "Incremental Subgradients for Constrained Convex Optimization: A Unified Framework and New Methods," SIAM J. on Optimization, Vol. 20, pp. 1547-1572.

[HeL89] Hearn, D. W., and Lawphongpanich, S., 1989. "Lagrangian Dual Ascent by Generalized Linear Programming," Operations Res. Letters, Vol. 8, pp. 189-196.

[HeM11] Henrion, D., and Malick, J., 2011. "Projection Methods for Conic Feasibility Problems: Applications to Polynomial Sum-of-Squares Decompositions," Optimization Methods and Software, Vol. 26, pp. 23-46.

[HeM12] Henrion, D., and Malick, J., 2012. "Projection Methods in Conic Optimization," In Handbook on Semidefinite, Conic and Polynomial Optimization, Springer, NY, pp. 565-600.

[Her09] Gabor, H., 2009. Fundamentals of Computerized Tomography: Image Reconstruction from Projection, (2nd ed.), Springer, NY.

[Hes69] Hestenes, M. R., 1969. "Multiplier and Gradient Methods," J. Opt. Th. and Appl., Vol. 4, pp. 303-320.

[Hes75] Hestenes, M. R., 1975. Optimization Theory: The Finite Dimensional Case, Wiley, NY.

[HiL93] Hiriart-Urruty, J.-B., and Lemarechal, C., 1993. Convex Analysis and Minimization Algorithms, Vols. I and II, Springer-Verlag, Berlin and NY.

[Hil57] Hildreth, C., 1957. "A Quadratic Programming Procedure," Naval Res. Logist. Quart., Vol. 4, pp. 79-85. See also "Erratum," Naval Res. Logist. Quart., Vol. 4, p. 361.

[HoK71] Hoffman, K., and Kunze, R., 1971. Linear Algebra, Pearson, Englewood Cliffs, NJ.

[HoL13] Hong, M., and Luo, Z. Q., 2013. "On the Linear Convergence of the Alternating Direction Method of Multipliers," arXiv preprint arXiv:1208.-3922.

[Hoh77] Hohenbalken, B. von, 1977. "Simplicial Decomposition in Nonlinear Programming," Math. Programming, Vol. 13, pp. 49-68.

[Hol74] Holloway, C. A., 1974. "An Extension of the Frank and Wolfe Method of Feasible Directions," Math. Programming, Vol. 6, pp. 14-27.

[IPS03] Iusem, A. N., Pennanen, T., and Svaiter, B. F., 2003. "Inexact Variants of the Proximal Point Algorithm Without Monotonicity," SIAM J. on Optimization, Vol. 13, pp. 1080-1097.

[IST94] Iusem, A. N., Svaiter, B. F., and Teboulle, M., 1994. "Entropy-Like Proximal Methods in Convex Programming," Math. of Operations Research, Vol. 19, pp. 790-814.

[IbF96] Ibaraki, S., and Fukushima, M., 1996. "Partial Proximal Method of Multipliers for Convex Programming Problems," J. of Operations Research Society of Japan, Vol. 39, pp. 213-229.

[IuT95] Iusem, A. N., and Teboulle, M., 1995. "Convergence Rate Analysis of Nonquadratic Proximal Methods for Convex and Linear Programming," Math. of Operations Research, Vol. 20, pp. 657-677.

[Ius99] Iusem, A. N., 1999. "Augmented Lagrangian Methods and Proximal Point Methods for Convex Minimization," Investigacion Operativa, Vol. 8, pp. 11-49.

[Ius03] Iusem, A. N., 2003. "On the Convergence Properties of the Projected Gradient Method for Convex Optimization," Computational and Applied Mathematics, Vol. 22, pp. 37-52.

[JFY09] Joachims, T., Finley, T., and Yu, C.-N. J., 2009. "Cutting-Plane Training of Structural SVMs," Machine Learning, Vol. 77, pp. 27-59.

[JRJ09] Johansson, B., Rabi, M., and Johansson, M., 2009. "A Randomized Incremental Subgradient Method for Distributed Optimization in Networked Systems," SIAM J. on Optimization, Vol. 20, pp. 1157-1170.

[Jag13] Jaggi, M., 2013. "Revisiting Frank-Wolfe: Projection-Free Sparse Convex Optimization," Proc. of ICML 2013.

[JiZ14] Jiang, B., and Zhang, S., 2014. "Iteration Bounds for Finding the ϵ-Stationary Points for Structured Nonconvex Optimization," arXiv preprint arXiv:1410.4066.

[JoY09] Joachims, T., and Yu, C.-N. J., 2009. "Sparse Kernel SVMs via Cutting-Plane Training," Machine Learning, Vol. 76, pp. 179-193.

[JoZ13] Johnson, R., and Zhang, T., 2013. "Accelerating Stochastic Gradient Descent Using Predictive Variance Reduction," Advances in Neural Information Processing Systems 26 (NIPS 2013).

[Joa06] Joachims, T., 2006. "Training Linear SVMs in Linear Time," International Conference on Knowledge Discovery and Data Mining, pp. 217-226.

[JuN11a] Juditsky, A., and Nemirovski, A., 2011. "First Order Methods for Nonsmooth Convex Large-Scale Optimization, I: General Purpose Methods," in Optimization for Machine Learning, by Sra, S., Nowozin, S., and Wright, S. J. (eds.), MIT Press, Cambridge, MA, pp. 121-148.

[JuN11b] Juditsky, A., and Nemirovski, A., 2011. "First Order Methods for Nonsmooth Convex Large-Scale Optimization, II: Utilizing Problem's Structure," in Optimization for Machine Learning, by Sra, S., Nowozin, S., and Wright, S. J. (eds.), MIT Press, Cambridge, MA, pp. 149-183.

[KaW94] Kall, P., and Wallace, S. W., 1994. Stochastic Programming, Wiley, Chichester, UK.

[Kac37] Kaczmarz, S., 1937. "Approximate Solution of Systems of Linear Equations," Bull. Acad. Pol. Sci., Lett. A 35, pp. 335-357 (in German); English transl.: Int. J. Control, Vol. 57, pp. 1269-1271, 1993.

[Kar84] Karmarkar, N., 1984. "A New Polynomial-Time Algorithm for Linear Programming," In Proc. of the 16th Annual ACM Symp. on Theory of Computing, pp. 302-311.

[Kel60] Kelley, J. E., 1960. "The Cutting-Plane Method for Solving Convex Programs," J. Soc. Indust. Appl. Math., Vol. 8, pp. 703-712.

[Kel99] Kelley, C. T., 1999. Iterative Methods for Optimization, Siam, Philadelphia, PA.

[Kib80] Kibardin, V. M., 1980. "Decomposition into Functions in the Minimization Problem," Automation and Remote Control, Vol. 40, pp. 1311-1323.

[Kiw04] Kiwiel, K. C., 2004. "Convergence of Approximate and Incremental Subgradient Methods for Convex Optimization," SIAM J. on Optimization, Vol. 14, pp. 807-840.

[KoB72] Kort, B. W., and Bertsekas, D. P., 1972. "A New Penalty Function Method for Constrained Minimization," Proc. 1972 IEEE Confer. Decision Control, New Orleans, LA, pp. 162-166.

[KoB76] Kort, B. W., and Bertsekas, D. P., 1976. "Combined Primal-Dual and Penalty Methods for Convex Programming," SIAM J. on Control and Optimization, Vol. 14, pp. 268-294.

[KoN93] Kortanek, K. O., and No, H., 1993. "A Central Cutting Plane

Algorithm for Convex Semi-Infinite Programming Problems," SIAM J. on Optimization, Vol. 3, pp. 901-918.

[Kor75] Kort, B. W., 1975. "Combined Primal-Dual and Penalty Function Algorithms for Nonlinear Programming," Ph.D. Thesis, Dept. of Engineering-Economic Systems, Stanford Univ., Stanford, Ca.

[Kra55] Krasnosel'skii, M. A., 1955. "Two Remarks on the Method of Successive Approximations," Uspehi Mat. Nauk, Vol. 10, pp. 123-127.

[KuC78] Kushner, H. J., and Clark, D. S., 1978. Stochastic Approximation Methods for Constrained and Unconstrained Systems, Springer-Verlag, NY.

[KuY03] Kushner, H. J., and Yin, G., 2003. Stochastic Approximation and Recursive Algorithms and Applications, Springer-Verlag, NY.

[LBB98] LeCun, Y., Bottou, L., Bengio, Y., and Haffner, P., 1998. "Gradient-Based Learning Applied to Document Recognition," Proceedings of the IEEE, Vol. 86, pp. 2278-2324.

[LJS12] Lacoste-Julien, S., Jaggi, M., Schmidt, M., and Pletscher, P., 2012. "Block-Coordinate Frank-Wolfe Optimization for Structural SVMs," arXiv preprint arXiv:1207.4747.

[LLS12] Lee, J., Sun, Y., and Saunders, M., 2012. "Proximal Newton-Type Methods for Convex Optimization," NIPS 2012.

[LLS14] Lee, J., Sun, Y., and Saunders, M., 2014. "Proximal Newton-Type Methods for Minimizing Composite Functions," arXiv preprint arXiv:1206.-1623.

[LLX14] Lin, Q., Lu, Z., and Xiao, L., 2014. "An Accelerated Proximal Coordinate Gradient Method and its Application to Regularized Empirical Risk Minimization," arXiv preprint arXiv:1407.1296.

[LLZ09] Langford, J., Li, L., and Zhang, T., 2009. "Sparse Online Learning via Truncated Gradient," In Advances in Neural Information Processing Systems (NIPS 2009), pp. 905-912.

[LMS92] Lustig, I. J., Marsten, R. E., and Shanno, D. F., 1992. "On Implementing Mehrotra's Predictor-Corrector Interior-Point Method for Linear Programming," SIAM J. on Optimization, Vol. 2, pp. 435-449.

[LMY12] Lu, Z., Monteiro, R. D. C., and Yuan, M., 2012. "Convex Optimization Methods for Dimension Reduction and Coefficient Estimation in Multivariate Linear Regression," Mathematical Programming, Vol. 131, pp. 163-194.

[LPS98] Larsson, T., Patriksson, M., and Stromberg, A.-B., 1998. "Ergodic Convergence in Subgradient Optimization," Optimization Methods and Software, Vol. 9, pp. 93-120.

[LRW98] Lagarias, J. C., Reeds, J. A., Wright, M. H., and Wright, P. E., 1998. "Convergence Properties of the Nelder-Mead Simplex Method in Low Dimensions," SIAM J. on Optimization, Vol. 9, pp. 112-147.

[LVB98] Lobo, M. S., Vandenberghe, L., Boyd, S., and Lebret, H., 1998. "Applications of Second-Order Cone Programming," Linear Algebra and Applications, Vol. 284, pp. 193-228.

[LWS14] Liu, J., Wright, S. J., and Sridhar, S., 2014. "An Asynchronous Parallel Randomized Kaczmarz Algorithm," Univ. of Wisconsin Report, arXiv preprint arXiv:1401.4780.

[LaD60] Land, A. H., and Doig, A. G., 1960. "An Automatic Method for Solving Discrete Programming Problems," Econometrica, Vol. 28, pp. 497-520.

[LaS87] Lawrence, J., and Spingarn, J. E., 1987. "On Fixed Points of Non-expansive Piecewise Isometric Mappings," Proc. London Math. Soc., Vol. 55, pp. 605-624.

[LaT85] Lancaster, P., and Tismenetsky, M., 1985. The Theory of Matrices, Academic Press, NY.

[Lan14] Landi, G., 2014. "A Modified Newton Projection Method for ℓ_1-Regularized Least Squares Image Deblurring," J. of Mathematical Imaging and Vision, pp. 1-14.

[LeL10] Leventhal, D., and Lewis, A. S., 2010. "Randomized Methods for Linear Constraints: Convergence Rates and Conditioning," Math. of Operations Research, Vol. 35, pp. 641-654.

[LeP65] Levitin, E. S., and Poljak, B. T., 1965. "Constrained Minimization Methods," Ž. Vyčisl. Mat. i Mat. Fiz., Vol. 6, pp. 787-823.

[LeS93] Lemaréchal, C., and Sagastizábal, C., 1993. "An Approach to Variable Metric Bundle Methods," in Systems Modelling and Optimization, Proc. of the 16th IFIP-TC7 Conference, Compiègne, Henry, J., and Yvon, J.-P., (Eds.), Lecture Notes in Control and Information Sciences 197, pp. 144-162.

[LeS99] Lee, D., and Seung, H., 1999. "Learning the Parts of Objects by Non-Negative Matrix Factorization," Nature, Vol. 401, pp. 788-791.

[LeS13] Lee, Y. T., and Sidford, A., 2013. "Efficient Accelerated Coordinate Descent Methods and Faster Algorithms for Solving Linear Systems," Proc. 2013 IEEE 54th Annual Symposium on Foundations of Computer Science (FOCS), pp. 147-156.

[LeW11] Lee, S., and Wright, S. J., 2011. "Approximate Stochastic Subgradient Estimation Training for Support Vector Machines," Univ. of Wisconsin Report, arXiv preprint arXiv:1111.0432.

[Lem74] Lemarechal, C., 1974. "An Algorithm for Minimizing Convex Functions," in Information Processing '74, Rosenfeld, J. L., (Ed.), North-Holland, Amsterdam, pp. 552-556.

[Lem75] Lemarechal, C., 1975. "An Extension of Davidon Methods to Nondifferentiable Problems," Math. Programming Study 3, Balinski, M., and Wolfe, P., (Eds.), North-Holland, Amsterdam, pp. 95-109.

[Lem89] Lemaire, B., 1989. "The Proximal Algorithm," in New Methods in Optimization and Their Industrial Uses, J.-P. Penot, (ed.), Birkhauser, Basel, pp. 73-87.

[LiM79] Lions, P. L., and Mercier, B., 1979. "Splitting Algorithms for the Sum of Two Nonlinear Operators," SIAM J. on Numerical Analysis, Vol. 16, pp. 964-979.

[LiP87] Lin, Y. Y., and Pang, J.-S., 1987. "Iterative Methods for Large Convex Quadratic Programs: A Survey," SIAM J. on Control and Optimization, Vol. 18, pp. 383-411.

[LiW14] Liu, J., and Wright, S. J., 2014. "Asynchronous Stochastic Coordinate Descent: Parallelism and Convergence Properties," Univ. of Wisconsin Report, arXiv preprint arXiv:1403.3862.

[Lin07] Lin, C. J., 2007. "Projected Gradient Methods for Nonnegative Matrix Factorization," Neural Computation, Vol. 19, pp. 2756-2779.

[Lit66] Litvakov, B. M., 1966. "On an Iteration Method in the Problem of Approximating a Function from a Finite Number of Observations," Avtom. Telemech., No. 4, pp. 104-113.

[Lju77] Ljung, L., 1977. "Analysis of Recursive Stochastic Algorithms," IEEE Trans. on Automatic Control, Vol. 22, pp. 551-575.

[LuT91] Luo, Z. Q., and Tseng, P., 1991. "On the Convergence of a Matrix-Splitting Algorithm for the Symmetric Monotone Linear Complementarity Problem," SIAM J. on Control and Optimization, Vol. 29, pp. 1037-1060.

[LuT92] Luo, Z. Q., and Tseng, P., 1992. "On the Convergence of the Coordinate Descent Method for Convex Differentiable Minimization," J. Optim. Theory Appl., Vol. 72, pp. 7-35.

[LuT93a] Luo, Z. Q., and Tseng, P., 1993. "On the Convergence Rate of Dual Ascent Methods for Linearly Constrained Convex Minimization," Math. of Operations Research, Vol. 18, pp. 846-867.

[LuT93b] Luo, Z. Q., and Tseng, P., 1993. "Error Bound and Reduced-Gradient Projection Algorithms for Convex Minimization over a Polyhedral Set," SIAM J. on Optimization, Vol. 3, pp. 43-59.

[LuT93c] Luo, Z. Q., and Tseng, P., 1993. "Error Bounds and Convergence Analysis of Feasible Descent Methods: A General Approach," Annals of Operations Research, Vol. 46, pp. 157-178.

[LuT94a] Luo, Z. Q., and Tseng, P., 1994. "Analysis of an Approximate Gradient Projection Method with Applications to the Backpropagation Algorithm," Optimization Methods and Software, Vol. 4, pp. 85-101.

[LuT94b] Luo, Z. Q., and Tseng, P., 1994. "On the Rate of Convergence of a Distributed Asynchronous Routing Algorithm," IEEE Trans. on Automatic Control, Vol. 39, pp. 1123-1129.

[LuT13] Luss, R., and Teboulle, M., 2013. "Conditional Gradient Algorithms for Rank-One Matrix Approximations with a Sparsity Constraint,"

SIAM Review, Vol. 55, pp. 65-98.

[LuY08] Luenberger, D. G., and Ye, Y., 2008. Linear and Nonlinear Programming, 3rd Edition, Springer, NY.

[Lue84] Luenberger, D. G., 1984. Introduction to Linear and Nonlinear Programming, 2nd Edition, Addison-Wesley, Reading, MA.

[Luo91] Luo, Z. Q., 1991. "On the Convergence of the LMS Algorithm with Adaptive Learning Rate for Linear Feedforward Networks," Neural Computation, Vol. 3, pp. 226-245.

[Luq84] Luque, F. J., 1984. "Asymptotic Convergence Analysis of the Proximal Point Algorithm," SIAM J. on Control and Optimization, Vol. 22, pp. 277-293.

[MRS10] Mosk-Aoyama, D., Roughgarden, T., and Shah, D., 2010. "Fully Distributed Algorithms for Convex Optimization Problems," SIAM J. on Optimization, Vol. 20, pp. 3260-3279.

[MSQ98] Mifflin, R., Sun, D., and Qi, L., 1998. "Quasi-Newton Bundle-Type Methods for Nondifferentiable Convex Optimization," SIAM J. on Optimization, Vol. 8, pp. 583-603.

[MYF03] Moriyama, H., Yamashita, N., and Fukushima, M., 2003. "The Incremental Gauss-Newton Algorithm with Adaptive Stepsize Rule," Computational Optimization and Applications, Vol. 26, pp. 107-141.

[MaM01] Mangasarian, O. L., Musicant, D. R., 2001. "Lagrangian Support Vector Machines," J. of Machine Learning Research, Vol. 1, pp. 161-177.

[MaS94] Mangasarian, O. L., and Solodov, M. V., 1994. "Serial and Parallel Backpropagation Convergence Via Nonmonotone Perturbed Minimization," Opt. Methods and Software, Vol. 4, pp. 103-116.

[Mai13] Mairal, J., 2013. "Optimization with First-Order Surrogate Functions," arXiv preprint arXiv:1305.3120.

[Mai14] Mairal, J., 2014. "Incremental Majorization-Minimization Optimization with Application to Large-Scale Machine Learning," arXiv preprint arXiv:1402.4419.

[Man53] Mann, W. R., 1953. "Mean Value Methods in Iteration," Proc. Amer. Math. Soc., Vol. 4, pp. 506-510.

[Mar70] Martinet, B., 1970. "Regularisation d' Inéquations Variationelles par Approximations Successives," Revue Fran. d'Automatique et Infomatique Rech. Opérationelle, Vol. 4, pp. 154-159.

[Mar72] Martinet, B., 1972. "Determination Approché d'un Point Fixe d'une Application Pseudo-Contractante. Cas de l'Application Prox," Comptes Rendus de l'Academie des Sciences, Paris, Serie A 274, pp. 163-165.

[Meh92] Mehrotra, S., 1992. "On the Implementation of a Primal-Dual Interior Point Method," SIAM J. on Optimization, Vol. 2, pp. 575-601.

[Mey07] Meyn, S., 2007. Control Techniques for Complex Networks, Cambridge Univ. Press, NY.

[MiF81] Mine, H., Fukushima, M. 1981. "A Minimization Method for the Sum of a Convex Function and a Continuously Differentiable Function," J. of Optimization Theory and Applications, Vol. 33, pp. 9-23.

[Mif96] Mifflin, R., 1996. "A Quasi-Second-Order Proximal Bundle Algorithm," Math. Programming, Vol. 73, pp. 51-72.

[Min62] Minty, G. J., 1962. "Monotone (Nonlinear) Operators in Hilbert Space," Duke J. of Math., Vol. 29, pp. 341-346.

[Min64] Minty, G. J., 1964. "On the Monotonicity of the Gradient of a Convex Function," Pacific J. of Math., Vol. 14, pp. 243-247.

[Min86] Minoux, M., 1986. Math. Programming: Theory and Algorithms, Wiley, NY.

[MoT89] Moré, J. J., and Toraldo, G., 1989. "Algorithms for Bound Constrained Quadratic Programming Problems," Numer. Math., Vol. 55, pp. 377-400.

[NBB01] Nedić, A., Bertsekas, D. P., and Borkar, V., 2001. "Distributed Asynchronous Incremental Subgradient Methods," Proc. of 2000 Haifa Workshop "Inherently Parallel Algorithms in Feasibility and Optimization and Their Applications," by D. Butnariu, Y. Censor, and S. Reich, Eds., Elsevier, Amsterdam.

[NJL09] Nemirovski, A., Juditsky, A., Lan, G., and Shapiro, A., 2009. "Robust Stochastic Approximation Approach to Stochastic Programming," SIAM J. on Optimization, Vol. 19, pp. 1574-1609.

[NSW14] Needell, D., Srebro, N., and Ward, R., 2014. "Stochastic Gradient Descent and the Randomized Kaczmarz Algorithm," arXiv preprint arXiv:1310.5715v3.

[NaT02] Nazareth, L., and Tseng, P., 2002. "Gilding the Lily: A Variant of the Nelder-Mead Algorithm Based on Golden-Section Search," Computational Optimization and Applications, Vol. 22, pp. 133-144.

[NaZ05] Narkiss, G., and Zibulevsky, M., 2005. "Sequential Subspace Optimization Method for Large-Scale Unconstrained Problems," Technion-IIT, Department of Electrical Engineering.

[NeB00] Nedić, A., and Bertsekas, D. P., 2000. "Convergence Rate of Incremental Subgradient Algorithms," Stochastic Optimization: Algorithms and Applications," S. Uryasev and P. M. Pardalos, Eds., Kluwer, pp. 263-304.

[NeB01] Nedić, A., and Bertsekas, D. P., 2001. "Incremental Subgradient Methods for Nondifferentiable Optimization," SIAM J. on Optimization, Vol. 12, pp. 109-138.

[NeB10] Nedić, A., and Bertsekas, D. P., 2010. "The Effect of Deterministic Noise in Subgradient Methods," Math. Programming, Ser. A, Vol. 125, pp. 75-99.

[NeC13] Necoara, I., and Clipici, D., 2013. "Distributed Coordinate Descent Methods for Composite Minimization," arXiv preprint arXiv:1312.5302.

[NeN94] Nesterov, Y., and Nemirovskii, A., 1994. Interior Point Polynomial Algorithms in Convex Programming, SIAM, Studies in Applied Mathematics 13, Philadelphia, PA.

[NeO09a] Nedić, A., and Ozdaglar, A., 2009. "Distributed Subgradient Methods for Multi-Agent Optimization," IEEE Trans. on Aut. Control, Vol. 54, pp. 48-61.

[NeO09b] Nedić, A., and Ozdaglar, A., 2009. "Subgradient Methods for Saddle-Point Problems," J. of Optimization Theory and Applications, Vol. 142, pp. 205-228.

[NeW88] Nemhauser, G. L., and Wolsey, L. A., 1988. Integer and Combinatorial Optimization, Wiley, NY.

[NeY83] Nemirovsky, A., and Yudin, D. B., 1983. Problem Complexity and Method Efficiency, Wiley, NY.

[Ned10] Nedić, A., 2010. "Random Projection Algorithms for Convex Set Intersection Problems," Proc. 2010 IEEE Conference on Decision and Control, Atlanta, Georgia, pp. 7655-7660.

[Ned11] Nedić, A., 2011. "Random Algorithms for Convex Minimization Problems," Math. Programming, Ser. B, Vol. 129, pp. 225-253.

[Nee10] Needell, D., 2010. "Randomized Kaczmarz Solver for Noisy Linear Systems," BIT Numerical Mathematics, Vol. 50, pp. 395-403.

[Nes83] Nesterov, Y., 1983. "A Method for Unconstrained Convex Minimization Problem with the Rate of Convergence $O(1/k^2)$," Doklady AN SSSR, Vol. 269, pp. 543-547; translated as Soviet Math. Dokl.

[Nes95] Nesterov, Y., 1995. "Complexity Estimates of Some Cutting Plane Methods Based on Analytic Barrier," Math. Programming, Vol. 69, pp. 149-176.

[Nes04] Nesterov, Y., 2004. Introductory Lectures on Convex Optimization, Kluwer Academic Publisher, Dordrecht, The Netherlands.

[Nes05] Nesterov, Y., 2005. "Smooth Minimization of Nonsmooth Functions," Math. Programming, Vol. 103, pp. 127-152.

[Nes12] Nesterov, Y., 2012. "Efficiency of Coordinate Descent Methods on Huge-Scale Optimization Problems," SIAM J. on Optimization, Vol. 22, pp. 341-362.

[Nev75] Neveu, J., 1975. Discrete Parameter Martingales, North-Holland, Amsterdam, The Netherlands.

[NoW06] Nocedal, J., and Wright, S. J., 2006. Numerical Optimization, 2nd Edition, Springer, NY.

[Noc80] Nocedal, J., 1980. "Updating Quasi-Newton Matrices with Limited Storage," Math. of Computation, Vol. 35, pp. 773-782.

[OBG05] Osher, S., Burger, M., Goldfarb, D., Xu, J., and Yin, W., 2005.

"An Iterative Regularization Method for Total Variation-Based Image Restoration," Multiscale Modeling and Simulation, Vol. 4, pp. 460-489.

[OJW05] Olafsson, A., Jeraj, R., and Wright, S. J., 2005. Optimization of Intensity-Modulated Radiation Therapy with Biological Objectives," Physics in Medicine and Biology, Vol. 50, pp. 53-57.

[OMV00] Ouorou, A., Mahey, P., and Vial, J. P., 2000. "A Survey of Algorithms for Convex Multicommodity Flow Problems," Management Science, Vol. 46, pp. 126-147.

[OrR70] Ortega, J. M., and Rheinboldt, W. C., 1970. Iterative Solution of Nonlinear Equations in Several Variables, Academic Press, NY.

[OvG14] Ovcharova, N., and Gwinner, J., 2014. "A Study of Regularization Techniques of Nondifferentiable Optimization in View of Application to Hemivariational Inequalities," J. of Optimization Theory and Applications, Vol. 162, pp. 754-778.

[OzB03] Ozdaglar, A. E., and Bertsekas, D. P., 2003. "Routing and Wavelength Assignment in Optical Networks," IEEE Trans. on Networking, Vol. 11, pp. 259-272.

[PKP09] Predd, J. B., Kulkarni, S. R., and Poor, H. V., 2009. "A Collaborative Training Algorithm for Distributed Learning," IEEE Transactions on Information Theory, Vol. 55, pp. 1856-1871.

[PaE10] Palomar, D. P., and Eldar, Y. C., (Eds.), 2010. Convex Optimization in Signal Processing and Communications, Cambridge Univ. Press, NY.

[PaT94] Paatero, P., and Tapper, U., 1994. "Positive Matrix Factorization: A Non-Negative Factor Model with Optimal Utilization of Error Estimates of Data Values," Environmetrics, Vol. 5, pp. 111-126.

[PaY84] Pang, J. S., Yu, C. S., 1984. "Linearized Simplicial Decomposition Methods for Computing Traffic Equilibria on Networks," Networks, Vol. 14, pp. 427-438.

[Paa97] Paatero, P., 1997. "Least Squares Formulation of Robust Non-Negative Factor Analysis," Chemometrics and Intell. Laboratory Syst., Vol. 37, pp. 23-35.

[Pan84] Pang, J. S., 1984. "On the Convergence of Dual Ascent Methods for Large-Scale Linearly Constrained Optimization Problems," Unpublished manuscript, The Univ. of Texas at Dallas.

[Pap81] Papavassilopoulos, G., 1981. "Algorithms for a Class of Nondifferentiable Problems," J. of Optimization Theory and Applications, Vol. 34, pp. 41-82.

[Pas79] Passty, G. B., 1979. "Ergodic Convergence to a Zero of the Sum of Monotone Operators in Hilbert Space," J. Math. Anal. Appl., Vol. 72, pp. 383-390.

[Pat93] Patriksson, M., 1993. "Partial Linearization Methods in Nonlinear

Programming," J. of Optimization Theory and Applications, Vol. 78, pp. 227-246.

[Pat98] Patriksson, M., 1998. "Cost Approximation: A Unified Framework of Descent Algorithms for Nonlinear Programs," SIAM J. Optimization, Vol. 8, pp. 561-582.

[Pat99] Patriksson, M.,1999. Nonlinear Programming and Variational Inequality Problems: A Unified Approach, Springer, NY.

[Pat01] Patriksson, M., 2001. "Simplicial Decomposition Algorithms," Encyclopedia of Optimization, Springer, pp. 2378-2386.

[Pat04] Patriksson, M., 2004. "Algorithms for Computing Traffic Equilibria," Networks and Spatial Economics, Vol. 4, pp. 23-38.

[Pen02] Pennanen, T., 2002. "Local Convergence of the Proximal Point Algorithm and Multiplier Methods Without Monotonicity," Math. of Operations Research, Vol. 27, pp. 170-191.

[Pfl96] Pflug, G., 1996. Optimization of Stochastic Models. The Interface Between Simulation and Optimization, Kluwer, Boston.

[PiZ94] Pinar, M., and Zenios, S., 1994. "On Smoothing Exact Penalty Functions for Convex Constrained Optimization," SIAM J. on Optimization, Vol. 4, pp. 486-511.

[PoJ92] Poljak, B. T., and Juditsky, A. B., 1992. "Acceleration of Stochastic Approximation by Averaging," SIAM J. on Control and Optimization, Vol. 30, pp. 838-855.

[PoT73] Poljak, B. T., and Tsypkin, Y. Z., 1973. "Pseudogradient Adaptation and Training Algorithms," Automation and Remote Control, Vol. 12, pp. 83-94.

[PoT74] Poljak, B. T., and Tretjakov, N. V., 1974. "An Iterative Method for Linear Programming and its Economic Interpretation," Matecon, Vol. 10, pp. 81-100.

[PoT80] Poljak, B. T., and Tsypkin, Y. Z., 1980. "Adaptive Estimation Algorithms (Convergence, Optimality, Stability)," Automation and Remote Control, Vol. 40, pp. 378-389.

[PoT81] Poljak, B. T., and Tsypkin, Y. Z., 1981. "Optimal Pseudogradient Adaptation Algorithms," Automation and Remote Control, Vol. 41, pp. 1101-1110.

[PoT97] Polyak, R., and Teboulle, M., 1997. "Nonlinear Rescaling and Proximal-Like Methods in Convex Optimization," Math. Programming, Vol. 76, pp. 265-284.

[Pol64] Poljak, B. T., 1964. "Some Methods of Speeding up the Convergence of Iteration Methods," Z. VyČisl. Mat. i Mat. Fiz., Vol. 4, pp. 1-17.

[Pol71] Polak, E., 1971. Computational Methods in Optimization: A Unified Approach, Academic Press, NY.

[Pol78] Poljak, B. T., 1978. "Nonlinear Programming Methods in the Presence of Noise," Math. Programming, Vol. 14, pp. 87–97.

[Pol79] Poljak, B. T., 1979. "On Bertsekas' Method for Minimization of Composite Functions," Internat. Symp. Systems Opt. Analysis, Benoussan, A., and Lions, J. L., (Eds.), Springer-Verlag, Berlin and NY, pp. 179-186.

[Pol87] Poljak, B. T., 1987. Introduction to Optimization, Optimization Software Inc., NY.

[Pol88] Polyak, R. A., 1988. "Smooth Optimization Methods for Minimax Problems," SIAM J. on Control and Optimization, Vol. 26, pp. 1274-1286.

[Pol92] Polyak, R. A., 1992. "Modified Barrier Functions (Theory and Methods)," Math. Programming, Vol. 54, pp. 177-222.

[Pow69] Powell, M. J. D., 1969. "A Method for Nonlinear Constraints in Minimizing Problems," in Optimization, Fletcher, R., (Ed.), Academic Press, NY, pp. 283-298.

[Pow73] Powell, M. J. D., 1973. "On Search Directions for Minimization Algorithms," Math. Programming, Vol. 4, pp. 193-201.

[Pow11] Powell, W. B., 2011. Approximate Dynamic Programming: Solving the Curses of Dimensionality, 2nd Ed., Wiley, NY.

[Pre95] Prekopa, A., 1995. Stochastic Programming, Kluwer, Boston.

[Psh65] Pshenichnyi, B. N., 1965. "Dual Methods in Extremum Problems," Kibernetika, Vol. 1, pp. 89-95.

[Pyt98] Pytlak, R., 1998. "An Efficient Algorithm for Large-Scale Nonlinear Programming Problems with Simple Bounds on the Variables," SIAM J. on Optimization, Vol. 8, pp. 532-560.

[QSG13] Qin, Z., Scheinberg, K., and Goldfarb, D., 2013. "Efficient Block-Coordinate Descent Algorithms for the Group Lasso," Math. Programming Computation, Vol. 5, pp. 143-169.

[RFP10] Recht, B., Fazel, M., and Parrilo, P. A., 2010. "Guaranteed Minimum-Rank Solutions of Linear Matrix Equations via Nuclear Norm Minimization," SIAM Review, Vol. 52, pp. 471-501.

[RGV14] Richard, E., Gaiffas, S., and Vayatis, N., 2014. "Link Prediction in Graphs with Autoregressive Features," J. of Machine Learning Research, Vol. 15, pp. 565-593.

[RHL13] Razaviyayn, M., Hong, M., and Luo, Z. Q., 2013. "A Unified Convergence Analysis of Block Successive Minimization Methods for Nonsmooth Optimization," SIAM J. on Optimization, Vol. 23, pp. 1126-1153.

[RHW86] Rumelhart, D. E., Hinton, G. E., and Williams, R. J., 1986. "Learning Internal Representation by Error Backpropagation," in Parallel Distributed Processing-Explorations in the Microstructure of Cognition, by Rumelhart and McClelland, (eds.), MIT Press, Cambridge, MA, pp. 318-362.

[RHW88] Rumelhart, D. E., Hinton, G. E., and Williams, R. J., 1988.

"Learning Representations by Back-Propagating Errors," in Cognitive Modeling, by T. A. Polk, and C. M. Seifert, (eds.), MIT Press, Cambridge, MA, pp. 213-220.

[RHZ14] Razaviyayn, M., Hong, M., and Luo, Z. Q., 2013. "A Unified Convergence Analysis of Block Successive Minimization Methods for Nonsmooth Optimization," SIAM J. on Optimization, Vol. 23, pp. 1126-1153.

[RNV09] Ram, S. S., Nedić, A., and Veeravalli, V. V., 2009. "Incremental Stochastic Subgradient Algorithms for Convex Optimization," SIAM J. on Optimization, Vol. 20, pp. 691-717.

[RNV10] Ram, S. S., Nedić, A., and Veeravalli, V. V., 2010. "Distributed Stochastic Subgradient Projection Algorithms for Convex Optimization," J. of Optimization Theory and Applications, Vol. 147, pp. 516-545.

[ROF92] Rudin, L. I., Osher, S., and Fatemi, E., 1992. "Nonlinear Total Variation Based Noise Removal Algorithms," Physica D: Nonlinear Phenomena, Vol. 60, pp. 259-268.

[RRW11] Recht, B., Re, C., Wright, S. J., and Niu, F., 2011. "Hogwild: A Lock-Free Approach to Parallelizing Stochastic Gradient Descent," in Advances in Neural Information Processing Systems (NIPS 2011), pp. 693-701.

[RSW13] Rao, N., Shah, P., Wright, S., and Nowak, R., 2013. "A Greedy Forward-Backward Algorithm for Atomic Norm Constrained Minimization, in Proc. 2013 IEEE International Conference on Acoustics, Speech and Signal Processing, pp. 5885-5889.

[RTV06] Roos, C., Terlaky, T., and Vial, J. P., 2006. Interior Point Methods for Linear Optimization, Springer, NY.

[RXB11] Recht, B., Xu, W., and Hassibi, B., 2011. "Null Space Conditions and Thresholds for Rank Minimization," Math. Programming, Vol. 127, pp. 175-202.

[RaN04] Rabbat, M. G., and Nowak, R. D., 2004. "Distributed Optimization in Sensor Networks," in Proc. Inf. Processing Sensor Networks, Berkeley, CA, pp. 20-27.

[RaN05] Rabbat M. G., and Nowak R. D., 2005. "Quantized Incremental Algorithms for Distributed Optimization," IEEE J. on Select Areas in Communications, Vol. 23, pp. 798-808.

[Ray93] Raydan, M., 1993. "On the Barzilai and Borwein Choice of Steplength for the Gradient Method," IMA J. of Numerical Analysis, Vol. 13, pp. 321-326.

[Ray97] Raydan, M., 1997. "The Barzilai and Borwein Gradient Method for the Large Scale Unconstrained Minimization Problem," SIAM J. on Optimization, Vol. 7, pp. 26-33.

[ReR13] Recht, B., and Ré, C., 2013. "Parallel Stochastic Gradient Algorithms for Large-Scale Matrix Completion," Math. Programming Computation, Vol. 5, pp. 201-226.

[Rec11] Recht, B., 2011. "A Simpler Approach to Matrix Completion," The J. of Machine Learning Research, Vol. 12, pp. 3413-3430.

[RiT14] Richtarik, P., and Takac, M., 2014. "Iteration Complexity of Randomized Block-Coordinate Descent Methods for Minimizing a Composite Function," Math. Programming, Vol. 144, pp. 1-38.

[RoS71] Robbins, H., and Siegmund, D. O., 1971. "A Convergence Theorem for Nonnegative Almost Supermartingales and Some Applications," Optimizing Methods in Statistics, pp. 233-257; see "Herbert Robbins Selected Papers," Springer, NY, 1985, pp. 111-135.

[RoW91] Rockafellar, R. T., and Wets, R. J.-B., 1991. "Scenarios and Policy Aggregation in Optimization Under Uncertainty," Math. of Operations Research, Vol. 16, pp. 119-147.

[RoW98] Rockafellar, R. T., and Wets, R. J.-B., 1998. Variational Analysis, Springer-Verlag, Berlin.

[Rob99] Robinson, S. M., 1999. "Linear Convergence of Epsilon-Subgradient Descent Methods for a Class of Convex Functions," Math. Programming, Ser. A, Vol. 86, pp. 41-50.

[Roc66] Rockafellar, R. T., 1966. "Characterization of the Subdifferentials of Convex Functions," Pacific J. of Mathematics, Vol. 17, pp. 497-510.

[Roc70] Rockafellar, R. T., 1970. Convex Analysis, Princeton Univ. Press, Princeton, NJ.

[Roc73] Rockafellar, R. T., 1973. "A Dual Approach to Solving Nonlinear Programming Problems by Unconstrained Optimization," Math. Programming, pp. 354-373.

[Roc76a] Rockafellar, R. T., 1976. "Monotone Operators and the Proximal Point Algorithm," SIAM J. on Control and Optimization, Vol. 14, pp. 877-898.

[Roc76b] Rockafellar, R. T., 1976. "Augmented Lagrangians and Applications of the Proximal Point Algorithm in Convex Programming," Math. of Operations Research, Vol. 1, pp. 97-116.

[Roc76c] Rockafellar, R. T., 1976. "Solving a Nonlinear Programming Problem by Way of a Dual Problem," Symp. Matematica, Vol. 27, pp. 135-160.

[Roc84] Rockafellar, R. T., 1984. Network Flows and Monotropic Optimization, Wiley, NY; republished by Athena Scientific, Belmont, MA, 1998.

[Rud76] Rudin, W., 1976. Real Analysis, McGraw-Hill, NY.

[Rup85] Ruppert, D., 1985. "A Newton-Raphson Version of the Multivariate Robbins-Monro Procedure," The Annals of Statistics, Vol. 13, pp. 236-245.

[Rus86] Ruszczynski, A., 1986. "A Regularized Decomposition Method for Minimizing a Sum of Polyhedral Functions," Math. Programming, Vol. 35, pp. 309-333.

[Rus06] Ruszczynski, A., 2006. Nonlinear Optimization, Princeton Univ. Press, Princeton, NJ.

[SBC91] Saarinen, S., Bramley, R. B., and Cybenko, G., 1991. "Neural Networks, Backpropagation and Automatic Differentiation," in Automatic Differentiation of Algorithms, by A. Griewank and G. F. Corliss, (eds.), SIAM, Philadelphia, PA, pp. 31-42.

[SBK64] Shah, B., Buehler, R., and Kempthorne, O., 1964. "Some Algorithms for Minimizing a Function of Several Variables," J. Soc. Indust. Appl. Math., Vol. 12, pp. 74-92.

[SBT12] Shah, P., Bhaskar, B. N., Tang, G., and Recht, B., 2012. "Linear System Identification via Atomic Norm Regularization," arXiv preprint arXiv:1204.0590.

[SDR09] Shapiro, A., Dentcheva, D., and Ruszczynski, A., 2009. Lectures on Stochastic Programming: Modeling and Theory, SIAM, Phila., PA.

[SFR09] Schmidt, M., Fung, G., and Rosales, R., 2009. "Optimization Methods for ℓ_1-Regularization," Univ. of British Columbia, Technical Report TR-2009-19.

[SKS12] Schmidt, M., Kim, D., and Sra, S., 2012. "Projected Newton-Type Methods in Machine Learning," in Optimization for Machine Learning, by Sra, S., Nowozin, S., and Wright, S. J., (eds.), MIT Press, Cambridge, MA, pp. 305-329.

[SLB13] Schmidt, M., Le Roux, N., and Bach, F., 2013. "Minimizing Finite Sums with the Stochastic Average Gradient," arXiv preprint arXiv:1309.-2388.

[SNW12] Sra, S., Nowozin, S., and Wright, S. J., 2012. Optimization for Machine Learning, MIT Press, Cambridge, MA.

[SRB11] Schmidt, M., Roux, N. L., and Bach, F. R., 2011. "Convergence Rates of Inexact Proximal-Gradient Methods for Convex Optimization," In Advances in Neural Information Processing Systems, pp. 1458-1466.

[SSS07] Shalev-Shwartz, S., Singer, Y., Srebro, N., and Cotter, A., 2007. "Pegasos: Primal Estimated Subgradient Solver for SVM," in ICML 07, New York, NY, pp. 807-814.

[SaT13] Saha, A., and Tewari, A., 2013. "On the Nonasymptotic Convergence of Cyclic Coordinate Descent Methods," SIAM J. on Optimization, Vol. 23, pp. 576-601.

[Sak66] Sakrison, D. T., 1966. "Stochastic Approximation: A Recursive Method for Solving Regression Problems," in Advances in Communication Theory and Applications, 2, A. V. Balakrishnan, ed., Academic Press, NY, pp. 51-106.

[Say14] Sayed, A. H., 2014. "Adaptation, Learning, and Optimization over Networks," Foundations and Trends in Machine Learning, Vol. 7, no. 4-5, pp. 311-801.

[ScF14] Schmidt, M., and Friedlander, M. P., 2014. "Coordinate Descent Converges Faster with the Gauss-Southwell Rule than Random Selection," Advances in Neural Information Processing Systems 27 (NIPS 2014).

[Sch82] Schnabel, R. B., 1982. "Determining Feasibility of a Set of Nonlinear Inequality Constraints," Math. Programming Studies, Vol. 16, pp. 137-148.

[Sch86] Schrijver, A., 1986. Theory of Linear and Integer Programming, Wiley, NY.

[Sch10] Schmidt, M., 2010. "Graphical Model Structure Learning with L1-Regularization," PhD Thesis, Univ. of British Columbia.

[Sch14a] Schmidt, M., 2014. "Convergence Rate of Stochastic Gradient with Constant Step Size," Computer Science Report, Univ. of British Columbia.

[Sch14b] Schmidt, M., 2014. "Convergence Rate of Proximal Gradient with General Step-Size," Dept. of Computer Science, Unpublished Note, Univ. of British Columbia.

[ShZ12] Shamir, O., and Zhang, T., 2012. "Stochastic Gradient Descent for Non-Smooth Optimization: Convergence Results and Optimal Averaging Schemes," arXiv preprint arXiv:1212.1824.

[Sha79] Shapiro, J. E., 1979. Mathematical Programming Structures and Algorithms, Wiley, NY.

[Sho85] Shor, N. Z., 1985. Minimization Methods for Nondifferentiable Functions, Springer-Verlag, Berlin.

[Sho98] Shor, N. Z., 1998. Nondifferentiable Optimization and Polynomial Problems, Kluwer Academic Publishers, Dordrecht, Netherlands.

[SmS04] Smola, A. J., and Scholkopf, B., 2004. "A Tutorial on Support Vector Regression," Statistics and Computing, Vol. 14, pp. 199-222.

[SoZ98] Solodov, M. V., and Zavriev, S. K., 1998. "Error Stability Properties of Generalized Gradient-Type Algorithms," J. Opt. Theory and Appl., Vol. 98, pp. 663-680.

[Sol98] Solodov, M. V., 1998. "Incremental Gradient Algorithms with Stepsizes Bounded Away from Zero," Computational Optimization and Applications, Vol. 11, pp. 23-35.

[Spa03] Spall, J. C., 2003. Introduction to Stochastic Search and Optimization: Estimation, Simulation, and Control, J. Wiley, Hoboken, NJ.

[Spa12] Spall, J. C., 2012. "Cyclic Seesaw Process for Optimization and Identification," J. of Optimization Theory and Applications, Vol. 154, pp. 187-208.

[Spi83] Spingarn, J. E., 1983. "Partial Inverse of a Monotone Operator," Applied Mathematics and Optimization, Vol. 10, pp. 247-265.

[Spi85] Spingarn, J. E., 1985. "Applications of the Method of Partial Inverses to Convex Programming: Decomposition," Math. Programming, Vol. 32, pp. 199-223.

[StV09] Strohmer, T., and Vershynin, R., 2009. "A Randomized Kaczmarz Algorithm with Exponential Convergence," J. Fourier Anal. Appl., Vol. 15, pp. 262-278.

[StW70] Stoer, J., and Witzgall, C., 1970. Convexity and Optimization in Finite Dimensions, Springer-Verlag, Berlin.

[StW75] Stephanopoulos, G., and Westerberg, A. W., 1975. "The Use of Hestenes' Method of Multipliers to Resolve Dual Gaps in Engineering System Optimization," J. Optimization Theory and Applications, Vol. 15, pp. 285-309.

[Str76] Strang, G., 1976. Linear Algebra and Its Applications, Academic Press, NY.

[Str97] Stromberg, A-B., 1997. Conditional Subgradient Methods and Ergodic Convergence in Nonsmooth Optimization, Ph.D. Thesis, Univ. of Linkoping, Sweden.

[SuB98] Sutton, R. S., and Barto, A. G., 1998. Reinforcement Learning, MIT Press, Cambridge, MA.

[TBA86] Tsitsiklis, J. N., Bertsekas, D. P., and Athans, M., 1986. "Distributed Asynchronous Deterministic and Stochastic Gradient Optimization Algorithms," IEEE Trans. on Automatic Control, Vol. AC-31, pp. 803-812.

[TBT90] Tseng, P., Bertsekas, D. P., and Tsitsiklis, J. N., 1990. "Partially Asynchronous, Parallel Algorithms for Network Flow and Other Problems," SIAM J. on Control and Optimization, Vol. 28, pp. 678-710.

[TVS10] Teo, C. H., Vishwanthan, S. V. N., Smola, A. J., and Le, Q. V., 2010. "Bundle Methods for Regularized Risk Minimization," The J. of Machine Learning Research, Vol. 11, pp. 311-365.

[TaP13] Talischi, C., and Paulino, G. H., 2013. "A Consistent Operator Splitting Algorithm and a Two-Metric Variant: Application to Topology Optimization," arXiv preprint arXiv:1307.5100.

[Teb92] Teboulle, M., 1992. "Entropic Proximal Mappings with Applications to Nonlinear Programming," Math. of Operations Research, Vol. 17, pp. 1-21.

[Teb97] Teboulle, M., 1997. "Convergence of Proximal-Like Algorithms," SIAM J. Optim., Vol. 7, pp. 1069-1083.

[Ter96] Terlaky, T. (Ed.), 1996. Interior Point Methods of Mathematical Programming, Springer, NY.

[Tib96] Tibshirani, R., 1996. "Regression Shrinkage and Selection via the Lasso," J. of the Royal Statistical Society, Series B (Methodological), Vol. 58, pp. 267-288.

[Tod01] Todd, M. J., 2001. "Semidefinite Optimization," Acta Numerica, Vol. 10, pp. 515-560.

[TsB87] Tseng, P., and Bertsekas, D. P., 1987. "Relaxation Methods for

Problems with Strictly Convex Separable Costs and Linear Constraints," Math. Programming, Vol. 38, pp. 303-321.

[TsB90] Tseng, P., and Bertsekas, D. P., 1990. "Relaxation Methods for Monotropic Programs," Math. Programming, Vol. 46, pp. 127-151.

[TsB91] Tseng, P., and Bertsekas, D. P., 1991. "Relaxation Methods for Problems with Strictly Convex Costs and Linear Constraints," Math. of Operations Research, Vol. 16, pp. 462-481.

[TsB93] Tseng, P., and Bertsekas, D. P., 1993. "On the Convergence of the Exponential Multiplier Method for Convex Programming," Math. Programming, Vol. 60, pp. 1-19.

[TsB00] Tseng, P., and Bertsekas, D. P., 2000. "An Epsilon-Relaxation Method for Separable Convex Cost Generalized Network Flow Problems," Math. Programming, Vol. 88, pp. 85-104.

[TsY09] Tseng, P. and Yun S., 2009. "A Coordinate Gradient Descent Method for Nonsmooth Separable Minimization," Math. Programming, Vol. 117, pp. 387-423.

[Tse91a] Tseng, P., 1991. "Decomposition Algorithm for Convex Differentiable Minimization," J. of Optimization Theory and Applications, Vol. 70, pp. 109-135.

[Tse91b] Tseng, P., 1991. "Applications of a Splitting Algorithm to Decomposition in Convex Programming and Variational Inequalities," SIAM J. on Control and Optimization, Vol. 29, pp. 119-138.

[Tse93] Tseng, P., 1993. "Dual Coordinate Ascent Methods for Non-Strictly Convex Minimization," Math. Programming, Vol. 59, pp. 231-247.

[Tse95] Tseng, P., 1995. "Fortified-Descent Simplicial Search Method," Report, Dept. of Math., Univ. of Washington, Seattle, Wash.; also in SIAM J. on Optimization, Vol. 10, 1999, pp. 269-288.

[Tse98] Tseng, P., 1998. "Incremental Gradient(-Projection) Method with Momentum Term and Adaptive Stepsize Rule," SIAM J. on Control and Optimization, Vol. 8, pp. 506-531.

[Tse00] Tseng, P., 2000. "A Modified Forward-Backward Splitting Method for Maximal Monotone Mappings," SIAM J. on Control and Optimization, Vol. 38, pp. 431-446.

[Tse01a] Tseng, P., 2001. "Convergence of Block Coordinate Descent Methods for Nondifferentiable Minimization," J. Optim. Theory Appl., Vol. 109, pp. 475-494.

[Tse01b] Tseng, P., 2001. "An Epsilon Out-of-Kilter Method for Monotropic Programming," Math. of Operations Research, Vol. 26, pp. 221-233.

[Tse04] Tseng, P., 2004. "An Analysis of the EM Algorithm and Entropy-Like Proximal Point Methods," Math. Operations Research, Vol. 29, pp. 27-44.

[Tse08] Tseng, P., 2008. "On Accelerated Proximal Gradient Methods for

Convex-Concave Optimization," Report, Math. Dept., Univ. of Washington.

[Tse09] Tseng, P., 2009. "Some Convex Programs Without a Duality Gap," Math. Programming, Vol. 116, pp. 553-578.

[Tse10] Tseng, P., 2010. "Approximation Accuracy, Gradient Methods, and Error Bound for Structured Convex Optimization," Math. Programming, Vol. 125, pp. 263-295.

[VKG14] Vetterli, M., Kovacevic, J., and Goyal, V. K., 2014. Foundations of Signal Processing, Cambridge Univ. Press, NY.

[VMR88] Vogl, T. P., Mangis, J. K., Rigler, A. K., Zink, W T., and Alkon, D. L., 1988. "Accelerating the Convergence of the Back-Propagation Method," Biological Cybernetics, Vol. 59, pp. 257-263.

[VaF08] Van Den Berg, E., and Friedlander, M. P., 2008. "Probing the Pareto Frontier for Basis Pursuit Solutions," SIAM J. on Scientific Computing, Vol. 31, pp. 890-912.

[Van01] Vanderbei, R. J., 2001. Linear Programming: Foundations and Extensions, Springer, NY.

[VeH93] Ventura, J. A., and Hearn, D. W., 1993. "Restricted Simplicial Decomposition for Convex Constrained Problems," Math. Programming, Vol. 59, pp. 71-85.

[Ven67] Venter, J. H., 1967. "An Extension of the Robbins-Monro Procedure," Ann. Math. Statist., Vol. 38, pp. 181-190.

[WDS13] Weinmann, A., Demaret, L., and Storath, M., 2013. "Total Variation Regularization for Manifold-Valued Data," arXiv preprint arXiv:1312.-7710.

[WFL14] Wang, M., Fang, E., and Liu, H., 2014. "Stochastic Compositional Gradient Descent: Algorithms for Minimizing Compositions of Expected-Value Functions," Optimization Online.

[WHM13] Wang, X., Hong, M., Ma, S., Luo, Z. Q., 2013. "Solving Multiple-Block Separable Convex Minimization Problems Using Two-Block Alternating Direction Method of Multipliers," arXiv preprint arXiv:1308.5294.

[WSK14] Wytock, M., Suvrit S., and Kolter, J. K., 2014. "Fast Newton Methods for the Group Fused Lasso," Proc. of 2014 Conf. on Uncertainty in Artificial Intelligence.

[WSV00] Wolkowicz, H., Saigal, R., and Vanderbergue, L., (eds), 2000. Handbook of Semidefinite Programming, Kluwer, Boston.

[WaB13a] Wang, M., and Bertsekas, D. P., 2013. "Incremental Constraint Projection-Proximal Methods for Nonsmooth Convex Optimization," Lab. for Information and Decision Systems Report LIDS-P-2907, MIT, to appear in SIAM J. on Optimization.

[WaB13b] Wang, M., and Bertsekas, D. P., 2013. "Convergence of Iterative Simulation-Based Methods for Singular Linear Systems," Stochastic

Systems, Vol. 3, pp. 38-95.

[WaB13c] Wang, M., and Bertsekas, D. P., 2013. "Stabilization of Stochastic Iterative Methods for Singular and Nearly Singular Linear Systems," Math. of Operations Research, Vol. 39, pp. 1-30.

[WaB14] Wang, M., and Bertsekas, D. P., 2014. "Incremental Constraint Projection Methods for Variational Inequalities," Mathematical Programming, pp. 1-43.

[Was04] Wasserman, L., 2004. All of Statistics: A Concise Course in Statistical Inference, Springer, NY.

[Wat92] Watson, G. A., 1992. "Characterization of the Subdifferential of Some Matrix Norms," Linear Algebra and its Applications, Vol. 170, pp. 33-45.

[WeO13] Wei, E., and Ozdaglar, A., 2013. "On the $O(1/k)$ Convergence of Asynchronous Distributed Alternating Direction Method of Multipliers," arXiv preprint arXiv:1307.8254.

[WiH60] Widrow, B., and Hoff, M. E., 1960. "Adaptive Switching Circuits," Institute of Radio Engineers, Western Electronic Show and Convention, Convention Record, Part 4, pp. 96-104.

[Wol75] Wolfe, P., 1975. "A Method of Conjugate Subgradients for Minimizing Nondifferentiable Functions," Math. Programming Study 3, Balinski, M., and Wolfe, P., (Eds.), North-Holland, Amsterdam, pp. 145-173.

[Wri97] Wright, S. J., 1997. Primal-Dual Interior Point Methods, SIAM, Philadelphia, PA.

[Wri14] Wright, S. J., 2014. "Coordinate Descent Algorithms," Optimization Online.

[XiZ14] Xiao, L., and Zhang, T., 2014. "A Proximal Stochastic Gradient Method with Progressive Variance Reduction," arXiv preprint arXiv:1403.-4699.

[Xia10] Xiao L., 2010. "Dual Averaging Methods for Regularized Stochastic Learning and Online Optimization," J. of Machine Learning Research, Vol. 11, pp. 2534-2596.

[YBR08] Yu, H., Bertsekas, D. P., and Rousu, J., 2008. "An Efficient Discriminative Training Method for Generative Models," Extended Abstract, the 6th International Workshop on Mining and Learning with Graphs (MLG).

[YGT93] Ye, Y., Guler, O., Tapia, R. A., and Zhang, Y., 1993. "A Quadratically Convergent O($\sqrt{nL}$)-Iteration Algorithm for Linear Programming," Math. Programming, Vol. 59, pp. 151-162.

[YNS10] Yousefian, F., Nedić, A., and Shanbhag, U. V., 2010. "Convex Nondifferentiable Stochastic Optimization: A Local Randomized Smoothing Technique," Proc. American Control Conference (ACC), pp. 4875-4880.

[YNS12] Yousefian, F., Nedić, A., and Shanbhag, U. V., 2012. "On Stochastic Gradient and Subgradient Methods with Adaptive Steplength Sequences," Automatica, Vol. 48, pp. 56-67.

[YOG08] Yin, W., Osher, S., Goldfarb, D., and Darbon, J., 2008. "Bregman Iterative Algorithms for ℓ_1-Minimization with Applications to Compressed Sensing," SIAM J. on Imaging Sciences, Vol. 1, pp. 143-168.

[YSQ14] You, K., Song, S., and Qiu, L., 2014. "Randomized Incremental Least Squares for Distributed Estimation Over Sensor Networks," Preprints of the 19th World Congress The International Federation of Automatic Control Cape Town, South Africa.

[Ye92] Ye, Y., 1992. "A Potential Reduction Algorithm Allowing Column Generation," SIAM J. on Optimization, Vol. 2, pp. 7-20.

[Ye97] Ye, Y., 1997. Interior Point Algorithms: Theory and Analysis, Wiley Interscience, NY.

[YuR07] Yu, H., and Rousu, J., 2007. "An Efficient Method for Large Margin Parameter Optimization in Structured Prediction Problems," Technical Report C-2007-87, Univ. of Helsinki.

[ZJL13] Zhang, H., Jiang, J., and Luo, Z. Q., 2013. "On the Linear Convergence of a Proximal Gradient Method for a Class of Nonsmooth Convex Minimization Problems," J. of the Operations Research Society of China, Vol. 1, pp. 163-186.

[ZLW99] Zhao, X., Luh, P. B., and Wang, J., 1999. "Surrogate Gradient Algorithm for Lagrangian Relaxation," J. Optimization Theory and Applications, Vol. 100, pp. 699-712.

[ZMJ13] Zhang, L., Mahdavi, M., and Jin, R., 2013. "Linear Convergence with Condition Number Independent Access of Full Gradients," Advances in Neural Information Processing Systems 26 (NIPS 2013), pp. 980-988.

[ZTD92] Zhang, Y., Tapia, R. A., and Dennis, J. E., 1992. "On the Superlinear and Quadratic Convergence of Primal-Dual Interior Point Linear Programming Algorithms," SIAM J. on Optimization, Vol. 2, pp. 304-324.

[Zal02] Zalinescu, C., 2002. Convex Analysis in General Vector Spaces, World Scientific, Singapore.

[Zan69] Zangwill, W. I., 1969. Nonlinear Programming, Prentice-Hall, Englewood Cliffs, NJ.

[Zou60] Zoutendijk, G., 1960. Methods of Feasible Directions, Elsevier Publ. Co., Amsterdam.

[Zou76] Zoutendijk, G., 1976. Mathematical Programming Methods, North Holland, Amsterdam.

INDEX

D

T

U

V

W

Y

Z